NATO ASI Series

Advanced Science Institutes Series

A series presenting the results of activities sponsored by the NATO Science Committee, which aims at the dissemination of advanced scientific and technological knowledge, with a view to strengthening links between scientific communities.

The Series is published by an international board of publishers in conjunction with the NATO Scientific Affairs Division

A Life Sciences **B Physics**	Plenum Publishing Corporation London and New York
C Mathematical and Physical Sciences **D Behavioural and Social Sciences** **E Applied Sciences**	Kluwer Academic Publishers Dordrecht, Boston and London
F Computer and Systems Sciences **G Ecological Sciences** **H Cell Biology** **I Global Environmental Change**	Springer-Verlag Berlin Heidelberg New York London Paris Tokyo Hong Kong Barcelona Budapest

PARTNERSHIP SUB-SERIES

1. Disarmament Technologies	Kluwer Academic Publishers
2. Environment	Springer-Verlag
3. High Technology	Kluwer Academic Publishers
4. Science and Technology Policy	Kluwer Academic Publishers
5. Computer Networking	Kluwer Academic Publishers

The Partnership Sub-Series incorporates activities undertaken in collaboration with NATO's Cooperation Partners, the countries of the CIS and Central and Eastern Europe, in Priority Areas of concern to those countries.

NATO-PCO DATABASE

The electronic index to the NATO ASI Series provides full bibliographical references (with keywords and/or abstracts) to about 50000 contributions from international scientists published in all sections of the NATO ASI Series. Access to the NATO-PCO DATABASE compiled by the NATO Publication Coordination Office is possible in two ways:

- via online FILE 128 (NATO-PCO DATABASE) hosted by ESRIN, Via Galileo Galilei, I-00044 Frascati, Italy.
- via CD-ROM "NATO Science & Technology Disk" with user-friendly retrieval software in English, French and German (© WTV GmbH and DATAWARE Technologies Inc. 1992).

The CD-ROM can be ordered through any member of the Board of Publishers or through NATO-PCO, Overijse, Belgium.

Series H: Cell Biology, Vol. 96

Springer
Berlin
Heidelberg
New York
Barcelona
Budapest
Hong Kong
London
Milan
Paris
Santa Clara
Singapore
Tokyo

Molecular Dynamics of Biomembranes

Edited by

Jos A. F. Op den Kamp

Centre for Biomembranes and Lipid Enzymology
Institute of Biomembranes
Utrecht University
Padualaan 8
3584 CH Utrecht, The Netherlands

Springer

Published in cooperation with NATO Scientific Affairs Division

Proceedings of the NATO Advanced Study Institute "Molecular Dynamics of Biomembranes", held in Cargèse, France, June 19–July 1, 1995

Library of Congress Cataloging-in-Publication Data applied for

Die Deutsche Bibliothek - CIP-Einheitsaufnahme

Molecular dynamics of biomembranes : [proceedings of the NATO Advanced Study Institute "Molecular Dynamics of Biomembranes" held in Cargèse, France, June 19 - July 1, 1995] / ed. by Jos A. F. op den Kamp. Publ. in cooperation with NATO Scientific Affairs Division. - Berlin ; Heidelberg ; New York ; Barcelona ; Budapest ; Hong Kong ; London ; Milan ; Paris ; Santa Clara ; Singapore ; Tokyo : Springer, 1996
(NATO ASI series : Ser. H, Cell biology ; Vol. 96)
ISBN-13: 978-3-642-64707-9
NE: Kamp, Jos A. F. op den [Hrsg.]; Advanced Study Institute Molecular Dynamics of Biomembranes <1995, Cargèse>; NATO: NATO ASI series / H

ISBN-13: 978-3-642-64707-9 e-ISBN-13: 978-3-642-61126-1
DOI: 10.1007/978-3-642-61126-1

Softcover reprint of the hardcover 1st edition 1996

Typesetting: Camera ready by authors/editors
Printed on acid-free paper
SPIN 10477136 31/3137 - 5 4 3 2 1 0

Preface

Protein insertion and translocation, intracellular traffic and sorting of membranes and their components, and lipid-protein interactions were the main topics of the Advanced Study Institute on "Molecular Dynamics of Membranes", which was held in June 1995 in Cargèse, Corsica, France.

The course, co-sponsored by NATO and FEBS, was the fifth in a series that started in 1987 and takes place every two years in the Institut d'Etudes Scientifiques in Cargèse. This Institute, ideally situated and fully equiped for this type of scientific meeting has greatly contributed to the great success of the courses. Of course, also the outstanding contributions of a large number of well known scientists and the enthousiastic participation of excellent graduate students and postdocs has given the "Cargèse Lectures on Biomembranes" a firm reputation in the scientific community.

The present proceedings is more than just a reflection on the information presented in the Course. First of all it contains a number of extensive reviews of specific areas of interest. Noteworthy are the articles dealing with:

- the general mechanisms of protein transport,
- the roles of invariant chain in antigen presentation,
- protein import and export in *E. coli*,
- protein folding and the role of chaperones,
- chloroplast and mitochondrial protein import,
- membrane traffic in general and during mitosis,

and with respect to membrane lipids:

- lipid domain formation,
- lipases: an extensive review about structure and properties,
- phospholipase A_2 and bioactive lipids,
- phospholipid transfer proteins,
- phospholipid localization and mobility and, finally,
- new strategies for protein reconstitution.

A wealth of information on important topics in membranology is presented here and extensive referencing to key publications complete these overviews.
Added to these main articles are more limited and specific summaries of recent developments in these fields, as well as short research papers. Together they offer a good representation of the interests of lecturers, postdocs and graduate students as presented during the course.

CONTENTS

CONTENTS

MECHANISMS INVOLVED IN CO- AND POSTTRANSLATIONAL PROTEIN TRANSPORT

Tom A. Rapoport, Ph.D.
Department of Cell Biology
Harvard Medical School
25 Shattuck Street
Boston, MA 02115
USA

The secretion and membrane insertion of proteins is a universal process occurring in living beings from bacteria to man. In prokaryotes, secretory proteins are transported directly across the plasma membrane or are inserted into it; in eukaryotes, they are initially translocated in an analogous process across the ER membrane, but are thereafter transported in vesicles to the plasma membrane. In all organisms, the translocation of proteins across and their integration into the membrane are initiated by hydrophobic signal sequences which are interchangeable; prokaryotic signal sequences can perform in eukaryotes and *vice versa*.

The transport of a protein across the membrane may occur during its synthesis (cotranslationally) or upon its completion (posttranslationally). In both cases, the process is started by a targeting phase (for a review, see Walter and Johnson, 1994). One mechanism of cotranslational targeting involves the signal recognition particle (SRP) which only recognizes signal sequences of nascent chains that are associated with the ribosome. Other targeting pathways involve cytosolic chaperones such as SecB, groEL and hsp 70, which may function either co- or posttranslationally and which maintain polypeptides in a translocation-competent structure.

The mechanism of the actual translocation process, that succeeds the targeting phase, also appears to differ depending on whether the polypeptide is transported co- or posttranslationally. The cotranslational mode requires the binding of the translating ribosome to the membrane and it is thought that the elongating nascent chain is transferred directly from the ribosome into the membrane (Blobel and Dobberstein, 1975). Therefore, the membrane binding of the ribosome may be necessary for the translocation process. In comparison, since the ribosome does not have a function in the posttranslational mode of translocation of proteins, other mechanisms of transport may be assumed. In both cases, however, it is believed that polypeptides are

NATO ASI Series, Vol. H 96
Molecular Dynamics of Biomembranes
Edited by Jos A. F. Op den Kamp

transferred across the membrane via a protein-conducting channel, formed at least in part from transmembrane proteins (Blobel and Dobberstein, 1975; Simon and Blobel, 1991).

In this brief review, I will focus on the actual process of protein translocation. I will discuss the membrane components involved in co- and posttranslational translocation processes and speculate about mechanisms.

COTRANSLATIONAL TRANSLOCATION OF PROTEINS

Cotranslational translocation occurs in every group of living beings in spite of the fact that its significance may vary. Whereas it seems to be the predominant mode in mammals, in prokaryotes and in *S. cerevisiae* many proteins may be transported posttranslationally.

The mechanism of cotranslational translocation has been best investigated in the mammalian system. The recent reconstitution of the translocation apparatus into proteoliposomes, using purified membrane components from dog pancreas microsomes (Görlich and Rapoport, 1993) indicates a remarkable simplicity of the system. The fundamental machinery appears to consist of only three components: the SRP- receptor, the Sec61p-complex and the TRAM protein. These components are sufficient for the translocation of all secretory proteins as well as for the insertion of all membrane proteins tested to date.

The SRP-receptor is probably only required for the targeting process. It is composed of two subunits that can both bind GTP (Tajima *et al.*, 1986; Connolly and Gilmore, 1989; Miller, *et al.*, 1993). The α–subunit makes contact with the ribosome/nascent chain/SRP- complex and its GTP-binding site is essential for the transfer of the nascent chain into the membrane (Rapiejko and Gilmore, 1992). It is still a mystery as to what the function of the J3 subunit is.

The Sec61p- complex is most likely the core component of the translocation site. It is made up of three subunits (Görlich and Rapoport, 1993). The large α-subunit was discovered as a homolog to Sec61p of *S . cerevisiae*, a protein found earlier in genetic screens for mutants defective in translocation (Deshaies and Schekman, 1987; Rothblatt *et al.*, 1989; Stirling *et al.*, 1992). Sec61α is predicted to span the membrane ten times. The β- and γ-subunits of the mammalian Sec61p-complex are small membrane proteins which are spanning the membrane once with a C-terminal hydrophobic tail. The γ-subunit is homologous to the SSS1 protein from *S. cerevisiae* and can effectively replace it in yeast cells (Hartmann *et al.*,1994).

Sec61α contacts polypeptide chains which are being transferred through the membrane (Müsch *et al.*, 1992; Görlich *et al.*, 1992; Sanders *et al.*, 1992). Photocrosslinking experiments have demonstrated that Sec61α is the primary neighbor of each of about 40 amino acids which follow the polypeptide segment located in the ribosome (Mothes *et al.*, 1994). Thus, Sec61α may be the main component of a putative protein-conducting channel.

In mammals, the Sec61p-complex is likely to mediate the binding of the ribosome during cotranslational translocation. The Sec61p-complex is closely associated with membrane-bound ribosomes after solubilization of ER membranes (Görlich *et al.*, 1992). This binding can be induced by the targeting of a nascent chain to the membrane and the isolated Sec61-ribosome complex is dissociated under conditions identical to those required for the release of ribosomes from the membranes. Sec61α is protected by the membrane-bound ribosome against the action of proteases and it exhibits the same extraordinary resistance as the ribosome- membrane linkage (Kalies et al, 1994) . Under physiological salt conditions, the Sec61p-complex accounts for the majority of binding sites for ribosomes lacking nascent chains. Additionally, since some proteins, like preprolactin, require only the presence of the SRP- receptor and the Sec61p-complex in proteoliposomes for their translocation (Görlich and Rapoport, 1993), it seems probable that the Sec61p-complex is sufficient for the binding of the translating ribosome. This conclusion is supported by recent data obtained with an SRP- independent system in which only the Sec61p-complex is required. All this data suggests that the Sec61p-complex is the receptor for ribosomes and makes it unlikely that other proposed proteins, such as p34 (Tazawa *et al.*, 1991) or p180 (Savitz and Meyer, 1990), play an essential part in the binding of ribosomes during translocation.

The ribosome appears to make a tight seal with the membrane, probably by forming numerous contacts with the cytosolic loops of Sec61α. The membrane-inserted nascent chain does not have access to proteases or iodide ions added to the cytosolic compartment (Connolly *et al.*, 1989; Crowley *et al.*, 1993). The close association of the ribosome and the Sec61p-complex suggests that the nascent chain is transferred directly from the channel in the ribosome into the protein-conducting channel in the membrane. Thus, the latter may be an extension of the ribosomal channel and the nascent chain may penetrate the membrane in a vectorial manner because there is only one exit from the extended channel. According to such a model, a pumping, pulling or pushing device would not be required.

The function of the third component of the mammalian translocation apparatus, the TRAM protein, is not completely understood. The TRAM protein is a multi-spanning membrane protein that contacts the nascent chain during early phases of its transfer through the membrane until its signal sequence is cleaved off (Görlich *et al.*, 1992). In the case of preprolactin, it has been shown to interact principally with the hydrophilic portion of the signal sequence preceding its hydrophobic core (High *et al.*, 1993). The presence of the TRAM protein is necessary for some proteins, like prepro-α-factor, and only stimulatory for others, like preprolactin (Görlich and Rapoport, 1993).

One may assume that the TRAM protein is involved in the membrane insertion of a nascent chain. A loop insertion, with one part of the hairpin being the hydrophobic core of the signal sequence, is thought to be the first step in protein translocation. For secretory proteins and some membrane proteins, the N-terminal end of the loop remains in the cytosol and the C-terminal part moves through the membrane. For other membrane proteins, the N-terminal part of the loop is transferred across the membrane producing the inverse orientation. By interacting with one of the polypeptide regions flanking the hydrophobic core of a signal sequence, TRAM may be involved in determining the orientation of a protein. It could also function during the insertion of polypeptide loops into the membrane in the case of multispanning proteins. This assumption would be consistent with the recent observation that TRAM is required for TRAM-dependent proteins at a step which follows the SRP-receptor - and Sec61p-complex - dependent membrane binding of the nascent chain for their productive insertion into the translocation site.

POSTTRANSLATIONAL TRANSLOCATION OF PROTEINS

Originally, the posttranslational mode of transport was studied in *E. coli*. Reconstitution experiments have demonstrated that the machinery required for the translocation of proteins across the cytoplasmic membrane consists of only two essential components: the peripheral membrane protein SecAp and the integral SecYp- complex.

SecAp is an ATPase (translocation ATPase) that receives the polypeptide chain from the cytosolic chaperone SecBp (Hartl *et al.*, 1990). It is believed to cycle between a membrane-inserted and free state and to thereby push the polypeptide chain across the membrane (Schiebel *et al.*, 1991; Economou and Wickner,1994).

The SecYp- complex is made up of SecYp, SecEp and SecGp (Brundage *et al.*, 1992). It is structurally related to the Sec61p-complex of eukaryotes. SecYp is homologous to the mammalian Sec61α and yeast Sec61p (Görlich *et al.*, 1992), and is also a neighbor of polypeptide chains crossing the membrane (Joly and Wickner, 1993). SecEp is predicted to span the membrane three times but only the C-terminal anchor is essential for its function (Schatz *et al.*, 1991), exactly the region that is similar in structure to Sec61p and SSS1p (Hartmann *et al.*, 1994). The structure of SecGp is unrelated to that of Sec61β.

In addition to these components, genetic screens have identified SecDp and SecFp, two multi-spanning membrane proteins which are required for viability of *E. coli* cells (Pogliano and Beckwith, 1994). These proteins may be involved in the maintenance of an electrochemical potential across the cytoplasmic membrane (Arkowitz and Wickner, 1994) or in the release of polypeptide chains from the translocation site into the periplasm (Matsuyama *et al.*,1993).

Posttranslational translocation in *S. cerevisiae* has been recently investigated in some detail. Yeast microsomes contain a trimeric Sec61p-complex that resembles that in mammals and the SecYp- complex in *E. coli*. It consists of Sec61p itself, Sbh1p and SSS1p, which are all homologous to the corresponding α -, β-, and γ-subunits of the mammalian Sec61p-complex (Panzner *et al.*, 1995). Both Sec61p and SSS1p are essential for the translocation of all proteins tested (Deshaies and Schekman, 1987; Esnault *et al.*, 1993). The trimeric Sec61p-complex is to some extent associated with ribosomes after solubilization of yeast microsomes and is therefore believed to be involved in cotranslational translocation (Panzner *et al.*, 1995). Yeast cells also contain a large Sec-complex which is composed of the components of the Sec61p-complex and all other membrane proteins found in genetic screens (Sec63p, Sec62p, Sec71p and Sec72p). When the Sec-complex was reconstituted into proteoliposomes, posttranslational translocation of prepro-α-factor was observed. However, a significant stimulation of the transport reaction was seen if Kar2p (BiP) was included into the lumen of the vesicles and ATP was added. With another translocation substrate, pro-OmpA, translocation was only observed in the presence of Kar2p and ATP. Thus, it seems that distinct membrane protein complexes are mediating the co- and posttranslational translocation processes.

What are the structure and functions of the additional components required for posttranslational transport? The Sec63, Sec71 and Sec62 proteins span the membrane 3, 1 and 2 times, respectively, Sec72p is a peripheral membrane protein and Kar2p (BiP) is a lumenal chaperone (Deshaies and Schekman, 1990; Feldheim *et al.*, 1992; Green *et al.*, 1992; Vogel *et al.*, 1990). Sec62p contacts polypeptides early during their translocation (Müsch *et al.*, 1992). Kar2p is believed to bind the polypeptide chain as it emerges in the ER lumen and to pull it across the membrane.

In conclusion, the Sec61/SecY-complex appears to be the core component in all known translocation pathways. For the cotranslational mode, it appears to cooperate with the SRV-receptor and TRAM and to bind directly to the ribosome. For the posttranslational pathway, different other proteins appear to associate with the Sec61/SecY- core. One function of these additional components must be to provide a driving force for vectorial translocation, either by pulling the polypeptide chain across the membrane, as proposed for Kar2p in the yeast system, or by pushing it across, as suggested for SecAp in *E. coli*. The additional components may also function in place of the ribosome to prohibit leakage of small molecules through the protein-conducting channel.

REFERENCES

Akimaru, J., Matsuyama, S.I., Tokuda, H. and Mizushima, S. (1991) Reconstitution of a Protein Translocation System Containing Purified SecY, SecE, and SecA from Escherichia coli. Proc. Natl. Acad.Sci. USA 88: 6545-6549.

Arkowitz, R.A. and Wickner, W. (1994) SecD and SecF are Required for the Proton Electrochemical Gradient Stimulation of Preprotein Translocation. EMBO J. 13: 954-963.

Blobel, G. and Dobberstein, B. (1975) Transfer of Proteins Across Membranes. I. Presence of Proteolytically Processed and Unprocessed Nascent Immunoglobulin Light Chains on Membrane-Bound Ribosomes of Murine Myeloma. J. Cell Biol. 67: 835-851.

Brodsky, J.L. and Schekman, R. (1993) A Sec63p-BiP Complex from Yeast Is Required for Protein Translocation in a Reconstituted Proteoliposome. J. Cell Biol.123: 1355-1363.

Brundage, L., Fimmel, C.J., Mizushima, S. and Wickner, W. (1992) SecY, SecE, and Band-1 Form the Membrane-Embedded Domain of *Escherichia coli* Preprotein Translocase. J. Biol Chem. 267: 4166-4170.

Brundage, L., Hendrick, J.P., Schiebel, E., Driessen, A.J.M. and Wickner, W. (1990) The Purified E. coli Integral Membrane Protein SecY/E Is Sufficient for Reconstitution of SecA-Dependent Precursor Protein Translocation. Cell 62: 649-657.

Connolly, T. and Gilmore, R. (1989) The Signal Recognition Particle Receptor Mediates the GTP-Dependent Displacement of SRP from the Signal Sequence of the Nascent Polypeptide. Cell 57: 599-610.

Connolly, T., Collins, P. and Gilmore, R. (1989) Access of Proteinase K to Partially Translocated Nascent Polypeptides in Intact and Detergent- Solubilized Membranes. J. Cell Biol. 108: 299-307.

Crowley, K.S., Reinhart, G.D. and Johnson, A.E. (1993) The Signal Sequence Moves Through a Ribosomal Tunnel into a Noncytoplasmic Aqueous Environment at the ER Membrane Early in Translocation. Cell 73: 1101-1115.

Deshaies, R.J. and Schekman, R. (1987) A Yeast Mutant Defective at an Early Stage in Import of Secretory Protein Precursors into the Endoplasmic Reticulum. J. Cell Biol. 105: 633-645.

Deshaies, R.J. and Schekman, R. (1990) Structural and Functional Dissection of Sec62P, a Membrane-Bound Component of the Yeast Endoplasmic Reticulum Protein Import Machinery. Mol. Cell Biol. 10: 6024-6035.

Deshaies, R.J., Sanders, S.L., Feldheim, D.A. and Schekman, R. (1991) Assembly of Yeast Sec Proteins Involved in Translocation into the Endoplasmic Reticulum into a Membrane-Bound Multisubunit Complex. Nature 349: 806-808.

Economou, A. and Wickner, W. (1994) SecA Promotes Preprotein Translocation by Undergoing ATP-Driven Cycles of Membrane Insertion and Deinsertion. Cell 78: 835-843.

Esnault, Y., Blondel, M.O., Deshaies, R.J., Schekman, R., Kepes, F. (1993) The Yeast Sssl Gene Is Essential for Secretory Protein Translocation and Encodes a Conserved Protein of the Endoplasmic Reticulum. EMBO J. 12: 4083-4094

Feldheim, D., Rothblatt, J. and Schekman, R. (1992) Topology and Functional Domains of Sec63p, an Endoplasmic Reticulum Membrane Protein Required for Secretory Protein Translocation. Mol. Cell Biol. 12: 3288-3296.

Gorlich, D., Hartmann, E., Prehn, S. and Rapoport, T.A. (1992) A Protein of the Endoplasmic Reticulum Involved Early in Polypeptide Translocation. Nature 357: 47-52.

Gorlich, D., Prehn, S., Hartmann, E., Kalies, K.U. and Rapoport, T.A. (1992) A Mammalian Homolog of Sec61p and SecYp Is Associated with Ribosomes and Nascent Polypeptides During Translocation. Cell 71: 489-503.

Gorlich, D., Rapoport, T.A. (1993) Protein Translocation into Proteoliposomes Reconstituted from Purified Components of the Endoplasmic Reticulum Membrane. Cell 75: 615-630

Green, N., Fang, H. and Walter, P. (1992) Mutants in Three Novel Complementation Groups Inhibit Membrane Protein Insertion into and Soluble Protein Translocation across the Endoplasmic Reticulum Membrane of Saccharomyces-Cerevisiae. J. Cell Biol. 116: 597-604.

Hartl, F.-U., Lecker, S., Schiebel, E., Hendrick, J.P. and Wickner, W. (1990) The Binding Cascade of SecB to SecA to SecY/E Mediates Preprotein Targeting to the *E. coli* Plasma Membrane. Cell 63: 269-279.

Hartmann, E., Sommer, T., Prehn, S., Gorlich, D., Jentsch, S. and Rapoport, T. A.(1994) Evolutionary Conservation of Components of the Protein Translocation Complex. Nature 367: 654-657.

High, S., Martoglio, B., Gorlich, D., Andersen, S.L.S., Ashford, A.J., Giner, A., Hartmann, E., Prehn, S., Rapoport, T.A., Dobberstein, B. and Brunner, J. (1993) Site-Specific Photocross-Linking Reveals That Sec61p and TRAM Contact Different Regions of a Membrane-Inserted Signal Sequence. J. Biol. Chem. 268: 26745-26751.

Joly, J.C. and Wickner, W. (1993) Translocation Can Drive the Unfolding of a Preprotein Domain. EMBO J. 12: 255-263.

Kalies, K.-U., Gorlich, D., and Rapoport, T.A. (1994) Binding of Ribosomes to the Rough Endoplasmic Reticulum-Mediated by the Sec61p-Complex. J. Cell Biol. 126: 925-934.

Matsuyama, S., Fujita, Y. and Mizushima, S. (1993) SecD Is Involved in the Release of Translocated Secretory Proteins from the Cytoplasmic Membrane of Escherichia coli. EMBO J. 12: 265-270.

Miller, J.D., Wilhelm, H., Gierasch, L., Gilmore, R. and Walter, P. (1993) GTP Binding and Hydrolysis by the Signal Recognition Particle During Initiation of Protein Translocation. Nature 336: 351-354.

Mothes, W., Prehn, S., and Rapoport, T.A. (1994) Systematic Probing of the Environment of a Translocating Secretory Protein During Translocation Through the ER Membrane. EMBO J. 13, No. 17: 3973-3982.

Musch, A., Wiedmann, M. and Rapoport, T.A. (1992) Yeast Sec Proteins Interact with Polypeptides Traversing the Endoplasmic Reticulum Membrane. Cell 69: 343-352.

Nishiyama, K., Mizushima, S. and Tokuda, H. (1993) A Novel Membrane Protein Involved in Protein Translocation Across the Cytoplasmic Membrane of *Escherichia coli.* EMBO J. 12: 3409-3415.

Panzner, S., Dreier, L., Hartmann, E., Kostka, S., and Rapoport, T.A. (1995) Posttranslational protein transport in yeast reconstituted with a purified complex of sec proteins and Kar2p. Cell 81, 1-20: in press.

Pogliano, J.A., Beckwith, J. (1994) SecD and SecF Facilitate Protein Export in *Escherichia coli.* EMBO J. 13: 554-561.

Rapiejko, P.J. and Gilmore, R. (1992) Protein Translocation Across the ER Requires a Functional GTP Binding Site in the Alpha-Subunit of the Signal Recognition Particle Receptor. J. Cell Biol. 117: 493-503.

Rapoport, T.A. (1992) Transport of Proteins Across the Endoplasmic Reticulum Membrane. Science 258: 931-936.

Rothblatt, J.A., Deshaies, R.J., Sanders, S.L., Daum, G. and Schekman, R. (1989) Multiple Genes Are Required for Proper Insertion of Secretory Proteins into the Endoplasmic Reticulum in Yeast. J. Cell Biol. 109: 2641-2652.

Sanders, S.L., Whitfield, K.M., Vogel, J.P., Rose, M.D. and Schekman, R.W. (1992) Sec61p and BiP Directly Facilitate Polypeptide Translocation into the ER. Cell 69: 353-365.

Savitz, A.J. and Meyer, D.I. (1990) Identification of a Ribosome Receptor in the Rough Endoplasmic Reticulum. Nature 346: 540-544

Schatz, P.J., Bieker, K.L., Ottemann, K.M., Silhavy, T.J. and Beckwith, J. (1991) One of Three Transmembrane Stretches Is Sufficient for the Functioning of the SecE Protein, a Membrane Component of the E-Coli Secretion Machinery. EMBO J. 10: 1749-1757.

Schiebel, E., Driessen, A.J.M., Hartl, F.-U. and Wickner, W. (1991) $D\mu H^+$ and ATP Function at Different Steps of the Catalytic Cycle of Preprotein Translocase. Cell 64: 927-939.

Simon, S.M. and Blobel, G. (1991) A Protein-Conducting Channel in the Endoplasmic Reticulum. Cell 65: 371-380.

Simon, S.M., Peskin, C.S. and Oster, G.F. (1992) What Drives the Translocation of Proteins? Proc. Natl. Acad. Sci. USA 89: 3770-3774.

Stirling, C.J., Rothblatt, J., Hosobuchi, M., Deshaies, R. and Schekman, R. (1992) Protein Translocation Mutants Defective in the Insertion of Integral Membrane Proteins into the Endoplasmic Reticulum. Mol. Biol. Cell 3:129-142.

Tajima, S., Lauffer, L., Rath, V.L. and Walter, P. (1986) The Signal Recognition Particle Receptor is a Complex that Contains Two Distinct Polypeptide Chains. J. Cell Biol. 103: 1167-1178.

Tazawa, S., Unuma, M., Tondokoro, N., Asano, Y., Ohsumi, T., Ichimura, T. and Sugano, H. (1991) Identification of a Membrane Protein Responsible for Ribosome Binding in Rough Microsomal Membranes. J. Biochem. Tokyo 109: 89-98.

Valentin, K. (1993) SecA Is Plastid-Encoded in a Red Alga Implications for the Evolution of Plastid Genomes and the Thylakoid Protein Import Apparatus. Mol. Gen. Genet. 236: 245-250.

Vogel, J.P., Misra, L.M. and Rose, M.D. (1990) Loss of Bip/Grp78 Function Blocks Translocation of Secretory Proteins in Yeast. J. Cell Biol. 110: 1885-1895.

Walter, P. and Johnson, A. E. (1994) Signal Sequence Recognition and Protein Targeting to the Endoplasmic Reticulum Membrane. Annu. Rev. Cell Biol. 10: 87-119.

Wickner, W.T. and Lodish, H.F. (1985) Multiple Mechanisms of Protein Insertion Into and Across Membranes. Science 230: 400-407.

GLYCOSYLATION MAPPING OF THE INTERACTION BETWEEN TOPOGENIC SEQUENCES AND THE ER TRANSLOCASE

Nilsson, IM., Whitley, P. and von Heijne, G.
Center for Molecular Biology, University of Heidelberg
D-6900 HEIDELBERG, GERMANY
Department of Biochemistry, University of Stockholm
S-106 91 STOCKHOLM, SWEDEN.

Many integral membrane proteins span the hydrophobic core of a membrane with one or more α-helical segments each consisting of about 20 hydrophobic amino acids. The orientation of the hydrophobic stretch in the membrane is determined by its flanking amino acids. In general there are more positively charged residues present in cytoplasmic loops than in extra-cytoplasmic ones (von Heijne, 1986). In both prokaryotic and eukaryotic cells, proteins are targeted for secretion by N-terminal signal sequences with a common basic design: a positively charged N-terminus, a central hydrophobic stretch, and a C-terminal cleavage region that serves as a recognition site for the signal peptidase enzyme (von Heijne, 1985). Signal sequences from prokaryotes and eukaryotes look very similar and are often functionally interchangeable. They are essential for the efficient and selective targeting of the nascent protein chains either to the endoplasmic reticulum (in eukaryotes) or to the cytoplasmic membrane (in prokaryotes) (Gierasch, 1989). Signal sequences also play a central role in the interaction with the translocation machinery of the cell and in the translocation of the polypeptide chains across the membrane. Proteins are co-translationally translocated across the endoplasmic reticulum (ER) membrane. In eukaryotes two kinds of topogenic sequences are important for assembly in ER membrane: Those that initiate translocation such as signal peptides (SPs) and signal-anchor sequences (SAs) and those that halt translocation, stop-transfer signals (STs). Signal peptides and signal anchor sequences differ in that SA sequences tend to have longer hydrophobic cores and lack a signal peptidase cleavage site (von Heijne, 1988).

The signal sequences of secretory and transmembrane proteins bind to the signal recognition particle (SRP) as they emerge from translating ribosomes. This interaction transiently arrests further elongation of the nascent chain. The SRP-nascent chain-ribosome complex is targeted to the ER membrane by binding to the heteromeric SRP receptor (docking protein). This interaction is dependent on the binding of GTP and results in the release of the signal peptide from SRP and continuation of translation (Walter, et al., 1984). The nascent chain then associates with the translocation machinery in the membrane (the "translocon") (Gilmore, 1993) and this machinery transports the growing chain into the lumen of the ER. The translocation machinery is assumed to be a protein-conducting channel formed from transmembrane proteins, such as Sec61p and TRAM (Rapoport, 1991). Other ER membrane

NATO ASI Series, Vol. H 96
Molecular Dynamics of Biomembranes
Edited by Jos A. F. Op den Kamp

proteins, such as the signal peptidase and the oligosaccharyl transferase, may also be included in the translocase (Görlich, et al., 1992).

The attachment of N-linked carbohydrate to eukaryotic secretory and membrane-bound proteins takes place during or soon after translocation of the nascent polypeptide across the ER membrane (Kaplan, et al., 1987). The transfer of high-mannose oligosaccharides from special lipid molecules (dolichol-carriers) to a signal for N-linked glycosylation, Asn-X-Thr/Ser, is catalysed by a membrane bound enzyme (oligosaccharyl transferase, OST). The oligosaccharyl transferase consist of Ribophorin I & II, and a third 48 kD protein (Kelleher, et al., 1992) and has its catalytic site exposed on the lumenal surface of the ER membrane. The characteristics of the acceptor sites are quite well understood (Gavel and von Heijne, 1990; Avanov, 1991), but little is known about the oligosaccharyl transferase itself. Previously it has been shown that there is a precise distance constraint that only allows glycosylation of acceptor sites when the Asn-residue is located at a minimum of 12-14 residues either at the C-terminal side or N-terminal side of a transmembrane segment (Nilsson and von Heijne, 1993).

Design Of Oligosaccharyl Transferase Substrate Molecules

We have chosen to use Leader peptidase (Lep), a protein normally found in the inner membrane of *Escherichia coli* (Fig.1), as a model protein to study the interactions between the ER translocase and SP/SA sequences. The insertion of Lep into the inner membrane of *E.coli* and into dog pancreas microsomes has been well characterized (Johansson, et al., 1993; von Heijne, 1994). Lep has two transmembrane segments H1 (residue 4-22) and H2 (residue 62-76), and both the N- and C-terminus are located in the periplasmic space in *E.coli.* When the protein is integrated into microsomes, using an *in vitro* transcription/translation system, the natural periplasmic segments face the lumen and allow N-linked glycosylation of Asn-Ser-Thr acceptor sites located either N- or C-terminally of transmembrane segments. The minimum distance for efficient glycosylation if the acceptor site is added to the C-terminal side of a transmembrane segment is 12-13 residues and if added to the N-terminal side 14-15 residues (Nilsson and von Heijne, 1993). If the potential glycosylation sites are situated any closer to transmembrane segments, presumably they are too close to the membrane to be recognised by the OST active site. Since the oligosaccharyl transferase is itself an integral part of the ER translocase it can be used as a fixed point of reference against which the position of SP and SA sequences in the translocase can be measured. Recently, we have found a sharp change in the behavior of signal sequences with regard to their minimum glycosylation distance as the length of the hydrophobic core increases from ~17 to ~20 residues. These and other observations suggest that SP and SA sequences may be positioned differently in the ER translocase (Martoglio, et al., 1995; Nilsson, et al., 1994).

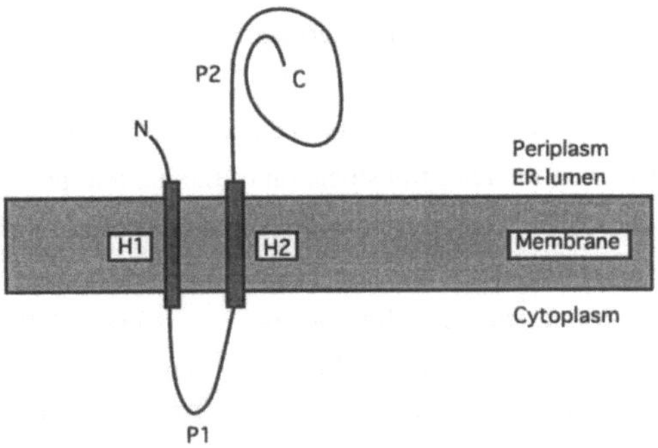

Figure 1. Orientation of the Lep in the inner membrane of *E.coli* and in microsomes. The two hydrophobic, transmembrane segments H1 and H2, the cytoplasmic loop P1, the large periplasmic (lumenal) domain P2.

Analysis Of Topogenic Polyleucine Trans-Membrane Regions

We have characterized polyleucine hydrophobic segments, containing a varying number of leucine residues at different positions in the model protein, in the ER membrane. Homopolymeric units of leucine have been used previously to replace the natural core segment of the cleavable signal peptide of alkaline phosphatase. A stretch of as few as 10 leucine residues was shown to function as a signal sequence and be efficiently cleaved. In *E.coli* however, a core increased to 20 leucine residues, although promoting translocation, was not cleaved and remained membrane bound (Chou and Kendall, 1990). Stop-transfer sequences interrupt the translocation of a polypeptide chain through the membrane and its efficiency is dependent on the hydrophobicity and the length of the inserted stretch. In the ER, more than 19 alanine residues are required for efficient stop-translocation, but as few as 9 leucine residues are sufficient (Kuroiwa, et al., 1991).

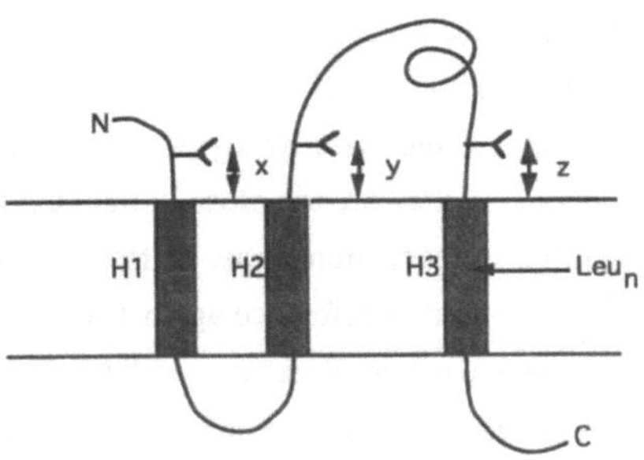

Figure 2. Polyleucine hydrophobic segments of varying length introduced at either H1, H2 or H3 in the model protein, Lep. n; number of leucines, x,y,z; number of amino acid residues to the nearest glycosylation site.

The synthetic polyleucine segments have been used to probe the location of a reverse SA sequence, a SA sequence, and a ST sequence in the ER translocase. The polyleucine transmembrane segments, containing a block of lysine residues at the cytoplasmic end of the stretch to fix this end to the cytoplasmic face of the membrane and three glutamine residues at the lumenal side, were introduced in the model protein. To determine the minimum glycosylation distance of different lengths of polyleucine segments (n=8-29) we introduced a tripeptide (Asn-Ser-Thr) at various distances (x, y, z=8-18) from the polyleucine segment (Fig.2). When the Asn-residue was placed 12 residues (in H2) C-terminally and 16 residues (in H1 & H3) N-terminally of the first glutamine residue on the lumenal side of the membrane, no glycosylation was observed if the number of leucines was ≤15 (Fig.3). However, when the distance to the acceptor site was longer, it was efficiently glycosylated. When the Asn-residue was placed 10 residues (in H2) C-terminally and 15 residues (in H1 & H3) N-terminally from the hydrophobic stretches of 20-29 leucine residues the acceptor site was glycosylated. In all cases except when the ST sequence (H3) is present, the P2-domain was translocated across the ER membrane as determined by its protection from externally added protease. The minimum glycosylation distance is thus different for topogenic signals with short (n≤15) and long (n≥20) hydrophobic segments irrespective of whether they initiate (SA) or halt (ST, R-SA) translocation.

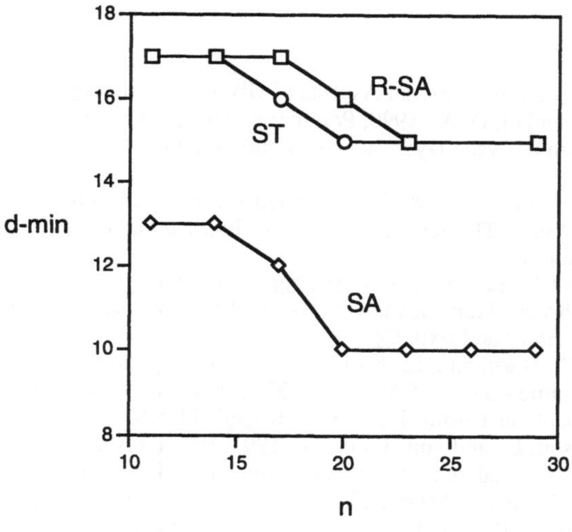

Figure 3. Determination of the minimum glycosylation distance d_{min} as a function of n, the number of leucines in H1, H2 and H3. Squares (H1), triangles (H2) and circles (H3) represent constructs that become glycosylated when expressed in the presence of microsomes; The line shows the variation of d_{min} with n. SA= signal-anchor sequence, ST= stop-transfer sequence and R-SA= reverse signal-anchor sequence.

Conclusion

Our current interpretation of the results presented above is that ST sequences and reverse SA sequences interact with the ER translocase in much the same way as SP and SA sequences. We hypothesize that the signal sequence is embedded within a hydrophobic part of the translocase, and that its apolar core adapts to this environment by adjusting its conformation such that it fills the putative channel. Thus, short apolar cores would be in an extended conformation and longer cores would be progressively more helical until they become long enough to completely fill the channel as a helix. This hypothesis is supported by an earlier observation that a signal peptide cleavage site placed at the C-terminal end of H2 is efficiently cleaved by the signal peptidase when the leucine stretch is ≤17 residue-long, but not when it is longer (Nilsson, et al., 1994). A possible interpretation, supported by crosslinking data (Martoglio, et al., 1995), is that long hydrophobic regions (n≥20) are not completely buried within the translocase, but are partly exposed to the lipid bilayer, whereas topogenic signals with shorter hydrophobic regions (n≤17) are positioned differently in the translocase and are less exposed to the lipids. To get further evidence as to how different topogenic sequences interact with the ER translocase, chemical crosslinking studies are being carried out.

References

Avanov, A. Y., 1991, Conformational Aspects of Glycosylation, Mol Biol-Engl Tr, 25:237.

Chou, M. M. , and Kendall, D. A., 1990, Polymeric Sequences Reveal a Functional Interrelationship Between Hydrophobicity and Length of Signal Peptides, J Biol Chem, 265:2873.

Gavel, Y. , and von Heijne, G., 1990, Sequence Differences Between Glycosylated and Non-Glycosylated Asn-X-Thr Ser Acceptor Sites - Implications for Protein Engineering, Protein Eng, 3:433.

Gierasch, L. M., 1989, Signal sequences, Biochemistry, 28:923.

Gilmore, R., 1993, Protein Translocation Across the Endoplasmic Reticulum - A Tunnel with Toll Booths at Entry and Exit, Cell, 75:589.

Görlich, D., Prehn, S., Hartmann, E., Kalies, K. U. , and Rapoport, T. A., 1992, A Mammalian Homolog of SEC61p and SECYp Is Associated with Ribosomes and Nascent Polypeptides During Translocation, Cell, 71:489.

Johansson, M., Nilsson, I. , and von Heijne, G., 1993, Positively Charged Amino Acids Placed Next to a Signal Sequence Block Protein Translocation More Efficiently in Escherichia-Coli Than in Mammalian Microsomes, Mol Gen Genet, 239:251.

Kaplan, H. A., Welply, J. K. , and Lennarz, W. J., 1987, Oligosaccharyl transferase: the central enzyme in the pathway of glycoprotein assembly, Biochim Biophys Acta, 906:161.

Kelleher, D. J., Kreibich, G. , and Gilmore, R., 1992, Oligosaccharyltransferase Activity is Associated with a Protein Complex Composed of Robophorins I and II and a 48 kd Protein, Cell, 69:55.

Kuroiwa, T., Sakaguchi, M., Mihara, K. , and Omura, T., 1991, Systematic Analysis of Stop-Transfer Sequence for Microsomal Membrane, J Biol Chem, 266:9251.

Martoglio, B., Hofmann, M. W., Brunner, J. , and Dobberstein, B., 1995, The protein-conducting channel in the membrane of the endoplasmic reticulum is open laterllay toward the lipid bilayer, Cell, 81:207.
Nilsson, I. , and von Heijne, G., 1993, Determination of the Distance Between the Oligosaccharyltransferase Active Site and the Endoplasmic Reticulum Membrane, J Biol Chem, 268:5798.
Nilsson, I., Whitley, P. , and von Heijne, G., 1994, The C-terminal ends of internal signal and signal-anchor sequences are positioned differently in the ER translocase, J Cell Biol, 126:1127.
Rapoport, T. A., 1991, Protein Transport Across the Endoplasmic Reticulum Membrane - Facts, Models, Mysteries, FASEB J, 5:2792.
von Heijne, G., 1985, Signal sequences. The limits of variation, J Mol Biol, 184:99.
von Heijne, G., 1986, The distribution of positively charged residues in bacterial inner membrane proteins correlates with the trans-membrane topology, EMBO J, 5:3021.
von Heijne, G., 1988, Transcending the impenetrable: how proteins come to terms with membranes, Biochim Biophys Acta, 947:307.
von Heijne, G., 1994, Membrane Proteins: From Sequence to Structure, Annu Rev Biophys Biomol Struct, 23:167.
Walter, P., Gilmore, R. , and Blobel, G., 1984, Protein translocation across the endoplasmic reticulum, Cell, 38:5.

The various roles of invariant chain in the act of antigen presentation

Tommy W. Nordeng*
Anne Simonsen * and
Oddmund Bakke
Div. Molecular Cell Biology,
Department of Biology,
University of Oslo,
0316 Oslo, Norway

Keywords / Abstract:
Invariant chain / MHC class II / intracellular transport / sorting signals / endosomes / peptide loading / antigen presentation

Foreign antigen are internalized by antigen presenting cells and processed to peptides presented in the context of major histocompatibility complex (MHC) class II molecules to $CD4^+$ T cells at the plasma membrane. Hence, class II molecules have to be sorted to endosomal compartments where they can meet and bind the antigenic peptides. The class II associated invariant chain contains sorting signals required for efficient class II accumulation in endosomes. Invariant chain also has several other features contributing to the immune system's specific combat of invaders.

Introduction

MHC class I and class II molecules bind peptides and present them to T-cells (Townsend and Bodmer, 1989; Harding and Unanue, 1990). Whereas class I molecules predominantly bind peptides generated from endogenously synthesized cytosolic proteins (Nuchtern et al. 1989), class II molecules present peptides derived from internalized degraded material (Lanzavecchia, 1987; Watts and Davidson, 1988).

Class II molecules are expressed on the cell surface as heterodimers of the transmembrane α and β chains (Kaufman et al. 1984). After synthesis and translocation into the

* The two first authors, T.W.N and A.S. have contributed equally to this paper.

NATO ASI Series, Vol. H 96
Molecular Dynamics of Biomembranes
Edited by Jos A. F. Op den Kamp

endoplasmic reticulum (ER), the α and β chains associate with a third transmembrane glycoprotein, the invariant chain (Ii) (Sung and Jones, 1981; Kvist et al. 1982), forming a nine subunit complex (Roche et al. 1991). Following subunit assembly, the αβ-Ii complex traverses the Golgi complex and accumulate in endosomal compartments (Bakke and Dobberstein, 1990; Lotteau et al. 1990; Pieters et al. 1991; Lamb et al. 1991), where Ii is proteolytically degraded (Blum and Cresswell, 1988) allowing class II molecules to separate into dimers and to bind peptides (Blum and Cresswell, 1988; Roche and Cresswell, 1990; Reyes et al. 1991; Roche and Cresswell, 1991). Peptide binding induces a conformational change of class II molecules, resulting in SDS resistant peptide-class II complexes (Sadegh-Nasseri and Germain, 1991; Wettstein et al. 1991; Germain and Hendrix, 1991; Neefjes and Ploegh, 1992). The class II molecules are subsequently transported to the cell surface where they present peptide to CD4 positive T-cells (for reviews see Germain and Margulies, 1993; Neefjes and Momburg, 1993; Cresswell, 1994).

Ii is found to stimulate antigen presentation (Stockinger et al. 1989; Peterson and Miller, 1990; Nadimi et al. 1991; Bertolino et al. 1991; Humbert et al. 1993a), possibly by regulating the intracellular transport of MHC class II molecules to a peptide loading compartment (for reviews see Sant and Miller, 1994; Schmid and Jackson, 1994) and/or by inhibiting premature peptide binding in the ER (Dodi et al. 1994; Long et al. 1994). A chondroitin sulphate form of Ii, found in association with class II molecules on the cell surface (Sant et al. 1985), is found to participate in the T cell interaction through CD 44 molecules on the T cell surface (Naujokas et al. 1993). Ii may also act as an inhibitor of various endosomal proteases (Katunuma et al. 1994) and thereby modulate the degradative capacity of the antigen processing cell. Finally, Ii is also found associated with MHC class I molecules in endosomes (Sugita and Brenner, 1995), and may thus be involved in the class I presentation of exogenous antigen as well. This review addresses the intracellular routing of Ii molecules and the various ways Ii influences routing of MHC molecules, antigen binding and antigen presentation.

Efficient transport of class II molecules out of ER depends on the association with Ii

Class II molecules are composed of two polymorphic transmembrane polypeptides, the α chain (35kD) and the β chain (27kD), that associate to form a noncovalent heterodimer (Kaufman et al. 1984). Whereas the luminal domains of class II molecules have an intrinsic ability to associate (Kjær-Nielsen et al. 1990; Wettstein et al. 1991), interactions by the transmembrane domains promote the formation of correctly assembled complexes (Cosson and Bonifacino, 1992). Class II molecules associate with Ii in the ER (Sung and Jones, 1981; Kvist et al. 1982), but the precise order of assembly is not clear. αβ dimers either assemble stepwise to pre-existing Ii trimers (Roche et al. 1991), or separate α and β chains associate with Ii trimers one at a time to form a nonameric complex (Lamb and Cresswell, 1992; Anderson and Cresswell, 1994).

Coordinate expression of class II and Ii is reported in human tissue (Quaranta et al. 1984; Volc-Platzer et al. 1984) and in tissue culture cells (reviewed in (Long, 1985). However, Ii has been observed in some cell lines with very low expression of class II (Accolla et al. 1985; Lenardo and Baltimore, 1989; Long et al. 1984)and *vice versa.* (Momburg et al. 1986). Expression of Ii and class II molecules can also be increased in a variety of cell lines when stimulated by cytokines (Collins et al. 1984; Kolk and Floyd-Smith, 1992; Brown et al. 1993; Pessara and Koch, 1990; Kolk and Floyd-Smith, 1993; Polla et al. 1986; Noelle et al. 1986). In antigen presenting cells (APC) Ii is usually produced in excess of class II molecules (Kvist et al. 1982; Nguyen and Humphreys, 1989), but Ii is not an absolute prerequisite for the formation of mature class II $\alpha\beta$ dimers (Miller and Germain, 1986; Sekaly et al. 1986) able to bind peptide antigen on the cell surface (Elliott et al. 1994).

Ii has a "chaperoning" role for class II molecules in their maturation process (Claesson-Welsh and Peterson, 1985; Layet and Germain, 1991; Schaiff et al. 1991; Anderson and Miller, 1992). Studies based on cross-linking of proteins suggest that class II and Ii molecules are released from the ER as a nine subunit complex (Roche et al. 1991). Thus, the binding of Ii aids, but is not an absolute requirement for, transport of class II molecules out of the ER (Anderson and Miller, 1992; Layet and Germain, 1991; Nijenhuis et al. 1994). Similarly, in the presence of class II, Ii is transported more efficiently from the ER to the Golgi, (Simonis et al. 1989; Lamb et al. 1991), indicating that exit from the ER is mutually facilitated by $\alpha\beta$-Ii assembly.

Calnexin, a ubiquitous ER phosphoprotein, has been found in association with unassembled subunits of multimeric complexes (Degen and Williams, 1991; Ahluwalia et al. 1992; Galvin et al. 1992; Hochstenbach et al. 1992; Schreiber et al. 1994) and viral (Hammond et al. 1994) and secretory monomeric (Ou et al. 1993) glycoproteins. Anderson and Cresswell (Anderson and Cresswell, 1994) have shown that calnexin associates, possibly cotranslationally, also with Ii and the α and β chains, and remains associated with the assembling $\alpha\beta$-Ii complex until the final class II subunit is added to form the nonameric complex. Dissociation of calnexin parallels egress of $\alpha\beta$-Ii from the ER. These results suggest that calnexin retains and stabilizes both free class II subunits and partially assembled class II-Ii complexes until the nonamer is complete. However, the molecular requirements for association of $\alpha\beta$-Ii complexes with calnexin remain uncertain as neither replacement of the transmembrane region of the DRβ subunit with a GPI-anchor or deglycosylation of the complex constituents abolish calnexin association (Arunachalam and Cresswell, 1995). In addition, class II molecules are found to aggregate with the ER resident chaperone BiP (Bole et al. 1986) when expressed in the absence of Ii (Bonnerot et al. 1994). Thus, it is likely that the $\alpha\beta$-Ii interaction mediates dissociation of these molecules from other chaperons known to bind and to retain misfolded or partially folded proteins in the ER (Fig. 1).

A number of forms of Ii exists, defined by the primary amino acid sequence. The major form is a glycoprotein of 31 - 33 kD (p33). In humans an alternative form of Ii, p35, containing an amino-terminal cytoplasmic extension of 16 residues, results from initiation of translation of an alternative AUG codon upstream of that used to generate p33 (O'Sullivan et al. 1987; Strubin et al. 1986). Additional species of the p33 and p35 forms of Ii result from alternative splicing of an additional exon giving rise to the p41 and p43 forms, respectively, containing a 64 amino acid insertion in the luminal portion (O'Sullivan et al. 1987; Strubin et al. 1986; Koch, 1988). The p35 and p43 forms of Ii are retained in the ER (Lotteau et al. 1990; Lamb et al. 1991) by a calnexin independent mechanism (Arunachalam and Cresswell, 1995) and a double arginine motif in the prolonged amino terminal segment of the cytosolic tail has been identified as the ER retention signal (Schutze et al. 1994). p35Ii inhibits presentation of endogenous antigen in a human fibroblast cell line (Dodi et al. 1994; Long et al. 1994) and *in vitro* studies utilizing isolated αβ–Ii trimers have shown that class II molecules only bind peptide in the absence of Ii (Roche and Cresswell, 1990; Newcomb and Cresswell, 1993; Ericson et al. 1994). Using antigen transgenic and Ii knock-out mice, (Bodmer et al. 1994) found Ii to limit the diversity of endogenous peptides bound to class II molecules. Although association of several peptides to class II alone not efficient in the ER (Ericson et al. 1994), probably due to the dependence of the low pH required for binding (Jensen, 1991), one of the functions of Ii seems to be prevention of premature peptide binding to class II molecules.

In the absence of association with class II molecules, human Ii forms a trimer, and an 18 kD fragment of Ii produced by proteinase K, lacking the cytosolic and the transmembrane region, is also trimeric (Marks et al. 1990). The region between amino acid 163 and 183 (exon 6) is essential for trimerization of Ii (Bijlmakers et al. 1994)(Gedde-Dahl, Freiswinkel, Staschewski, Schenk, Koch and Bakke, submitted). Together these data suggest that trimerization of Ii is a function of the carboxy-terminal luminal region of the molecule only. However, recent data indicates that this region of Ii is not required for maintenance of the nonameric complex (Amigorena et al. 1995). The association of p33/p41Ii with p35/p43Ii in trimers and higher molecular weight aggregates may contribute to the retention of the p33/p41 forms in the ER (Marks et al. 1990; Lamb and Cresswell, 1992), where they undergo degradation to smaller, amino-terminally cleaved forms (Nguyen and Humphreys, 1989; Marks et al. 1990). In contrast, when not complexed with the p35/p43 forms, a fractions (5-20%) of human p33Ii move through the Golgi complex and accumulate in vesicular structures (Bakke and Dobberstein, 1990; Lotteau et al. 1990; Simonsen et al. 1993).

The cytosolic tail of Ii contains several sorting signals

After release from the ER, the αβ-Ii nonamers are transported through the Golgi complex to the endocytic pathway where Ii is degraded and the class II molecules liberated to bind peptides. The intracellular route used by class II-Ii to the endosomal compartments is still under debate. The class II-Ii complex could either be sorted directly from the trans-Golgi network (TGN) to an endosomal compartment (Neefjes et al. 1990; Peters et al. 1991; Odorizzi et al. 1994; Bénaroch et al. 1995), or indirectly via the plasma membrane by rapid internalization (Roche et al. 1993; Bremnes et al. 1994; Odorizzi et al. 1994) (Fig. 1). Several reports indicate that class II-Ii complexes move from the TGN to early endosomes (for review see Germain and Margulies, 1993; Neefjes and Momburg, 1993; Cresswell, 1994) and then accumulate in late endosomal compartments distinct from terminal, dense lysosomes, which show virtually no class II content (Harding et al. 1990; Peters et al. 1991; Harding and Geuze, 1993). From the above we must conclude that class II-Ii complexes may reach endosomes by dual pathways (Odorizzi et al. 1994; Bénaroch et al. 1995) and it has been reported that other membrane proteins, including the lysosomal associated membrane protein LAMP1 (Carlsson and Fukuda, 1992), the lysosomal membrane glycoprotein lgp120 (lgp-A) (Harter and Mellman, 1992), and the mannose-6-phosphate/insulin-like growth factor-II receptor (Johnson and Kornfeld, 1992), are able to reach the endocytic pathway by two different routes from the TGN; one direct and one via the plasma membrane. Class II molecules expressed in the absence of Ii may also localize in endosomes (Salamero et al. 1990; Simonsen et al. 1993; Humbert et al. 1993b; Pinet et al. 1995), possibly by recycling from the plasma membrane. However, for efficient sorting of newly synthesized class II molecules to endosomes, association with Ii is required (Lotteau et al. 1990; Simonsen et al. 1993).

Class I molecules have been found to co-localize with Ii in endosomes in HeLa cells (Sugita and Brenner, 1995) and Ii is able to associate with class I molecules in the ER (Cerundolo et al. 1992; Sugita and Brenner, 1995). This phenomenon may serve as an explanation for the presentation of exogenous peptides by class I molecules (Carbone and Bevan, 1990; Rock et al. 1990; Pfeifer et al. 1993; Kovacsovics-Bankowski et al. 1993).

The cytosolic tail of Ii has been shown to contain information for endosomal sorting (Bakke and Dobberstein, 1990; Lotteau et al. 1990; Simonsen et al. 1993; Pieters et al. 1993). Two di-leucine based motifs, leucine-isoleucine (LI) in position 7 and 8 and methionine-leucine (ML) in position 16 and 17 in the cytosolic tail are independently sufficient for endosomal localization of Ii and for rapid internalization of Ii from the plasma membrane ($t\ 1/2 < 1$ min)(Bremnes et al. 1994). This could explain the observed low steady state level of Ii on the cell surface. Two-dimensional nuclear magnetic resonance spectroscopy (2D-NMR) studies on a peptide corresponding to the cytosolic tail of Ii show that the LI motif is located within a regular α-helix (Motta, Bremnes, Castiglione, Morelli, Frank, Saviano, Bakke, submitted). This prediction was supported by biological data showing that neighbouring recidues to LI in the secondary structure could abolish internalization whereas recidues opposite to this were mutated without influencing the sorting, making this combined motif a putative signal patch. Data from Arneson and Miller (1995) show furthermore that multimers of the Ii cytosolic tail may be required for efficient endosomal localization of class II-Ii complexes.

We have studied the sorting of class II and Ii in polarized Madin Darby Canine Kidney (MDCK) epithelial cells (Simonsen, Stang, Bremnes, Røe, Prydz and Bakke, submitted). It has been suggested that non-polarized and polarized cells have common sorting pathways and sorting machinery (Matter and Mellman, 1994), and in fact APC, as B cells, become transiently polarized upon interaction with T cells. Thus, analyzing the sorting of these molecules in polarized cells might add information about these processes in non-polarized cells as well. Moreover, tissue-epithelial cells have been shown to express class II molecules and present antigen to intra-epithelial lymphocytes present on the vascular side of the epithelium (for review see Brandtzaeg et al. 1988). We found that Ii is required for efficient targeting of the class II molecules to the basolateral surface, the side of the epithelia facing the vascular space. Both di-leucine based motifs are individually sufficient for basolateral targeting of Ii. In addition, a separate novel basolateral signal is located within the 10 membrane-proximal residues of the Ii cytosolic tail (Simonsen, Stang, Bremnes, Røe, Prydz and Bakke, submitted) and class II molecules may also posess sorting information for basolateral distribution.

Is the pathway via the plasma membrane just a recover mechanism to retrieve Ii missorted to the cell surface, or is there some immunological significance of more than one entry for class II-Ii to the endocytic pathway? Obviously, co-internalization of antigen and class II-Ii complexes into the same endocytic vesicle could ensure the meeting of class II molecules and antigenic peptides at a proper processing stage of easily degradable antigen. Indeed, antigen processing and peptide association to class II have been demonstrated in early endosomes (McCoy et al. 1993a; 1993b; Gagliardi et al. 1994). Cells expressing class II molecules in the absence of Ii or together with a truncated Ii lacking the endosomal sorting signals have been shown to present certain antigen very efficiently (Anderson et al. 1993; Nijenhuis et al. 1994; Pinet et al. 1994), suggesting that recycling class II molecules reach endosomes containing degraded antigen (Long et al. 1993; Nijenhuis et al. 1994; Pinet et al. 1995). On the other hand, direct sorting of class II-Ii complexes to a late endosomal compartment could prevent occupancy of all available class II molecules by peptides from easily degradable antigen. Alternative entry levels of class II-Ii complexes to the endocytic pathway may thus promote the presentation of a larger spectrum of antigenic peptides, regardless of the vulnerability of the endocytosed antigen to protease activity. The multiple sorting signals in the Ii cytosolic tail might be involved in the fine tuning of the intracellular transport of the class II-Ii complex, like sorting between endosomal populations or retention in some endosomal maturation stage.

Class II molecules are retained in the antigen processing pathway

Newly synthesized class II molecules are retained in the endocytic pathway for 1-3 hours before they appear at the cell surface (Neefjes et al. 1990). In endosomes Ii is sequentially degraded by proteases (Blum and Cresswell, 1988; Marks et al. 1990; Pieters et al. 1991; Newcomb and Cresswell, 1993; Nguyen and Humphreys, 1989; Xu et al. 1994), an event required for peptide-class II complex formation (Roche and Cresswell, 1991; Neefjes and Ploegh, 1992; Daibata et al. 1994; Maric et al. 1994; Amigorena et al. 1995). Degradation of Ii may a rate limiting step in the transport of class II molecules through the endocytic pathway, as inhibition of Ii degradation by protease inhibitors or by lysosomotrophic agents delays the surface appearance of class II molecules (Neefjes and Ploegh, 1992; Loss and Sant, 1993; Amigorena et al. 1995).

By subcellular fractionation of a human fibroblast cell line stably transfected with Ii and HLA-DR, we have shown that material endocytosed in the fluid phase is retained in early endosomes together with Ii and class II molecules (Gorvel et al. 1995). Ii has been found to accumulate in an unusual cohort of intracellular large vesicular structures (LVS) at high level of expression in transfected cells (Romagnoli et al. 1993; Pieters et al. 1993; Stang et al. 1993) and in a processing defect cell line (Riberdy et al. 1994), and the rate of endocytic flow seems to be lowered from this compartment (Romagnoli et al. 1993; Gorvel et al. 1995).
In conclusion, these findings suggest that Ii regulates movement of endocytosed antigen and of newly synthesized class II-Ii complexes in the endosomal pathway. This may enhance endosomal fusion and mixing so that internalized antigen and class II molecules will reside in the same maturing endocytic vesicle. Such a mechanism might contribute to the efficiency of peptide capture by class II when antigen is limiting. The phenotypical alterations induced by Ii, however, are so far only seen in transfected cells. The effect of Ii is thus a biological phenomena, but it still remains to see whether the same or similar mechanisms are active in "professional" antigen presenting cells as B cells and macrophages. Interestingly, however, is the observation that isolated Langerhans cells expressing Ii are found to contain large lucent acidic vacuoles with the characteristics of early endosomes (Kämpgen et al. 1991; Pure et al. 1990; Stössel et al. 1990).

Processing events in the endocytic pathway

In general, for class II presentation, antigen must be internalized by endocytosis for subsequent processing within endosomes/lysosomes. The processing includes both proteolysis, apparently mediated by a spectrum of different proteases (Vidard et al. 1992), and disulphide reduction (Jensen, 1991; Hampl et al. 1992). The disulphide reduction appears to be mediated in high-density, lysosome-like compartments (Collins et al. 1991). Immunoelectron microscopy has illustrated that lysosome-like compartments in some cells contain high levels of class II molecules (Peters et al. 1991; Harding and Geuze, 1992). Processing of endocytosed antigen is blocked at 18°C (Harding and Unanue, 1990), also suggesting a requirement for lysosomal or late endosomal function, as transport from early endosomes to later compartments is blocked at this temperature. Furthermore, liposome-encapsuled antigen are efficiently processed only after lysosomal targeting (Harding et al. 1991a; Harding et al. 1991b). These observations suggest a role for lysosomes in class II antigen processing.

During intracellular transport of Class II-Ii complexes, Ii is sequentially degraded from the luminal C-terminal side, but is still associated with class II molecules (Blum and Cresswell, 1988; Nguyen and Humphreys, 1989; Marks et al. 1990; Pieters et al. 1991; Newcomb and Cresswell, 1993; Xu et al. 1994) in a nonameric complex (Amigorena et al. 1995). Processing of Ii is necessary to achieve peptide loading on mature class II molecules (Roche and Cresswell, 1991; Neefjes and Ploegh, 1992; Daibata et al. 1994). Although it is established that processing of Ii and exogenous antigen involves proteolysis, it has been difficult to resolve the involvement of the specific proteases (for review see Berg et al. 1995). Both Cathepsin D and Cathepsin B are necessary for Ii degradation (Mizuochi et al. 1994; Daibata et al. 1994) and it seems that an aspartyl protease may be involved in the early steps of Ii degradation, whereas a cysteine protease catalyzes the final steps (Maric et al. 1994). Both cysteine proteases (Cathepsin B and L) (Takahashi et al. 1989; Neefjes and Ploegh, 1992; Mizuochi et al. 1994) and the aspartyl protease Cathepsin D (Diment, 1990; Mizuochi et al. 1994) are involved in processing of exogenous antigen. Ii has sequence similarity with the cystatin family of cysteine protease inhibitors, and Ii has recently been demonstrated to inhibit the enzymatic activity of Cathepsin L and H, whereas Cathepsin B was not inhibited (Katunuma et al. 1994). Thus, another role of Ii in antigen presentation may be to modulate endosomal degradation. This function of Ii may also explain the Ii mediated endosomal accumulation/retention, as altered proteolytic activity influences the rate of endocytic flow (Neefjes and Ploegh, 1992; Zachgo et al. 1992).

The peptide loading compartments

Lysosomal compartments may be heterogeneous, and a degradative compartment containing high levels of class II may represent an earlier compartment than terminal lysosomes (Harding and Geuze, 1992). Attempts have been made to describe the peptide loading compartment as distinct class II-positive vesicular structures, referred to as MHC class II Compartment (MIIC) (Peters et al. 1991; West et al. 1994; Kleijmeer et al. 1994; Tulp et al. 1994), Class II Vesicle (CIIV) (Amigorena et al. 1994; Escola et al. 1995) compartments containing floppy or compact class II molecules (Qiu et al. 1994); commonly referred to as Compartment for Peptide Loading (CPL) (Schmid and Jackson, 1994) The formation of such a compartment, proposed to resemble transcytotic endosomes in polarized cells (Barroso and Sztul, 1994; Apodaca et al. 1994) and the class II containing transport vesicles in Mel Juso cells

(Zachgo et al. 1992), is suggested to be induced by the expression of class II itself (Calafat et al. 1994). However, Ii, class II and the human leukocyte antigen HLA-DM (see later for description) has been shown to be the minimally required components to reconstitute an operational MIIC in nonantigen-presenting cells (Karlsson et al. 1994). Recent data show that SDS-stable class II molecules are present in MIIC (Escola et al. 1995) and CIIV (Escola et al. 1995) compartments, suggesting that peptide-class II complexes could be formed therein. A compartment, distinct from MIIC-like compartments, that contain Ii and "floppy" SDS resistant class II molecules receptive to peptide binding, but not containing processed antigen, has also been described (Qiu et al. 1994).

Class II molecules have also been localized to endosomes earlier in the pathway (Pieters et al. 1991; Harding and Geuze, 1992) and several studies have shown that antigen can be processed and associated to class II in early endosomal compartments (McCoy et al. 1993a; Gagliardi et al. 1994; Amigorena et al. 1995). Moreover, retrograde traffic between terminal lysosomes and late endosomes (Jahraus et al. 1994) could enable binding of terminally processed antigen without entry of class II molecules into terminal lysosomes. Based on the different characteristics of the peptide loading compartments described and the different degradative requirements for antigen processing reported, we conclude that peptide loading may occur at various steps of the endocytic pathway (Escola et al. 1995) dependent on the cell specific degradative capacity of endosomal populations and the vulnerability of disparate antigen to proteolysis.

Peptide loading

The analysis of mutant B-cell lines expressing class II molecules unable to present exogenous antigen (Mellins et al. 1990), led to the search of genes mapped to the class II region of the MHC locus (Mellins et al. 1991; Ceman et al. 1992; Riberdy and Cresswell, 1992; Malnati et al. 1993; Ceman et al. 1994). The two genes HLA-DMA and HLA-DMB, encoding a heterodimer, HLA-DM, has been shown to play a critical regulatory role in class II-restricted antigen presentation, probably by promoting endosomal peptide loading of class II molecules (Morris et al. 1994; Fling et al. 1994; Sloan et al. 1995) (Fig. 1). Cell lines lacking HLA-DM are defect in the presentation of intact protein antigen, but not of exogenously supplied peptides. Moreover, the class II molecules lack the characteristics of mature, peptide-loaded molecules as SDS stability and recognition by conformation-specific antibodies. The major fraction of class II molecules in such cells lack a wild type repertoire of endogenous peptides. Instead, they are associated with a nested set of invariant chain-derived peptides, CLIP, encompassing residue 80-103 of the Ii luminal domain (Riberdy et al. 1992; Sette et al. 1992).

The ability of Ii to interact with class II and interfere with peptide loading has been mapped to Ii exon 3, the region encoding CLIP (Freisewinkel et al. 1993; Bijlmakers et al.

1994; Romagnoli and Germain, 1994). Moreover, this region is essential for the promoting effect of Ii on class II folding and ER to Golgi transport (Romagnoli and Germain, 1994), suggesting functional similarities between the CLIP-dependent effects of Ii on class II post synthetic behaviour and the effects of peptide binding on class I molecules in the ER. Recent data suggest that CLIP binds to class II in an analogous fashion as conventional antigenic peptides and actually occupy the class II peptide binding site (Chicz et al. 1993; Sette et al. 1995; Malcherek et al. 1995). CLIP has been found stably associated with several class II alleles (Rudensky et al. 1991; Hunt et al. 1992; Chicz et al. 1992; 1993) and the affinity of the interaction is controlled by polymorphic residues in the class II chains (Sette et al. 1995). Furthermore, structural changes in CLIP can modulate class II binding in an allele-dependent manner (Sette et al. 1995). CLIP can be released from class II molecules at pH 5, in contrast to conventional natural ligands, which are irreversibly bound (Malcherek et al. 1995). However, introduction of allele-specific anchor residues in CLIP generate a CLIP variant that is stable in SDS as allele-specific ligands (Malcherek et al. 1995). Thus, CLIP seem to be designed for promiscuous binding in the groove of many class II molecules by taking advantage of one or more supermotifs.

A role for HLA-HLA-DM in the displacement of CLIP from newly synthesized class II molecules is proposed (Brooks et al. 1994; Stebbins et al. 1995; Sloan et al. 1995) and different hypothesis suggest the role(s) of HLA-DM: 1) a class II chaperone, mediating either a trafficking step or a folding event necessary for intracellular peptide binding, 2) a shuttle to deliver peptides to a compartment for association with class II molecules, and 3) a "CLIP-sink" enabling removal of CLIP without the use of peptidases. The latter possibility is less likely, as HLA-DM was found to increase the amount of SDS stable class II molecules also in the absence of Ii (Monji et al. 1994) and CLIP association with HLA-DM has not been observed. The murine equivalent of HLA-DM, H-2M, has been shown to associate with Ii during synthesis (Monji et al. 1994), suggesting that Ii prevents H-2M from exerting its function before reaching the endosomal system. Monji et al. (1994) also found that HLA-DM physically interact with class II molecules during their functional maturation, suggesting that HLA-DM may direct the class II molecules into a low pH compartment where CLIPs can be removed nonenzymatically and cognate peptides with higher affinity for class II could bind. HLA-DM has been found to accumulate together with class II molecules in typical CPL structures (Sanderson et al. 1994; Karlsson et al. 1994). Also a 10-kD Ii fragment containing CLIP associates with SDS-labile class II molecules in CIIV, and removal of this fragment is followed by formation of SDS-stable class II molecules and their appearance at the plasma membrane (Amigorena et al. 1995).

Transport of peptide-class II complexes to the surface

Peptide loading is shown to promote a conformational change in class II molecules, converting them to an SDS resistant "compact" form (Sadegh-Nasseri and Germain, 1991; Wettstein et al. 1991; Germain and Hendrix, 1991; Neefjes and Ploegh, 1992). Degradation of the peptide binding block (Ii) and processing of antigen have to parallel peptide binding and transport of peptide-class II complexes to the plasma membrane. If the class II molecules themselves were to contain the sorting signals, some mechanism would have to ensure a change in sorting information as a response to peptide binding. Such mechanisms exists for instance for receptors on the cell surface which are autophosphorylated upon ligand binding, thereby activating internalization signals (Honegger et al. 1987; Glenney et al. 1988; Chen et al. 1989). Activation of several lysosomal proteases are also coupled to proteolytic cleavage and thus entry into a proteolytic compartment (Neufeld, 1991).

In class II antigen presentation, both removal of the peptide binding block and the change of transport route seem to be combined by a single event; Ii degradation. When associated, Ii may mask a putative sorting signal in class II molecules (Pinet et al. 1995; Chervonsky et al. 1994). As Ii is degraded, antigen can bind and class II molecules may be liberated to enter a compartment for recycling to the plasma membrane. The conformational change of class II molecules observed upon peptide binding (Sadegh-Nasseri and Germain, 1991; Wettstein et al. 1991; Germain and Hendrix, 1991; Neefjes and Ploegh, 1992) could also mediate a change in sorting destination. It appears that peptide-class II complexes are formed within an intracellular compartment using nascent class II (Davidson et al. 1991; Lanzavecchia et al. 1992), but a second pathway independent of invariant chain and independent of newly synthesized class II molecules has also been reported (Pinet et al. 1994; Pinet et al. 1995). Endocytosis and recycling of class II molecules have been demonstrated (Harding and Unanue, 1989; Davis and Cresswell, 1990; Reid and Watts, 1990), but the functional significance of this is unclear.

Peptide-class II complexes are transported to the cell surface within 4 hours after endocytosis of the antigen (Qiu et al. 1994). The route taken to the plasma membrane is not known, but recent data suggest that multivesicular class II containing compartments are able to fuse directly with the plasma membrane, liberating the internal vesicles (exosomes) to present antigen in a cell-independent fashion (Raposo et al. 1995).

Fig 1

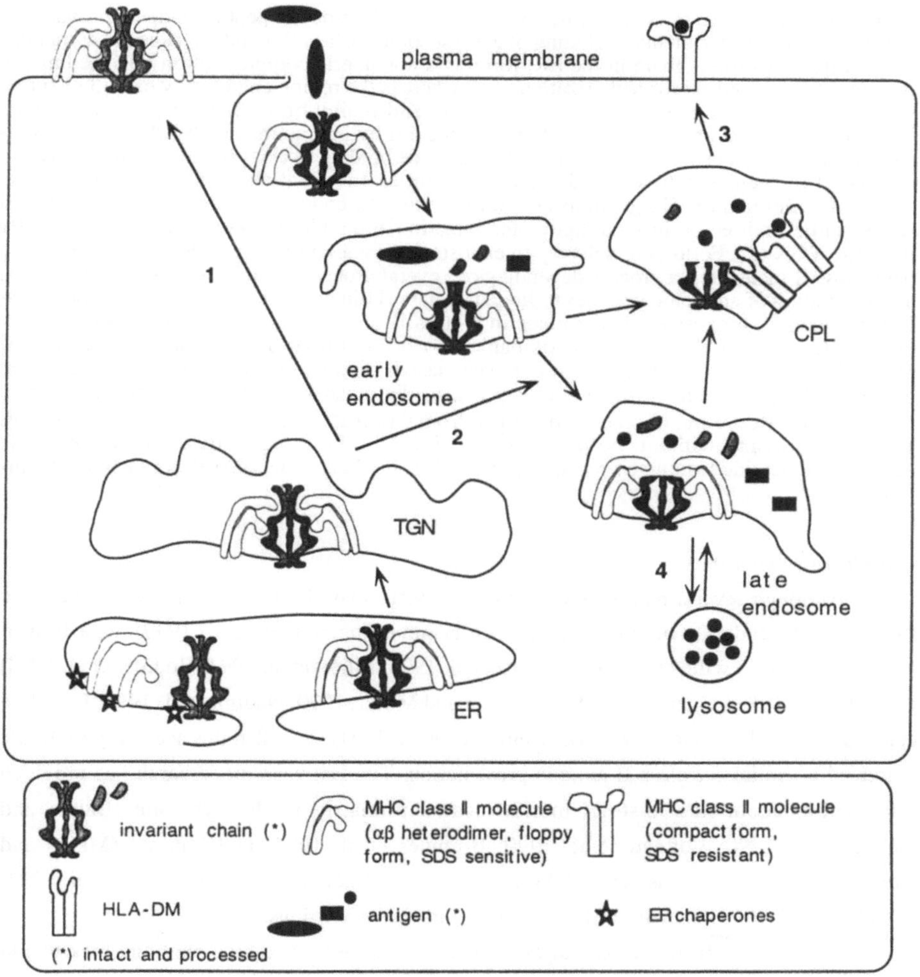
plasma membrane
1
2
3
4
early
endosome
CPL
TGN
late
endosome
ER
lysosome
invariant chain (*)
MHC class II molecule
(αβ heterodimer, floppy
form, SDS sensitive)
MHC class II molecule
(compact form,
SDS resistant)
HLA-DM
antigen (*)
ER chaperones
(*) intact and processed

Model for the class II processing pathway, the assembly and intracellular transport of class II molecules

Exogenous antigen are internalized by endocytosis and degraded to peptide antigen, primarily in late endocytic compartments, including phagolysosomes. Class II α and β chains are associated with trimeric invariant chain in the ER, forming a nonameric complex. Exon 6 in the luminal domain of Ii is required for trimerization of Ii, whereas the region encoded by exon 3 of Ii (the CLIP region) is involved in class II association and inhibition of class II peptide binding. Partially assembled complexes are retained in the ER by interactions with ER chaperones, as calnexin and BiP. Thus, Ii prevents premature peptide binding to class II and stimulates ER exit of class II molecules with a peptide binding potential. Invariant chain-class II nonamers are transported through the Golgi complex and may enter the endocytic pathway by two different routes; either indirectly through rapid internalization at the plasma membrane (1) or directly from the TGN (2). Following delivery to endosomes, invariant chain is cleaved and removed from class II molecules, a process depending on several proteases and HLA-DM. Ii degradation allows binding of antigenic peptides to liberated class II dimers. The formation of peptide-class II complexes is suggested to take place in a class II enriched specialized Compartment for Peptide Loading (CPL) and is accompanied by a conformational change in the class II molecules, resulting in SDS resistant compact class II dimers. Peptide-class II complexes are transported, by an undetermined pathway (3), to the plasma membrane for presentation to $CD4^+$ T cells. Recycling of processed antigen from terminal lysosomes (4) could aid extensive antigen processing without the presence of class II in these organelles. The last possible pathway for peptide loading by recycling MHC class II (Pinet et al. 1995)is not included in this figure.

Concluding remarks

Although several reports have shown that MHC class II antigen presentation may also function independently of Ii (for review see Hämmerling and Moreno, 1990), data from Ii knock-out mice (Viville et al. 1993; Bikoff et al. 1993; Elliott et al. 1994; Bodmer et al. 1994) and cell lines (Stockinger et al. 1989; Peterson and Miller, 1990; Nadimi et al. 1991; Bertolino et al. 1991; Humbert et al. 1993a; van Kemenade et al. 1994) show that this accessory molecule is indeed essential for class II antigen presentation. The last years of research has provided information about how distinct motifs located primarily in the cytosolic (Bakke and Dobberstein, 1990; Lotteau et al. 1990; Bremnes et al. 1994) and luminal (Marks and Cresswell, 1986; Naujokas et al. 1993; Peterson and Miller, 1992; Freisewinkel et al. 1993; Bertolino et al. 1995) domains of Ii contribute to its many functions.

Exit of class II molecules out of the ER in absence of a peptide binding block would eventually lead to presentation of peptides present in the biosynthetic compartments. Ii prevents premature peptide binding to class II molecules, probably by occupying the peptide binding groove, a function shown to depend on the luminal CLIP-containing region of Ii. Moreover, p35/43 Ii, containing an ER retention signal in the cytosolic tail, associate with p33/41 Ii and αβ-chains and might, together with ER-chaperones, retain class II-Ii molecules until properly folded complexes is obtained. Thus, both luminal and cytosolic domains of Ii participate to ensure that only class II molecules with an antigenic peptide binding potential are allowed to travel the roads to a peptide binding compartment. The cytosolic tail of Ii also contains signals for sorting of the class II molecules to, and probably retention in, an endosomal compartment. The Ii mediated endosomal retention observed in Ii-transfected fibroblasts (Romagnoli et al. 1993; Gorvel et al. 1995) may be a mechanism to increase the presentation potential of APC as antigen presentation has been demonstrated to be stimulated as the Ii/class II ratio is increased (Bertolino et al. 1991). Degradation and removal of Ii liberates free class II molecules able to bind antigenic peptides and might reveal signals for a change in the sorting destination of the class II molecules.

Since the immunologist and cell biologist in the late 1980´s realised that MHC class II was found in endosomal compartments (Guagliardi et al. 1990; Pieters et al. 1991) and that Ii was needed for efficient presentation (Stockinger et al. 1989) and contained sorting information for endosomal compartments (First presented at a previous NATO/FEBS conference in this series, Cargese, 1989, as published in Bakke and Dobberstein, 1990) there has been a great effort by many groups to solve the mechanisms behind antigen presenting via MHC class II. It was known that MHC class I and class II did in general present peptides from endogenous and exogenous sources and based on the emerging new data this difference was to a large degree postulated to be caused by the MHC associated Ii (Koch et al. 1989; Long, 1989). As reviewed here, the proposed influence of Ii on MHC class II, like blocking peptide binding, sorting to endosomes, aiding transport and altering conformation have all been experimentally proven and we have now entered the stage where the aim is to understand the more detailed mechanism behind MHC class II antigen presentation.

As discussed in this review many of the published reports do not agree and some aspects are confusing even for people working with the problems. Fortunately, the interest in the field has also lead to important findings that have broadened our understanding considerably. Among these the major "milestones" are the discovery of CLIP and of HLA-DM that may catalyse the release of CLIP from class II, thereby changing class II plus peptide into an SDS resistant conformation. In addition to this, an area that has not been dealt with in this review, the structure determination of MHC (for review see Bjorkman and Burmeister, 1994) has been of great help to understand MHC-peptide interactions, and model and visualize the molecular details. Other questions like the definition of the peptide loading compartment, the proteolytic machinery involved in antigen processing and where this take place and the intracellular route taken by MHC II is still not settled. The future will show whether there is a simple explanation to these questions or if nature use for instance a multitude of proteolytic machineries and many peptide loading compartments and intracellular pathways to cope with the high number of antigens that has to be presented for an organism to survive.

References

Accolla RS, Carra G and Guardiola J (1985) Reactivation by a trans-acting factor of human major histocompatibility complex Ia gene expression in interspecies hybrids between an Ia-negative human B-cell variant and an Ia-positive mouse B-cell lymphoma. Proc. Natl. Acad. Sci. USA 82:5145-5149

Ahluwalia N, Bergeron JJ, Wada I, Degen E and Williams DB (1992) The p88 molecular chaperone is identical to the endoplasmic reticulum membrane protein, calnexin. J. Biol. Chem. 267:10914-10918

Amigorena S, Drake JR, Webster P and Mellman I (1994) Transient accumulation of new class II MHC molecules in a novel endocytic compartment in B lymphocytes. Nature 369:113-120

Amigorena S, Webster P, Drake J, Newcomb J, Cresswell P and Mellman I (1995) Invariant chain cleavage and peptide loading in major histocompatibility complex class II vesicles. J. Exp. Med. 181:1729-1741

Anderson KS and Cresswell P (1994) A role for calnexin (IP90) in the assembly of class II MHC molecules. EMBO J. 13 (3):675-682

Anderson MS and Miller J (1992) Invariant chain can function as a chaperone protein for class II major histocompatibility complex molecules. Proc. Natl. Acad. Sci. USA 89:2282-2286

Anderson MS, Swier K, Arneson L and Miller J (1993) Enhanced antigen presentation in the absence of the invariant chain endosomal localization signal. J. Exp. Med. 178:1959-1969

Apodaca G, Katz LA and Mostov KE (1994) Receptor-mediated transcytosis of IgA in MDCK cells is via apical recycling endosomes. J. Cell Biol. 125:67-86

Arneson LS and Miller J (1995) Efficient endosomal localization of major histocompatibility complex class II-invariant chain complexes requires multimerization of the invariant chain targeting sequence. J. Cell Biol. 129:1217-1228

Arunachalam B and Cresswell P (1995) Molecular requirements for the interaction of class II major histocompatibility complex molecules and invariant chain with calnexin. J. Biol. Chem. 270:2784-2790

Bakke O and Dobberstein B (1990) MHC class II-associated invariant chain contains a sorting signal for endosomal compartments. Cell 63:707-716

Barroso M and Sztul ES (1994) Basolateral to apical transcytosis in polarized cells is indirect and involved BFA and trimeric G protein sensitive passage through the apical endosome. J. Cell Biol. 124:83-100

Berg T, Gjoen T and Bakke O (1995) Physiological functions of endosomal proteolysis. Biochem. J. 307:313-326

Bertolino P, Forquet F, Pont S, Koch N, Gerlier D and Rabourdin-Combe C (1991) Correlation between invariant chain expression level and capability to present antigen to MHC class II-restricted T cells. Int. Immunol. 3:435-443

Bertolino P, Staschewski M, Trescol-Biémont M-C, Freisewinkel IM, Schenck K, Chrétien I, Forquet F, Gerlier D, Rabourdin-Combe C and Koch N (1995) Deletion of a C-terminal sequence of the class II-associated invariant chain abrogates invariant chains oligomer formation and class II antigen presentation. J. Immunol. 154:5620-5629

Bijlmakers ME, Benaroch P and Ploegh HL (1994) Mapping functional regions in the lumenal domain of the class II-associated invariant chain. J. Exp. Med. 180:623-629

Bikoff EK, Huang LY, Episkopou V, van Meerwijk J, Germain RN and Robertson EJ (1993) Defective major histocompatibility complex class II assembly, transport, peptide acquisition, and CD4+ T cell selection in mice lacking invariant chain expression. J. Exp. Med. 177:1699-1712

Bjorkman PJ and Burmeister WP (1994) Structures of two classes of MHC molecules elucidated: Crucial differences and similarities. Curr. Opin. Struct. Biol. 4:852-856

Blum JS and Cresswell P (1988) Role for intracellular proteases in the processing and transport of class II HLA antigens. Proc. Natl. Acad. Sci. USA 85:3975-3979

Bodmer H, Viville S, Benoist C and Mathis D (1994) Diversity of endogenous epitopes bound to MHC class II molecules limited by invariant chain. Science 263:1284-1286

Bole DG, Hendershot LM and Kearney JF (1986) Posttranslational association of immunoglobulin heavy chain binding protein with nascent heavy chains in nonsecreting and secreting hybridomas. J. Cell Biol. 102:1558-1566

Bonnerot C, Marks MS, Cosson P, Robertson EJ, Bikoff EK, Germain RN and Bonifacino JS (1994) Association with BiP and aggregation of class II MHC molecules synthesized in the absence of invariant chain. EMBO J. 13:934-944

Brandtzaeg P, Sollid LM, Thrane PS, Kvale D, Bjerke K, Scott H, Kett K and Rognum TO (1988) Lymphoepithelial interactions in the mucosal immune system. [Review]. Gut 29:1116-1130

Bremnes B, Madsen T, Gedde-Dahl M and Bakke O (1994) An LI and ML motif in the cytoplasmic tail of the MHC-associated invariant chain mediate rapid internalization. J. Cell Sci. 107:2021-2032

Brooks AG, Campbell PL, Reynolds P, Gautam AM and McCluskey J (1994) Antigen presentation and assembly by mouse I-A^k class II molecules in human APC containing deleted or mutated *HLA DM* genes. J. Immunol. 153:5382-5392

Brown AM, Wright KL and Ting JP-Y (1993) Human major histocompatibility complex class II-associated invariant chain gene promoter. Functional analysis and *in vivo* protein/DNA interactions of constitutive and IFN-gamma-induced expression. J. Biol. Chem. 268:26328-26333

Bénaroch P, Yilla M, Raposo G, Ito K, Miwa K, Geuze HJ and Ploegh HL (1995) How MHC class II molecules reach the endocytic pathway. EMBO J. 14:37-49

Calafat J, Nijenhuis M, Janssen H, Tulp A, Dusseljee S, Wubbolts R and Neefjes J (1994) Major histocompatibility complex class II molecules induce the formation of endocytic MIIC-like structures. J. Cell Biol. 126:967-977

Carbone FR and Bevan MJ (1990) Class I-restricted processing and presentation of exogenous cell-associated antigen in vivo. J. Exp. Med. 171:377-387

Carlsson SR and Fukuda M (1992) The lysosomal membrane glycoprotein lamp-1 is transported to lysosomes by two alternative pathways. Arch. Biochem. Biophys. 296:630-639

Ceman S, Rudersdorf R, Long EO and Demars R (1992) MHC class II deletion mutant expresses normal levels of transgene encoded class II molecules that have abnormal conformation and impaired antigen presentation ability. J. Immunol. 149:754-761

Ceman S, Petersen JW, Pinet V and Demars R (1994) Gene required for normal MHC class II expression and function is localized to approximately 45 kb of DNA in the class II region of the MHC. J. Immunol. 152:2865-2873

Cerundolo V, Elliott T, Elvin J, Bastin J and Townsend A (1992) Association of the human invariant chain with H-2 Db class I molecules. Eur. J. Immunol. 22:2243-2248

Chen WS, Lazar CS, Lund KA, Welsh JB, Chang CP, Walton GM, Der CJ, Wiley HS, Gill GN and Rosenfeld MG (1989) Functional independence of the epidermal growth factor receptor from a domain required for ligand-induced internalization and calcium regulation. Cell 59:33-43

Chervonsky AV, Gordon L and Sant AJ (1994) A segment of the MHC class II beta chain plays a critical role in targeting class II molecules to the endocytic pathway. Int. Immunol. 6:973-982

Chicz RM, Urban RG, Lane WS, Gorga JC, Stern LJ, Vignali DA and Strominger JL (1992) Predominant naturally processed peptides bound to HLA-DR1 are derived from MHC-related molecules and are heterogeneous in size. Nature 358:764-768

Chicz RM, Urban RG, Gorga JC, Vignali DA, Lane WS and Strominger JL (1993) Specificity and promiscuity among naturally processed peptides bound to HLA-DR alleles. J. Exp. Med. 178:27-47

Claesson-Welsh L and Peterson PA (1985) Implications of the invariant gamma-chain on the intracellular transport of class II histocompatibility antigens. J. Immunol. 135:3551-3557

Collins T, Korman AJ, Wake CT, Boss JM, Kappes DJ, Fiers W, Ault KA, Gimbrone MA,Jr., Strominger JL and Pober JS (1984) Immune interferon activates multiple class II major histocompatibility complex genes and the associated invariant chain gene in human endothelial cells and dermal fibroblasts. Proc. Natl. Acad. Sci. USA 81:4917-4921

Collins DS, Unanue ER and Harding CV (1991) Reduction of disulfide bonds within lysosomes is a key step in antigen processing. J. Immunol. 147:4054-4059

Cosson P and Bonifacino JS (1992) Role of transmembrane domain interactions in the assembly of class II MHC molecules. Science 258:659-662

Cresswell P (1994) Assembly, transport, and function of MHC class II molecules [Review]. Annu. Rev. Immunol. 12:259-293

Daibata M, Xu M, Humphreys RE and Reyes VE (1994) More efficient peptide binding to MHC class II molecules during cathepsin B digestion of I_i than after I_i release. Mol. Immunol. 31:255-260

Davidson HW, Reid PA, Lanzavecchia A and Watts C (1991) Processed antigen binds to newly synthesized MHC class II molecules in antigen-specific B lymphocytes. Cell 67:105-116

Davis JE and Cresswell P (1990) Lack of detectable endocytosis of B lymphocyte MHC class II antigens using an antibody-independent technique. J. Immunol. 144:990-997

Degen E and Williams DB (1991) Participation of a novel 88-kD protein in the biogenesis of murine class I histocompatibility molecules. J. Cell Biol. 112:1099-1115

Diment S (1990) Different roles for thiol and aspartyl proteases in antigen presentation of ovalbumin. J. Immunol. 145:417-422

Dodi AI, Brett S, Nordeng TW, Sidhu S, Batchelor RJ, Lombardi G, Bakke O and Lechler RI (1994) The invariant chain inhibits presentation of endogenous antigens by a human fibroblast cell line. Eur. J. Immunol. 24:1632-1639

Elliott EA, Drake JR, Amigorena S, Elsemore J, Webster P, Mellman I and Flavell RA (1994) The invariant chain is required for intracellular transport and function of major histocompatibility complex class II molecules. J. Exp. Med. 179:681-694

Ericson ML, Sundström M, Sansom DM and Charron DJ (1994) Mutually exclusive binding of peptide and invariant chain to major histocompatibility complex class II antigens. J. Biol. Chem. 269:26531-26538

Escola J-M, Grivel J-C, Chavrier P and Gorvel J-P (1995) Different endocytic compartments are involved in the tight association of class II molecules with processed hen egg lysozyme and ribonuclease A in B cells. J. Cell Sci. 108:2337-2345

Fling SP, Arp B and Pious D (1994) HLA-DMA and -DMB genes are both required for MHC class II/peptide complex formation in antigen-presenting cells. Nature 368:554-558

Freisewinkel IM, Schenck K and Koch N (1993) The segment of invariant chain that is critical for association with major histocompatibility complex class II molecules contains the sequence of a peptide eluted from class II polypeptides. Proc. Natl. Acad. Sci. USA 90:9703-9706

Gagliardi M-C, Nisini R, Benvenuto R, De Petrillo G, Michel M-L and Barnaba V (1994) Soluble transferrin mediates targeting of hepatitis B envelope antigen to transferrin receptor and its presentation by activated T cells. Eur. J. Immunol. 24:1372-1376

Galvin K, Krishna S, Ponchel F, Frohlich M, Cummings DE, Carlson R, Wands JR, Isselbacher KJ, Pillai S and Ozturk M (1992) The major histocompatibility complex class I antigen-binding protein p88 is the product of the calnexin gene. Proceedings of the National Academy of Sciences of the United States of America 89:8452-8456

Germain RN and Hendrix LR (1991) MHC class II structure, occupancy and surface expression determined by post-endoplasmic reticulum antigen binding. Nature 353:134-139

Germain RN and Margulies DH (1993) The biochemistry and cell biology of antigen processing and presentation. [Review]. Annu. Rev. Immunol. 11:403-450

Glenney JR,Jr., Chen WS, Lazar CS, Walton GM, Zokas LM, Rosenfeld MG and Gill GN (1988) Ligand-induced endocytosis of the EGF receptor is blocked by mutational inactivation and by microinjection of anti-phosphotyrosine antibodies. Cell 52:675-684

Gorvel J-P, Escola J-M, Stang E and Bakke O (1995) Invariant chain induces a delayed transport from early to late endosomes. J. Biol. Chem. 270:2741-2746

Guagliardi LE, Koppelman B, Blum JS, Marks MS, Cresswell P and Brodsky FM (1990) Co-localization of molecules involved in antigen processing and presentation in an early endocytic compartment. Nature 343:133-139

Hammond C, Braakman I and Helenius A (1994) Role of N-linked oligosaccharide recognition, glucose trimming, and calnexin in glycoprotein folding and quality control. Proc. Natl. Acad. Sci. USA 91:913-917

Hampl J, Gradehandt G, Kalbacher H and Rude E (1992) In vitro processing of insulin for recognition by murine T cells results in the generation of A chains with free CysSH. J. Immunol. 148:2664-2671

Harding CV and Unanue ER (1989) Antigen processing and intracellular Ia. Possible roles of endocytosis and protein synthesis in Ia function. J. Immunol. 142:12-19

Harding CV and Unanue ER (1990) Low-temperature inhibition of antigen processing and iron uptake from transferrin: deficits in endosome functions at 18 degrees C. Eur. J. Immunol. 20:323-329

Harding CV, Unanue ER, Slot JW, Schwartz AL and Geuze HJ (1990) Functional and ultrastructural evidence for intracellular formation of major histocompatibility complex class II-peptide complexes during antigen processing. Proc. Natl. Acad. Sci. USA 87:5553-5557

Harding CV, Collins DS, Kanagawa O and Unanue ER (1991a) Liposome-encapsulated antigens engender lysosomal processing for class II MHC presentation and cytosolic processing for class I presentation. J. Immunol. 147:2860-2863

Harding CV, Collins DS, Slot JW, Geuze HJ and Unanue ER (1991b) Liposome-encapsulated antigens are processed in lysosomes, recycled, and presented to T cells. Cell 64:393-401

Harding CV and Geuze HJ (1992) Class II MHC molecules are present in macrophage lysosomes and phagolysosomes that function in the phagocytic processing of Listeria monocytogenes for presentation to T cells. J. Cell Biol. 119:531-542

Harding CV and Geuze HJ (1993) Immunogenic peptides bind to class II MHC molecules in an early lysosomal compartment. J. Immunol. 151:3988-3998

Harding CV and Unanue ER (1990) Cellular mechanisms of antigen processing and the function of class I and II major histocompatibility complex molecules. [Review]. Cell Regul. 1:499-509

Harter C and Mellman I (1992) Transport of the lysosomal membrane glycoprotein lgp120 (lgp-A) to lysosomes does not require appearance on the plasma membrane. J. Cell Biol. 117:311-325

Hochstenbach F, David V, Watkins S and Brenner MB (1992) Endoplasmic reticulum resident protein of 90 kilodaltons associates with the T- and B-cell antigen receptors and major histocompatibility complex antigens during their assembly. Proceedings of the National Academy of Sciences of the United States of America 89:4734-4738

Honegger AM, Dull TJ, Felder S, Van Obberghen E, Bellot F, Szapary D, Schmidt A, Ullrich A and Schlessinger J (1987) Point mutation at the ATP binding site of EGF receptor abolishes protein-tyrosine kinase activity and alters cellular routing. Cell 51:199-209

Humbert M, Bertolino P, Forquet F, Rabourdin-Combe C, Gerlier D, Davoust J and Salamero J (1993a) Major histocompatibility complex class II-restricted presentation of secreted and endoplasmic reticulum resident antigens requires the invariant chains and is sensitive to lysosomotropic agents. Eur. J. Immunol. 23:3167-3172

Humbert M, Raposo G, Cosson P, Reggio H, Davoust J and Salamero J (1993b) The invariant chain induces compact forms of class II molecules localized in late endosomal compartments. Eur. J. Immunol. 23:3158-3166

Hunt DF, Michel H, Dickinson TA, Shabanowitz J, Cox AL, Sakaguchi K, Appella E, Grey HM and Sette A (1992) Peptides presented to the immune system by the murine class II major histocompatibility complex molecule I-Ad. Science 256:1817-1820

Hämmerling GJ and Moreno J (1990) The function of the invariant chain in antigen presentation by MHC class II molecules [news]. [Review]. Immunol. Today 11:337-340

Jahraus A, Storrie B, Griffiths G and Desjardins M (1994) Evidence for retrograde traffic between terminal lysosomes and the prelysosomal/late endosome compartment. J. Cell Sci. 107:145-157

Jensen PE (1991) Enhanced binding of peptide antigen to purified class II major histocompatibility glycoproteins at acidic pH. J. Exp. Med. 174:1111-1120

Johnson KF and Kornfeld S (1992) The cytoplasmic tail of the mannose 6-phosphate/insulin-like growth factor-II receptor has two signals for lysosomal enzyme sorting in the Golgi. J. Cell Biol. 119:249-257

Karlsson L, Péléraux A, Lindstedt R, Liljedahl M and Peterson PA (1994) Reconstitution of an operational MHC class II compartment in nonantigen-presenting cells. Science 266:1569-1573

Katunuma N, Kakegawa H, Matsunaga Y and Saibara T (1994) Immunological significances of invariant chain from the aspect of its structural homology with the cystatin family. FEBS Lett. 349:265-269

Kaufman JF, Auffray C, Korman AJ, Shackelford DA and Strominger J (1984) The class II molecules of the human and murine major histocompatibility complex. [Review]. Cell 36:1-13

Kjær-Nielsen L, Perera JD, Boyd LF, Margulies DH and McCluskey J (1990) The extracellular domains of MHC class II molecules determine their processing requirements for antigen presentation. J. Immunol. 144:2915-2924

Kleijmeer MJ, Oorschot VMJ and Geuze HJ (1994) Human resident Langerhans cells display a lysosomal compartment enriched in MHC class II. J. Invest. Dermatol. 103:516-523
Kolk DP and Floyd-Smith G (1992) The HXY box regulatory element modulates expression of the murine IA antigen-associated invariant chain in L fibroblasts. DNA Cell Biol. 11:745-754

Koch N (1988) Posttranslational modifications of the Ia-associated invariant protein p41 after gene transfer. Biochemistry 27:4097-4102

Koch N, Lipp J, Pessara U, Schenck K, Wraight C and Dobberstein B (1989) MHC class II invariant chains in antigen processing and presentation. [Review]. Trends Biochem. Sci. 14:383-386

Kolk DP and Floyd-Smith G (1993) Induction of the murine class-II antigen-associated invariant chain by TNF-alpha is controlled by an NF-kappa B-like element. Gene 126:179-185

Kovacsovics-Bankowski M, Clark K, Benacerraf B and Rock KL (1993) Efficient major histocompatibility complex class I presentation of exogenous antigen upon phagocytosis by macrophages. Proceedings of the National Academy of Sciences of the United States of America 90:4942-4946

Kvist S, Wiman K, Claesson L, Peterson PA and Dobberstein B (1982) Membrane insertion and oligomeric assembly of HLA-DR histocompatibility antigens. Cell 29:61-69

Kämpgen E, Koch N, Koch F, Stoger P, Heufler C, Schuler G and Romani N (1991) Class II major histocompatibility complex molecules of murine dendritic cells: synthesis, sialylation of invariant chain, and antigen processing capacity are down-regulated upon culture. Proc. Natl. Acad. Sci. USA 88:3014-3018

Lamb CA, Yewdell JW, Bennink JR and Cresswell P (1991) Invariant chain targets HLA class II molecules to acidic endosomes containing internalized influenza virus. Proc. Natl. Acad. Sci. USA 88:5998-6002

Lamb CA and Cresswell P (1992) Assembly and transport properties of invariant chain trimers and HLA-DR-invariant chain complexes. J. Immunol. 148:3478-3482

Lanzavecchia A (1987) Antigen uptake and accumulation in antigen-specific B cells. [Review]. Immunol. Rev. 99:39-51

Lanzavecchia A, Reid PA and Watts C (1992) Irreversible association of peptides with class II MHC molecules in living cells. Nature 357:249-252

Layet C and Germain RN (1991) Invariant chain promotes egress of poorly expressed, haplotype-mismatched class II major histocompatibility complex A alpha A beta dimers from the endoplasmic reticulum/cis-Golgi compartment. Proc. Natl. Acad. Sci. USA 88:2346-2350

Lenardo MJ and Baltimore D (1989) NF-kappa B: a pleiotropic mediator of inducible and tissue-specific gene control. [Review]. Cell 58:227-229

Long EO, Mach B and Accolla RS (1984) Ia-negative B-cell variants reveal a coordinate regulation in the transcription of the HLA class II gene family. Immunogenetics 19:349-353

Long EO (1985) In search of a function for the invariant chain associated with Ia antigens. [Review]. Surv. Immunol. Res. 4:27-34

Long, E.O, P.A. Roche, M.S. Malnati, T. LaVaute and V. Pinet, 1993, Multiple pathways of antigen processing for MHC class II-restricted presentation, in: Antigen processing and presentation, eds. R.E. Humphreys and S.K. Pierce (Academic Press)

Long EO(1989) Intracellular traffic adn antigen processing. Immunol. Today 10:232-234

Long EO, LaVaute T, Pinet V and Jaraquemada D (1994) Invariant chain prevents the HLA-DR-restricted presentation of a cytosolic peptide. J. Immunol. 153:1487-1494

Loss GE,Jr. and Sant AJ (1993) Invariant chain retains MHC class II molecules in the endocytic pathway. J. Immunol. 150:3187-3197

Lotteau V, Teyton L, Peleraux A, Nilsson T, Karlsson L, Schmid SL, Quaranta V and Peterson PA (1990) Intracellular transport of class II MHC molecules directed by invariant chain. Nature 348:600-605

Malcherek G, Gnau V, Jung G, Rammensee H-G and Melms A (1995) Supermotifs enable natural invariant chain-derived peptides to interact with many major histocompatibility complex-class II molecules. J. Exp. Med. 181:527-536

Malnati MS, Ceman S, Weston M, Demars R and Long EO (1993) Presentation of cytosolic antigen by HLA-DR requires a function encoded in the class II region of the MHC. J. Immunol. 151:6751-6756

Maric MA, Taylor MD and Blum JS (1994) Endosomal aspartic proteinases are required for invariant-chain processing. Proc. Natl. Acad. Sci. USA 91:2171-2175

Marks MS and Cresswell P (1986) Invariant chain associates with HLA class II antigens via its extracytoplasmic region. J. Immunol. 136:2519-2525

Marks MS, Blum JS and Cresswell P (1990) Invariant chain trimers are sequestered in the rough endoplasmic reticulum in the absence of association with HLA class II antigens. J. Cell Biol. 111:839-855

Matter K and Mellman I (1994) Mechanisms of cell polarity: Sorting and transport in epithelial cells. Curr. Opin. Cell Biol. 6:545-554

McCoy KL, Gainey D, Inman JK and Stutzman R (1993a) Antigen presentation by B lymphoma cells: Requirements for processing of exogenous antigen internalized through transferrin receptors. J. Immunol. 151:4583-4594

McCoy KL, Noone M, Inman JK and Stutzman R (1993b) Exogenous antigens internalized through transferrin receptors activate CD4+ T cells. J. Immunol. 150:1691-1704

Mellins E, Smith L, Arp B, Cotner T, Celis E and Pious D (1990) Defective processing and presentation of exogenous antigens in mutants with normal HLA class II genes. Nature 343:71-74

Mellins E, Kempin S, Smith L, Monji T and Pious D (1991) A gene required for class II-restricted antigen presentation maps to the major histocompatibility complex. J. Exp. Med. 174:1607-1615

Miller J and Germain RN (1986) Efficient cell surface expression of class II MHC molecules in the absence of associated invariant chain. J. Exp. Med. 164:1478-1489

Mizuochi T, Yee S-T, Kasai M, Kakiuchi T, Muno D and Kominami E (1994) Both cathepsin B and cathepsin D are necessary for processing of ovalbumin as well as for degradation of class II MHC invariant chain. Immunol. Lett. 43:189-193

Momburg F, Koch N, Moller P, Moldenhauer G, Butcher GW and Hammerling GJ (1986) Differential expression of Ia and Ia-associated invariant chain in mouse tissues after in vivo treatment with IFN-gamma. J. Immunol. 136:940-948

Monji T, McCormack AL, Yates JR and Pious D (1994) Invariant-cognate peptide exchange restores class II dimer stability in HLA-DM mutants. J. Immunol. 153:4468-4477

Morris P, Shaman J, Attaya M, Amaya M, Goodman S, Bergman C, Monaco JJ and Mellins E (1994) An essential role for HLA-DM in antigen presentation by class II major histocompatibility molecules. Nature 368:551-554

Nadimi F, Moreno J, Momburg F, Heuser A, Fuchs S, Adorini L and Hammerling GJ (1991) Antigen presentation of hen egg-white lysozyme but not of ribonuclease A is augmented by the major histocompatibility complex class II-associated invariant chain. Eur. J. Immunol. 21:1255-1263

Naujokas MF, Morin M, Anderson MS, Peterson M and Miller J (1993) The chondroitin sulfate form of invariant chain can enhance stimulation of T cell responses through interaction with CD44. Cell 74:257-268

Neefjes JJ, Stollorz V, Peters PJ, Geuze HJ and Ploegh HL (1990) The biosynthetic pathway of MHC class II but not class I molecules intersects the endocytic route. Cell 61:171-183

Neefjes JJ and Ploegh HL (1992) Inhibition of endosomal proteolytic activity by leupeptin blocks surface expression of MHC class II molecules and their conversion to SDS resistance alpha beta heterodimers in endosomes. EMBO J. 11:411-416

Neefjes JJ and Momburg F (1993) Cell biology of antigen presentation. [Review]. Curr. Opin. Immunol. 5:27-34

Neufeld EF (1991) Lysosomal storage diseases. [Review]. Annu Rev Biochem 60:257-280

Newcomb JR and Cresswell P (1993) Structural analysis of proteolytic products of MHC class II-invariant chain complexes generated in vivo. J. Immunol. 151:4153-4163

Nguyen QV and Humphreys RE (1989) Time course of intracellular associations, processing, and cleavages of Ii forms and class II major histocompatibility complex molecules. J. Biol. Chem. 264:1631-1637

Nijenhuis M, Calafat J, Kuijpers KC, Janssen H, de Haas M, Nordeng TW, Bakke O and Neefjes JJ (1994) Targeting major histocompatibility complex class II molecules to the cell surface by invariant chain allows antigen presentation upon recycling. Eur. J. Immunol. 24:873-883

Noelle RJ, Kuziel WA, Maliszewski CR, McAdams E, Vitetta ES and Tucker PW (1986) Regulation of the expression of multiple class II genes in murine B cells by B cell stimulatory factor-1 (BSF-1). J. Immunol. 137:1718-1723

Nuchtern JG, Bonifacino JS, Biddison WE and Klausner RD (1989) Brefeldin A implicates egress from endoplasmic reticulum in class I restricted antigen presentation. Nature 339:223-226

O'Sullivan DM, Noonan D and Quaranta V (1987) Four Ia invariant chain forms derive from a single gene by alternate splicing and alternate initiation of transcription/translation. J. Exp. Med. 166:444-460

Odorizzi CG, Trowbridge IS, Xue L, Hopkins CR, Davis CD and Collawn JF (1994) Sorting signals in the MHC class II invariant chain cytoplasmic tail and transmembrane region determine trafficking to an endocytic processing compartment. J. Cell Biol. 126:317-330

Ou WJ, Cameron PH, Thomas DY and Bergeron JJ (1993) Association of folding intermediates of glycoproteins with calnexin during protein maturation. Nature 364:771-776

Pessara U and Koch N (1990) Tumor necrosis factor alpha regulates expression of the major histocompatibility complex class II-associated invariant chain by binding of an NF-kappa B-like factor to a promoter element. Mol. Cell. Biol. 10:4146-4154

Peters PJ, Neefjes JJ, Oorschot V, Ploegh HL and Geuze HJ (1991) Segregation of MHC class II molecules from MHC class I molecules in the Golgi complex for transport to lysosomal compartments [see comments]. Nature 349:669-676

Peterson M and Miller J (1990) Invariant chain influences the immunological recognition of MHC class II molecules. Nature 345:172-174

Peterson M and Miller J (1992) Antigen presentation enhanced by the alternatively spliced invariant chain gene product p41. Nature 357:596-598

Pfeifer JD, Wick MJ, Roberts RL, Findlay K, Normark SJ and Harding CV (1993) Phagocytic processing of bacterial antigens for class I MHC presentation to T cells. Nature 361:359-362

Pieters J, Horstmann H, Bakke O, Griffiths G and Lipp J (1991) Intracellular transport and localization of major histocompatibility complex class II molecules and associated invariant chain. J. Cell Biol. 115:1213-1223

Pieters J, Bakke O and Dobberstein B (1993) The MHC class II associated invariant chain contains two endosomal targeting siganls within its cytoplasmic tail. J. Cell Sci. 106:831-846

Pinet V, Malnati MS and Long EO (1994) Two processing pathways for the MHC class II-restricted presentation of exogenous influenza virus antigen. J. Immunol. 152:4852-4860

Pinet V, Vergelli M, Martin R, Bakke O and Long EO (1995) Antigen presentation mediated by recycling of surface HLA-DR molecules. Nature 375:603-606

Polla BS, Poljak A, Ohara J, Paul WE and Glimcher LH (1986) Regulation of class II gene expression: analysis in B cell stimulatory factor 1-inducible murine pre-B cell lines. J. Immunol. 137:3332-3337

Pure E, Inaba K, Crowley MT, Tardelli L, Witmer-Pack MD, Ruberti G, Fathman G and Steinman RM (1990) Antigen processing by epidermal Langerhans cells correlates with the level of biosynthesis of major histocompatibility complex class II molecules and expression of invariant chain. J. Exp. Med. 172:1459-1469

Qiu Y, Xu X, Wandinger-Ness A, Dalke DP and Pierce SK (1994) Separation of subcellular compartments containing distinct functional forms of MHC class II. J. Cell Biol. 125:595-605

Quaranta V, Majdic O, Stingl G, Liszka K, Honigsmann H and Knapp W (1984) A human Ia cytoplasmic determinant located on multiple forms of invariant chain (gamma, gamma 2, gamma 3). J. Immunol. 132:1900-1905

Raposo G, Nijman H, Stoorvogel W, Leijendekker RL, Harding CV, Melief CJM and Geuze HJ (1995) B lymphocytes secrete antigen presenting vesicles. Eur. J. Cell Biol. Supplement:153(Abstract)

Reid PA and Watts C (1990) Cycling of cell-surface MHC glycoproteins through primaquine-sensitive intracellular compartments. Nature 346:655-657

Reyes VE, Lu S and Humphreys RE (1991) Cathepsin B cleavage of Ii from class II MHC alpha- and beta-chains. J. Immunol. 146:3877-3880

Riberdy JM and Cresswell P (1992) The antigen-processing mutant T2 suggests a role for MHC-linked genes in class II antigen presentation [published erratum appears in J Immunol 1992 Aug 1;149(3):1113]. J. Immunol. 148:2586-2590

Riberdy JM, Newcomb JR, Surman MJ, Barbosa JA and Cresswell P (1992) HLA-DR molecules from an antigen-processing mutant cell line are associated with invariant chain peptides. Nature 360:474-477

Riberdy JM, Avva RR, Geuze HJ and Cresswell P (1994) Transport and intracellular distribution of MHC class II molecules and associated invariant chain in normal and antigen-processing mutant cell lines. J. Cell Biol. 125:1225-1237

Roche PA and Cresswell P (1990) Invariant chain association with HLA-DR molecules inhibits immunogenic peptide binding. Nature 345:615-618

Roche PA and Cresswell P (1991) Proteolysis of the class II-associated invariant chain generates a peptide binding site in intracellular HLA-DR molecules. Proc. Natl. Acad. Sci. USA 88:3150-3154

Roche PA, Marks MS and Cresswell P (1991) Formation of a nine-subunit complex by HLA class II glycoproteins and the invariant chain. Nature 354:392-394

Roche PA, Teletski CL, Stang E, Bakke O and Long EO (1993) Cell surface HLA-DR-invariant chain complexes are targeted to endosomes by rapid internalization. Proc. Natl. Acad. Sci. USA 90:8581-8585

Rock KL, Gamble S and Rothstein L (1990) Presentation of exogenous antigen with class I major histocompatibility complex molecules. Science 249:918-921

Romagnoli P, Layet C, Yewdell J, Bakke O and Germain RN (1993) Relationship between invariant chain expression and major histocompatibility complex class II transport into early and late endocytic compartments. J. Exp. Med. 177:583-596

Romagnoli P and Germain RN (1994) The CLIP region of invariant chain plays a critical role in regulating major histocompatibility complex class II folding, transport, and peptide occupancy. J. Exp. Med. 180:1107-1113

Rudensky AY, Preston-Hurlburt P, Hong SC, Barlow A and Janeway CA,Jr. (1991) Sequence analysis of peptides bound to MHC class II molecules. Nature 353:622-627

Sadegh-Nasseri S and Germain RN (1991) A role for peptide in determining MHC class II structure. Nature 353:167-170

Salamero J, Humbert M, Cosson P and Davoust J (1990) Mouse B lymphocyte specific endocytosis and recycling of MHC class II molecules. EMBO J. 9:3489-3496

Sanderson F, Kleijmeer MJ, Kelly A, Verwoerd D, Tulp A, Neefjes JJ, Geuze HJ and Trowsdale J (1994) Accumulation of HLA-DM, a regulator of antigen presentation, in MHC class II compartments. Science 266:1566-1569

Sant AJ, Cullen SE and Schwartz BD (1985) Biosynthetic relationships of the chondroitin sulfate proteoglycan with Ia and invariant chain glycoproteins. J. Immunol. 135:416-422

Sant AJ and Miller J (1994) MHC class II antigen processing: Biology of invariant chain [Review]. Curr. Opin. Immunol. 6:57-63

Schaiff WT, Hruska KA,Jr., Bono C, Shuman S and Schwartz BD (1991) Invariant chain influences post-translational processing of HLA-DR molecules. J. Immunol. 147:603-608

Schmid SL and Jackson MR (1994) Immunology: Making class II presentable. Nature 369:103-104

Schreiber KL, Bell MP, Huntoon CJ, Rajagopalan S, Brenner MB and McKean DJ (1994) Class II histocompatibility molecules associate with calnexin during assembly in the endoplasmic reticulum. Int. Immunol. 6:101-111

Schutze M-P, Peterson PA and Jackson MR (1994) An N-terminal double-arginine motif maintains type II membrane proteins in the endoplasmic reticulum. EMBO J. 13:1696-1705

Sekaly RP, Tonnelle C, Strubin M, Mach B and Long EO (1986) Cell surface expression of class II histocompatibility antigens occurs in the absence of the invariant chain. J. Exp. Med. 164:1490-1504

Sette A, Ceman S, Kubo RT, Sakaguchi K, Appella E, Hunt DF, Davis TA, Michel H, Shabanowitz J, Rudersdorf R and et al (1992) Invariant chain peptides in most HLA-DR molecules of an antigen-processing mutant. Science 258:1801-1804

Sette A, Southwood S, Miller J and Appella E (1995) Binding of major histocompatibility complex class II to the invariant chain-derived peptide, CLIP, is regulated by allelic polymorphism in class II. J. Exp. Med. 181:677-683

Simonis S, Miller J and Cullen SE (1989) The role of the Ia-invariant chain complex in the posttranslational processing and transport of Ia and invariant chain glycoproteins. J. Immunol. 143:3619-3625

Simonsen A, Momburg F, Drexler J, Hämmerling GJ and Bakke O (1993) Intracellular distribution of the MHC class II molecules and the associated invariant chain (Ii) in different cell lines. Int. Immunol. 5, No. 8:903-917

Sloan VS, Cameron P, Porter G, Gammon M, Amaya M, Mellins E and Zaller DM (1995) Mediation by HLA-DM of dissociation of peptides from HLA-DR. Nature 375:802-805

Stang E, Long EO and Bakke O (1993) Intracellular localization of MHC class II molecules and associated invariant chain. J. Cell. Bioch. Supplement 17C:29(Abstract)

Stebbins CC, Loss GE,Jr., Elias CG, Chervonsky A and Sant AJ (1995) The requirement for DM in class II-restricted antigen presentation and SDS-stable dimer formation is allele and species dependent. J. Exp. Med. 181:223-234

Stockinger B, Pessara U, Lin RH, Habicht J, Grez M and Koch N (1989) A role of Ia-associated invariant chains in antigen processing and presentation. Cell 56:683-689

Strubin M, Berte C and Mach B (1986) Alternative splicing and alternative initiation of translation explain the four forms of the Ia antigen-associated invariant chain. EMBO J. 5:3483-3488

Stössel H, Koch F, Kämpgen E, Stöger P, Lenz A, Heufler C, Romani N and Schuler G (1990) Disappearance of certain acidic organelles (endosomes and Langerhans cell granules) accompanies loss of antigen processing capacity upon culture of epidermal Langerhans cells. J. Exp. Med. 172:1471-1482

Sugita M and Brenner MB (1995) Association of the invariant chain with major histocompatibility complex class I molecules directs trafficking to endocytic compartments. J. Biol. Chem. 270:1443-1448

Sung E and Jones PP (1981) The invariant chain of murine Ia antigens: its glycosylation, abundance and subcellular localization. Mol. Immunol. 18:899-913

Takahashi H, Cease KB and Berzofsky JA (1989) Identification of proteases that process distinct epitopes on the same protein. J. Immunol. 142:2221-2229

Townsend A and Bodmer H (1989) Antigen recognition by class I-restricted T lymphocytes. [Review]. Annu. Rev. Immunol. 7:601-624

Tulp A, Verwoerd D, Dobberstein B, Ploegh HL and Pieters J (1994) Isolation and characterization of the intracellular MHC class II compartment. Nature 369:120-126

van Kemenade FJ, Baars P, Hooibrink B, van Lier RAW and Miedema F (1994) Enhanced antigen-specific human T-cell proliferation by co-expression of the invariant chain on DR1+ fibroblasts. Eur. J. Immunol. (in press)

Vidard L, Rock KL and Benacerraf B (1992) Diversity in MHC class II ovalbumin T cell epitopes generated by distinct proteases. J. Immunol. 149:498-504

Viville S, Neefjes J, Lotteau V, Dierich A, Lemeur M, Ploegh H, Benoist C and Mathis D (1993) Mice lacking the MHC class II-associated invariant chain. Cell 72:635-648

Volc-Platzer B, Majdic O, Knapp W, Wolff K, Hinterberger W, Lechner K and Stingl G (1984) Evidence of HLA-DR antigen biosynthesis by human keratinocytes in disease. J. Exp. Med. 159:1784-1789

Watts C and Davidson HW (1988) Endocytosis and recycling of specific antigen by human B cell lines. EMBO J. 7:1937-1945

West MA, Lucocq JM and Watts C (1994) Antigen processing and class II MHC peptide-loading compartments in human B-lymphoblastoid cells. Nature 369:147-151

Wettstein DA, Boniface JJ, Reay PA, Schild H and Davis MM (1991) Expression of a class II major histocompatibility complex (MHC) heterodimer in a lipid-linked form with enhanced peptide/soluble MHC complex formation at low pH. J. Exp. Med. 174:219-228

Xu M, Capraro GA, Daibata M, Reyes VE and Humphreys RE (1994) Cathepsin B cleavage and release of invariant chain from MHC class II molecules follow a staged pattern. Mol. Immunol. 31:723-731

Zachgo S, Dobberstein B and Griffiths G (1992) A block in degradation of MHC class II-associated invariant chain correlates with a reduction in transport from endosome carrier vesicles to the prelysosome compartment. J. Cell Sci. 103:811-822

Acknowledgements

The work of this research group has been supported by the Norwegian Cancer Society and the Norwegian Research Council. In addition we have received grants from Novo Nordisk Foundation, Family Blix Foundation, Anders Jahre Foundation and Nansen Foundation.

Progress towards the identification of secretion signals in a protein transported in a folded state across a lipid bilayer.

Nathalie Sauvonnet and Anthony P. Pugsley.
Unité de Génétique Moléculaire (CNRS URA1149)
Institut Pasteur
25, rue du Dr. Roux
75724 Paris
France.
Telephone: 33.1.45688494. Telefax: 33.1.45868960. E-mail: max@pasteur.fr

Targeting signals.

One of the basic principles underlying current studies of transmembrane protein traffic is that protein sorting is signal mediated. According to experimentally verified dogma, proteins without targeting signals remain in the cell compartment in which they are synthesised whereas proteins with targeting signals are recognised by receptors in the membrane to which they are targeted or by circulating receptors which then dock at the surface or the target organelle or membrane (Pugsley 1989).

In most of the extensively characterised examples of transmembrane protein translocation (protein import into mitochondria, chloroplasts and the endoplasmic reticulum and protein export across the bacterial cytoplasmic membrane), linear targeting sequences located near the N-terminus of a polypeptide are removed shortly after translocation is initiated, either because they no longer serve any useful purpose or because they would be deleterious to protein function or sorting after translocation. In all of these cases, the proteins are unfolded during translocation so that they cross the membrane through protein-lined channels in strict linear order, starting at the N-terminus (Pugsley 1989).

There are some exceptions, however. For example, most imported peroxisomal proteins and most of the proteins which are exported across the bacterial cytoplasmic membrane via the so-called ABC protein secretion pathway have C-terminal targeting signals which are not removed after translocation (Pugsley 1989). Other exceptions include proteins imported into nuclei. These proteins, which pass through the nuclear

NATO ASI Series, Vol. H 96
Molecular Dynamics of Biomembranes
Edited by Jos A. F. Op den Kamp

pore in a fully-folded state, often have internal, unprocessed targeting signals (Pugsley 1989).

The bacterial General Secretory Pathway

A further example of the latter phenomenon is provided by the general secretory pathway (GSP) in Gram-negative bacteria. The GSP is composed of two independent steps. The first, signal peptide-dependent step, leads exoproteins across the cytoplasmic membrane via the same pathway (usually called Sec) as that used by proteins to destined for the periplasm or the outer membrane. A few, selected proteins are then translocated from the periplasm across the outer membrane via the second step in the GSP (Pugsley 1993). A substantial body of evidence indicates that secreted exoproteins adopt considerable tertiary and even quaternary structure during their transit through the periplasm. Furthermore, it appears that correct folding is crucial to their recognition by the GSP (Bortoli-German and others 1994; Hardie and others ; Hirst and Holmgren 1987; Pugsley 1992; Shevchik and others in press).

The identification and characterisation of the secretion signal(s) in GSP exoproteins would represent a considerable step towards understanding how this system works. For example, their identification would allow us to use both biochemical and genetical methods to identify the cognate receptor component(s) of the GSP. Furthermore, the fact that proteins secreted by the GSP are subject to catalysed folding including disulphide bond formation in the periplasm before they are released from the cells makes this pathway ideally suited to the secretion of heterologous proteins. It would therefore be desirable to fuse secretion signals from GSP exoproteins to other proteins such as toxins, hormones, protective antigens and enzymes to promote their secretion in a correctly-folded, fully active form. We therefore decided to try to locate the secretion signal in a polypeptide, pullulanase, which is secreted by this pathway.

Pullulanase

Pullulanase is an amylolytic protein produced by *Klebsiella oxytoca*, a Gram-negative bacterium closely related to *Escherichia coli*. Transfer of the structural gene together with about 16 kb of flanking DNA into *E. coli* resulted in correct synthesis, processing and secretion of pullulanase in this bacterium (d'Enfert and others 1987). The DNA flanking the structural gene includes 14 genes which are essential for pullulanase secretion. They are all uniquely involved in the second step in the GSP. One unusual feature of pullulanase which is relevant to the present discussion is that it is a lipoprotein: processing of the N-terminal signal peptide is accompanied by fatty acylation of the N-terminal cysteine residue of the mature polypeptide (Pugsley and others 1986). The fatty acids anchor the protein in the cell envelope so that it faces the periplasm during transit between the two steps in the secretion pathway, and so that it faces the outside milieu once its secretion has been completed (Pugsley and others 1986).

Earlier studies in our laboratory by Michael Kornacker (Kornacker and Pugsley 1990) and Isabelle Poquet (unpublished data) showed that the insertion of 2 or 4 amino acids at different positions in the mature segment of pullulanase did not affect its secretion. This suggested that although folding of pullulanase could be shown to occur and might even be necessary for transport across the outer membrane, minor sequence changes could be tolerated throughout the polypeptide without affecting secretion. Furthermore, Michael Kornacker showed that the C-terminal 256 amino acids of the 1071 amino acid pullulanase polypeptide could be replaced by the mature sequence of periplasmic ß-lactamase without dramatic effects on secretion efficiency. Our recent studies have been based on the use of ß-lactamase as a reporter protein to locate a secretion signal in pullulanase.

Working models for the pullulanase secretion signal.

At the start of this study, we considered three models for the location and organisation of the secretion signal in the pullulanase polypeptide. In the simplest of these models, the secretion signal is envisaged as a short linear sequence of amino acids within a defined sector of the polypeptide (Fig. 1). This model is compatible with the failure of the 2-4 amino acid inserts to affect secretion and with the secretion of the pullulanase-ß-lactamase hybrid, since none of these changes appeared to alter or remove this peptide. Although the sequence of the signal would be independent of amino acids located elsewhere in the protein, its correct presentation might still depend on the correct folding of the complete protein. This would explain why intramolecular disulphide bond formation and oligomerisation are both required for the secretion of at least some proteins by the GSP.

This model predicts that progressive removal of segments of the pullulanase polypeptide outside of the signal might not affect its secretion and that it might be possible to transplant it into other periplasmic proteins to promote their secretion. Such a signal would be similar to the nuclear localisation signals identified in proteins imported into nuclei. However, sequence comparisons of the unrelated proteins secreted by the GSP in other bacteria (unlike *K. oxytoca*, which secretes only pullulanase, other Gram-negative bacteria secrete up to 10 or more different proteins by the GSP) failed to identify any homologous segments that might represent a conserved secretion signal.

The second model (Fig. 1) predicts the presence of multiple secretion signals in the pullulanase polypeptide. It is based on the observation that some forms of pullulanase which have been truncated near the C-terminus are not secreted whereas other which have either longer or shorter deletions are secreted (unpublished observations). We interpreted these observations to indicate that progressive deletion from the C-terminus of the polypeptide removed first a secretion signal and then a segment of the polypeptide whose secretion depended uniquely on the presence of the signal that had been deleted. Complete removal of this segment would restore secretion but further deletion would remove yet another secretion signal and hence abolish secretion, and so-on. This model is technically difficult to test because most pullulanase truncates are highly unstable and therefore difficult to analyse.

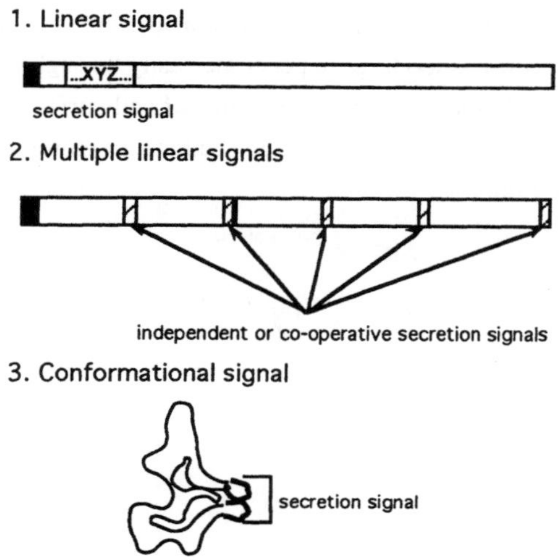

Fig. 1. Three speculative models for the organisation of secretion signals in the pulllanase polypeptide.

In the third model (Fig. 1), secretion is determined by a patch signal composed of a few amino acids from different segments of the pullulanase polypeptide that are brought together only in the final or near-final folded state of the polypeptide. This model is fully compatible with all of the experimental data available: small insertions would not necessarily affect the conformation of pullulanase sufficiently to prevent the correct juxtapositioning of key amino acids in the signal, and neither the key amino acids nor segments of pullulanase required for their correct positioning are located in the C-terminal 256 amino acids. This model predicts that more radical changes to the pullulanase polypeptide should prevent secretion signal presentation because they would cause incorrect folding.

Pullulanase-reporter protein hybrids.

One of the disadvantages of working with pullulanase is that it is not released into the growth medium. This means that the levels of pullulanase secretion are more difficult to assess than with other, totally soluble proteins secreted by the same pathway. Furthermore, it is difficult to perform kinetic experiments on protein secretion. These technical difficulties can be overcome to some extent by determining whether or not the protein is cleaved by exogenous proteases such as proteinase K. However, it is difficult to quantify protease accessibility accurately. We therefore adopted a second approach which relies on the use of pullulanase-reporter hybrids (Fig. 2). ß-lactamase is an almost perfect reporter protein for this purpose since one of its substrates, the chromogenic ß-lactam antibiotic nitrocefin, is too large to penetrate easily through the constitutive pores in the *E. coli* outer membrane. Thus, measurements of cell surface ß-lactamase activity combined with protease accessibility provide an accurate measurement of pullulanase-ß-lactamase hybrid secretion.

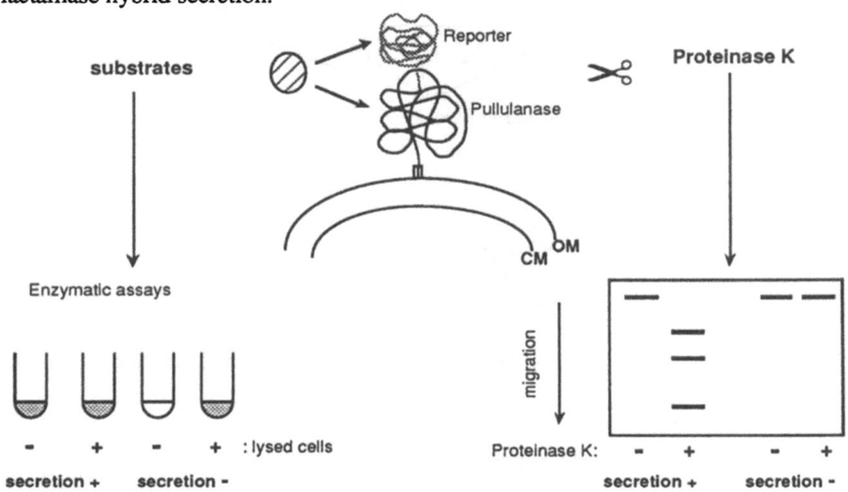

Fig. 2. Tests for secretion of pullulanase-reporter protein hybrids.

The impetus for carrying out this part of the study came from Michael Kornacker's observation that a pullulanase-ß-lactamase hybrid was secreted. We also constructed other hybrids in which the same 256 amino acids of pullulanase were replaced by periplasmic maltose binding protein from *E. coli*, periplasmic alkaline phosphatase from *E. coli*, extracellular levan sucrase from the Gram-positive bacterium *Bacillus subtilis*, and the extracellular *Erwinia chrysanthemi* enzyme endoglucanase Z or its 60 amino acid cellulase binding domain (endoglucanase Z is secreted by the GSP in basically the same was as pullulanase except that it is not fatty acylated). Only the cellulose binding domain was secreted (Fig. 3).

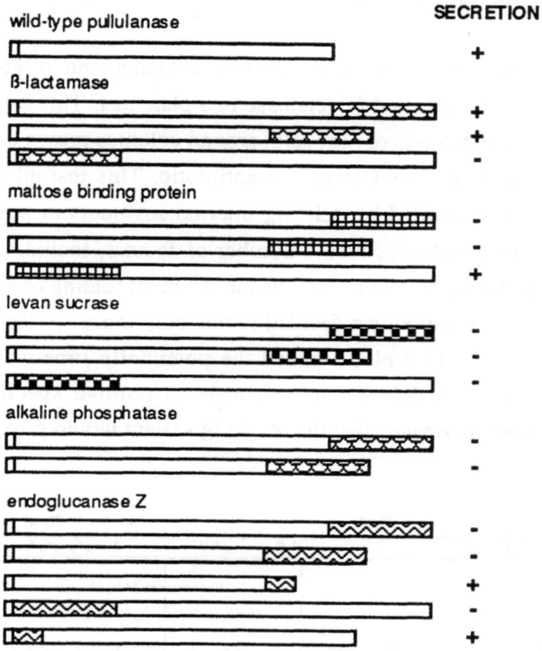

Fig. 3. Secretion of pullulanase-reporter hybrid proteins in *E. coli* expressing the pullulanase-specific secretion genes cloned from *K. oxytoca*; In each case, the reporter protein is schematized as a shaded bar. The small box at the left-hand-side of the representation of the pullulanase polypeptide indicates the signal peptide. The figure is not drawn to scale.

One possible reason for the failure to secrete most of these hybrids could be that the 256 amino acid segment of pullulanase which is missing from these hybrids contains information that is essential for their secretion. We therefore constructed a new series of hybrids in which the same reporters (except the cellulose binding domain) were fused to the extreme C-terminal end of a complete pullulanase polypeptide. Once again, only the pullulanase-ß-lactamase hybrid was secreted (Fig. 3). Finally, we reversed the order of the pullulanase and reporter segments in these hybrids so that the reporter was now fused near the N-terminal end of the pullulanase polypeptide. This time, the hybrid formed with ß-lactamase was not secreted, but the hybrids formed with maltose binding protein and with the cellulase binding domain were secreted (Fig. 3).

In some respects, there results were disappointing since they indicated that pullulanase-mediated reporter protein secretion was apparently affected by the size, position and nature of the reporter. Nevertheless, we decided to exploit the pullulanase ß-lactamase hybrids in an attempt to locate a secretion signal in pullulanase.

Deletion strategy to locate a pullulanase secretion signal.

In the first step of this new strategy, we developed a simple screen that allowed us to detect clones which secreted ß-lactamase in a plate test. This test is based on the formation of zones of hydrolysis of nitrocefin around colonies grown on agar plates and then overlaid with soft agar containing the antibiotic. This test allows us to screen collections of mutations created by deleting internal segments of pullulanase from a pullulanase-ß-lactamase hybrid. Since a number of factors, including reading frame changes, mRNA instability, protein instability and incorrect folding or aggregation might prevent hybrid protein secretion, we decided to test only those clones which scored as positive in the plate test. This also avoided the potentially time-wasting and costly exercise of using directed deletion mutagenesis to remove specific segments of pullulanase without any assurance that the resulting variant would provide interpretable data.

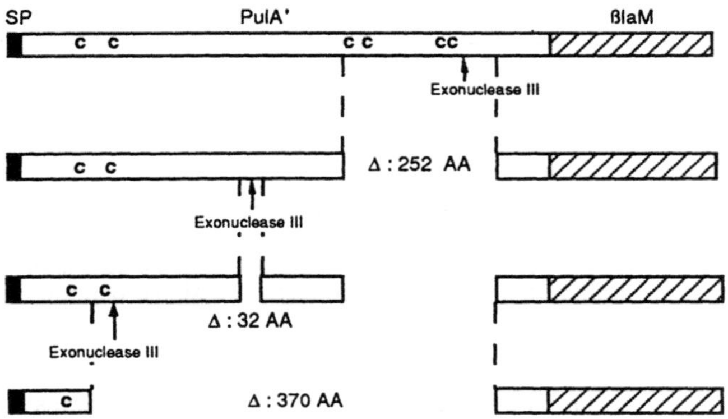

Fig. 4. Detection of potential secretion signals in the pullulanase polypeptide by progressive deletion with exonuclease III. SP, signal peptide; PulA, pullulanase; ßlaM, ß-lactamase; C, cysteine.

For the rest our analyses, we used the pullulanase-ß-lactamase hybrid that lacked the non-essential C-terminal 256 residues of pullulanase. In the strategy employed, a plasmid carrying of this pullulanase-ß-lactamase gene fusion was linearised at a unique restriction site in the pullulanase segment and subjected to bidirectional deletion with exonuclease III. The resulting clones which still secreted ß-lactamase were analysed by restriction mapping and then DNA sequencing. The clone which contained the largest deletion which did not interfere with secretion was then subjected to a further round of deletion analysis, starting at a different site. The results of this analysis are presented in Fig. 4. The current status of this work is that the smallest hybrid protein that is still secreted contains two blocks of the pullulanase polypeptide, each of approximately 80 amino acids. We are currently trying to delete each of these segments in turn to determine whether they are both required for secretion, and to further reduce the overall size of the pullulanase component of the hybrid. Even if these strategies fail, we should be able to subject the relatively small DNA fragment which codes for these two blocks of sequences to random PCR mutagenesis to identify mutations which prevent secretion.

Re-examination of the models for the secretion signal.

How do these data fit with the three models for the secretion signal? They are clearly compatible with the presence of one or 2 secretion signals located in well-defined regions of the pullulanase polypeptide (models 1 and 2). Intriguingly, one of the two segments is located at the extreme N-terminus of the pullulanase polypeptide in a region which is totally devoid of any sequence homology with other amylases. On the other hand, the second, internal segment is close to the centre of a ca 70 kDa domain of pullulanase which contains sequences found in all other amylases. It would be interesting to determine whether the same regions of the pullulanase polypeptide are identified if a completely different strategy is employed to create the deletions. For example, is it possible to delete the first ca 80 amino acids from an otherwise complete pullulanase-ß-lactamase hybrid without affecting its secretion? Such a strategy might allow us to determine whether pullulanase has multiple secretion signals which can function independently in appropriate contexts (model 2).

Do the data exclude the third model for the secretion signal? One of the interesting aspects of the results is that the shortest secreted variant identified to date retains only one of the 6 unmodified cysteine initially present in the pullulanase polypeptide. This is important in the context of the demonstrated requirement for disulphide bond formation in secretion of wild-type pullulanase (Pugsley 1992) since it means that the disulphide bond itself does not form part of the signal(s) that we have identified. Perhaps the disulphide bond is required for another, as yet unidentified secretion signal, or perhaps its role is to prevent misfolding and hence masking of the identified signals in the complete polypeptide chain? Nevertheless, we cannot completely rule out the possible existence of a patch signal since the shortest secretable hybrid still contains two segments of pullulanase which might, by chance, be correctly positioned to form a the secretion signal in an otherwise radically different polypeptide context.

These and other questions concerning the identity of the secretion signals in pullulanase should be answered by further mutation analysis and by crystallographic analysis of the pullulanase polypeptide.

Acknowledgements. This work was supported in part by European Union grant number ERBCHRX-CT92-0018 to the Molecular Dynamics of Membrane Biogenesis Network.

References.

Bortoli-German I, Brun E, Py B, Chippeaux M, Barras F (1994) Periplasmic disulphide bond formation is essential for cellulase secretion by the plant pathogen *Erwinia chrysanthemi*. Mol Microbiol.11:545-553.

d'Enfert C, Ryter A, Pugsley AP (1987) Cloning and expression in *Escherichia coli* of the *Klebsiella pneumoniae* genes for production, surface localization and secretion of the lipoprotein pullulanase. EMBO J 6:3531-3538.

Hardie KR, Schulze A, Parker MW, Buckley JT *Vibrio sp.* secrete proaerolysin as a folded dimer without the need for disulphide bond formation. Mol Microbiol in press.

Hirst TR, Holmgren J (1987) Conformation of protein secreted across bacterial outer membanes : a study of enterotoxin translocation from *Vibrio cholerae*. Proc Natl Acad Sci USA 84:7418-7422

Kornacker MG, Pugsley AP (1990) The normally periplasmic enzyme ß-lactamase is specifically and efficiently translocated through the *Escherichia coli* outer membrane when it is fused to the cell surface enzyme pullulanase. Mol Microbiol 4:1101-1109

Pugsley AP (1989) Protein targeting. Academic Press, San Diego, USA

Pugsley AP (1992) Translocation of a folded protein across the outer membrane via the general secretory pathway in *Escherichia coli*. Proc Natl Acad Sci USA 89:12058-12062

Pugsley AP (1993 The complete general secretory pathway in gram-negative bacteria. Microbiol Rev 57:50-108

Pugsley AP, Chapon C, Schwartz M (1986) Extracellular pullulanase of *Klebsiella pneumoniae* is a lipoprotein. J Bacteriol 166:1083-1088

Shevchik VE, Bortoli-German I, Robert-Baudouy J, Robinet S, Barras F, Condemine G Differential effect of *dsbA* and *dsbC* mutations on extracellular enzyme secretion in *Erwinia chrysanthemi*. Mol Microbiol in press

E. coli preprotein translocase: A 6 stroke engine with 2 fuels and 2 piston rods

William Wickner and Marilyn Rice Leonard
Department of Biochemistry
Dartmouth Medical School
Hanover, N.H. 03755-3844 U.S.A.

The elucidation of the pathway for preprotein transit across the plasma membrane of *E. coli* is a fine example of the synergy between genetic, biochemical, and physiological approaches to study a common problem. Early physiological studies of secretion by Ito and by Randall established important boundary parameters; secretion requires metabolic energy and is not coupled to ongoing polypeptide chain growth. Genetic efforts in the labs of Beckwith and Silhavy were focused on obtaining *prl* supressor mutants with enhanced capacities to export proteins with mutant leader sequences and *sec* mutants which were temperature sensitive in the export process itself. In our lab, the biochemical studies of this era focused on the sec-independent M13 procoat protein and on the isolation of leader peptidase.

With the development in 1984 of in vitro reactions of translocation of large, sec-dependent preproteins, an enzymological approach became possible for the study of large protein secretion. Preproteins were purified in denaturant and, though competent for translocation when diluted, required chaperones to stabilize their assembly-competent structure. Chaperone cascades are now emerging from the mists; bacterial SRP and trigger factor, GroELS, the heat shock DnaKJ/GrpE system, and the export-dedicated SecB may all play overlapping roles (to different extents for different preproteins) in delivering proteins to the membrane.

In vitro translocation reactions circa 1987 used pure soluble components, ATP, and purified membrane vesicles. However, attempts to solubilize inner membrane vesicles (IMV's) in detergent

NATO ASI Series, Vol. H 96
Molecular Dynamics of Biomembranes
Edited by Jos A. F. Op den Kamp

and reconstitute proteoliposomes which were active for translocation met with repeated failure. Only later did the reason become apparent through the "retrospectoscope" (the most powerful scientific instrument ever devised!)- SecA is exquisitely sensitive to common detergents.

We observed that membranes became dependent on a cytosolic factor when the membranes were urea-treated to remove peripheral membrane proteins. Sizing on glycerol gradients showed the factor to be of a size which fit with that predicted for SecA by the elegant studies of Don Oliver. He generously shared a precious SecA overproducer strain, which allowed the immediate identification of SecA as the relevant component. At the same time, following the observations by Tai that membranes could be inactivated for translocation by azido-ATP plus light, we found that pure SecA could restore these inactivated membranes to full translocation activity. Thus, we were delighted, but not suprised, to find that pure SecA is an ATPase. In contrast, we were very suprised indeed to find that the ATPase activity of SecA is strongly stimulated by simultaneous interactions with membranes and with a preprotein such as proOmpA. This "translocation ATPase" activity provided a rapid, colorimetric assay for each of the components of translocation, and thus transformed our work. Arnold Driessen then came to the lab and discovered that a mixture of detergent with lipids and glycerol was able to solubilize the integral membrane component(s) in a form which could be reconstituted into translocation-competent proteoliposomes. An eager and capable crew, led by Lorna Brundage, then fractionated this extract. To our delight, just one integral membrane complex was needed. It was shown to be comprised of three polypeptide subunits- SecY, SecE, and a third subunit which was given the rather neutral name band 1. Band 1, though the object of calumny and scorn for a year or two, was firmly bound with the other two subunits, and the "reverse genetics" of this protein in the lab of Professor Tokuda in Japan showed that it merits the designation SecG. With purified components and antibodies available, Ulrich Hartl and Stewart Lecker found that preproteins are guided to the membrane by a cascade of binding events. Many preproteins are stabilized for translocation through formation of a stoichiometric complex with a SecB tetramer. As shown through the elegant studies of Randall, SecB captures slow-folding mature domains, and the folding of at least some preprotein's mature domains is actually slowed by the presence of the leader sequence. Though much of SecA is cytosolic, and cytosolic SecA has affinity for the preprotein/chaperone complex, membrane-bound SecA is activated for these binding affinities. SecA binds to the membrane both by its affinity for acidic phospholipids and for the SecYEG heterotrimeric, membrane-embedded domain. When so bound, SecA is activated to bind the preprotein/chaperone complexes by virtue of its recognition of the SecB chaperone, the preprotein leader region, and the preprotein mature domain. The latter specificity is especially suprising, and may again reflect an evolutionary constraint against rapid and tight folding for proteins which must be translocated. This has been underscored by studies of Beckwith and Silhavy's groups in which they find that "leaderless" exported proteins are still very selectively translocated across the plasma membrane in certain "*prl*" mutants in the secY gene.

Translocation can be reconstituted with proteoliposomes bearing even a single acidic phospholipid species, pure SecYEG, SecA, ATP, and a pure complex of proOmpA and SecB. When a light-driven proton "pump" is also incorporated into these proteoliposomes, light dramatically accelerates and enhances the translocation process. With this chemically defined, if complex, reaction, true initial rates and translocation yields are obtained which are within a factor of two of those found with the starting membranes or even the intact cell. Translocation, whether in vivo or in a reconstituted reaction, thus uses two energy sources, ATP and the membrane protonmotive force. Though this reconstitution was satisfying, it still left us in the dark as to the mechanism whereby preproteins were actually delivered by translocase (SecA + SecYEG) across the membrane.

Elmar Schiebel gained great insight into the steps of translocation by studying intermediates of translocation in which a preprotein chain is kinetically "caught" with only a part of its sequence spanning the membrane. He noted that intermediates of all length of translocated portion showed a common feature of translocating an extra ~25 residues upon the addition of AMP-PNP, suggesting that ATP drives translocation through the energy of its binding to the SecA subunit of translocase and that translocation occurs in "quanta"!! Elmar found that he could also use urea to remove functional SecA from translocation intermediates. The intermediates, spanning the membrane via SecYEG, could not undergo ATP-driven translocation unless fresh SecA was added back, showing that the translocation site remained functional. Most strikingly, the imposition of a membrane potential drove a rapid and efficient completion of translocation in the absence of functional SecA, showing that SecA was not some obligate translocation "pore". Furthermore, the addition of SecA and AMP-PNP completely blocked the potential-driven completion of translocation, showing that the two energy sources act at distinct phases or steps of the catalytic cycle. It is still not at all clear (at least to us!) how the potential drives translocation, for each simple and simplistic experiment which might suggest a mechanism (e.g., electrophoresis of the ionized amino acyl sidechains) has fallen victim to a control experiment.

We have, however, found how SecA can promote translocation. Anastassios ("Tassos") Economou checked that SecA he had labeled with radioactive iodine was still active by using it "carrier-free" in a translocation reaction. Not only did he find that this SecA supported the translocation of radioactive proOmpA into the lumen of vesicles, but that a 30kDa domain of the SecA also became inaccessable to the protease. This domain actually translocates, becoming accessable from the lumenal (periplasmic) side of the vesicles, as shown in careful experiments of proteolysis of intact and disrupted vesicles. Its membrane "insertion" requires ATP, proOmpA, and membranes with SecYEG. The genetic co-overexpression of SecYEG creates not only additional sites for translocation but additional sites for SecA

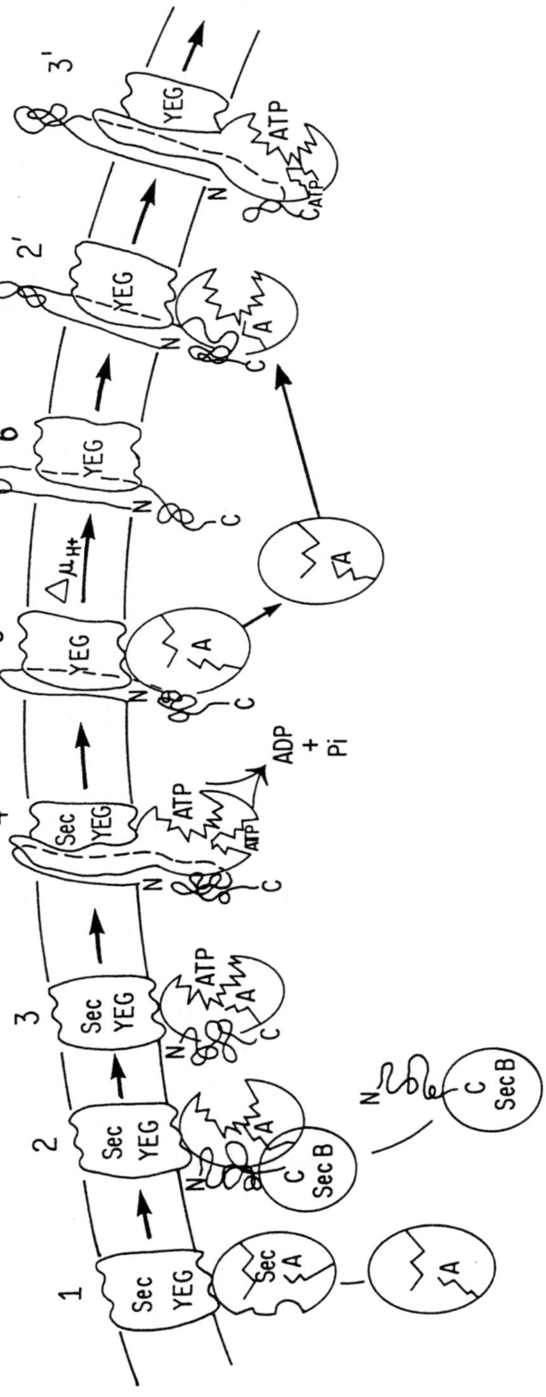
1
2
3
4
5
6
2'
3'
Sec
YEG
A
Sec B
ATP
ADP
+
Pi
N
C
Δμ H+

insertion. Insertion is only temporary; addition of excess, nonradioactive SecA almost immediately displaced even the membrane-inserted radioactive SecA in these experiments. Furthermore, the deinsertion reaction required additional input of ATP! Recently, in collaboration with Don Oliver, we have found that the insertion and deinsertion steps use the two ATP binding sites on SecA in distinct fashion.

Similar findings of inserted SecA that is accessable from the periplasmic surface of the membrane were made by Kim and Oliver. They found that SecD and SecF (and/or other nearby genes) can strongly influence the fraction of the SecA in the inserted state. Our recent in vitro studies are in agreement with their findings, and suggest that SecD/F may stabilize the inserted state of SecA. Further studies, including the purification of the functional protein(s), will be essential to illuminate the function performed by SecD/F at this stage of the translocation reaction.

Our current understanding of the translocation process is summarized in the Figure, one in a (hopefully!) never-ending series of cartoons which reflect our growing understanding of this translocation reaction. Future questions include the dynamics of the other subunits; epitope-tagging studies suggest that SecY and SecE subunits remain stably associated for generations, while SecG is more dynamic. The relationship between the membrane potential and SecA movements will bear study. It will also be fascinating to determine which parts of the apparently polar SecA protein are inserted into the membrane in contact with SecYEG, which are in contact with the lipid fatty acyl phase, and which are in contact with the periplasmic surface. The lateral "escape" of integral membrane proteins into the lipid phase is also a central mystery of this system. Finally, the basis whereby some proteins, like M13 procoat protein, can assemble without translocase is an enduring puzzle.

The Tol/PAL and TonB systems : two envelope-spanning protein complexes involved in colicin import in *E. coli.*

Emmanuelle Bouveret, Claude Lazdunski and Hélène Bénédetti,
Laboratoire d'Ingénierie et Dynamique des Systèmes Membranaires,
CNRS, 31, chemin Joseph Aiguier, BP 71,
13402 Marseille cedex 20,
FRANCE

Mutants in *tolA*, *B*, *Q*, and *R* genes have been isolated on the basis of their tolerance to bacterial toxins (colicins) and their resistance to the infection of filamentous phages (M13, fd, and f1) (Davies and Reeves, 1975a, 1975b ; Nagel de Zwaig and Luria, 1967). These genes form a cluster at 16,8 min on the chromosomal map of *E. coli*. *tol* mutants are hypersensitive to detergents and to certain drugs, and they release periplasmic proteins into the growth medium (Nagel de Zwaig and Luria, 1967). Mutations in a contiguous gene, *pal*, which encodes the outer membrane Peptidoglycan Associated Lipoprotein (PAL), generate a similar phenotype (Fognini-Lefebvre *et al.*, 1987). This suggests that the Tol/PAL proteins are involved in maintaining the integrity of the outer membrane of E. *coli*. However, the exact physiological role of the Tol/PAL system has not yet been elucidated.

After colicins have been released from producing bacteria, they bind to specific receptors at the surface of sensitive bacteria, translocate through the cell envelope and kill them. However, colicins are too large to freely pass through the outer membrane barrier. The toxins and a variety of bacteriophages have taken advantage of resident envelope proteins to enter bacteria. Group B colicins (including B, D, Ia, Ib, M, and 5) and phages T1 and Φ80 have parasitized TonB and its associated proteins ExbB and ExbD. The role of these proteins, in association with specific receptors of the outer membrane, is to promote the active transport of vitamin B12 and iron siderophores through the outer membrane. Group A colicins (A, E1 to E9, K, L, N, and 10) and phages f1, fd, and M13 have parasitized the Tol proteins. The sequences and the localizations of the Tol/PAL proteins and the proteins of the TonB system have been determined. In addition, the topologies of the inner membrane anchored proteins have been elucidated. Recently, it has been demonstrated that both systems form a complex and the interactions involved between the different components are on the way to be characterized.

The present paper reviews these recent data with emphasis on the Tol proteins and the more recent contribution from the authors' laboratory. We also report the recent progress in the understanding of the translocation step of colicins and describe the approach we are now following to try to fully elucidate the translocation mechanism of group A colicins. Here, we report and discuss recent progresses in these studies.

Description of the components of the Tol/PAL and TonB systems

TonB system: The TonB system is composed of three proteins : TonB (242 aa), ExbB (244 aa), and ExbD (141 aa) (Postle and Good, 1983 ; Eick-Helmerich and Braun, 1989). *tonB* is unlinked

NATO ASI Series, Vol. H 96
Molecular Dynamics of Biomembranes
Edited by Jos A. F. Op den Kamp

to the *exb* locus which consists of *exbB* and *exbD* genes. TonB and ExbD are anchored to the inner membrane by a single transmembrane domain while ExbB is an integral membrane protein with three transmembrane segments. The transmembrane domains of TonB and ExbD are localized in their N-terminal part and the remainder of the proteins protrudes into the periplasmic space (Kampfenkel and Braun, 1992a ; Roof *et al.*, 1991). TonB suppressors of outer membrane vitamin B12 or iron siderophores receptor mutants have been isolated (Schöffler and Braun, 1989 ; Bell *et al.*, 1990 ; Braun *et al.*, 1991). Furthermore, *in vivo* crosslinking studies have shown that TonB physically interacts with one of these receptors : FepA (Skare *et al.*, 1993). These lines of evidence indicate that TonB spans the entire periplasm and interacts with high affinity outer membrane receptors. This is consistent with the role of TonB as an energy transducer coupling the cytoplasmic membrane proton motive force to the transport of vitamin B12 and iron siderophores across the outer membrane. A 33-residue proline-rich peptide segment composed of $(Glu\text{-}Pro)_n$- $(Lys\text{-}Pro)_m$ repeat motif and predicted to assume an extended rigid conformation has been shown to provide a physical extension sufficient for TonB to span the periplasm (Larsen *et al.*, 1993). The TonB C-terminal domain consists of three amphiphilic structures (β-strand, α-helix, β-strand). The C-terminal β-strand has been shown to be essential for TonB function (Anton and Heller, 1991) and might interact with the receptors. Mutations in *exbB* or *exbD* decrease TonB activity (Gutermann and Dann, 1973; Hantke and Zimmerman, 1981) and its stability (Fischer *et al.*, 1989; Skare and Postle, 1991). However, the role of ExbB in energy transduction is not restricted to its effect on TonB turnover because ExbB remains essential for efficient energy transduction in *OmpT* strains where TonB levels are stabilized (Skare *et al.*, 1993).

Tol/PAL system : The complete sequence of the Tol/PAL cluster has been determined (Sun and Webster, 1987; Levengood and Webster, 1989 ; Lazzaroni *et al.*, 1989). Each gene can be independently transcribed but it has been recently shown that they are organized in two operons : *tolQ*, *R*, and *A* operon and *tolB*, *pal*, and *orf2* operon (*orf2* encodes a putative protein of 262 aa) (Lazzaroni, unpublished results). TolQ, R, A, B, and PAL are proteins of respectively 230, 142, 421, 431, and 173 aa.

TolQ and TolR are 75% homologous with ExbB and ExbD, respectively (Eick-Helmerich and Braun, 1989). Moreover, they have been found to have similar topologies : TolQ is an integral membrane protein with three transmembrane domains with its N-terminus facing the periplasm space (Bourdineaud *et al.*, 1989 ; Vianney *et al.*, 1994; Kampfenkel and Braun, 1992b). TolR spans the inner membrane only once through a single hydrophobic region (Muller *et al.*, 1993; Kampfenkel and Braun, 1992b). Furthermore, some functional cross-reactivity has been found to exist between TolQ/R and ExbB/D (Braun, 1989).

TolA is anchored to the inner membrane by a 20-aa N-terminal hydrophobic region. Its central domain (residues 34 to 264) is periplasmic and very rich in Ala, Lys, Glu, and Asp residues, since it contains the unit $ED(K)_{1-2}(A)_{2-4}$ repeated ten times. This repeated motif might contribute to strongly stabilize α-helical structures (Levengood and Webster, 1989) and might serve to provide an extended conformation of TolA. This structure might allow TolA, like TonB, to span the periplasm (Webster, 1991). The TolA central domain seems to play no role in colicin translocation because cells producing the corresponding TolA mutant are almost equally sensitive to group A colicins than wild type cells. Conversely, this domain seems to be important in maintaining the cell envelope integrity (Derouiche *et al.*, 1995). The C-terminal domain of TolA is involved in colicin import since mutations in this domain block the translocation step of colicins and since this domain has been shown to interact with colicins A and E1 *in vitro* (Bénédetti *et al.*, 1991).

TolB is a periplasmic protein. It is synthesized in a precursor form and its 21-aa N-terminal signal sequence is cleaved upon its SecA-dependent export. A low amount of the protein has been shown to be peripherally membrane associated (Isnard *et al.*, 1994).

PAL is an outer membrane lipoprotein. It is synthesized as a precursor with a N-terminal signal sequence containing a consensus clivage site for the lipoprotein signal peptidase. Before the signal sequence is cleaved, the lipoprotein is maturated by acylation of the first cystein residue of the mature sequence. This lipid moiety is responsible for the integration of the protein into the outer membrane. PAL has been shown to be strongly but not covalently associated with the peptidoglycan (Mizuno, 1979). A positively charged region of its C-terminus might be involved in the interaction (Lazzaroni and Portalier, 1992). *pal* mutants exhibit the same phenotype as *tol* mutants with respect to leakiness and drug sensitivity (Lazzaroni and Portalier, 1981 ; Lazzaroni *et al.*, 1989). However, they are not tolerant to group A colicins nor to filamentous phages.

Proteins of the TonB and Tol/PAL systems form a complex linking the inner and outer membranes of *E. coli*.

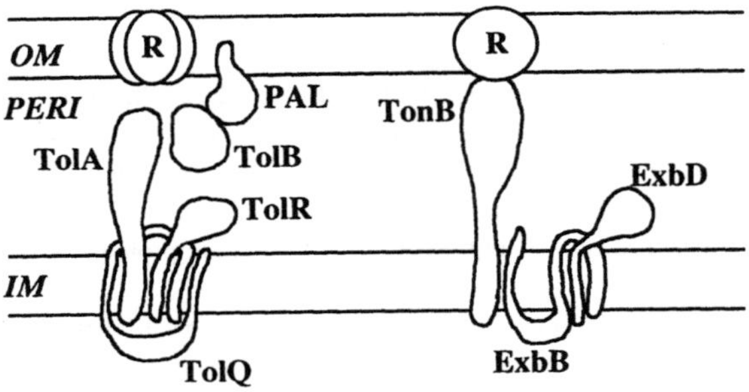

Figure 1 : The Tol/PAL and TonB systems

Different lines of evidence favor an interaction between TonB and high affinity outer membrane receptors of iron siderophores and vitamin B12. All receptors whose activities depend on TonB share a consensus sequence of 5 to 6 aa called the TonB-box. When mutated in this box, the receptors still bind to their ligand but cannot transport them through the outer membrane. This common structure in the receptors is expected to be recognized by the TonB protein. Indeed, mutations in TonB are able to suppress mutations in the TonB-box of BtuB, FhuA, and Cir receptors (Heller *et al.*, 1988 ; Schöffler and Braun; 1989 ; Bell *et al.*, 1990). Furthermore, a pentapeptide corresponding to a consensus TonB-box has been shown to inhibit TonB activity when added to the external medium (Tuckman and Osburne, 1992). Recently, *in vivo* cross-linking studies have provided direct evidence that TonB can physically interact with the FepA receptor (Skare *et al.*, 1993).

Beside its interactions with proteins of the outer membrane, TonB has also been shown to interact with ExbB. The first evidence that favoured such an interaction were indirect and were inferred from the TonB unstability in *exbB/exbD* mutants (Fischer *et al.*, 1989; Skare and Postle, 1991). Karlsson *et al.* (1993) have also demonstrated that the requirement of ExbB and ExbD proteins for phage Φ80 sensitivity was specified by the TonB transmembrane domain. This result suggested that a functional interaction might occur between the N-terminal domain of TonB and ExbB and ExbD proteins inside the membrane. Direct pieces of evidence arose from *in vivo* cross-linking experiments (Skare *et al.*, 1993). This type of experiments confirmed that TonB and ExbB interact inside the membrane and that the first transmembrane domain of ExbB might be involved in this interaction (Jaskula *et al.*, 1994 ; Larsen *et al.*, 1994).

Until recently, the idea that the proteins of the Tol/PAL system formed a complex was only supported by indirect evidence. First, the TolA transmembrane domain was found to specify the dependence on TolQ and TolR (Karlsson *et al.*, 1993). By analogy with the TonB system, this result suggested that TolA and TolQ/TolR interacted via their transmembrane segments. Second, Tol/PAL proteins were shown to co-fractionate with contact sites between the inner and outer membranes (Bourdineaud *et al.*, 1989 ; Leduc *et al.*, 1992 ; Guihard *et al.*, 1993a). When colicin A was added to the cells, the relative amount of Tol proteins in contact sites was increased. This suggested that TolA, B, Q, and R formed a complex with a given stoechiometry (Guihard *et al.*, 1993a). The first direct evidence that the Tol proteins interacted came from genetic studies involving TolR suppressors of a TolQ protein mutated in its third transmembrane domain (Lazzaroni *et al.*, 1995). *In vivo* cross-linking experiments further demonstrated that TolA transmembrane domain interacts with TolQ and TolR (Derouiche *et al.*, 1995). Finally, combining this later technique to co-immunoprecipitation experiments, we showed that TolB interacted with PAL (Bouveret *et al.*, 1995). Inside the inner membrane, the transmembrane domain of TolR therefore interacts with the transmembrane domain of TolA and the third transmembrane domain of TolQ. Genetic evidence favor an intramolecular interaction between the first and third transmembrane segments of TolQ (Lazzaroni *et al.*, 1995). Although we do not know which portion of TolQ interacts with the TolA transmembrane domain, figure 1 presents a hypothetical model of the Tol/PAL complex. Because TolB cofractionates with the other Tol proteins in the contact sites, it might interact with one of these proteins. Indeed, preliminary results indicate that TolB might interact with TolA (Bénédetti, unpublished data). Therefore, TolB and PAL would connect TolA from the inner membrane to the outer membrane.

Like the TonB system, the Tol system spans the whole cell envelope and connects the inner and outer membranes of *E. coli*. In the case of the TonB system, the role of this connection is to allow an energy transduction of the proton motive force of the cytoplasmic membrane to high-affinity outer membrane receptors. However, the mechanism of this transduction is not yet clear. According to cross-linking experiments, other not yet identified proteins might be involved in energy transduction (Skare *et al.*, 1993). It has recently been proposed that the outer membrane receptor proteins function as gated pores (Rutz *et al.*, 1992 ; Killmann *et al.*, 1993). The current hypothesis is that the proton motive force would drive a conformational change in TonB that would then open the gate and allow ligands to move through the pores (Rutz *et al.*, 1992). Cycles of conformational changes in TonB would result in active pumping of ligands. ExbB and ExbD might play a role in reenergizing TonB. In contrast, in the case of the Tol/PAL system, the physiological role of the contact sites formed is not yet known.

The translocation step of colicins

Little is known about the mechanism of translocation of group B colicins. Most of them have parasitized TonB-dependent receptors of the outer membrane; however, a recently identified TonB-dependent colicin (colicin 5) has been shown to use Tsx which is TonB-independent and has no TonB-box (Bradley and Howard, 1992). Group B colicins might interact with TonB since all of them have a TonB-box (Köck *et al.*, 1987 ; Schramm *et al.*, 1988 ; Ross *et al.*, 1989). In the case of colicins B and M, the TonB-box and those of their receptors have been shown to be important for the translocation step (Mende and Braun, 1990 ; Pilsl *et al.*, 1993). These data suggest that TonB interacts with colicins B and M and their receptors during their translocation.

Although the translocation step of Tol-dependent colicins remains unclear by many aspects, much progress has been accomplished with regard to our understanding of the mechanisms involved. Most of the data come from the studies of pore-forming colicins that kill the cells by depolarizing the inner membrane and by inducing a phosphate efflux (Guihard *et al.*, 1993b), which results in the depletion of the cytoplasmic ATP. These colicins also induce an efflux of cytoplasmic K^+ which can be measured. The kinetics of this efflux have been extensively studied in the case of colicin A (Bourdineaud *et al.*, 1990). It has been shown that the K^+ efflux is preceded by a lag time of a few tens of seconds, which corresponds to the time required for colicin A to bind to its receptor and to be translocated through the envelope (Bourdineaud *et al.*, 1990). Under defined conditions, it is possible to discriminate between the two steps and to analyze the translocation step independently of the receptor binding step. This technique allowed to demonstrate that the translocation step is accelerated if colicin A is urea-denaturated (Bénédetti *et al.*, 1992a), suggesting that the unfolding of the toxin accelerates its translocation.

Trypsin has been shown to induce the closing of the colicin A pore while having access neither to the periplasmic space nor to the cytoplasmic membrane of the sensitive cells (Bénédetti *et al.*, 1992a). Furthermore, a disulfide bond-engineered colicin A mutant able to be translocated but unable to open its pore has been shown to compete with the receptor binding and the translocation steps of wild type colicin A (Duché *et al.*, 1994). These data indicate that colicin A maintains an extended conformation accross the cell envelope and is still in contact with its receptor and its translocation machinery even when its pore has formed in the inner membrane. The unfolding of the colicin would be triggered very early in the process by the binding to the receptor (Duché *et al.*, 1994). These data agree with fractionnation experiment results which showed that colicin cofractionnates with the Tol proteins in contact sites (Guihard *et al.*, 1993a).

Group A and B colicins have been divided into three domains. A specific function has been assigned to each of them. The central domain is involved in receptor binding, the N-terminal domain is required for translocation and the C-terminal domain carries the lethal activity (pore formation, RNase or DNase). Hybrid colicins containing various combinations of group A and group B colicin domains have been constructed (Géli and Lazdunski, 1992; Bénédetti *et al.*, 1992b). The translocation pathway they follow appears to be defined by their sole N-terminal domain. Therefore, the N-terminal domains of colicins contain all the information needed for the translocation step. Indeed the TonB-box is found in the N-terminus of all group B colicins. The N-terminal domains of colicin A and E1 have been shown to interact *in vitro* with the C-terminal domain of TolA. This interaction seems to reflect an interaction occuring *in vivo* during the translocation step of these colicins because a TolA mutant unable to promote colicin translocation is also unable to interact *in vitro* with the N-terminal domain of colicin A and E1 (Bénédetti *et al.*, 1991). A TolA-box has recently been identified in the N-terminus of all Tol-dependent colicins (Pilsl and Braun, 1995). This TolA-box is a conserved glycin-rich pentapeptide region. At present, there is no evidence that this motif might allow colicins to interact with TolA. However, it seems to be important for colicin translocation because a colicin E3 (RNase type) mutant in one residue of the box (E3 (Ser37Phe)) binds normally to its receptor but is unable to be translocated (Mock and Schwarz, 1980; Escuyer and Mock, 1987). For the moment, it is not known whether this colicin E3 mutant binds or not to TolA since *in vitro* conditions used to test the colicin A or E1 interaction with TolA do not work for interaction of RNase- or DNase-type colicins with TolA (Bénédetti, unpublished results). This difference might simply reflect a lower affinity in the interaction or suggest a distinct mechanism of translocation between colicins having an inner membrane target and colicins having a cytoplasmic target. However, the translocation mechanisms might be somehow similar because a hybrid colicin having the N-terminal domain of colicin A and the C-terminal domain of colicin E3 has been shown to be active on sensitive cells (Bénédetti *et al.*, 1992b). This interaction with TolA seems necessary but not sufficient for the translocation step since a colicin A mutant deleted from aa 15 to 30 and with an intact TolA-box still binds with TolA *in vitro* but is unable to translocate through the cell envelope (Bénédetti *et al.*, 1991 ; Baty *et al.*, 1988). Interactions with other Tol proteins might be necessary for translocation to proceed.

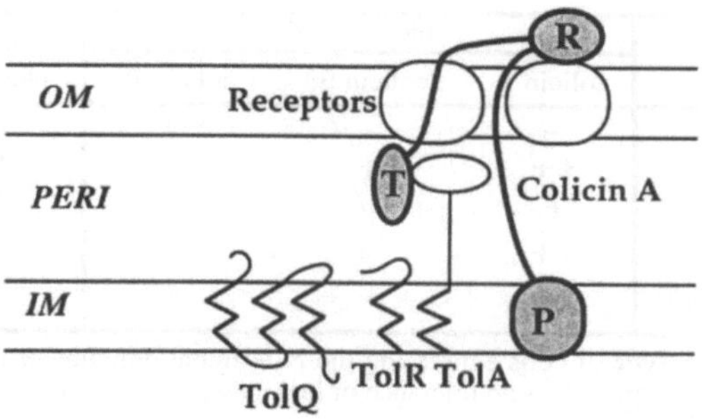

figure 2 : model of the insertion of the colicin A in *E. coli* envelope.
T: translocation domain; R: receptor binding domain; P: pore forming domain.

To get new insights on the mechanism of translocation of group A colicins, we developed a system allowing us to study the interaction of the N-terminal domains of group A colicins with the translocation system *in vivo*. For this purpose, the approach chosen, described in the following paragraph, was to produce the N-terminal domain of different colicins in the periplasm of the bacteria.

How to use periplasm-targeted N-terminal domains of colicins to study the colicin translocation step ?

If the N-terminal domains of colicins are exported to the the periplasm of bacteria and interact with at least one component of the group A colicin translocation system, cells should become tolerant to these colicins coming from the outside. Indeed, such a competition has already been observed in the following cases : (i) Cells infected by Tol-dependent phage f1 are tolerant to group A colicins (Zinder , 1973). (ii) When the N-terminal domain of protein G3p from phage f1 is produced in the periplasm, the bacteria become tolerant to colicins (Boeke *et al.*, 1982). (iii) The same phenotype is also induced when the C-terminal domain of TolA is produced in the periplasm (Levengood-Freyermuth *et al.*, 1993). In the case of the N-terminal domains of colicins, this approach might allow us to increase the copy number of colicin N-terminal domains in the periplasm as compared to cells treated with exogenous colicins. This factor might facilitate the characterization of the interactions between the N-terminal domains of colicins and the envelope proteins.

By using the plasmid Pin2 (Ghrayeb *et al.*, 1984), plasmids were engineered to encode the N-terminal domain of colicins A, E3, B, A(Δ15-30), or E3(Ser37Phe) (Baty *et al.*, 1988 ; Mock and Schwarz, 1980 ; Escuyer and Mock, 1987) joined to the signal sequence of OmpA. Synthesis of these proteins (induced by IPTG) should lead to the secretion, into the periplasm, of soluble processed N-terminal domains. Cell fractionation and imunocytochemical experiments confirmed the correct exportation of these domains into the periplasm. Cells which produce these constructs were tested for (i) their tolerance to colicins, (ii) their leakiness, and (iii) their sensitivity to deoxycholate. Results are summarized in the table 1:

N-terminal of colicin	tolerance to			leaky phenotype
	colicin A	colicin E3	colicin B	
B	-	-	++	-
A	++	+	-	++
E3	+	+/-	-	++
A(Δ15-30)	++	+	-	++
E3(Ser37Phe)	-	-	-	-

Table 1 : Phenotype of cells expressing the N-terminal domains of colicins in their periplasmic space. **-** no tolerance or leakiness; **+** tolerance or leakiness

(i) Cells producing the N-terminal domain of colicin A are tolerant to colicin A and E3 but not to colicin B while cells producing the N-terminal domain of colicin B are still sensitive to colicins A and E3 and tolerant to colicin B. This result proves that the N-terminal domain of colicin A interacts, in the periplasm, with components specifically involved in the uptake of group A colicins. Interestingly, cells which produce the N-terminal domain of colicin E3 are only partially tolerant to colicins A and E3.

(ii) When the N-terminal domains of colicins A or E3 were produced in the periplasm of the cells, the release of β-lactamase, alcaline phosphatase and RNase I in the medium was triggered. This leaky phenotype was not induced when the colicin B N-terminal domain was produced.

(iii) None of the cells was rendered sensitive to deoxycholate.

Points (i) and (ii) can be related to the phenotype of *tol* mutants. Because cells which express the N-terminal domain of colicins A or E3 present common phenotypes with *tol* mutants, these N-terminal domains may interact with one or more Tol proteins and thus render these proteins inoperative for their physiological function in the cell.

The results obtained with the two colicin mutants are noteworthy. First, whereas colicin

A(Δ15-30) is not active on sensitive cells, its N-terminal domain still provokes tolerance to exogenous colicins A or E3 and leakiness of the bacteria. Therefore, this domain still competes with the N-terminal domain of the entire colicins for the interaction with Tol proteins or other proteins of the cell envelope. Nevertheless, this interaction is not sufficient to promote the colicin translocation because colicin A(Δ15-30) is unable to reach its target. In the case of the colicin E3 mutant (Ser37Phe), the phenotype induced by the production of the N-terminal domain in the periplasm was drastically different. Cells were as sensitive to colicins as wild type cells and above all, they were not leaky at all. Therefore, this N-terminal domain does not seem to compete with the N-terminal domain of entire colicins coming from the outside for the interaction with the Tol proteins or other proteins of the cell envelope. In this case, the inactivity of the entire colicin E3 mutant can be correlated to the lack of the competition we detect by our system. It is of interest to note that colicin E3(Ser37Phe) is mutated in the supposed TolA-box whereas the deletion in colicin A (Δ15-30) does not include this TolA-box.

We would also like to use this approach to identify the components which interact with the N-terminal domains of colicins. *In vivo* cross-linking experiments with formaldehyde on cells producing the N-terminal domain of colicins in their periplasm have shown that the N-terminal domains of colicins A and E3 interact with yet unidentified proteins distinct from the Tol proteins. Furthermore, when total cell proteins are immunoprecipitated, after *in vivo* cross-linking, with an antibody directed against TolB, the N-terminal domain of colicin E3 is co-immunoprecipitated. This result suggests that the N-terminal domain of colicin E3 interacts with TolB in the periplasm.

Another way to identify interactions between the N-terminal domains of colicins and the Tol proteins is to do affinity experiments with purified proteins. TolA and TolB proteins were purified on a Nickel column after a Histidine-tagging. The N-terminal domains of colicins A, E3, A(Δ15-30), and E3(Ser37Phe) were also purified. Preliminary results with Affigel beads suggest that TolB interacts with the N-terminal domain of colicin E3 and that TolB interacts with TolA. We intend to confirm the existence of these interactions and to measure their affinity constants by using the technique of surface plasmon resonance.

Conclusion

The TonB and Tol/PAL systems define contact sites between the inner and outer membranes of *Escherichia coli*. The TonB-dependent sites are used for the transport of specific compounds (vitamin B12 and iron siderophores) which are too large to freely diffuse across outer membrane porins. The physiological role of the Tol-dependent contact sites is unknown. They may be involved in a yet unidentified transport function. *tol* and *pal* mutants are affected in the integrity of their outer membrane. This observation supports the hypothesis that the Tol system might be involved in the export and correct assembly of components of the outer membrane. This hypothesis is currently tested. However, the perturbation in the outer membrane organization might be explained differently. A mutation in a component of the Tol system might break the link between the inner membrane and the peptidoglycan (via PAL which is directly associated to it) thereby destabilizing the envelope. The two contact site systems have been parasitized by colicins to promote their entry into cells. Considerable progress have been accomplished with regard to our understanding of the mechanisms involved. We know that the toxins have to unfold completely and to interact with components of these translocation sites. Our efforts are now aimed at identifying the proteins involved in these interactions and at understanding the dynamics of translocation. A better knowledge of the structure of the proteins of the Tol/PAL and TonB systems and of the N-terminal domains of colicins, and the use of periplasm-targeted N-terminal domain of colicins, should help us to further investigate the molecular mechanisms underlying import into sensitive cells.

Acknowledgements

We are grateful to A. Rigal and D. Espesset for careful reading of the manuscript. We are particularly indepted to A. Rigal who actively participated in the work described. We also thank R. Dérouiche and R. Lloubes for helpful discussion and M. Gavioli for technical assistance. This work was supported by the CNRS and by the Fondation pour la Recherche Médicale.

REFERENCES

Anton M and Heller KJ (1991) Functional analysis of a C-terminally altered TonB protein of *Escherichia coli* Gene 105:23-29.

Baty D, Frenette M, Lloubès R, Géli V, Howard SP, Pattus F, Lazdunski C (1988) Functional domains of colicin A. Mol. Microbiol. 2, 807-811.

Bell PE, Nau CD, Brown JT, Konisky J and Kadner RJ. (1990) Genetic suppression demonstrates interaction of Ton B protein with outer membrane transport proteins in *Escherichia coli*. J. Bacteriol. 172:3826-3829.

Bénédetti H, Lazdunski C and Lloubès R (1991). Protein import into *Escherichia coli*: colicins A and E1 interact with a component of their translocation system. EMBO J. 10, 1989-1995.

Bénédetti H, Lloubès R, Lazdunski C and Letellier L (1992a) Colicin A unfolds during its translocation in *Escherichia coli* cells and spans the whole cell envelope when its pore has formed. EMBO J. 11: 441-447.

Bénédetti H, Letellier L, Lloubes R, Géli V, Baty D and Lazdunski C (1992b) Study of the import mechanisms of colicins through protein engineering and K+ efflux kinetics. In Bacteriocins, Microcins and lantibiotics, Nato ASI Series Vol. H65, pp. 215-223 R. James, C. Lazdunski, F. Pattus (eds). Springer Verlag,Berlin, Heidelberg.

Boeke JD, Model P and Zinder ND (1982) Effects of bacteriophage f1 gene III protein on the host cell membrane. Mol. Gen. Genet. 186:185-192.

Bourdineaud JP, Howard SP and Lazdunski C (1989) Localization and assembly onto the *Escherichia coli* envelope of a protein required for entry of colicin A. J. Bacteriol. 171:2458-2465.

Bourdineaud JP, Boulanger P, Lazdunski C and Letellier L (1990) In vivo properties of colicin A : channel activity is voltage dependent but translocation may be voltage independent. Proc. Natl. Acad. Sci. USA. 87, 1037-1041.

Bouveret E, Derouiche R, Rigal A, Lloubès R, Lazdunski C and Bénédetti H (1995) Peptidoglycan-associated lipoprotein-TolB interaction J. Biol. Chem. 270:11071-11077.

Bradley DE and Howard SP (1992) A new colicin that absorbs to outer-membrane protein Tsx is dependent on the tonB instead of the tolQ membrane transport system. J. Gen. Microbiol. 135:1857-1863.

Braun V (1989) The structurally related *exbB* and *tolQ* genes are interchangeable in conferring tonB-dependent colicin, bacteriophage and albomycin sensitivity. J. Bacteriol. 171: 6387-6390.

Braun V, Günter K and Hantke K (1991). Transport of iron across the outer membrane. Biol. Metals 4: 14-22.

Davies JK and Reeves P (1975a) Genetic of resistance to colicins in *Escherichia coli* K-12. Cross-resistance among colicins of group B. J. Bacteriol. 123:96-101.

Davies JK and Reeves P (1975b) Genetic of resistance to colicins in *Escherichia coli* K-12. Cross-resistance among colicins of group A. J. Bacteriol. 123:102-117.

Derouiche R, Bénédetti H, Lazzaroni JC, Lazdunski C and Lloubès R (1995) Protein complex within *Escherichia coli* inner membrane - TolA N-terminal domain interacts with TolQ and TolR proteins J. Biol. Chem. 270:11078-11084

Duché D, Baty D, Chartier M, Letellier L (1994) Unfolding of colicin A during its translocation through the *Escherichia coli* envelope as demonstrated by disulfide bond engineering J. Biol. Chem. 269:24820-24825.

Eick-Helmerich K and Braun V. (1989) Import of biopolymers into *Escherichia coli*: nucleotide sequences of the *exbB* and *exbD* genes are homologous to those of the *tolQ* and *tolR* genes, respectively. J. Bacteriol. 171: 5117-5127.

Escuyer V and Mock M. (1987) DNA sequence analysis of three missense mutations affecting colicin E3 bactericical activity. Mol. Microbiol. 1:82-85.

Fisher E, Günter K and Braun V (1989) Involvement of ExbB and TonB in transport across the outer membrane of *Escherichia coli*: phenotypic complementation of *exb* mutants by overexpressed tonB and physical stabilization of TonB by ExbB. J. Bacteriol. 171:5127-5134.

Fognini-Lefebvre N, Lazzaroni JC and Portalier R (1987) *tolA*, *tolB* and *excC*, three cistrons involved in the control of pleiotropic release of periplasmic proteins by *Escherichia coli* K12. Mol. Gen. Genet. 209:391-395.

Géli V and Lazdunski C (1992) An α-helical hairpin as a specific determinant in protein-protein interaction occurring in *Escherichia coli* colicin A and colicin B immunity systems. J. Bacteriol. 174, 6432-6437.

Ghrayeb J, Kimura H, Takahara M, Hsiung H, Masui Y and Inouye, M. (1984) Secretion cloning vectors in *Escherichia coli*. EMBO J. 3:2437-2442.

Guihard G, Boulanger P, Bénédetti H, Lloubès R, Besnard M and Letellier L (1993a) Colicin A and the Tol proteins involved in its translocation are preferentially located in the contact sites between the inner and outer membranes of *Escherichia coli* cells. J. Biol. Chem. 269:5874-5880.

Guihard G, Bénédetti H, Besnard M and Letellier L (1993b) Phosphate efflux through the channels formed by colicins and phage T5 in *Escherichia coli* cells is responsible for the fall in cytoplasmic ATP. J. Biol. Chem. 268:17775-17780.

Guterman SK and Dann L (1973) Excretion of enterochelin by *exbA* and *exbB* mutants of *Escherichia coli*. J. Bacteriol. 114:1225-1230.

Hantke K and Zimmermann L (1981) The importance of the *exbB* gene for vitamin B12 and ferric iron transport. FEMS Microbiol. Lett. 12:31-35.

Heller K, Kadner RJ, and Günther K (1988) Suppression of the btuB451 mutation by mutations of the *tonB* gene suggests a direct interaction between TonB and Ton-dependent receptor proteins in the outer membrane of *Escherichia coli*. Gene 64:147-153.

Isnard M, Rigal A, Lazzaroni JC, Lazdunski C and Lloubès R (1994) Mutation and localization of the TolB protein required for colicin import. J. Bacteriol. 176:6392-6396.

Jaskula JC, Letain TE, Roof SK, Skare JT and Postle K (1994) Role of the TonB amino terminus in energy transduction between membranes. J. Bacteriol. 175: 2326-2338.

Kampfenkel K and Braun V (1992a) Membrane topology of the *Escherichia coli* ExbB protein. J. Bacteriol. 174: 5485-5487.

Kampfenkel K and Braun V (1992b) Topology of the ExbB protein in the cytoplasmic membrane of *Escherichia coli*. J. Biol. Chem. 268:6050-6057.

Karlsson M, Hannavy K and Higgins CF (1993) A sequence-specific function for the N-terminal signal-like sequence of the TonB protein. Mol. Microbiol. 8:379-388.

Killmann H, Benz R and Braun V (1993) Conversion of the FhuA transport protein into a diffusion channel through the outer membrane of *Escherichia coli*. EMBO J. 12:3007-3016.

Köck J, Ölschläger T, Kamp RM and Braun V (1987) Primary structure of colicin M, an inhibitor of murein biosynthesis. J. Bacteriol. 169:3358-3361.

Larsen RA, Thomas MG, Wood GE and Postle K (1994) Partial suppression of an *Escherichia coli* TonB transmembrane domain mutation (DV17) by a missense mutation in ExbB. Mol. Microbiol. 13:627-640.

Larsen RA, Wood GE and Postle K (1993) The conserved proline-rich motif is not essential for energy transduction by *Escherichia coli* TonB protein. Mol. Microbiol. 10 : 943-953.

Lazzaroni JC and Portalier R (1981) Genetic and biochemical characterization of periplasmic-leaky mutants of *Escherichia coli* K-12. J. Bacteriol. 145:1351-1358.

Lazzaroni JC and Portalier R (1992) The *excC* gene of *Escherichia coli* K-12 required for cell envelope integrity encodes the pepidoglycan-associated lipoprotein (PAL). Mol. Microbiol. 6: 735-742.

Lazzaroni JC, Fognini-Lefebvre N and Portalier R (1989) Cloning of *excC* and *excD* genes involved in the release of periplasmic proteins by *Escherichia coli* K-12. Mol. Gen. Genet. 218:460-464.

Lazzaroni JC, Vianney A, Popot JL, Benedetti H, Samatey F, Lazdunski C, Portalier R and Geli V (1995) Transmembrane alpha-helix interactions are required for the functional assembly of the

Escherichia coli Tol complex J. Mol. Biol. 246:1-7
Leduc M., Ishidate K, Shakibai N and Rothfield L (1992) Interactions of *Escherichia coli* membrane lipoproteins with the murein sacculus. J. Bacteriol. 174:7982-7988.
Levengood SK and Webster RE (1989) Nucleotide sequences of the *tolA* and *tolB* genes and localization of their products, components of a multistep translocation system in *Escherichia coli* J.Bacteriol. 171, 6600-6609.
Levengood-Freyermuth SK, Click EM and Webster RE (1993) Role of the carboxyterminal domain of TolA in protein import and the integrity of the outer membrane. J. Bacteriol. 175: 222-228.
Mende J and Braun V (1990) Import-defective colicin B derivatives mutated in the TonB box. Mol. Microbiol. 4:1523-1533.
Mizuno T (1979) A novel peptidoglycan-associated lipoprotein found in the cell envelopes of *Pseudomonas aeruginosa* and *Escherichia coli*. J. Biochem. 86:991-1000.
Mock M and Schwarz M (1980) Mutations which affect the structure and activity of colicin E3. J. Bacteriol. 142:384-390.
Müller MM, Vianney A, Lazzaroni JC, Webster RE and Portalier R (1993) Membrane topology of the *Escherichia coli* TolR protein required for cell envelope integrity. J. Bacteriol. 175:6059-6061.
Nagel de Zwaig R and Luria SE (1967) Genetics and physiology of colicin-tolerant mutants of *Escherichia coli*. J. Bacteriol. 94:1112-1123.
Pilsl H and Braun V (1995) Novel colicin 10: Assignment of four domains to TonB- and TolC-dependent uptake via the Tsx receptor and to pore formation Mol. Microbiol. 16:57-67
Pilsl H, Glaser C, Gross P, Killman H, Olschläger T and Braun V (1993) Domains of colicin M involved in uptake and activity. Mol. Gen. Genet. 240:103-112.
Postle K, and Good RF (1983) DNA sequence of the *Escherichia coli tonB* gene. Proc. Natl. Acad. Sci. USA. 80:5235-5239.
Roof SK, Allard JD, Bertrand KP and Postle K (1991) Analysis of *Escherichia coli* TonB membrane topology by use of PhoA fusions. J. Bacteriol. 173:5554-5557.
Ross U, Harkness RE and Braun V (1989) Assembly of colicin genes from a few DNA fragments. Nucleotide sequence of colicin D. Mol. Microbiol. 3: 891-902.
Rutz JM, Liu J, Lyons JA, Gotanson J, Amstrong SK, Mc Intosh MA, Feix JB and Klebba PE (1992) Formation of a gated channel by a ligand-specific transport protein in the bacterial membrane. Science 258:471-475.
Schoffler H and Braun V (1989) Transport across the outer membrane of *Escherichia coli* K12 via the FhuA receptor is regulated by the TonB protein of the cytoplasmic membrane. Mol. Gen. Genet. 217:378-383.
Schramm E, Mende J, Braun V and Kamp R (1987) Nucleotide sequence of the colicin B activity gene cba: consensus pentapeptide among TonB dependent colicins and receptors. J. Bacteriol. 169:3350-3357.
Skare JT, Ahmer BMM, Seachord CL, Darveau RP and Postle K (1993) Energy transduction between membranes. TonB, a cytoplasmic membrane protein, can be chemically cross-linked in vitro to the outer membrane receptor FepA. J. Biol. Chem. 268:16302-16308.
Skare JT and Postle K (1991) Evidence for a TolB-dependent energy transduction complex in *Escherichia coli*. Mol. Microbiol. 5:2883-2890.
Sun TP and Webster RE (1987) Nucleotide sequence of a gene cluster involved in entry of E colicins and single stranded DNA of infecting filamentous bacteriophages into *Escherichia coli*. J. Bacteriol. 169, 2667-2674.
Tuckman M and Osburne MS (1992) *In vivo* inhibition of TonB-dependent processes by a TonB box consensus pentapeptide. J. Bacteriol. 174:320-323.
Vianney A, Lewin TM, Beyer Jr WF, Lazzaroni JC, Portalier R and Webster RE (1994) Membrane topology and mutational analysis of the TolQ protein of *Escherichia coli* required for the uptake of macromolecules and cell envelope integrity. J. Bacteriol. 176:822-829.
Webster RE (1991) The *tol* gene products and the import of macromolecules into *Escherichia coli*. Mol. Microbiol. 5:1005-1011.
Zinder ND (1973) Resistance to colicin E3 and K induced by infection with bacteriophage f1. Proc. Natl. Acad. Sci. USA. 70:3160-3164.

In vitro assembly of outer membrane protein PhoE of *E. coli.*

Carmen Jansen, Hans de Cock, Patrick Van Gelder and Jan Tommassen
Department of Molecular Cell Biology
Institute of Biomembranes
Utrecht University
Padualaan 8
3584 CH Utrecht
The Netherlands

Introduction

The cell envelope of Gram-negative bacteria, such as *Escherichia coli*, contains a double membrane. The inner membrane (IM) surrounds the cytoplasm and the outer membrane (OM) protects the cell by forming a barrier for harmful compounds, such as bile salts and antibiotics. The periplasm is located between the two membranes and contains the peptidoglycan. The OM is an asymmetric bilayer. The inner monolayer contains phospholipids and the outer monolayer contains lipopolysaccharides (LPS) (Lugtenberg and van Alphen, 1983). The OM contains various proteins (OMPs), including pore-forming proteins, which allow the passage of small hydrophilic molecules with molecular weights up to about 600 Da (Benz and Bauer, 1988). In contrast to inner membrane proteins, which span the membrane by hydrophilic α-helices, OMPs span the membrane by amphipathic ß-strands. After translocation across the IM, OMPs have to fold and insert into the OM. These late steps in OMP biogenesis are the subject of our study. The porin PhoE, whose synthesis is induced when bacteria are grown under phosphate starvation (Overbeeke and Lugtenberg, 1980), is used as a model protein. The PhoE protein forms general diffusion pores with a selectivity for anions (Benz *et al.*, 1985). The crystal structure of PhoE has been resolved (Cowan *et al.*, 1992). Each monomer of this trimeric protein forms a 16-stranded antiparallel ß-barrel with hydrophobic amino acid residues exposed to the lipids and to the subunit interface.

NATO ASI Series, Vol. H 96
Molecular Dynamics of Biomembranes
Edited by Jos A. F. Op den Kamp

In vivo assembly

Intragenic information required for assembly and insertion

OMPs have a hydrophobic exterior only after folding into the ß-barrel structure. Therefore, it is likely that they insert into the OM in a folded configuration. Hence, it is expected that mutations, perturbating the tertiary structure, will affect the assembly process. Indeed, large deletions in PhoE (removing over 30 amino acid residues) prevented OM incorporation and resulted in periplasmic accumulation of the mutant proteins (Bosch *et al.*, 1986). Although the overall ß-structure of the porin seems to be important for OM assembly and insertion, it remains a possibility that, in addition to the tertiary structure, specific targeting signals are required for OM insertion. Comparison of the last 10 amino acid residues of a variety of OMPs of Gram-negative bacteria revealed a consensus-motif with, as most pronounced feature, the presence of a conserved phenylalanine at the carboxy terminus. Substitution of this residue of PhoE by other amino acid residues interfered with correct assembly *in vivo* (Struyvé *et al.*, 1991). Therefore, this C-terminal Phe might represent (part of) a targeting signal. Furthermore, a glycine residue, corresponding to Gly-144 in PhoE protein, seemed to be conserved in various OMPs (Nikaido and Wu, 1984). The substitution of Gly-144 in PhoE by Leu affected the efficiency of OM incorporation *in vivo*. However, further *in vitro* analysis revealed that the mutation interfered with the correct folding of the protein rather than with targeting (de Cock *et al.*, 1991). This conclusion was underscored by inspection of the crystal structure of PhoE, which suggested that the torsion angle at the position of Gly-144 could be accomodated by leucine only with difficulty.

Extragenic information required for assembly and insertion

Another factor that could direct the targeting of OMPs to the correct membrane is lipopolysaccharide. LPS is specifically present in the OM and could therefore discriminate the OM from the IM as the target membrane for OMP insertion. Consistent with an important role of LPS in the targeting of OMPs is the observation that mutants producing defective LPS molecules ('deep rough' mutants) have reduced amounts of OMPs in the OM (Nikaido and Vaara, 1985). Furthermore, the antibiotic cerulenin, which inhibits fatty acid synthesis and, consequently, also LPS synthesis, interferes with the assembly of porin OmpF (Bolla *et al.*, 1988).

Assembly of *in vitro* synthesized PhoE

To study the molecular mechanism of folding, assembly and insertion of the porins, *in vitro* assays have been developed in our laboratory. Mature PhoE protein, lacking the signal sequence, but extended at the N-terminus with two residues, Met and Ser, was synthesized *in vitro* in a radioactive form using an S135 lysate of *E. coli*. After translation, the folding of the protein was studied in immunoprecipitation experiments using PhoE-specific monoclonal antibodies (mAbs) that recognize conformational epitopes (de Cock *et al.*, 1990). In these assays, a folded monomeric form of PhoE protein with a higher electrophoretic mobility during SDS-polyacrylamide gel electrophoresis (SDS-PAGE) than the fully denatured protein was detected. This heat-modifiable folded monomer (M_r 31 kDa) was converted to the denatured monomer (M_r 38 kDa) when heated at 56 °C in 2% SDS. The trimeric unit of PhoE has an apparent molecular weight of approximately 70 kDa and is heat-stable up to about 70 °C in the presence of 2% SDS. Making use of these characteristics, one can visualize the folded state of PhoE protein by gel electrophoresis.

It turned out that *in vitro* synthesized PhoE formed folded monomers and trimers upon the addition of detergents (Van Gelder *et al.*, 1994). A remarkable specificity of phenyl-containing detergents, such as Triton X-100, was observed. Moreover, Triton X-100 appeared to affect the folding of PhoE in a concentration-dependent manner. More folded PhoE was obtained with increasing detergent concentrations, reaching a maximum at 2% Triton X-100. Addition of outer membranes could increase the rate of trimerization, probably by serving as 'concentration points' for the PhoE molecules (Van Gelder *et al.*, 1994). However, in the absence of Triton X-100, outer membranes or LPS were inactive in folding of PhoE. In contrast an assembly-intermediate of OmpA (imp-OmpA), accumulated in overproducing *E. coli* strains, could be converted *in vitro* into the folded heat-modifiable form by the addition of LPS (Freudl *et al.*, 1986), showing a direct effect of LPS on the folding of OMPs. On the other hand, purified denatured OmpA could also be refolded *in vitro* and inserted into a lipid bilayer in the absence of LPS (Dornmair *et al.*, 1990).

Monomeric OmpF, secreted by sphaeroplasts (Sen and Nikaido, 1990), as well as purified OmpF denatured in 6 M guanidium hydrochloride (Eisele and Rosenbusch, 1990) could be renatured into trimers in the presence of SDS and lipids or detergents. In these experiments, SDS, which has a surface negative charge density comparable with LPS, might mimic LPS in the assembly process. Trimerization of *in vitro* synthesized OmpF porin could be induced by adding outer membranes, LPS or a mixture of an anionic and a nonionic detergent (Sen and Nikaido, 1991).

Taken together, it seems that LPS, either directly or indirectly, is involved in the trimerization process. However, it remains to be elucidated in which way LPS exactly acts on the assembly process.

Assembly of purified PhoE

The *in vitro* transcription/translation system described above showed that assembly of OMPs *in vitro* is possible. A disadvantage, however, is that the components responsible for the assembly cannot readily be identified in this system. Therefore, an *in vitro* system is being developed in our laboratory, in which the folding and assembly of purified PhoE protein can be studied. An advantage of this *in vitro* system with purified components is that the role of LPS, lipids and periplasmic proteins can be studied. Next to the wild-type PhoE protein, also mutant PhoE proteins that are defective in folding and/or assembly *in vivo* will be studied in this system. The development of the *in vitro* system is, at the time of writing this manuscript, still in progress. We will describe the results obtained so far.

Purification

To purify the PhoE protein two procedures have been applied. The first procedure involves the extraction of the porin from isolated cell envelopes. Cell envelopes were heated in 2% SDS to 60 °C in order to remove almost all proteins present in, or associated to, the cell envelope. PhoE remains associated with the peptidoglycan layer at this stage, and can be dissociated upon heating to 100 °C. The protein was finally dissolved in a buffer with 8 M urea. This stock, designated OM-PhoE, contains LPS as a major contaminant (Figure 1, lane 2). To study the possible role of LPS in the biogenesis process, a PhoE preparation devoid of LPS was essential. Therefore, we cloned the part of the *phoE* gene encoding the mature domain of the protein (i.e., without the signal sequence) behind the T7 promoter, resulting in plasmid pCJ2. In strain BL21(DE3), containing a chromosomal copy of the T7 RNA polymerase gene under control of the *lac*UV5 promoter, mature PhoE was expressed upon addition of isopropyl-ß-D-thiogalactopyranoside (IPTG) (Figure 2, lane 2) and accumulated in cytoplasmic inclusion bodies. These inclusion bodies could be isolated by various centrifugation steps after sonication (Figure 2, lane 3-5). The protein, designated cib-PhoE (cytoplasmic inclusion bodies), was finally dissolved in a buffer with 8 M urea, and the preparation contained only minor LPS contamination (Figure 1, lane 1). N-terminal amino acid sequencing revealed that the methionine, encoded by the start-codon, was removed *in vivo* for about 90%. Both protein stocks, OM-PhoE and cib-PhoE, were used for refolding studies.

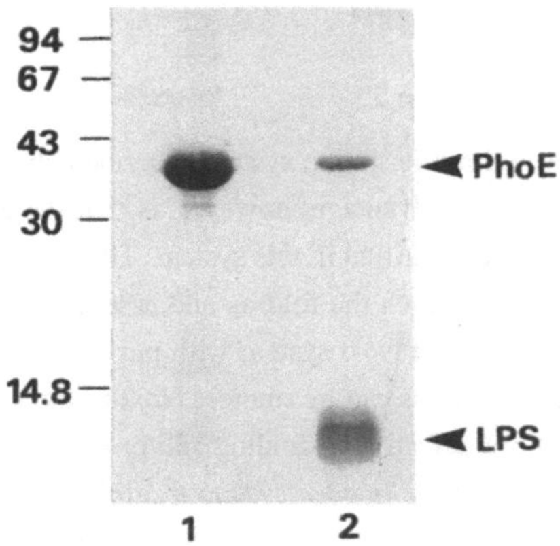

Figure 1. SDS-PAGE gel (14%) containing wild-type PhoE protein, isolated from either cytoplasmic inclusion bodies (cib-PhoE, lane 1) or from cell envelopes (OM-PhoE, lane 2). The gel was stained both with Coomassie Brilliant Blue (CBB) and with silver, to visualize proteins and LPS, respectively. The positions of molecular weight standard proteins are indicated (kDa).

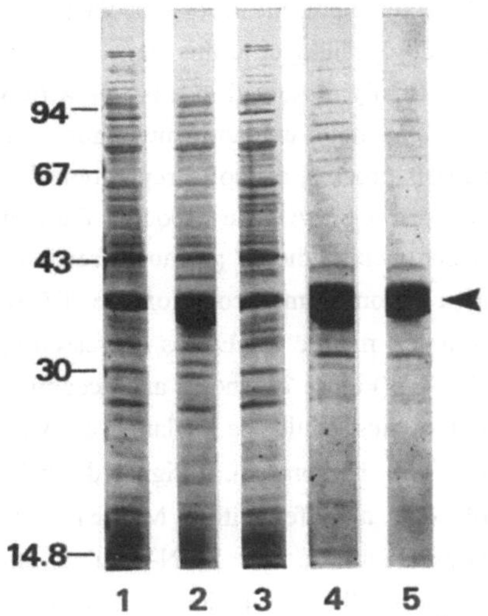

Figure 2. Isolation procedure of cib-PhoE. Strain BL21(DE3) containing pCJ2 was grown in L-broth. Just before induction with IPTG, a whole cell sample was analysed by SDS-PAGE (lane 1). Three hours after induction (lane 2), cells were harvested and sonicated. After sonication, the cytoplasmic inclusion bodies, containing PhoE, were pelleted (lane 4); the supernatant contained almost all other bacterial proteins (lane 3). The pellet was dissolved in a buffer containing 8 M urea. A final ultra centrifugation step was performed to remove residual membrane fragments. Lane 5 contains the purified wild-type PhoE protein (indicated by an arrow). The proteins were stained with CBB. Positions of molecular weight standard proteins are indicated (kDa).

Refolding

The denatured protein was refolded by rapid dilution in a refolding buffer containing Tris or Hepes, detergent and residual amounts of urea. This mixture was incubated overnight at room temperature. Refolded proteins were analysed by SDS-PAGE where folded monomers and trimers can be distinguished from denatured protein because of their different migration behaviour (see section 'assembly of *in vitro* synthesized PhoE').

In the first trials, the OM-PhoE protein refolded with an efficiency of about 50%. Both folded monomers and trimers were detected (Figure 3, panel A). This result was obtained using a refolding buffer containing Hepes pH 6.5, 2% Triton N-101 and 0.3 M urea.

The first refolding trials with the cib-PhoE protein did not reach, until now, the efficiency levels obtained with the OM-PhoE protein. Some trimers, but no folded monomers were detected after refolding of cib-PhoE (Figure 3, panel B). The difference in folding efficiency is not simply due to the lower amount of LPS present in the cib-PhoE preparation, since addition of LPS to concentrations present in the OM-PhoE preparation did not drastically increase the refolding efficiency (Figure 4).

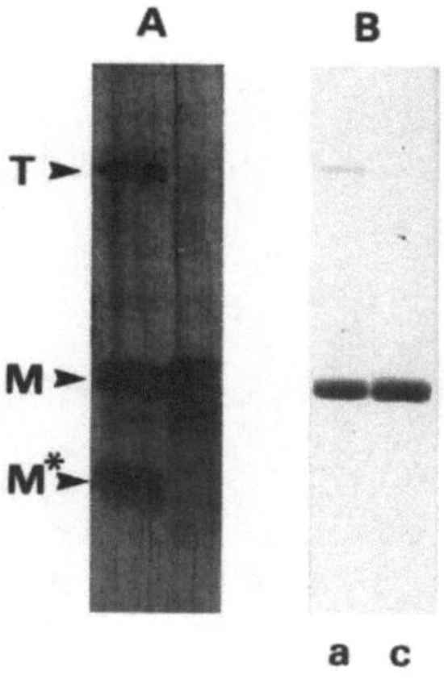

Figure 3. Refolding experiment of OM-PhoE (panel A) and cib-PhoE (panel B). OM-PhoE was refolded in a buffer containing Hepes pH 6.5, 2% Triton N-101 and 0.3 M urea. Cib-PhoE was refolded in a buffer containing Tris pH 8.0, 2% Triton N-101 and 0.1 M urea. In both refolding mixtures, the final protein concentration was 0.1 mg/ml. The refolding mixtures were incubated overnight at room temperature. After refolding, the mixtures were divided into two parts and incubated in sample buffer for 10 min at room temperature (lanes a) or at 100 °C (lanes c) prior to SDS-PAGE. T, M and M* indicate trimers, denatured monomers and folded monomers of PhoE protein, respectively. Panel A is silver stained, whereas panel B is CBB stained.

Another observation concerned the difference in refolding efficiency obtained with various detergents. Triton N-101 was slightly more efficient in refolding as compared with Triton X-100 (Figure 4). The detergent octylpolyoxy ethylene (OPOE) showed an intermediate refolding efficiency (not shown). Also the concentration of the detergent is of importance. A concentration of 2% is favoured above one of 0.6% for example, consistent with the results found in the *in vitro* transcription/translation system (van Gelder *et al.*, 1994). Remarkably, as compared with Triton X-100, Triton N-101 caused a slightly faster migration of the denatured monomer during SDS-PAGE (Figure 4). Possibly, Triton N-101 affects the conformation of the monomer, resulting in more efficient trimerization.

Parameters like protein concentration, pH and temperature will have to be varied in a systematic way in a search for better refolding efficiencies. Another explanation for the low folding efficiency of cib-PhoE may be the presence of an inhibiting component in the protein preparation, which may be removed by further purification steps.

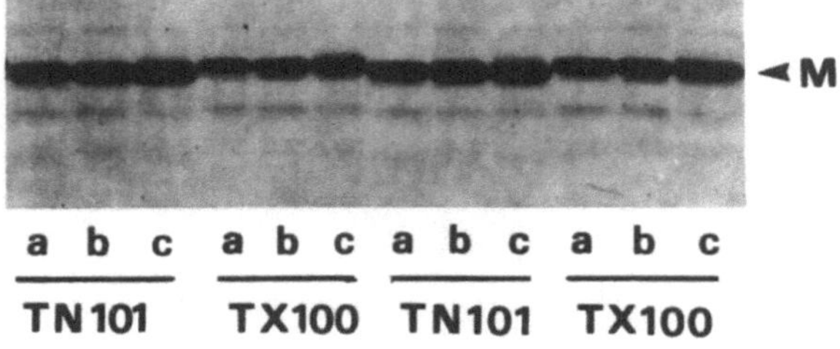

Figure 4. Refolding experiment of cib-PhoE in absence or presence of LPS.
Cib-PhoE was refolded in a buffer containing Hepes pH 6.5 with either Triton X-100 or Triton N-101 (as indicated) and 0.6 M urea. Incubations were performed overnight at room temperature with or without LPS (as indicated). The final protein concentration in the refolding mixture was 0.1 mg/ml. After refolding, the mixtures were divided into three parts and incubated in sample buffer for 10 min at room temperature (lanes a), 56 °C (lanes b) or 100 °C (lanes c) prior to SDS-PAGE. T and M indicate trimers and denatured monomers of PhoE protein, respectively. The proteins were visualized with CBB.

Concluding remarks

Once a high refolding efficiency is obtained with wild-type PhoE protein, PhoE mutants, defective in folding and/or assembly *in vivo* can be used in the new *in vitro* system with purified components. Other, biophysical techniques such as circular dichroism and tryptophan fluorescence, can be used to assess the stage in which the assemby of these mutants is affected. By comparing all the studies, one can gain insight into the biogenesis process of outer membrane proteins in more detail.

References

Benz, R. and Bauer, K. (1988) Permeation of hydrophilic molecules through the outer membrane of Gram-negative bacteria. Review on bacterial porins. Eur. J. Biochem. 176: 1-19.

Bolla, J.M., Lazdunski, C. and Pagès, J.M. (1988) The assembly of the major outer membrane protein OmpF of *Escherichia coli* depends on lipid synthesis. EMBO J. 7: 3595-3599.

Bosch, D., Leunissen, J., Verbakel, J., de Jong, M., van Erp, H. and Tommassen, J. (1986) Periplasmic accumulation of truncated forms of outer membrane PhoE protein of *Escherichia coli* K-12. J. Mol. Biol. 189: 449-455.

Cowan, S.W., Schirmer, T., Rummel, G., Steiert, M., Ghosh, R., Paupit, R.A., Jansonius, J.N. and Rosenbusch, J.P. (1992) Crystal structures explain functional properties of two *E. coli* porins. Nature 358:727-733.

De Cock, H., Hendriks, R., de Vrije, T. and Tommassen, J. (1990) Assembly of an *in vitro* synthesized *Escherichia coli* outer membrane porin into its stable trimeric configuration. J. Biol. Chem. 265: 4646-4651.

De Cock, H., Quaedvlieg, N., Bosch, D., Scholten, M. and Tommassen, J. (1991) Glycine-144 is required for efficient folding of outer membrane protein PhoE of *E. coli* K-12. FEBS Lett. 279: 285-288.

Dornmair, K., Kiefer, H. and Jähnig, F. (1990) Refolding of an integral membrane protein. OmpA of *Escherichia coli*. J. Biol. Chem. 265: 18907-18911.

Eisele, J.L. and Rosenbusch J.P. (1990) *In vitro* folding and oligomerization of a membrane protein. Transition of a bacterial porin from random coil to native conformation. J. Biol. Chem. 265: 10217-10220.

Freudl, R., Schwarz, H., Stierhof, Y., Gamon, K., Hindennach, I. and Henning, U. (1986) An outer membrane protein (OmpA) of *Escherichia coli* K-12 undergoes a conformational change during export. J. Biol. Chem. 261: 11355-11361.

Lugtenberg, B. and van Alphen, L. (1983) Molecular architecture and functioning of the outer membrane of *Escherichia coli* and other Gram-negative bacteria. Biochim. Biophys. Acta 737: 51-115.

Nikaido, H. and Vaara, M. (1985) Molecular basis of bacterial outer membrane permeability. Microbiol. Rev. 49: 1-32.

Nikaido, H. and Wu, H. (1984) Amino acid sequence homology among the major outer membranes proteins of *Escherichia coli*. Proc. Natl. Acad. Sci. USA 81: 1048-1052.

Sen, K. and Nikaido, H. (1990) *In vitro* trimerization of OmpF porin secreted by spheroplasts of *Escherichia coli*. Proc. Natl. Acad. Sci. USA 87: 743-747.

Sen, K. and Nikaido, H. (1991) Trimerization of an *in vitro* synthesized OmpF porin of *Escherichia coli* outer membrane. J. Biol. Chem. 266: 11295-11300.

Struyvé, M., Moons, M. and Tommassen, J. (1991) Carboxy-terminal phenylalanine is essential for the correct assembly of a bacterial outer membrane protein. J. Mol. Biol. 218: 141-148.

Tommassen, J. (1987) Biogenesis and membrane topology of outer membrane proteins in *Escherichia coli*. In: Membrane Biogenesis, Op den Kamp, J.A.F. (Ed.) Springer-Verlag, Berlin, pp.351-373.

Van Gelder, P., de Cock, H. and Tommassen, J. (1994) Detergent-induced folding of the outer membrane protein PhoE, a pore protein induced by phosphate limitation. Eur. J. Biochem. 226: 783-787.

PROTEIN FOLDING IN THE CELL: THE ROLE OF MOLECULAR CHAPERONES

Franz-Ulrich Hartl
Program of Cellular Biochemistry & Biophysics
Memorial Sloan-Kettering Cancer Center
1275 York Avenue
New York, NY 10021
USA

The problem of protein folding in the cell

Understanding the mechanisms and pathways of protein folding constitutes a problem of fundamental biological significance. It has been known for more than three decades that all the information required for the acqisition of the native state is contained in the linear amino acid sequence of the polypeptide chain. Proteins are capable of spontaneous folding in the test-tube, at least under carefully chosen conditions, and this has led to the view that also within cells newly-synthesized polypeptides reach their native state in an essentially spontaneous reaction. Only more recently has it been realized that this is not generally the case. Cells contain a complex machinery of proteins, folding catalysts and so-called molecular chaperones, which mediate folding in the cytosol as well as within subcellular compartments such as mitochondria, chloroplasts and the endoplasmic reticulum (Hendrick and Hartl, 1993; Hartl et al., 1994; Hartl and Martin, 1995). Molecular chaperones, mostly constitutively expressed stress proteins, play a preeminent role in these processes and are the main focus of this chapter.

When considering the enormous difficulties often encountered in the renaturation of unfolded proteins *in vitro*, it is surprsing that folding in the cell occurs with remarkably high efficiency. Unfolded polypeptides have the general tendency to misfold and aggregate due to their exposure of hydrophobic surfaces to solvent. This property would be expected to be most pronounced in the cellular solution which is characterized by considerable molecular crowding due to the very high concentration of total protein and other macromolecules (~340 g/l) and by a high concentration of newly-synthesized, folding polypeptide chains (30-50 µM) (Zimmerman and Trach, 1991). Nevertheless, the yield of protein folding *in vivo* is generally high, reaching almost 100% in certain cases. This can be attributed to the action of molecular chaperones, helper proteins which interact with non-native polypeptides and prevent unproductive off-pathway reactions during folding (Ellis, 1987; Hendrick and Hartl, 1993). It

NATO ASI Series, Vol. H 96
Molecular Dynamics of Biomembranes
Edited by Jos A. F. Op den Kamp

is now generally believed that molecular chaperones function by shielding the hydrophobic sequences or surfaces exposed by conformational intermediates on the pathways of protein folding or unfolding. They do not recognize a consensus sequence motif and therefore have the ability to prevent the incorrect intra- and inter-molecular folding and association of many different proteins. The Hsp70s and chaperonins then promote correct folding by repeatedly binding and releasing their substrate proteins regulated by ATP binding and hydrolysis. The molecular chaperones do not typically function as catalysts of protein folding and increase the yield of a folding reaction rather than its speed. Once a protein has reached its native state, it no longer presents hydrophobic surfaces for chaperone binding. However, exposure to elevated tempertaure or other forms of cellular stress may cause the partial or complete unfolding of proteins, leading to their renewed interaction with chaperones. Many, but not all, molecular chaperones are indeed heat-shock or stress proteins as they are expressed at increased levels under conditions of cellular stress which may structurally destabilize folded proteins.

A pathway of chaperone-assisted protein folding

Essential functions in protein folding have been established for the members of the heat-shock protein 70 (Hsp70) and the so-called "chaperonins" (Hartl et al., 1994). Chaperones of the Hsp70 class and their regulatory cofactors bind nascent polypeptide chains on ribosomes and prevent their premature (mis)folding and aggregation until at least a domain capable of forming a stable structure is synthesized. For many proteins, completion of folding requires the subsequent interaction with one of the large ring-shaped protein complexes of the chaperonin family, which is comprised of the GroEL-like proteins in eubacteria, mitochondria, and chloroplasts, and the TRiC (TCP-1 ring complex) family in eukaryotic cytosol and archaea. The chaperonins bind partially folded polypeptide in their central cavity and promote folding by ATP-dependent cycles of release and rebinding. Such a pathway has been described for proteins which fold following import into mitochondria (Ostermann et al., 1989; Kang et al., 1990; Manning-Krieg et al., 1991; Langer et al., 1992a) and also exists in the cytosol of bacteria and eukaryotic cells (Langer et al., 1992a; Frydman et al., 1994) (Figure 1). It is still unclear what fraction of newly-synthesized polypeptide chains have to use this pathway in order to reach the native state. Evidence has been provided that about 20-30% of the cytosolic proteins in *E. coli* fail to fold correctly when the chaperonin, GroEL, is defective (Horwich et al., 1993). In the following sections I will discuss the main functional properties of the Hsp70s and the chaperonins and their respective roles in protein folding.

Mechanism of the Hsp70 chaperone system

The members of the Hsp70 class of chaperones bind short, extended peptide segments of seven or eight residues which are enriched in hydrophobic amino acids (Flynn et al. 1991; Blond-Elguindi et al. 1993). For full function the Hsp70s depend on the regulation by further proteins. The Hsp70 homologue of *E. coli*, DnaK, cooperates with the chaperone protein DnaJ and the nucleotide exchange factor GrpE, proteins of about 40 kDa and 20 kDa, respectively (Georgopoulos, 1992).

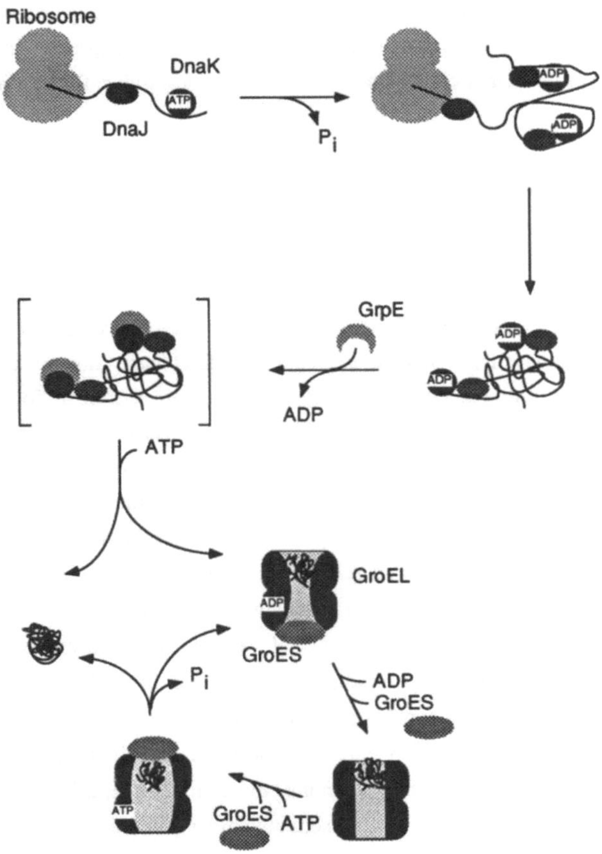

Figure 1: Model for the pathway of chaperone-mediated protein folding in the bacterial cytosol. The polypeptide chain emerging from the ribosome interacts with DnaJ and DnaK (Hsp70) (Hendrick et al. 1993; Gaitanaris et al. 1994). Direct interaction between DnaK and DnaJ in the

Both DnaK and DnaJ associate co-translationally with nascent polypeptide chains (Hendrick et al. 1993; Gaitanaris et al. 1994; Kudlicki et al. 1994), whereby DnaJ seems to facilitate the loading of DnaK onto the elongating chain (Figure 1). This is due to the direct interaction of DnaJ with DnaK which accelerates the hydrolysis of DnaK-bound ATP (Liberek et al. 1991) and stabilises the ADP-state of DnaK, the high affinity state for unfolded polypeptide (Palleros et al. 1993). As a result, a stable ternary complex consisting of polypeptide substrate, DnaJ and ADP-bound DnaK is formed (Langer et al. 1992a; Szabo et al. 1994). GrpE then functions as a nucleotide exchange factor in dissociating the ADP bound to DnaK. Subsequent ATP binding to DnaK triggers the release of the polypeptide substrate (Szabo et al., 1994), allowing its transfer to the bacterial chaperonin GroEL (Langer et al. 1992a). At least *in vitro*, the folding of certain proteins may be achieved through ATP-dependent cycles of binding and release to DnaK and DnaJ alone (Schröder et al., 1993; Szabo et al., 1994). *In vivo*, the primary function of the Hsp70 system may be in maintaining the polypeptide chain in an unfolded but non-aggregated state, competent for folding by the chaperonin. The eukaryotic cytsosol contains several DnaJ homologues (Caplan et al., 1993), but a structural or functional equivalent of GrpE has not yet been identified in this compartment. The recent analysis of the Hsp70 ATPase cycle in the eukaryotic cytosol suggests that a GrpE homologue may not be required since ADP-dissociation from Hsp70 is not a rate-limiting step in the cycle (Ziegelhoffer et al., 1995; Höhfeld et al., 1995).

Mechanism of the Chaperonin System

In contrast to the Hsp70 and DnaJ proteins, which function as monomers or dimers, the best studied chaperonin, GroEL, is a large complex consisting of two stacked rings of seven identical ~60 kDa subunits, forming a cylinder with a central cavity (Hohn et al. 1979; Hendrix 1979; Langer et al. 1992b; Saibil et al. 1993; Braig et al. 1994). GroEL cooperates with the cofactor GroES, a single heptameric ring of 10 kDa subunits that binds to GroEL and increases the cooperativity of ATP hydrolysis in the GroEL ring system. While this regulation is not required for the ATP-dependent release of bound protein from GroEL per se, with many substrate proteins it is necessary to couple protein release with productive folding (Martin et al., 1991b). Under physiological conditions, binding of GroES to either end of the GroEL cylinder reduces the affinity of the opposite end for binding a second GroES, resulting in the formation of

---

presence of ATP leads to the formation of a ternary complex between nascent chain, DnaK and DnaJ in which DnaK is in the ADP-state. This complex dissociates upon the GrpE-dependent dissociation of ADP and the binding (not hydrolysis) of ATP to DnaK (Szabo et al. 1994). The protein may reach the native state by multiple rounds of interaction with the DnaK, DnaJ, GrpE system or upon transfer to the chaperonins GroEL and GroES for final folding (Langer et al. 1992a). A model of the mechanism of GroEL/GroES in folding is shown in Figure 2.

asymmetrical GroEL:GroES complexes (Langer et al. 1992b; Saibil et al. 1993; Chen et al. 1994; Engel et al., 1995; Hayer-Hartl et al., 1995). This negative cooperativity of GroES binding is decreased at high concentrations of Mg^{2+} (15-50 mM) and at elevated pH (pH 7.7-8.0), conditions which allow the symmetrical binding of two GroES to GroEL (Llorca et al. 1994; Schmidt et al. 1994). A detailed electron microscopic and biochemical analysis of GroES binding to GroEL as well as a kinetic analysis of the GroEL-GroES reaction cycle by surface plasmon resonance failed to demonstrate the functional significance of these so-called "football" structures, and established the significance of asymmetry in the GroEL:GroES complex (Engel et al., 1995; Hayer-Hartl et al., 1995).

The interaction between GroEL and GroES is highly dynamic, both in the absence and presence of substrate polypeptide. Asymmetrical binding of GroES stabilizes the seven subunits in the adjacent GroEL toroid in a tight ADP state, resulting in a 50% inhibition of the GroEL ATPase (Martin et al. 1993; Todd et al. 1993; Hayer-Hartl et al., 1995). Binding and hydrolysis of ATP in the opposite GroEL toroid causes the transient release of this ADP and of GroES. Polypeptide binding stimulates the ATPase activity of GroEL (Martin et al. 1991; Jackson et al. 1993). The unfolded polypeptide binds initially to the GroEL ring that is not covered by GroES (Figure 2). This facilitates the dissociation of GroES from GroEL (Martin et al., 1993; Hayer-Hartl et al., 1995). Upon ATP binding, GroES may then reassociate with the GroEL toroid that contains the substrate protein, inducing its release into the GroEL cavity (Martin et al. 1993).

Interestingly, binding of GroES causes a significant outwards movement of the apical domains of the GroEL subunits, creating an enclosed, dome-shaped space with a maximum height and width of about 70 Å (Figure 2) (Chen et al. 1994). GroES may initially contact the outer surface of the GroEL cylinder (Fenton et al., 1994), triggering further domain movement, thus transiently displacing the polypeptide substrate into the cavity for folding (Hartl 1994). This mechanism has now been demonstrated directly with a dihydrofolate reductase (DHFR) fusion protein as the substrate protein that binds the folate analog methotrexate in its native state (Mayhew et al., 1995). Upon GroES binding, approximately 50 % of the GroEL-associated DHFR is displaced into the GroEL cavity by GroES. About one third of this protein reaches the native state while being enclosed underneath the hood of GroES and subsequently leaves the chaperonin upon ATP-dependent GroES release (Figure 2). Although not every protein may be able to fold to completion in association with GroEL, at least partial folding is thus likely to occur in a shielded microenvironment, before the polypeptide emerges from the chaperonin cavity into the bulk solution (Martin et al. 1993; Mayhew et al., 1995). Multiple rounds of GroES-dependent polypeptide release into the GroEL cavity and rebinding may be necessary for completion of folding (Martin et al. 1991; Weissman et al., 1994). The (re)binding of kinetically trapped folding intermediates to multiple sites on GroEL is thought to result in partial unfolding and structural rearrangement. This would prepare the not yet folded polypeptide for another folding trial in the GroEL cavity.

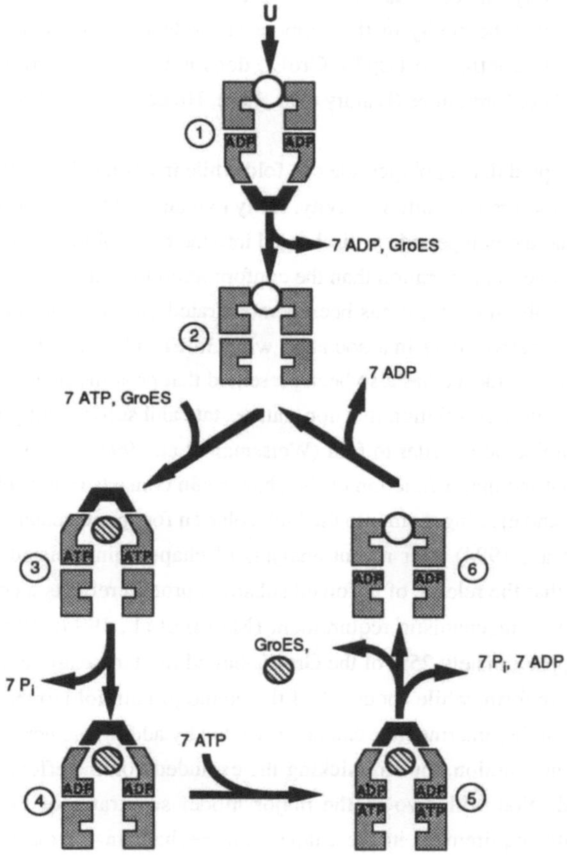

Figure 2: Model for the folding of dihydrofolate reductase (DHFR) by GroEL and GroES. ADP, the seven subunits in a GroEL ring that are bound to GroES in a high-affinity ADP state; ATP, the seven subunits in a GroEL ring in the ATP-bound state; U, unfolded polypeptide substrate as compact folding intermediate (open sphere). 1, Polypeptide binding facilitates dissociation of tightly bound ADP and GroES. 2, Polypeptide transiently bound in nucleotide-free ring. 3, ATP binds and GroES reassociates with the polypeptide-containing ring of GroEL. DHFR protein is released into the GroEL cavity where it folds to the native state (hatched sphere).(GroES reassociation with the free GroEL ring is not shown). The seven subunits of GroEL that interact with GroES hydrolyze ATP. This step may be required for the folding of certain polypeptide

GroEL binds its substrate in the conformation of a compact, yet flexible, molten globule-state which exposes hydrophobic surfaces to solvent (Martin et al. 1991; Hayer-Hartl et al. 1994; Robinson et al. 1994). Mutational analysis of GroEL in conjunction with the structural analysis by X-ray crystallography indeed suggested the presence of a complementary hydrophobic binding surface that lines the cavity of the cylinder at the level of the apical GroEL domains (Fenton et al. 1994). In contrast to Hsp70, GroEL does not seem to recognize short peptide sequences in extended conformations (Landry et al. 1992; Hlodan et al., 1995).

It is not generally accepted that a polypeptide can fold while in association with the inner surface of the chaperonin or upon release into its cavity. Early evidence obtained with rhodanese as the substrate suggested that the polypeptide is discharged into the bulk solution in a conformation that is significantly less prone to aggregation than the conformation initially bound by the chaperonin (Martin et al. 1991). Subsequently, it has been demonstrated that proteins such as barnase and DHFR may reach their native states in association with GroEL (Gray and Fersht 1993; Mayhew et al., 1995). In contrast, evidence has also been presented that proteins such as rhodanese can be released from GroEL into free solution in a non-native state and subsequently have to re-bind to another chaperonin molecule in order to fold (Weissman et al. 1994). These latter observations led to the proposal that the major function of the chaperonin consists in unfolding unproductive folding intermediates and ejecting them into the bulk solution for spontaneous folding (Weissman et al., 1994; Todd et al., 1994). Our recent analysis of chaperonin-assisted folding reactions suggested, however, that the release of unfolded substrate protein reflects a certain leakiness of the system rather than a mechanistic requirement (Martin et al., 1995). We find that in each ATPase cycle only approximately 25% of the GroEL-bound rhodanese are released into the bulk solution in a non-native form, while about 5% of the bound protein fold to the native state. This premature release of folding intermediate can be prevented by adding molecular crowding agents such as Ficoll 70 to the solution, thus mimicking the excluded volume effects prevailing in the cytosol. We conclude that with two of the major model substrates of GroEL, DHFR and rhodanese, there is no requirement in the chaperonin mechanism for the release of actively unfolded protein species into the bulk solution. Rather, folding of these proteins to the native state or to a conformation committed to reach the native state occurs in the protected environment of the chaperonin cavity through repeated cycles of folding and structural rearrangement.

---

substrates but is not necessary for the folding of DHFR. 4, GroES bound stably to GroEL subunits in the ADP state, enclosing the folded DHFR in the GroEL cavity. 5, ATP binding and hydrolysis in the opposite GroEL ring leads to GroES dissociation and folded DHFR emerges into the bulk solution. 6, Incompletely folded protein is recaptured before it leaves GroEL and may undergo unfolding to allow structural rearrangement in preparation for another folding cycle in the GroEL cavity.

References

Blond-Elguindi, S., Cwirla, S.E., Dower, W.J., Lipshutz, R.J., Sprang, S.R.,Sambrook, J.F. & Gething, M.-J.H. (1993) Affinity panning of a library of peptides displayed on bacteriophages reveals the binding specificity of BiP.Cell 75, 717-728

Braig, K., Otwinowski, Z., Hegda, R., Boisvert, D., Joahimiak, A., Horwich, A.L. & Sigler, P.B. (1994) The crystal structure of GroEL at 2.8 Å resolution. Nature, in press

Caplan, A. J., Cyr, D. & Douglas, M.G. (1993) Eukaryotic homologues of*Escherichia coli* dnaJ: A diverse protein family that functions with HSP70 stress proteins. Mol. Biol. Cell 4, 555-563

Chen, S., Roseman, A.M., Hunter, A.S., Wood, S.P., Burston, S.G., Ranson, N.A., Clarke, A.R. & Saibil, H.R. (1994) Location of a folding protein and shape changes in GroEL-GroES complexes imaged by cryo-electron microscopy. Nature 371, 261-264

Ellis, J. (1987) Proteins as molecular chaperones. Nature 328, 378-379

Engel, A., Hayer-Hartl, M., Hartl, F.-U. (1995) Symmetrical vs. asymmetrical GroEL-GroES chaperonin complexes: Evaluation of functional significance. Science 269, 832-836

Fenton, W.A., Kashi, Y., Furtak, K. & Horwich, A.L. (1994) Functional analysis of the chaperonin GroEL: Identification of residues required for polypeptide binding and release. Nature, in press

Flynn GC, Pohl J, Flocco MT, Rothman JE (1991) Peptide-binding specificity of the molecular chaperone BiP. Nature 353:726-730

Frydman, J., Nimmesgern, E., Ohtsuka, K. & Hartl, F.U. (1994) Folding of nascent poypeptide chains in a high molecular mass assembly with molecular chaperones. Nature 370, 111-117

Gaitanaris, G.A., Vysokanov, A., Gottesman, M. & Gragerov, A. (1994) *Escherichia coli* chaperones are associated with nascent polypeptide chains and promote the folding of l repressor. Mol. Microbiol., in press

Georgopoulos, C. (1992) The emergence of the chaperone machines. Trends in Biochem. Sci. 17, 295-299

Gray, T. E. & Fersht, A.R. (1993) Refolding of barnase in the presence of GroE. J. Mol. Biol. 232, 1197-1207

Hartl, F.-U., Hlodan, R. & Langer, T. (1994) Molecular chaperones in protein folding: the art of avoiding sticky situations. Trends in Biochem. Sci. 19, 20-25

Hartl, F.U. (1994) Protein folding: Secrets of a double-doughnut. Nature 371, 557-559.

Hartl, F.U. and Martin, J. (1995) Molecular chaperones in cellular protein folding. Curr. Op. Struct. Biol. 5, 92-102

Hayer-Hartl, M., Martin, J., Hartl, F.-U. (1995) The asymmetrical interaction of GroEL and GroES in the chaperonin ATPases cycle of assisted protein folding. Science 269, 836-841

Hayer-Hartl, M.K., Ewbank, J.J., Creighton, T.E. & Hartl, F.U. (1994) Conformational specificity of the chaperonin GroEL for the compact folding intermediates of a-lactalbumin. EMBO J. 13, 3192-3202

Hendrick, J.P. and Hartl, F. U. (1993) Molecular chaperone functions of heat-shock proteins. Annu. Rev. Biochem. 62, 349-384

Hendrick, J. P., Langer, T., Davis, T.A., Hartl, F.U. & Wiedmann, M. (1993) Control of folding and membrane translocation by binding of the chaperone DnaJ to nascent polypeptides. Proc. Natl. Acad. Sci. U.S.A. 90, 10216-10220

Hendrix, R.W. (1979) Purification and properties of GroE, a host protein involved in bacteriophage assembly. J. Mol. Biol. 129, 375-392

Hlodan, R., Pempst, P. and Hartl, F.U. (1995) Binding of defined regions of a polypeptide to GroEL and its implications for chaperonin-mediated protein folding. Nature Struct. Biol. 2, 587-595

Höhfeld, J., Minami, Y. and Hartl, F.U. (1995) Hip, a new cochaperone involved in the eukaryotic Hsc70/Hsp40 reaction cycle. Submitted

Hohn T, Hohn B, Engel A, Wurtz M (1979) Isolation and characterization of the host protein groE involved in bacteriophage lambda assembly. J Mol Biol 129:359-373

Horwich, A.L., Low, K.B., Fenton, W.A., Hirshfield, I.N., and Furtak, K. (1993) Folding in vivo of bacterial cytoplasmic proteins: role of GroEL. Cell 74, 909-917

Jackson, G. S., Staniforth, R.A., Halsall, D.J., Atkinson, T., Holbrook, J.J., Clarke, A.R. & Burston, S.G. (1993) Binding and hydrolysis of nucleotides in the chaperonin catalytic cycle: Implications for the mechanism of assisted protein folding. Biochemistry 32, 2554-2563

Kang PJ, Ostermann J, Shilling J, Neupert W, Craig EA, Pfanner N (1990) Requirement of hsp70 in the mitochondrial matrix for translocation and folding of precursor proteins. Nature 348:137-143

Kudlicki, W., Odom, O.W., Kramer, G. & Hardesty, B. (1994) Activation and release of enzymatically inactive, full-length rhodanese that is bound to ribosomes as peptidyl-tRNA. J. Biol. Chem. 269, 16549-16553

Landry, S. J., Jordan, R., McMacken, R. & Gierasch, L.M. (1992) Different conformations for the same polypeptide bound to chaperones DnaK and GroEL. Nature 355, 455-457

Langer T, Lu C, Echols H, Flanagan J, Hayer MK, Hartl FU (1992a) Successive action of molecular chaperones DnaK, DnaJ and GroEL along the pathway of assisted protein folding.Nature 356:683-689

Langer T, Pfeifer G, Martin, J, Baumeister W, Hartl FU (1992b) Chaperonin-mediated protein folding: GroES binds to one end of the GroEL cylinder, which accommodates the protein substrate within its central cavity. EMBO J. 11: 4757-4766

Liberek, K., Marszalek, J., Ang, D., Georgopoulos, C. & Zylicz, M. (1991)*Escherichia coli* DnaJ and GrpE heat shock proteins jointly stimulate ATPase activity of DnaK. Proc. Natl. Acad. Sci. U.S.A. 88, 2874-2878

Llorca, O., Marco, S., Carrascosa, J.L. & Valpuesta, J.M. (1994) The formation of symmetrical GroEL-GroES complexes in the presence of ATP. FEBS Lett. 345, 181-186

Manning-Krieg UC, Scherer PE, Schatz G (1991) Sequential action of mitochondrial chaperones in protein import into the matrix. EMBO J 10:3273-3280

Martin J, Langer T, Boteva R, Schramel A, Horwich AL, Hartl FU (1991b) Chaperonin-mediated protein folding at the surface of groEL through a 'molten globule'-like intermediate.Nature 352:36-42

Martin, J., Mayhew, M. and Hartl, F.-U. (1995) Successive rounds of protein folding by GroEL without intermittent release of unfolded polypeptides into the bulk solution. Submitted

Martin, J., Mayhew, M., Langer, T. & Hartl, F-U. (1993) The reaction cycle of GroEL and GroES in chaperonin-assissted protein folding. Nature 366, 228-233

Mayhew, M., Da Silva, A., Erdjument-Bromage, H., Tempst, P. and Hartl, F.U. (1995) Protein folding in the central cavity of the GroEL-GroES chaperonin complex. Submitted

Ostermann, J., Horwich, A.L., Neupert, W. & Hartl, F.U. (1989) Protein folding in mitochondria requires complex formation with hsp60 and ATP hydrolysis. Nature 341, 125-130

Palleros, D. R., Reid, K.L., Shi, L., Welch, W.J. & Fink, A.L. (1993) ATP-induced protein-HSP70 complex dissociation requires K^+ but not ATP hydrolysis. Nature 365, 664-666

Robinson, C.V., Groß, M., Eyles, S.J., Ewbank, J.E., Mayhew, M., Hartl, F.U., Dobson, C.M. & Radford, S. (1994) Hydrogen exchange protection in GroEL-bound a-lactalbumin detected by mass spectrometry. Submitted

Saibil, H. R., Zheng, D., Roseman, A.M., Hunter, A.S., Watson, G.M.F., Chen, S., auf der Mauer, A., O'Hara, B.P., Wood, S.P., Mann, N.H., Barnett, L.K. & Ellis, R.J. (1993) ATP induces large quarternary rearrangements in a cage-like chaperonin structure. Current Biology 3, 265-273

Schmidt, M., Rutkat, K., Rachel, R., Pfeifer, G., Jaenicke, R., Viitanen, P., Lorimer, G. & Buchner, J. (1994) Symmetric complexes of GroE chaperonins as part of the functional cycle. Science 265, 656-659

Schröder, H., Langer, T., Hartl, F.U. & Bukau, B. (1993) DnaK, DnaJ and GrpE form a cellular chaperone machinery capable of repairing heat-induced protein damage. EMBO J. 12, 4137-4144

Szabo, A., Langer, T., Schröder, H., Flanagan, J., Bukau, B. & Hartl, F.U. (1994) The ATP hydrolysis-dependent reaction cycle of the *Escherichia coli* Hsp70 system - DnaK, DnaJ and GrpE. Proc. Natl. Acad. Sci. U.S.A. 91, in press

Todd, M.J., Viitanen, P.V. & Lorimer, G.H. (1993) Hydrolysis of adenosine 5'-triphosphate by Escherichia coli GroEL: Effects of GroES and potassium ion. Biochemistry 32, 8560-8567

Weissman, J.S., Kashi, Y., Fenton, W.A. & Horwich, A.L. (1994) GroEL-mediated protein folding proceeds by multiple rounds of binding and release of nonnative forms. Cell 78, 693-702

Ziegelhoffer, T., Lopez-Buesa, P., and Craig, E. A. (1995). The dissociation of ATP from hsp70 of Saccharomyces cerevisiae is stimulated by both Ydj1p and peptide substrates. J. Biol. Chem. 270, 10412-10419

Zimmerman SB, Trach SO (1991) Estimation of macromolecule concentrations and excluded volume effects for the cytoplasm of *Escherichia coli*. J Mol Biol 222:599-620

Thermodynamics of the membrane insertion process of the M13 procoat protein, a lipid bilayer traversing protein comprising a leader sequence

Martin Eisenhawer‡, Mark Soekarjo‡, Andreas Kuhn§ and Horst Vogel‡
‡ Swiss Federal Institute of Technology Lausanne (EPFL)
CH-1015 Lausanne, Switzerland
§ University of Karlsruhe; D-76128 Karlsruhe, Germany

Proteins may use several different modes to insert into a membrane. *In vitro*, some small proteins are known to insert spontaneously into liposomes (Rietveld & de Kruijff, 1984; Geller & Wickner 1985) or membrane vesicles (Müller & Zimmermann, 1988), while others require the addition of cellular factors (Müller & Blobel, 1984) and high energy phosphates (Chen & Tai 1985; Waters & Blobel 1986; Eilers et al. 1987 and Schlenstedt & Zimmermann 1987). *In vivo*, insertion modes vary as far as oligatory coupling to protein synthesis (Rothblatt & Meyer 1986 and Zimmermann and Meyer 1986) and involvement of cellular secretory functions are concerned.

The M13 coat protein is synthesized with a 23 amino acid long signal sequence in the cytoplasm of *E. coli* and is then inserted into the plasma membrane. The membrane insertion process of this so-called M13 procoat protein has been extensively studied genetically and biochemically (Kuhn & Troschel, 1992). There are experimental indications that the binding of the protein to the negatively charged membrane surface occurs via the positively charged amino acid residues at the amino and carboxy termini of the procoat protein (Gallusser & Kuhn, 1990). The subsequent partitioning of the two hydrophobic regions into the membrane results in an U-like configuration. Biophysical investigations have shown that at least part of the predicted transmembrane regions adopt an α-helical conformation (Thiaudière et al., 1993). The topology of this intermediate has been investigated using procoat proteins with either N-terminal or C-terminal extensions (Kuhn et al., 1986; Kuhn, 1987). It was also found that the N- and C-termini do not leave the cytoplasm during the insertion process and only the negatively charged central region traverses the membrane in accordance with the "positive-inside" rule (von Heijne, 1992). After membrane insertion, the signal sequence is cleaved off by the *E. coli* leader peptidase which has its enzymatic activity at the periplasmic face of the cytoplasmic membrane (Dalbey, 1991; Dalbey & von Heijne, 1992). The membrane insertion process of the procoat protein is apparently independent of any "helper " protein complex, such as *E. coli* translocase

NATO ASI Series, Vol. H 96
Molecular Dynamics of Biomembranes
Edited by Jos A. F. Op den Kamp

and other factors like the ffh protein (Kuhn, unpublished results). In this context it is interesting to note that newly synthesized M13 procoat protein is capable of insertion into liposomes from the aqueous phase (Geller & Wickner, 1985) suggesting that the insertion process is mainly driven by hydrophobic interactions. This makes quantitative investigation of the membrane insertion of this protein attractive.

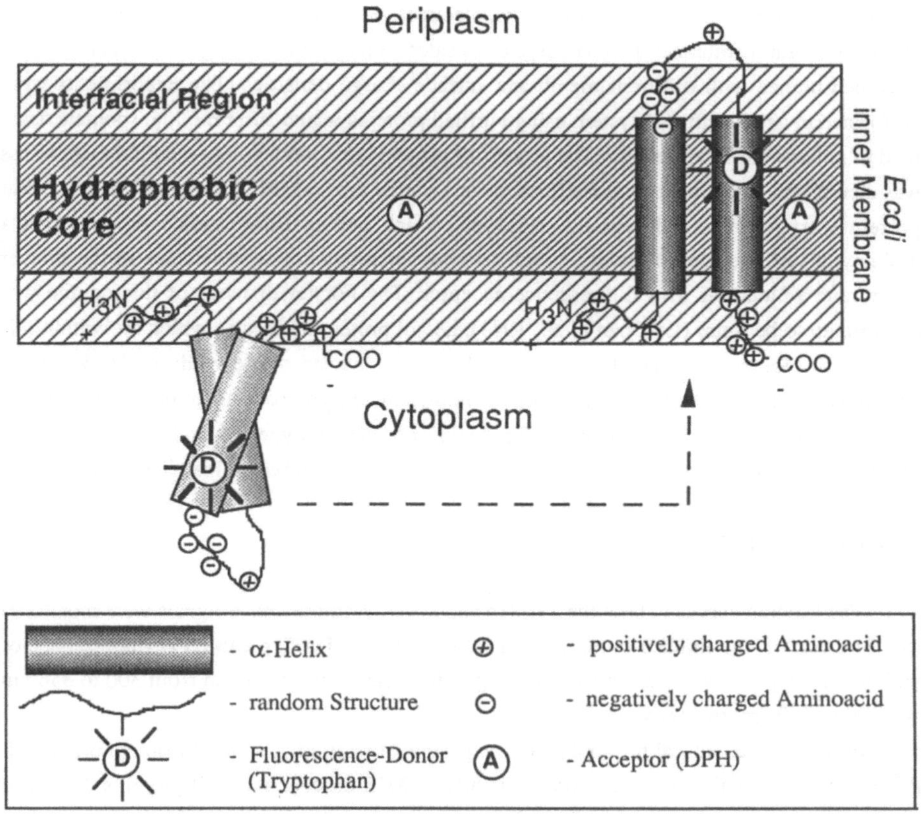

Figure 1: Insertion model for M13 Procoat Protein. The initial electrostatic interaction is followed by hydrophobic partitioning.

We determined experimentally the free energy change ΔG^o occuring during the spontaneous transfer of the procoat protein from the aqueous to the lipid bilayer phase. These data will serve as a basis for a thermodynamic model describing the membrane insertion of procoat protein. In order to distinguish the energetic contributions which arise from the leader sequence and from the mature protein part, parallel experiments were also performed with the coat protein as well as with the procoat mutant protein OM30R. The OM30R mutant differs from the wild type procoat protein in having a Val -> Arg mutation at position +30, inserting a positive charge into the hydrophobic stretch of the mature part of the protein. This mutant protein is known from *in vivo* experiments to interact with the membrane, but not to be translocated across the *E. coli* plasma membrane. The interest in the present context is to learn whether or not this changed biological property is correlated with a change of the partitioning into lipid bilayer membranes. Furthermore, electrostatic contributions of the procoat protein interaction with membranes were taken into consideration by investigating the protein insertion into membranes with different membrane surface charge densities. The experiments were performed in a reconstituted system using purified proteins which, after mixing, spontaneously bind to unilamellar lipid vesicles. The process of protein incorporation into the lipid membranes was then observed by measuring the fluorescence energy transfer (FET) between the (pro-) coat protein's tryptophan and tyrosine residues, which act as fluorescence donors, and a diphenylhexatriene (DPH) moiety coupled to phosphatidylcholine molecules in the bilayer membrane which acts as a fluorescence acceptor. The Förster distance of this couple was determined to be 4 nm (Le Doan et al., 1983), i.e. all protein molecules which are within this distance or closer to the membrane will lead to FET. Instead of the wildtype M13 procoat protein we used in the course of this work the non-cleavable procoat mutant protein H5 (with a Phe at position -3) which inserts into the lipid bilayer as the wildtype protein but can be isolated by a simple procedure in chemical quantities as was shown elsewhere (Thiaudière et al., 1993).

Analysis of the fluorescence data

The titration curves of both the Tyr/Trp and the DPH fluorescence signals were monitored as a function of the lipid-to-protein molar ratio, L/P. The Tyr/Trp and DPH fluorescence of individual samples was measured as the integral intensity of the emission from 300 to 356 nm and 390 to 450 nm, respectively.

FET from the natural tryptophan of the proteins to a lipid attached chromophore, DPH-PC, was used to evaluate the membrane affinity of M13 procoat and coat proteins. FET was characterized as the increase of the DPH fluorescence due to the close proximity of Tyr and Trp residues to the DPH-PC molecules in the membrane. Upon irradiation at 280 nm DPH is directly exited to some extent; therefore background fluorescence of DPH is observed in absence of FET. Spectra were corrected for intrinsic DPH fluorescence by substracting the spectrum of a sample which contained the appropriate amount of lipids, but no protein. The DPH fluorescence intensity, F, is taken to be the integral of the corresponding difference spectra from 390 to 450 nm. Measuring the FET-induced decrease of the Tyr/Trp fluorescence yielded identical information on the membrane partitioning. However, the experimental scatter of the data was higher than for DPH fluorescence for reasons we do not yet understand. We assume that at the moment of sample injection the protein equilibrates between the aqueous and the lipid bilayer. All protein molecules which are not lipid-associated and do not contribute to the DPH fluorescence will be included in a term called 'free' protein concentration, c_f.

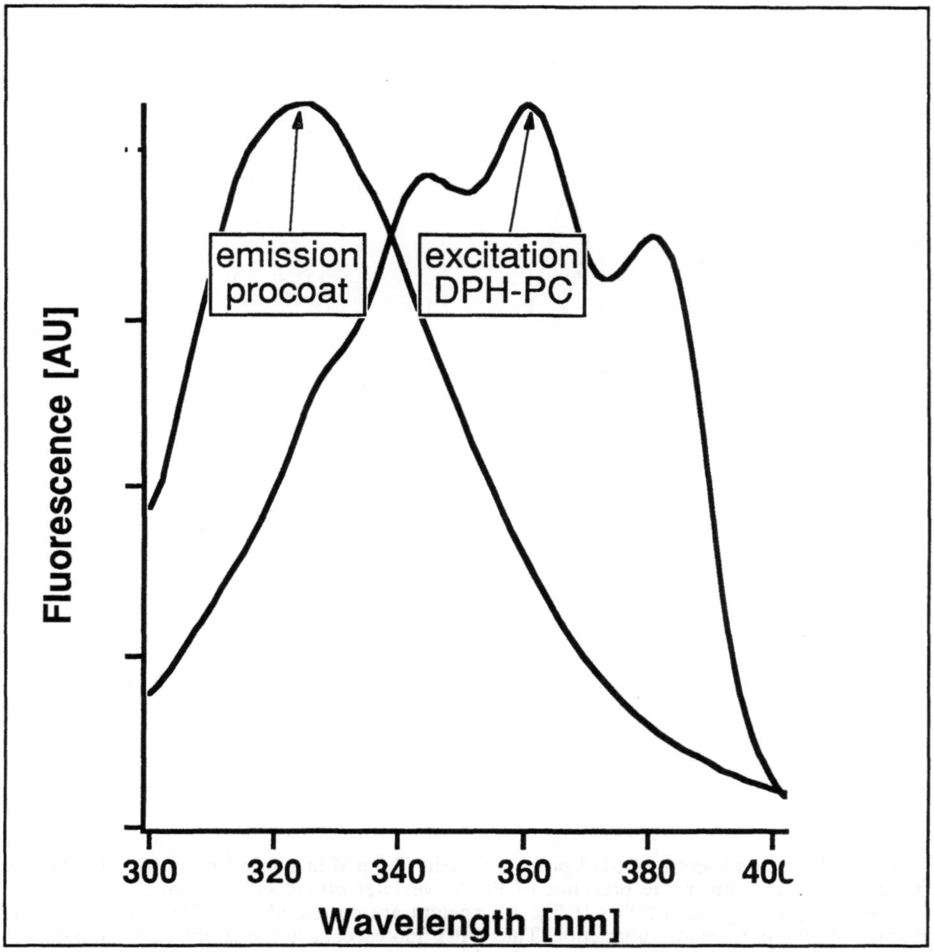

Figure 2: Tyr/Trp emission spectrum of M13 procoat protein (exitation at 280 nm) and diphenylhexatriene (DPH) exitation spectrum of DPH-PC (emission at 430 nm). Due to the overlap between the two spectra fluorescence energy transfer occurs between the Tyr and Trp residues of vesicle-associated proteins which act as donors and lipid-anchored DPH moieties which act as acceptors.

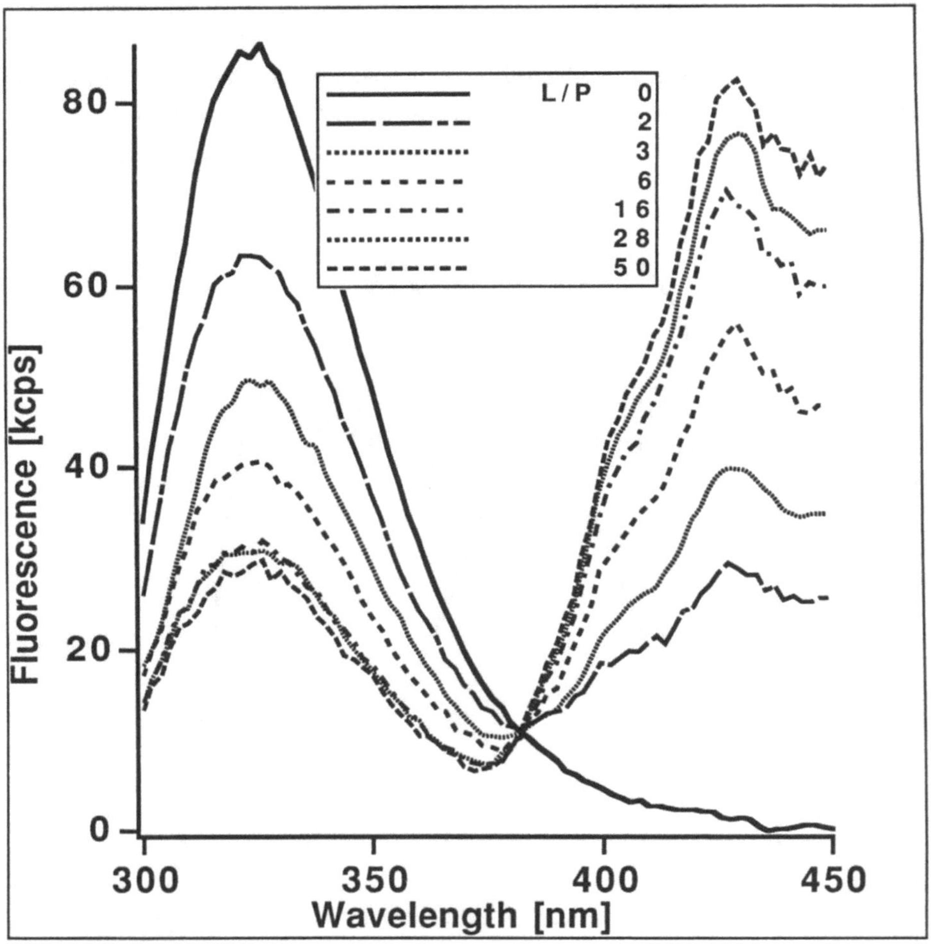

Figure 3: Emission spectra of M13 procoat protein (0.5 μM in 1 mM Tris buffer, pH 7.4, 23 ^{o}C) exited at 280 nm in the presence of POPC vesicles doped with 2 mol % DPH-PC at increasing lipid to protein ratios (L/P). All spectra are corrected for DPH fluorescence in absence of protein by blank subtraction. Thus DPH fluorescence in the samples containing lipid vesicles is due only to FET between the Tyr/Trp residues of vesicle associated proteins and lipid-bound DPH moieties.

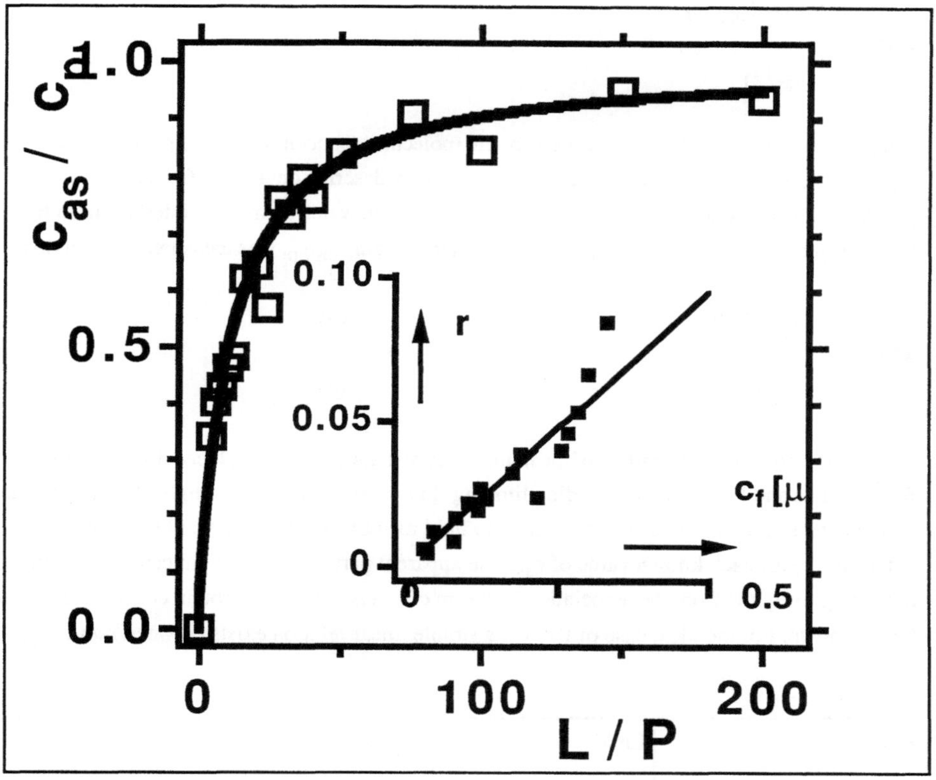

Figure 4: The fraction of coat protein associated to POPC lipid vesicles relative to the total peptide concentration, c_{as}/c_p, plotted as a function of increasing lipid-to-protein ratios L/P. c_{as} was determined from the FET induced increase of DPH-PC fluorescence of the data shown and described in Fig. 3, according to eqs (3) and (4). The insert shows a plot of r, the molar ratio of lipid-associated protein molecules per lipid molecule, as a function of the free protein concentration c_f. The straight line corresponds to a least squares linear fit of an ideal protein partitioning into the lipid bilayers according to eq (1), yielding a partition coefficient $\Gamma = 1.0 \times 10^5\,M^{-1}$.

The association of polypeptides to lipid membranes has been quantitatively evaluated in the literature in terms of a partition equilibrium between the aqueous and the lipid bilayer phase for several amphipatic polypeptides (Pawlak et al., 1994 and refs. therein). FET measurements have already been used by others to detect protein association to lipid membranes (Vaz et al., 1977; Tamm & Bartoldus, 1988). Here we apply this method to M13 procoat protein and M13 coat protein using the formalism of Schwarz et al. (1986). An apparent partition coefficient Γ_{app} is defined as

$$r \cdot f(r) = \Gamma_{app} \cdot c_f \quad (1)$$

with

$$r = c_{as} / c_l \quad (2)$$

being the molar ratio of lipid-associated protein molecules (concentration c_{as}) per lipid molecule (c_l). $f(r)$, the activity coefficient as a function of r, may describe deviations from ideal partitioning ($f = 1$) due to protein-protein interaction at the water-membrane interface or in the lipid bilayer. The molar fraction of lipid associated protein, c_{as}/c_p, was measured by applying

$$c_{as} / c_p = F / F_\infty \quad (3)$$

where

$$c_p = c_{as} + c_f \quad (4)$$

is the total concentration of protein, F represents the FET induced increase in DPH fluorescence and F_∞ the corresponding limiting fluorescence intensity at high L/P ratios for the case where all protein molecules are bound to lipid membranes. By using eq (4) it is possible to determine c_f for each known value of c_{as}. The apparent partition coefficient can be determined according to eq. (1) from the association isotherm of r versus the free protein concentration c_f if $f(r)$ is known. For the ideal case of $f(r) = 1$ a simple linear relation exists between r and Γ_{app}.

Protein	Lipid	$\Gamma_{app} \times 10^{-5}$ M^{-1}	$-\Delta G^o$ [kcal/mol]
H5 procoat protein	POPC	6.5	10.4
	POPC/POPG 1/1	>10	>10.7
OM30R procoat protein	POPC	0.3	8.6
M13 coat protein	POPC	1.0	9.3
	POPC/POPG 1/1	3.5	10.1
	DPPC	no spontaneous insertion	

Table 1 : Apparent partition coefficients Γ_{app} and values for the free energy of transfer ΔG^o of coat and procoat proteins from water to different lipid membranes at 23 °C[a)]

a) The partition coefficients Γ_{app} are defined using the protein/lipid molar ratio as the calculated as $\Delta G° = -RT \ln\gamma_{MF}$ using the mole-fraction partition coefficient γ_{MF} which for sufficiently high dilution of the proteins in the membranes is given by $\gamma_{MF} = \Gamma_{app}/\bar{V}_W$ with the partial molar volume of water $\bar{V}_W = 18$ ml/mol.[3] This form has been recently applied for the calculation of the partitioning of small peptides to lipid bilayers (Terzi et al., 1994).

Conclusions

1. M13 (pro-)coat inserts spontaneously into fluid lipid membranes, showing an ideal partitioning of the protein between the aqueous and the membrane phase. The partition coefficients for POPC membranes are higher for procoat than for coat, but still in the same order of magnitude. Procoat seems to span the membrane by two hydrophobic helices.

2. The membrane insertion depends on the membrane surface charge. The partition coefficient for coat and procoat for binding to membranes with a net negative surface charge is higher than for pure POPC membranes. This might be due to electrostatic interactions between the positively charged amino acids of the proteins and the negative charges of POPG.

3. The OM30R mutant protein is known from in vivo experiments to bind only to the membrane interface. It shows a distinctly lower partitioning for POPC membranes than H5 procoat protein, but still in the same order of magnitude. From this we conclude that the major part of the free energy change during membrane interaction already occurs at the membrane interface.

4. Preliminary experiments have been performed with synthetic M13 signal peptide: In contrast to entire procoat protein the signal peptide only binds to negatively charged lipid membranes such as POPG but not to POPC. This clearly indicates that the isolated signal peptide behaves significantly different as the entire procoat protein.

5. The presented procedure might be generally applicable to probe the membrane insertion process of hydrophobic proteins into lipid bilayers. This is independent of the particular structure of the protein at or in the lipid membrane and holds for partitioning to both the lipid headgroup region and the hydrocarbon core.

References

Chen, L. & Tai, P. C. (1985) *Proc. Natl. Acad. Sci. USA 82*, 4384-4388

Crooke, E. & Wickner, W. (1987) *Proc. Natl. Acad. Sci. USA 84*, 5216-5220

Dalbey, R.E. (1991) *Mol. Microbiol. 5*, 2855-2860.

Dalbey, R.E. & von Heijne, G. (1992) *Trends Biochem. Sci. 17*, 474-478.

Deber, C. M., Khan, A. R., Li, Z., Joensson, C., Glibowicka, M. & Wang, J. (1993) *Proc. Natl. Acad. Sci. USA 90*, 11648-11652.

Eilers, M., Oppliger, W. & Schatz, G. (1987) *EMBO J. 6*, 1073-1077

Geller, B.L. & Wickner, W. (1985) *J. Biol. Chem. 260*, 13281-13285.

Gallusser, A. & Kuhn, A. (1990) *EMBO J. 9*, 2723-2729.

Jähnig, F. (1983) *Proc. Natl. Acad. Sci. USA 80*, 3691-3695.

Kuhn, A. (1987) *Science 238*, 1413-1415.

Kuhn, A., Wickner, W. & Kreil, G. (1986 a) *Nature 322*, 335-339.

Kuhn, A., Kreil, G. & Wickner, W. (1986 b) *EMBO J. 5*, 3681-3685.

Kuhn, A. & Wickner, W. (1985) *J. Biol. Chem. 260*, 15914-15918.

Kuhn, A. & Troschel, D. (1992) in *Membrane Biogenesis and Protein Targeting* (Neupert, W. & Lill, R., Eds.) pp 33-47, Elsevier, New York.

Le Doan, T., Tagasuki, M., Aragon, I., Boudet, G., Montenay-Garestier, T. & Helene, C. (1983) *Biochim. Biophys. Acta 735*, 259-270.

Müller, G. & Zimmermann, R. (1988) *EMBO J. 7*, 639-648

Müller, M. & Blobel, G.(1984) *Proc. Natl. Acad. Sci. USA 81*, 7421-7425

Pawlak, M., Kuhn, A. & Vogel, H. (1994) *Biochemistry 33*, 283-290.

Pawlak, M., Meseth, U., Dhanapal, B., Mutter, M. & Vogel, H. (1994) *Protein Science, 3*, 1788-1805.

Rietveld, A. & de Kruijff, B. (1984) *J. Biol. Chem. 259*, 6704-6707

Rothblatt, J. A. & Meyer, D. I. (1986) *EMBO J. 5*, 1031-1036

Schlenstedt, G. & Zimmermann, R. (1987) *EMBO J. 6*, 699-703

Schwarz, G., Stankowski, S. & Rizzo, V. (1986) *Biochim. Biophys. Acta 861*, 141-151.

Tanford, C. (1980) *The Hydrophobic Effect, 2nd edition*, New York, John Wiley.

Tamm, L. K. & Bartoldus, I. (1988) *Biochemistry 27*, 7453-7458.

Terzi, E., Hölzemann, G. & Seelig, J. (1994) *Biochemistry 33*, 7434-7441.

Thiaudière, E., Soekarjo, M., Kuchinka, E., Kuhn, A. & Vogel, H. (1993) *Biochemistry 32*, 12186-12196.

Vaz, W. L. C., Kaufman, K. & Nicksch, A. (1977) *Anal. Biochemistry 83*, 385-393.

Von Heijne, G. (1992) *J. Mol. Biol. 225*, 487-494.

Waters, M. G. & Blobel, G. (1986) *J. Cell. Biol. 102*, 1543-1550

Zimmermann, R. & Meyer, D. I. (1986) *Trends Biochem. Sci. 11*, 512-515

Lipid-protein interactions in chloroplast protein import

Ben de Kruijff*, Rien Pilon [1], Ron van 't Hof [2] & Rudy Demel

* Corresponding author: Department of Biochemistry of Membranes, Center for Biomembranes and Lipid Enzymology, Institute of Biomembranes, Utrecht University, Padualaan 8, 3584 CH Utrecht, The Netherlands. fax: +31-30-2522478, E mail: dekruijff@chem.ruu.nl

1) Present adress: Department of Molecular and Cell Biology, University of California, Berkeley, USA

2) Present adress: Center for Protein Technology, TNO/WAU, The Netherlands

NATO ASI Series, Vol. H 96
Molecular Dynamics of Biomembranes
Edited by Jos A. F. Op den Kamp

Table of contents

1 Summary

Chloroplasts rely for their biogenesis on import of cytosolically synthesized proteins. These proteins carry N-terminal transit sequences which are responsible for the organelle specific import. The hypothesis is presented that specific interactions between the transit sequence and chloroplast specific lipids are involved in envelope passage of the precursor proteins. Experimental evidence obtained in model systems on the transit sequence-lipid interaction in support of this hypothesis is reviewed. The specificity of the transit sequence-lipid interaction and its consequences for the structure of the interacting components are translated into a model for the early steps in chloroplast proteins import. The model predicts that the combination of specific lipid-protein and protein-protein interactions are responsible for an efficient import process.

1.1 Abbreviations

apoFd	apoferredoxin
C_{12}-maltose	dodecyl maltose
C_{12}-PN	dodecyl phosphocholine
DGDG	digalactosyldiglyceride
DOPG	dioleoyl phosphatidylglycerol
$\Delta\Pi$	surface pressure increase
Fd	Ferredoxin
H_{II}	inverted hexagonal phase
IM	inner membrane
MGDG	monogalactosyldiglyceride
OM	outer membrane
Π	surface pressure
Π_i	initial surface pressure
PC	phosphatidylcholine
PG	phosphatidylglycerol
P-glycol	dodecyl phosphoglycol
PI	phosphatidylinositol
preFd	precursor of ferredoxin
SQDG	sulfoquinovosyldiglyceride
trFd	transit peptide of ferredoxin
WT	wild type

2. Introduction

Proteins function at distinct sites in a cell, yet they are primarily synthesized in one compartment, the cytosol. (For a concise collection of reviews see (Neupert and Lill,1992)). This implies that newly synthesized proteins have to be transported to their final destination. Protein transport often involves membrane passage which is a fascinating event because large often polar polypeptide chains of variable composition have to move through an ultra-thin membrane without compromising the other essential barrier function of that membrane.

Within nucleated cells such as plant cells several main protein transport routes exist (Fig.1). Cytosolically synthesized proteins directed into the endoplasmic reticulum enter the secretion pathway. Vesicular transport can carry there proteins further to compartments such as the Golgi system, the plasma membrane, the vacuole and the outside world (Rapoport, 1992; Rothman, 1994).

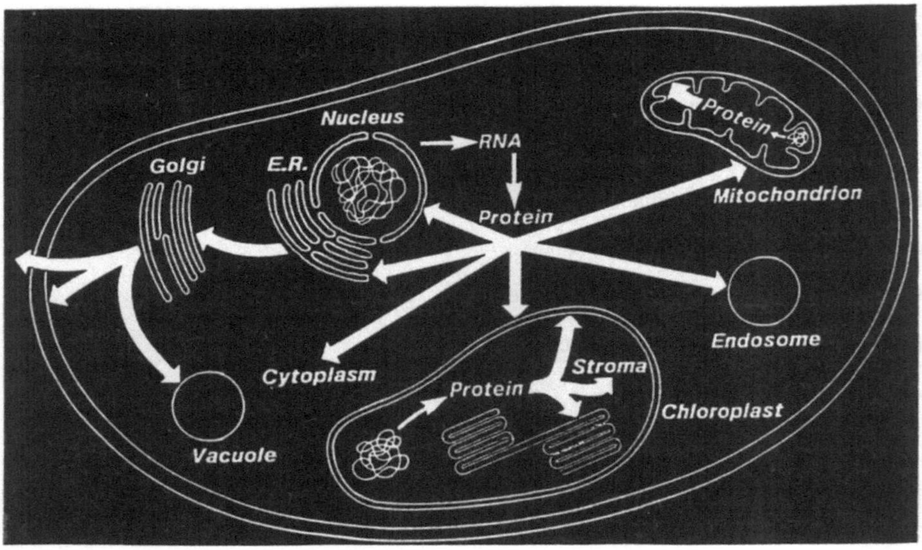

Fig. 1: A schematic representation of a plant cell in which the main protein transport routes are indicated.

Newly synthesized proteins can also be directed into the nucleus or stay in the cytosol. Of particular interest for this article is the transport route into plastids like chloroplasts, which are the most prominent organelles found in the plant cell. Chloroplasts are complex structures bounded by an envelope consisting of an outer and an inner membrane (Douce and Joyard, 1990). Within the stroma the thylakoid membrane system is present, which harbors the photosynthetic apparatus (Hall and Rao, 1987). Chloroplasts contain their own genome and have the capacity to synthesize proteins (Ellis, 1981; Sugiura, 1992). However, the coding capacity of the DNA in chloroplasts is much too limited to account for all chloroplast proteins (Umesono and Ozeki, 1987). Therefore, chloroplast biogenesis strictly depends on import of protein synthesized in the cytosol. The same is true for mitochondria which also have a genome of insufficient coding capacity and therefore depend on protein import for growth (Douglas et al., 1986). Within the chloroplast extensive protein sorting processes have to take place because each intra-organellar compartment has its own specific set of proteins coming both from the cytosol as well as from synthesis within the stroma. It is the aim of this contribution to briefly describe the current insight into protein import into chloroplasts with special emphasis on protein passage through the envelope and to give an overview of our own research on the role specific chloroplast lipids might play in this process.

At some points a comparison will be made to mitochondrial protein import because these routes show similarities and only in plant cells operate simultaneously and yet with high fidelity. For recent comprehensive reviews on chloroplast protein import the reader is referred to (De Boer and Weisbeek, 1991; Keegstra, 1989).

2.1 Protein import into chloroplasts

The envelope forms the barrier for proteins imported into the chloroplast stroma. There appears to be one general import pathway for such proteins. The key features of this pathway are illustrated in Fig. 2 which pictures the import of Ferredoxin (Fd) which protein fulfills its function in electron transport during photosynthesis on the stromal site of the thylakoid

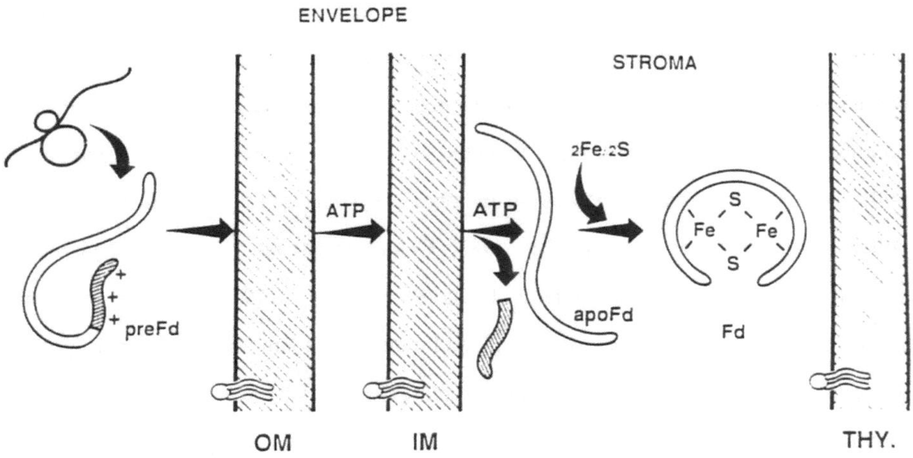

Fig. 2: The general chloroplast import pathway illustrated for ferredoxin.

membrane system (Hall and Rao, 1977). Fd follows the general import route and is synthesized in the cytosol as a precursor preFerredoxin (preFd) carrying an N-terminal extension called the transit sequence (Smeekens et al., 1985). This transit sequence contains all information for chloroplast specific import. For instance gene fusion experiments revealed that it can direct foreign proteins into the chloroplast (Smeekens et al., 1987). Both chloroplast and mitochondrial precursor proteins are posttranslationally imported into the organelle (Hay et al., 1984; Umesono and Ozeki, 1987). *In vitro* experiments using isolated chloroplasts revealed that preFd is efficiently imported without the aid of cytosolic factors (Pilon et al., 1992). This might be a special property of this relatively small and polar precursor protein. More hydrophobic precursors seem to require cytosol proteinaceous factors for import (Waegemann et al., 1990). Such helper proteins might have a chaperone type of function to maintain (or acquire) a translocation competent conformation, alternatively or in addition they could have a targeting function. PreFd is by itself an unstructured protein (Pilon et al., 1992) and therefore might bypass involvement of cytosolic proteins. Envelope passage can be dissected into several steps. Binding of the precursor to the outer surface of the outer membrane (OM) is dependent on the transit sequence (Friedman and Keegstra, 1989) and requires low concentration of ATP (Olsen et al., 1989) and proteinaceous components on the envelope (Cornwell and Keegstra, 1987). Pretreatment of chloroplasts with proteases strongly reduces precursor binding and import (Friedman and Keegstra, 1989). What happens to the precursor between the outer surface of the outer membrane and the inner surface of the inner membrane (IM) is largely a mystery. Envelope passage of the precursor requires ATP which appears to be consumed in the stroma (Theg et al., 1989). Most likely several proteinaceous components are involved. Several of these proteins were recently identified mainly via cross linking and affinity purification approaches (Hirsch et al, 1994; Kessler et al., 1994; Perry and Keegstra, 1994; Schnell et al., 1994). Reactive sulfhydryls are involved in import of preFd (Pilon et al., 1992) and Cu^{2+} appears to be a selective import inhibitor because of its ability to oxidize these sulfhydryls (Seedorf and Soll, 1995). Once the precursor emerges from the inside of the inner membrane it becomes processed. A specific protease removes the transit sequence and liberates the mature protein into the stroma (Robinson and Ellis, 1984). The transit peptide is rapidly digested and it appears that the resulting products are efficiently exported out of the organelle by an at yet unknown mechanism (Van 't Hof and De Kruijff, 1995). Once in the stroma the mature proteins can undergo further reactions finally resulting in the active molecule. ApoFd generated from the precursor is still an unfolded protein which matures upon coupling of a 2 iron - 2 sulfur cluster onto 4 cysteines in the protein therefore folding the protein into the compact and functional Fd molecule (Pilon et al., 1992). Proteins destined for the thylakoid systems also initially follow this general import route but subsequent sorting has to take place which occurs in the stroma and which is directed by an additional targeting signal located between the stromal targeting signal and the mature protein (Hageman et al., 1990).

The molecular mechanism of protein transport across the envelope membranes is virtually unknown but appears to involve specific recognition between the precursor and the target organelle and membrane insertion followed by membrane passage most likely through flexible tunnel-like structures. In analogy with other protein transport systems it is believed that chloroplast proteins move across the envelope as unfolded or loosely folded (Eilers and Schatz, 1986; Muller and Zimmermann, 1988) structures passing the envelope at sites of close contact between the outer and inner membrane (Schnell et al., 1990). It is assumed that each membrane has its own translocation machinery but that these can be coupled for efficient translocation.

The similarities between mitochondrial and chloroplast protein import are often emphasized. However, at least two obvious differenced between these routes exist. The targeting signals are very different (see further) and a second difference is the requirement of a membrane potential across the inner membrane for import of mitochondrial proteins (Gasser et al., 1982; Schleyer et al., 1982).

2.2 Chloroplast lipids

The different membranes within a nucleated cell have in general a specific protein and lipid composition. This specificity is most apparent for membrane proteins. A given protein is restricted to a specific membrane. However, a membrane lipid is often found in different cellular membranes but often in different concentrations. There are some exceptions on this general rule. For instance, cardiolipin is an abundant lipid found only in mitochondria (Hovius et al., 1990). Even more striking examples can be found for chloroplasts. Their membranes contain several lipid classes and species found nowhere else in the plant cell. These include the abundant galacto lipids monogalactosyldiglyceride (MGDG) and digalactosyldiglyceride (DGDG) and the sulfolipid sulfoquinovosyldiglyceride (SQDG) (Douce et al., 1984). For chemical structures of MGDG and SQDG see Fig. 3. As a consequence chloroplast membranes have a relatively low phospholipid content as compared to non-chloroplast membranes in the plant cell. Fig. 4 schematically illustrates the lipid composition of the outer and inner membrane of pea chloroplasts. In particular the inner membrane is very rich in galactolipids and has a composition which is similar that of thylakoids. Zwitteronic phosphatidylcholine (PC) is an abundant lipid in the outer membrane. The main anionic lipids in the envelope are SQDG, phosphatidylglycerol (PG) and phosphatidylinositol (PI). Several chloroplast lipid classes have in addition a special fatty acid composition and distribution (Douce et al., 1984).

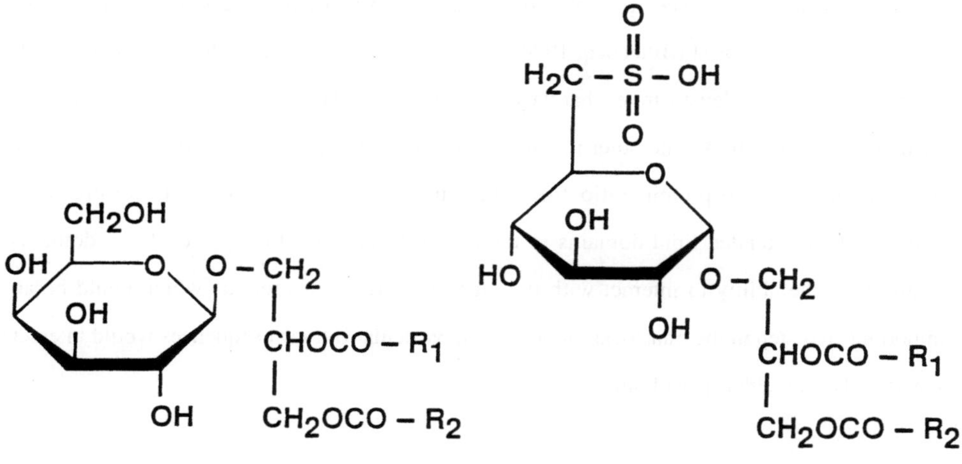

MGDG SQDG

Fig. 3: The chemical structures of MGDG and SQDG.

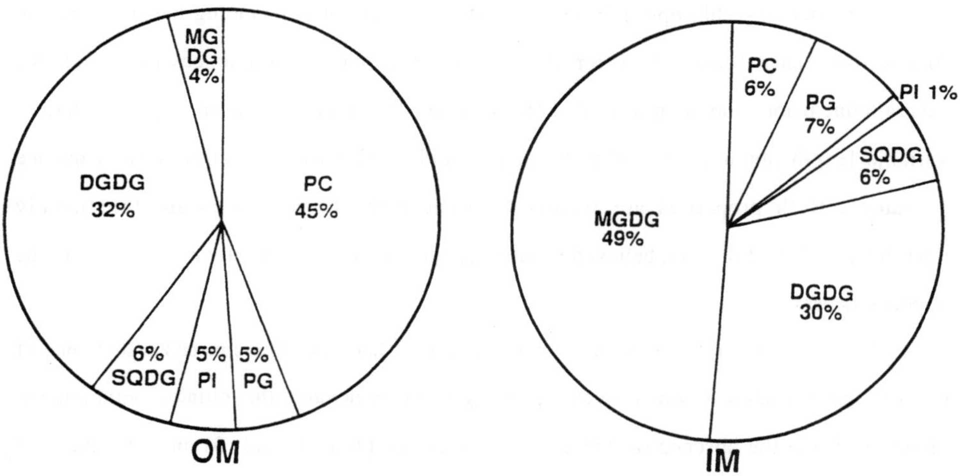

Fig. 4: The membrane lipid composition of the outer and inner membrane of pea chloroplast (Douce et al, 1984).

Only little is known about the topology of the membrane lipids across the envelope membranes. PC in the outer membrane is preferentially localized in the outer leaflet (Dorne et al., 1985). Galactolipid specific antibodies are able to recognize the galactolipids at the surface of the chloroplast (Billecocq, 1974). This not only demonstrates that these latter lipids are exposed but also demonstrates that they are accessible to large molecules. In this context it is important to recall that the outer membrane of both chloroplasts and mitochondria have a much higher lipid-to-protein ratio then other intracellular membranes. Therefore, it is expected that extended lipid domains in these membranes face the cytosol. Such domains might have the ability to interact with the incoming precursor proteins which could be an important first step in the translocation process in particular if transit sequences would interact specifically with chloroplast lipids.

3. Structure and function of transit sequences

Understanding chloroplast protein import will require understanding the structure and function of transit sequences. From the vast number of transit sequences now available several conclusions can be drawn (Von Heijne et al., 1989). Firstly, transit sequences have a variable length ranging from 30 to 80 amino acids. Secondly, they show little sequence similarity. Both properties are surprising because transit sequences are functionally interchangeable and they are believed to undergo specific interactions at various stages in the import process.

The characteristic features that transit sequences share are (1) a consensus sequence of the first amino acids MA which is believed to signal the removal of the initiator methionine in the cytosol; (2) the absence of negatively charged amino acids and (3) an enrichment of hydroxylated and small hydrophobic amino acids (Von Heijne et al., 1989). Three regions can be recognized: a N-terminal region of ± 15 amino acids that is uncharged. A middle region of variable length which lacks charges but which is enriched in hydroxylated amino acids and a

C-terminal region which contains positively charged amino acids. In the central region usually one or more prolines are found (Von Heijne et al., 1989).

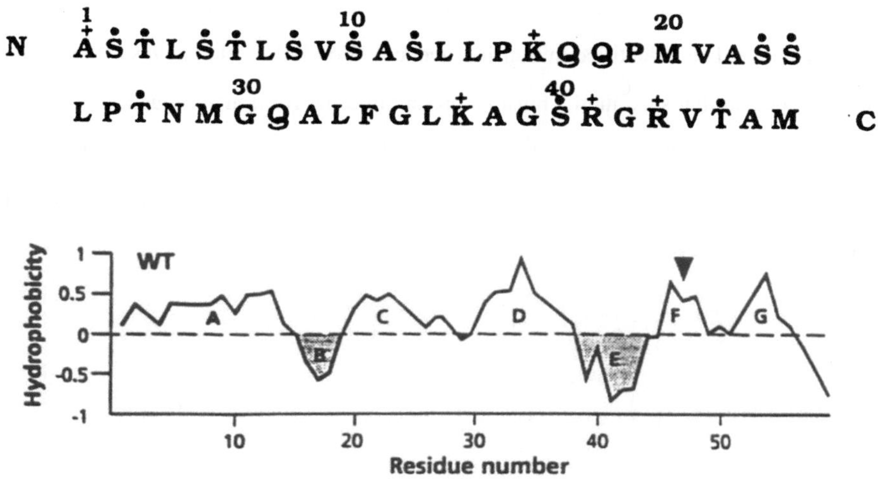

Fig. 5: The transit sequence of preFd form S.pratensis and its hydrophobicity profile (Pilon et al., 1995). The arrow indicates the position corresponding to the processing site. The dots indicate hydroxylated residues.

The 47-amino acid transit sequence of preFd from *S.pratensis* is shown in Fig. 5. The hydrophobicity profile identifies four more hydrophobic regions (A,C,D,F) and two more polar regions (B and E). This amphiphilicity which is a hallmark of most targeting sequences makes the transit sequence prone to interact with membrane lipids. Theoretical analysis of secondary structure preferences of transit sequences (Von Heijne and Nishikawa, 1991) as well as circular dichroism measurements on corresponding peptides (Endo et al., 1992; Muller and Zimmermann, 1988) (transit peptides) in solution strongly suggest that transit sequences are structurally flexible polypeptides with little preference to adopt a defined structure (see also section 4.4.1).

3.1 Transit sequence domain structure and function

In contrast to other targeting signals such as those functional in the secretion pathway (Gierash, 1989) and in mitochondrial protein import (Roise and Schatz, 1988) no specific motifs have yet been identified in transit sequences (Von Heijne et al., 1989). Nevertheless, it can be expected that such motifs must be present in order for transit sequences to fulfill their functions. A substitution and deletion analysis of the ferredoxin transit sequence has given some insights into the importance of the various domains of the sequence for the different steps in the translocation process.

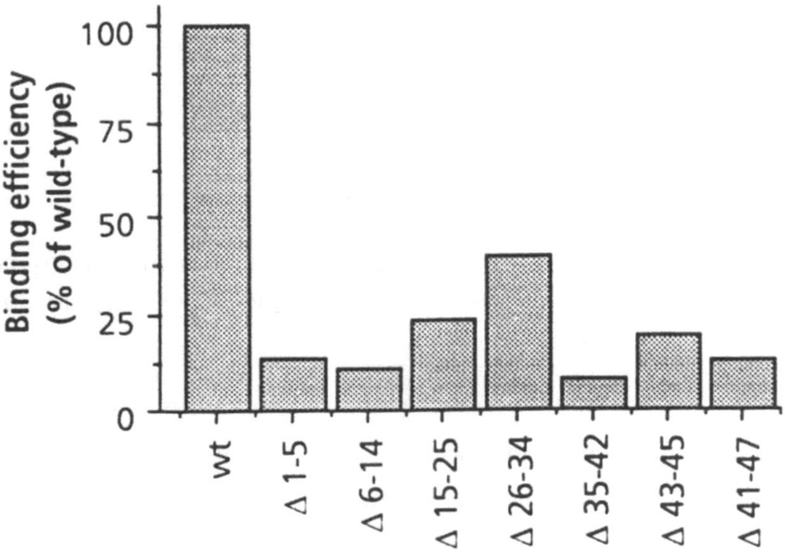

Fig. 6: Binding of wild type preFd and some transit sequence deletion mutants to chloroplasts. For further details see (Pilon et al., 1995). Deletions are indicated by Δ followed by the sequence deleted.

The N- and C-terminus appear to be important for targeting because small deletions in these regions strongly interfere with chloroplast binding as illustrated for some deletion mutants in Fig. 6. Large deletions in the central region are less harmful for binding yet the entire transit sequence gives maximal binding. It is not surprising that impaired binding will result in impaired import. Indeed, deletions in the N- and C-terminus of the transit sequence also greatly reduce the import of the precursor (Fig. 7). This approach identified a third region from approximately residues 15-25 that was important for translocation but less for initial binding. Amino acids 26-38 constitute a region which is less essential for *in-vitro* import.

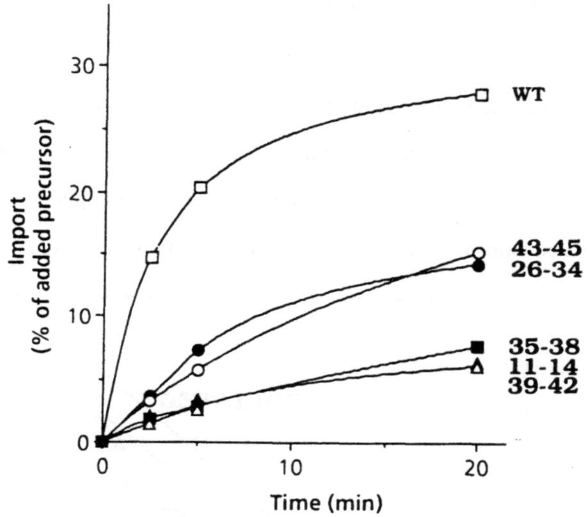

Fig. 7: Time course of import of some deletion mutants of preFd into chloroplasts. For further details see (Pilon et al., 1995).

This region is rich in α-helix breaking residues and might act as a flexible linker between functional domains. Deletions in the C-terminus strongly interfere with processing which is expected because the protease will have to recognize the amino acids around the cleavage site (Pilon et al., 1995). A mechanistic attractive possibility for functioning of transit sequences would be that they could open up channels through which the precursor could pass the envelope membranes. Some support for this hypothesis is obtained. Analysis of the electrical resistance of giant chloroplasts patch-clamped in the whole cell configuration revealed that the transit sequence greatly increased the envelope conductance most likely by opening translocation pores (Bulychev et al., 1994). Import of preFd does not require cytosolic factors. Therefore, the transit sequence will have to specifically recognize components of the envelope membranes. Likely candidates are proteinaceous receptors as well as the membrane lipids as will become clear in section 4.

3.2 Import of the transit peptide

One way to dissect the relative importance of the mature and transit part of a precursor protein for import is to analyze in a comparitive way the import of a precursor and the corresponding transit peptide. The chemically synthesized transit peptide of ferredoxin (trFd) associates with isolated chloroplasts and becomes protease protected (Fig. 8).

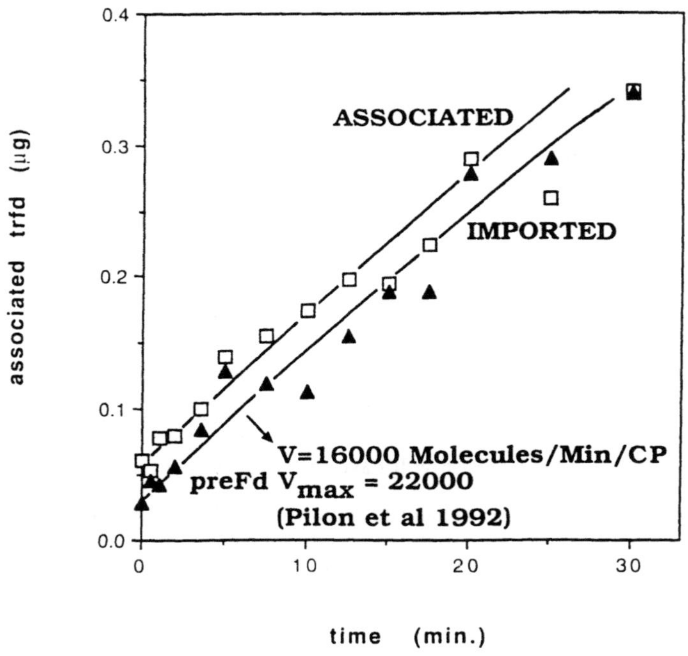

Fig. 8: Time course of association and import of trfd into chloroplasts. For details see (Van 't Hof and De Kruijff, 1995).

Protease protection is the result of its import into the organelle via the general import pathway (Van 't Hof and De Kruijff, 1995). The rate of import of the transit peptide can be estimated to be 16.000 molecules per minute per chloroplast which is close to the maximal rate of import of the intact precursor (Pilon et al., 1992). This suggests that for this precursor the rate limiting step is the interaction of the transit sequence with the machinery.

Like in the case of the precursor, ATP stimulates import of the transit peptide (Fig. 9).

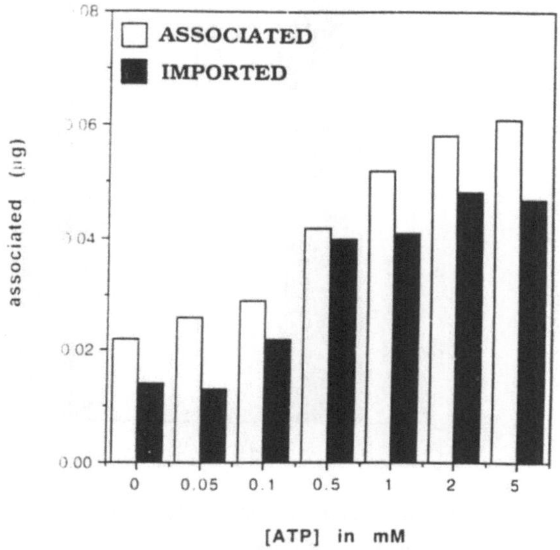

Fig. 9: ATP dependency of trfd import into chloroplasts. For details see (Van 't Hof and De Kruijff).

This ATP requirement for membrane translocation of a targeting signal is unique among the various cellular transport routes and supports the view that protein import into chloroplasts has very distinct features. In contrast to import of preFd, import of trFd is not inhibited by protease pretreatment of the chloroplasts (Fig. 10). The most simple explanation for this observation is that protease sensitive protein(s) on the outside of the outer membrane are involved in passage of the mature part of the precursor. Apparently no surface exposed receptor proteins are present for binding of the transit sequence. Instead it appears that the transit sequence either interacts with the outer membrane lipids or with protease insensitive components of the outer membrane. The first possibility requires that the transit sequence interacts with the outer membrane lipids.

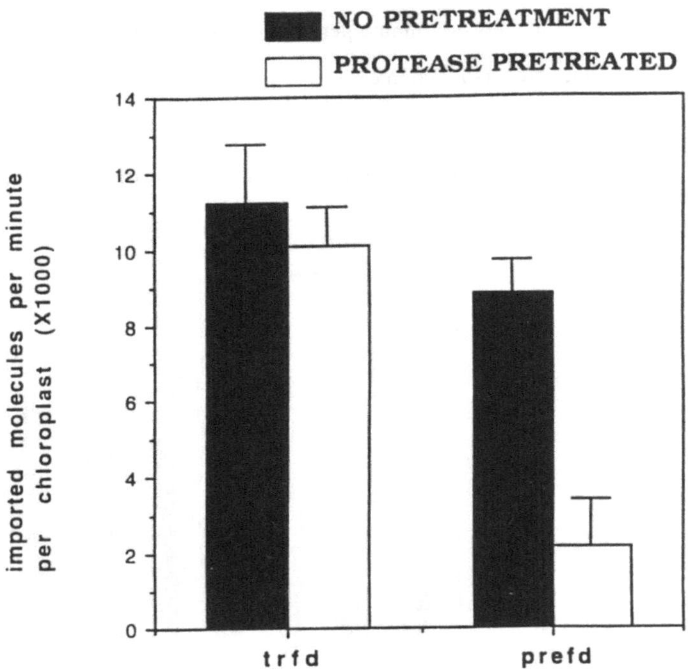

Fig. 10: Effect of thermolysin pretreatment of chloroplasts on trfd and preFd import. Mean values and the standard deviation of three individual experiments are shown. For details see (Van 't Hof and De Kruijff, 1995).

4. Transit sequence-lipid interactions

4.1 Monolayer insertion

Lipid extracts from biological membranes when spread at the air-water interface form monomolecular layers which are useful membrane models to analyze the possible affinity of a protein for lipids.

When pure preFd is injected underneath such a monolayer, prepared from the total lipids of the chloroplast outer membrane, the surface pressure rapidly increases and soon reaches a stable level (Fig. 11). This demonstrates that the precursor has an interaction with its target lipids and that this results in insertion of (part of) the precursor in between the lipid molecules thereby increasing the packing density which corresponds to the change in surface pressure.

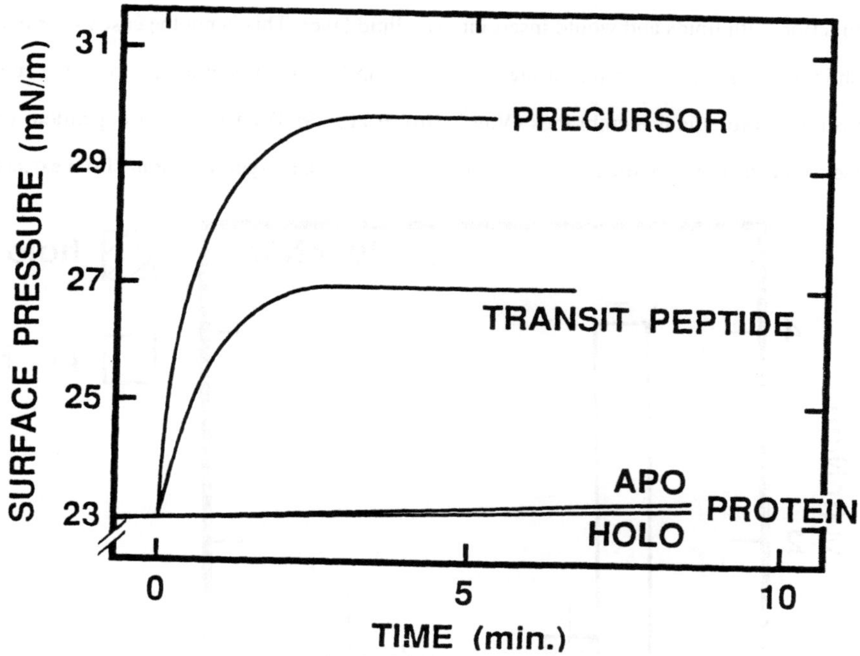

Fig. 11: Surface pressure changes induced by injection of precursor preFd, the corresponding transit peptide, apo Fd and holo Fd underneath a monolayer of the total lipid extract of the chloroplast outer envelope membrane. For details see (Van 't Hof et al., 1994).

The figure also demonstrates that this is not a general or aspecific effect of any protein. Injection of the unfolded apoFd or the folded holo protein does not cause a surface pressure increase at this initial surface pressure despite the fact that they have an identical polypeptide sequence as the mature part of the precursor. This strongly suggests that the lipid insertion of the precursor is due to the transit sequence. Indeed, injection of the corresponding transit peptide also causes a surface pressure increase which however is somewhat lower suggesting that the interaction of the transit sequence with lipids is enhanced by the mature part of the precursor.

4.2 Lipid specificity

It could now be argued that the transit sequence itself has an aspecific mode of interaction with lipids and would insert into any lipid layer. This is not the case as for instance is shown in Fig. 12. At an initial pressure of 30 mN/m which is assumed to be the relevant pressure of biological membranes (Van 't Hof et al., 1994) an insertion dependent on the presence of the transit sequence can be observed only in the target membrane lipid extract.

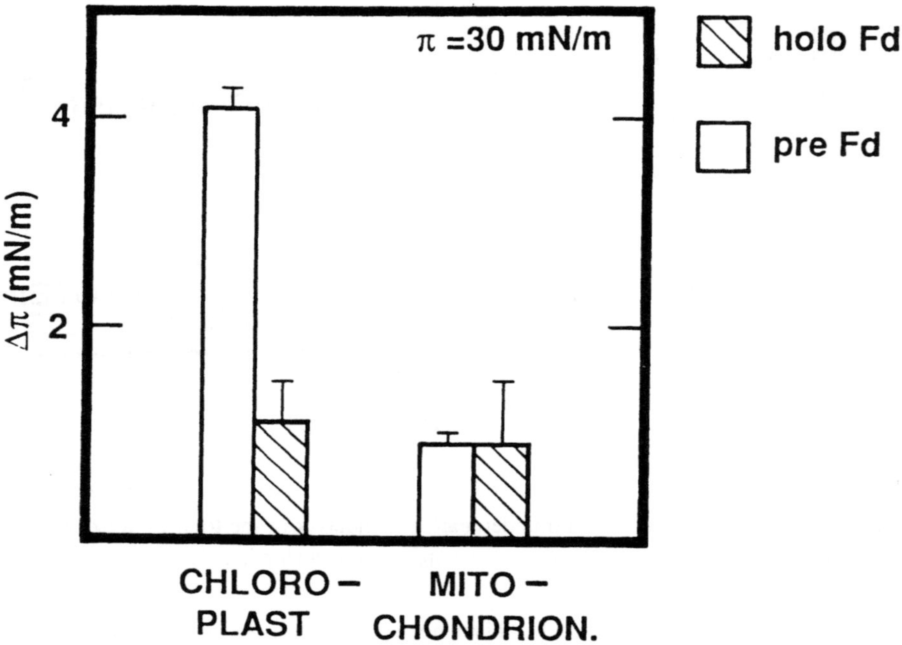

Fig. 12: Surface pressure increases induced by preFd and holo Fd when injected underneath lipid monolayers prepared from lipid extracts of chloroplast and mitochondrial outer membranes. The initial surface pressure was 30 mN/m. For details see (Van 't Hof et al., 1994).

No transit sequence dependent insertion occurs when the precursor is injected underneath a monolayer of the lipid extract of mitochondrial membranes. The mitochondrial lipid extract was chosen for such a comparison because protein import in mitochondria also occurs posttranslationally and is mediated by the presequence an N-terminal extension on mitochondrial precursor protein which is also an amphiphatic targeting sequence. If specific targeting sequence-lipid interactions would occur between the two systems then these could contribute, next to other factors, to organelle specific targeting.

A comparison of monolayer insertion of a typical presequence peptide and the transit peptide of Fd into their target lipid extracts is shown in Fig. 13 in the form of a Πi-$\Delta\Pi$ plot. With increasing initial surface pressure (Πi) the peptide induced surface pressure increase ($\Delta\Pi$) becomes less due to a closer lipid packing.

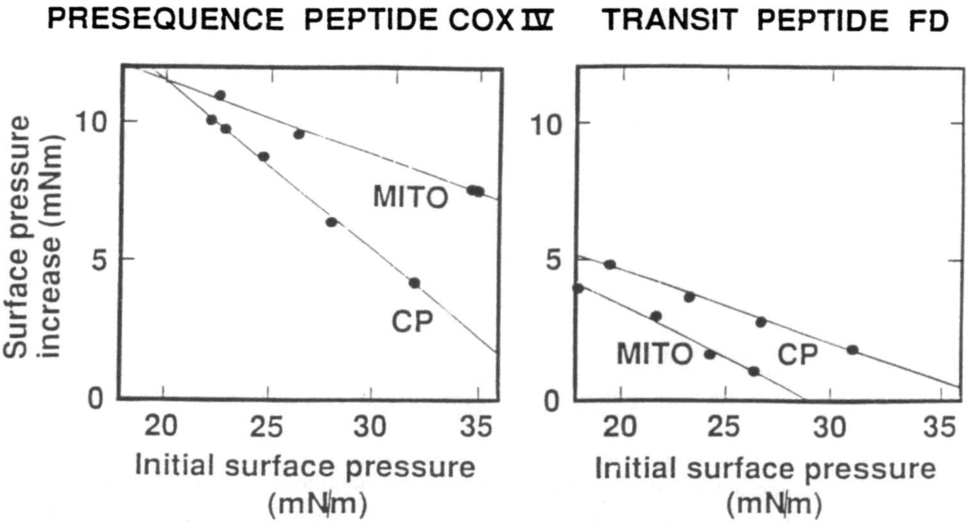

Fig. 13: Surface pressure increases induced by injection of $P_{25}L18W$ peptide (an analogue of the presequence of cytochrome oxidase subunit IV from yeast in which leu-18 is replaced by tryptophan) and trfd underneath monolayers of lipid extracts of chloroplast and mitochondrial outer membranes. The initial surface pressure is changed in the range indicated and the maximal polypeptide induced surface pressure increase is measured. For details see (Van 't Hof et al., 1994).

Like for the precursor, also the transit peptide preferentially inserts into its target lipids (right panel). Extrapolation of the Πi-$\Delta\Pi$ plot to $\Delta\Pi$=0 provides the limiting insertion pressure which is around 36 mN/m for insertion of the transit peptide into the chloroplast outer membrane extract. At the "physiological" surface pressure of around 30 mN/m no insertion can detected in the mitochondrial lipid extract. The presequence peptide has a much stronger capacity to insert into lipid layers because the resulting surface pressure increases are much higher (left panel). This might be due to the stronger amphiphilicity of the presequence peptide. Also the presequence peptide preferentially inserts into its target lipid extract, suggesting that indeed targeting sequence-lipid interactions could contribute to targeting of the precursor to the correct organelle. The preferential association of the targeting sequence with the lipid extract from the target membrane could be the result of either the overall properties of the lipid mixture or the presence of specific lipid classes. A way to discriminate between these possibilities is to analyze the peptide-lipid interactions for the various main lipid classes found in the outer membranes (Fig. 14).

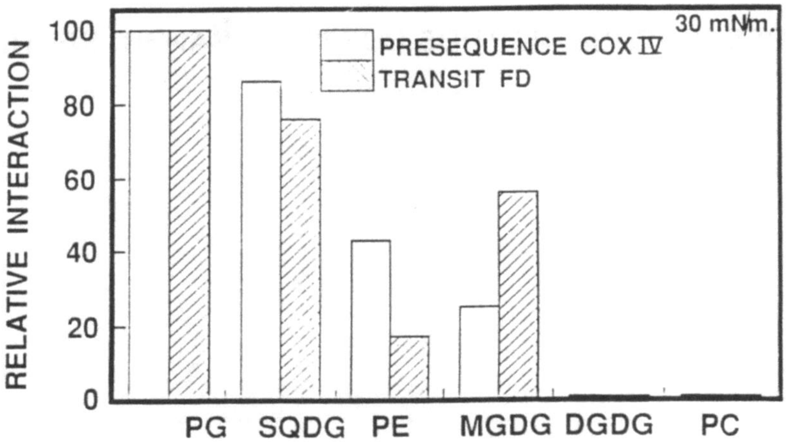

Fig. 14: The interaction of targeting peptides with monolayers prepared from several individual lipid classes. The dioleoyl species of PG, PE and PC are compared to SQDG, MGDG and DGDG from chloroplasts and the resulting surface pressure increases observed at an initial surface pressure of 30 mN/m are normalized to the value observed for PG. For details see the legend of Fig. 13 and (Van 't Hof et al., 1994).

The transit peptide inserts most efficiently in monolayers of PG, SQDG and MGDG suggesting that these lipid classes are mainly responsible for insertion into the target lipid extract and that the transit sequence-lipid interaction has electrostatic (the transit peptide is positively charged, whereas PG and SQDG are negatively charged) and other contributions. The strong interaction with the neutral MGDG is particularly intriguing. The transit peptide does not show an interaction with the related DGDG nor with PC which is zwitteronic but also has no charge overall. Comparing the overall pattern of lipid insertion of the transit and the presequence peptide learns that the anionic lipids are the strongest determinants for lipid insertion and that PE (which is an abundant lipid in mitochondria) and MGDG contribute most to the specific insertion of the peptides into the target lipid extract.

Peptides corresponding to the various parts of the transit sequence of the precursor of the small subunit of ribulose-1,5-biphosphate carboxylase/oxygenase display a similar lipid specific insertion as for preFd (Van 't Hof et al., 1991) which suggests that this is a general property of transit sequences. The interaction of preFd with MGDG showed some unexpected features. MGDG from chloroplasts is highly unsaturated and model membranes made from this lipid will be in a fluid state. Yet preFd also preferentially interacts with MGDG in the gel state (Demel et al., 1995) which indicates the involvement of specific head group interactions of which hydrogen bonding between the galactose group and the hydroxylated amino acids in the transit sequence are an obvious possibility. Surprisingly, the conformation of the sugar moiety and the sugar-glycerol linkage is of little influence to the preFd-monolayer interaction (Demel et al., 1995). However, methylation of position 3 of the galactose largely inhibits the interaction (Demel et al, 1995).

The mode of insertion of preFd into MGDG is also different from that in PG. In the glycolipid the penetrated domain of the precursor occupies an area of 650Å^2 when interaction was initiated at an initial surface pressure of the monolayer of 20 mN/m (Van 't Hof and De Kruijff, 1995). Despite the stronger interaction with PG (compare Fig. 14) the inserted domain only amounts to 400Å^2. This demonstrates that larger parts of the precursor insert into the glycolipid layer. The same specific affinity of the transit sequence for anionic lipids and MGDG was also observed for lipid bilayers in the form of large unilamellar vesicles (Van 't Hof and De Kruijff, 1995). That study also established that the preFd-bilayer interaction did not lead to a loss of barrier function of the bilayer and that the affinity of preFd for the lipid extract is less than the affinity of a precursor for chloroplasts.

A final intriguing aspect of the preFd-MGDG interaction is shown in Fig. 15. Very low concentrations of MGDG in PC monolayers already greatly enhance insertion of the transit

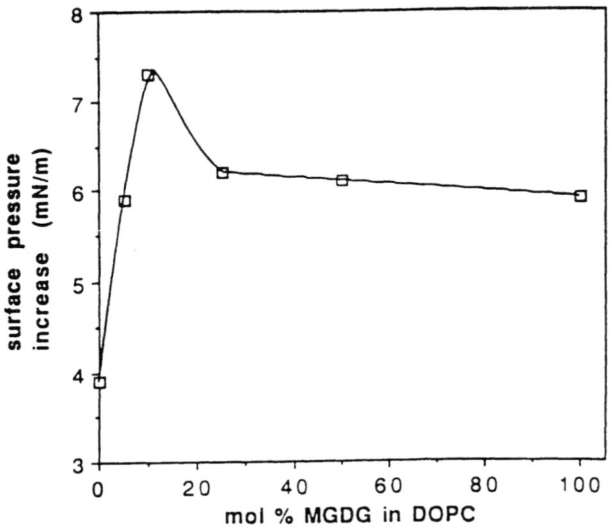

Fig. 15: Effect of the MGDG concentration in DOPC (dioleoyl phosphatidylcholine) monolayers on the interaction with trfd. The initial surface pressure was 20 mN/m. Reproduced with permission from (Van 't Hof et al., 1994). See that reference for further details.

peptide. The outer membrane of the chloroplast contains such low concentrations of this lipid which could be sufficient for efficient and specific insertion of the transit sequence between the lipids. This result suggests that the head group of MGDG could function as an insertion site for the transit sequence. In the section 4.4.2 it will be illustrated that mixed PG-MGDG systems have specific packing properties. The precise nature of the transit sequence-MGDG interaction remains to be determined.

4.3 Localization of insertion domains

In a search to localize the domains within the transit sequence which are involved in lipid insertion a selection of transit sequence deletion mutants of preFd were expressed in *E.coli* and purified (Pilon et al., 1995). The functionality of the resulting proteins was tested via their ability to compete for import of *in-vitro* synthesized wild type precursor into chloroplasts. The results obtained were fully consistent with the import studies of the *in-vitro* synthesized deletion mutants (Pilon et al., 1995) as they were summarized in section 3.1. The precursors were analyzed for their ability to insert into monolayers of either a total lipid extract of the outer membrane of chloroplasts, MGDG or DOPG (Fig. 16).

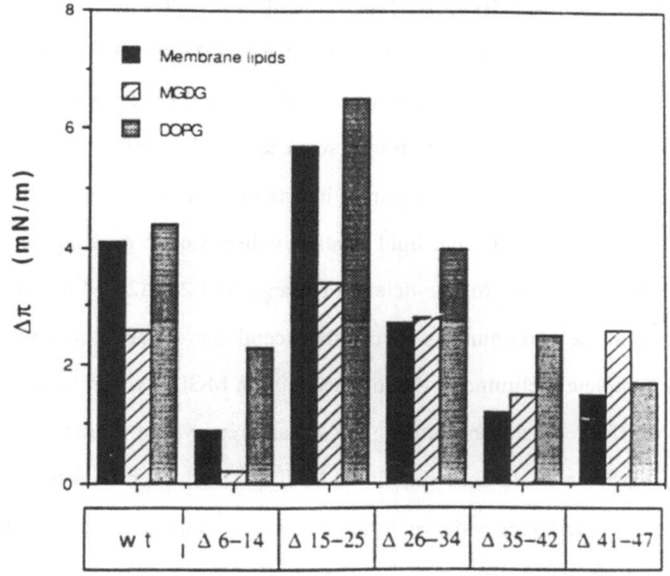

Fig. 16: Surface pressure increases caused by the injection of wild type (WT) preFd and a series of transit sequence deletion mutants of this precursor underneath monolayers of the indicated composition. The surface pressure increases at an initial surface pressure of 28 mN/m were determined from $\Pi_i / \Delta\Pi$ diagrames.

Most deletions cause a decrease in insertion ability into the lipid extract. This is not due to just a shortening of the transit sequence because the largest deletion tested (Δ15-25) inserts even more efficient then the wild type protein. This already suggests that specific domains of the transit sequence are responsible for insertion into the target lipid extract. The limiting surface pressure for insertion of WT preFd is around 35 mN/m (Van 't Hof et al., 1994) which is close to the value observed for the transit peptide (see Fig. 13). This is increased to 38 mN/m for Δ15-25 but is decreased to around 30 mN/m for the N-terminal (Δ6-14) and C-terminal (Δ41-47) deletion mutants. This implies that these mutants would be unable to penetrate a biological membrane. Because these mutants also have almost completely lost the ability to bind to chloroplasts (Fig. 6) the obvious suggestion is that these domains are involved in initial binding of the precursor to the surface of the chloroplast.

Interestingly Δ15-25 which is completely unable to import (Pilon et al., 1995) but which still has considerable activity to bind to chloroplasts also interacts strongly with lipids suggesting that the 15-25 domain is involved in translocation but less in binding. Analysis of the insertion of the deletion mutants into the MGDG and PG monolayers allowed to identify the regions responsible within the transit sequence for the preferential interaction with these lipids (Fig. 16). The wild type precursor inserts both in the PG and the MGDG monolayer with a preference for the former lipid as already discussed in previous sections. This pattern of specificity is also seen for the deletion mutants Δ15-25, Δ26-34 and Δ35-42. The N- and most C-terminal deletion mutants are clearly exceptional. Removal of amino acid sequence 6-14 nearly completely eliminates the interaction with MGDG which identifies this region as a putative MGDG recognition domain. In contrast, removal of amino acid residues 41 to 47 shifts the insertion preference mainly to MGDG which identifies this C-terminal region, which contains one lysine and two arginine residues, as the putative domain interacting with the negatively charged lipids.

4.4 Structural consequences

Up till now the transit-sequence-lipid interaction has only been described in terms of overall properties of the system without specifying the consequences of the interaction for the structure of the system. Insight into possible structural changes is likely to provide insight into the function of these interactions for the import process. Both aspects of the structure of the transit sequence and the structure of the lipids will be considered.

4.4.1 Transit sequence

The transit peptide of ferredoxin is like the complete precursor a relatively unstructured polypeptide as can be inferred from a circular dichroism analysis of solutions of these polypeptides (Pilon et al., 1992). Upon interaction with detergent micelles and membrane lipids in small unilamellar vesicles the transit sequence undergoes a two-state random coil-helix transition (Horniak et al., 1993). The CD spectra can be deconvoluted to estimate the α-helix content. Fig. 17 illustrates some typical results for a selection of detergents studied.

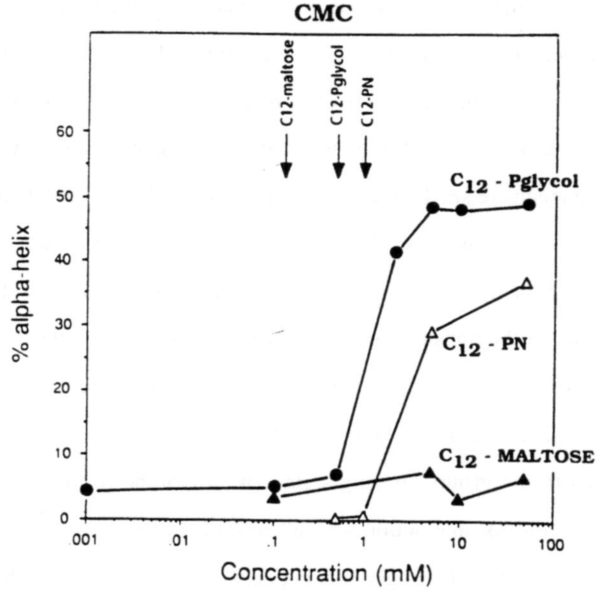

Fig. 17: Effect of different detergents on the α-helix content of trFd in buffer. The arrows indicate the critical micelle concentrations (cmc) of the detergents used and the α-helix is estimated from circular dichroism spectra. For details see (Horniak et al., 1993).

All detergents had an identical acyl chain but differed in their head group. The phosphoglycol (Pglycol) head group mimicks the head group of PG, C_{12}-PN has like PC a phosphocholine head group. C_{12}-maltose carries a sugar head group which models the MGDG head group. The figure shows that α-helix formation is induced by C_{12}-Pglycol and to a lesser extent also by C_{12}-PN. This random coil-helix transition thus requires the presence of a lipid-water interface. C_{12}-maltose micelles do not increase the α-helix content. Anionic detergents like C_{12}-Pglycol or lipids like PG and SQDG are the strongest α-helix promoters. They can convert up to 50% of the polypeptide into a helical conformation. The location of the induced α-helical parts within the sequence is not known. Speculations on its location are hampered by lack of clear predictions on the secondary structure. Secondary structure prediction programs fail to identify regions with high propensity to form defined secondary structures in the transit sequence of ferredoxin (Pilon et al., 1992) as well as in other transit sequences (Von Heijne and Nishikawa, 1991). The only precise information on transit sequence structure comes from a 2D-^{1}H NMR analysis of the transit peptide of ferredoxin from *chlamydomonas reinhardtii* (Lancelin et al., 1994). In water this transit peptide is also largely in a random coil conformation but in a more apolar environment the α-helix content is increased. The α-helix was found to be localized in the N-terminus. Transit sequences from this primitive organism are small and have characteristics of both mitochondrial presequences and higher plant transit sequences. Therefore, it can be questioned whether this result can be extrapolated to the transit sequence of ferredoxin from *S.pratensis* . But if this is the case an interesting property of the transit sequence is revealed (Fig. 18). A helical conformation of the first 12 amino acids of the transit sequence of ferredoxin would generate an amphipatic helix of special design. One side of the helix would be hydrophobic whereas the other would be formed by the hydroxyl group carrying serine and threonine residues. Such a helix would have a strong potential for interaction with glycolipids because the hydroxyl rich site of the helix could participate in the hydrogen binding network of the glycolipid head groups. Such a putative amphipathic helix can also be identified in the N-terminus (residues 6-13) of the small subunit of ribulose-1,5-biphosphate carboxylase/oxygenase. Whether this is a more general property of transit sequences is unclear. The observed structural flexibility of transit sequences is consistent with the view that transit sequences function in import by acquiring different structures in various steps of the envelope translocation process. Future research will have to give a more precise picture of the conformation of transit sequences.

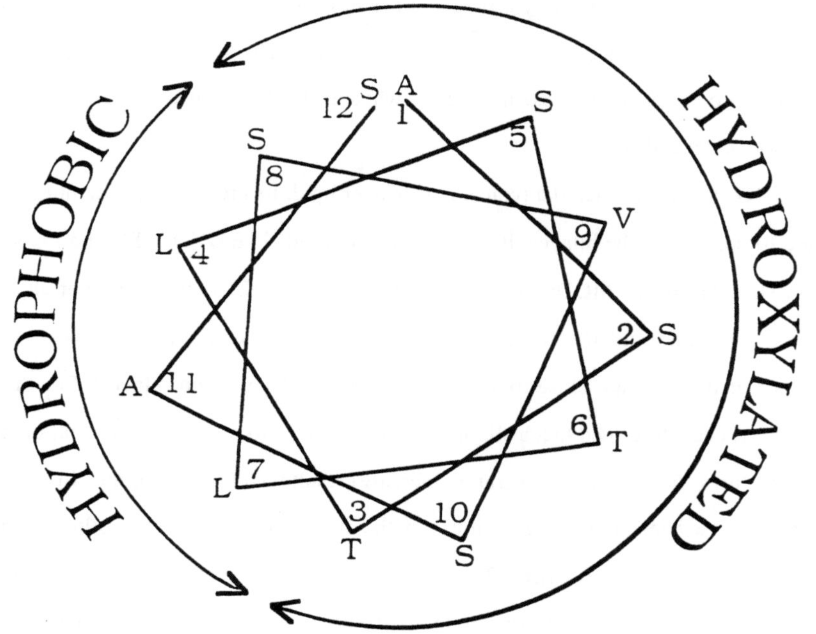

Fig. 18: Helical wheel representations of the 1-12 sequence of the transit sequence of ferredoxin.

4.4.2 Lipids

Given the specific interaction between the transit sequence and MGDG it is useful to consider in more detail the properties of this lipid and then to discuss the consequences of the transit sequence-MGDG interaction for the structure of the lipid. MGDG from plants is highly enriched in linolenic (18:3) acid. This poly-unsaturation makes that MGDG/water systems will be in a disordered (fluid) state at all temperatures. The galactose head group is small and appears to be oriented such that it is extended away from the membrane surface (Howard and Prestigard, 1995). Moreover, the head groups have a strong tendency to interact intermolecularly via hydrogen bonding. These properties are responsible for the phase properties of this lipid. In aqueous dispersion MGDG from plants does not organize in bilayers but instead forms an inverse hexagonal phase (H_{II} phase) (Shipley and Green 1973).

This phase consists of hexagonally arranged tubes in which the lipid head groups face inward and line the narrow aqueous channels present in the tubes. The geometry of the phase is such that the headgroups can be in close interaction whereas the unsaturated acyl chain can occupy a much more extended area.

MGDG can be considered a typical non-bilayer lipid. Its presence in a membrane can be expected to destabilize the bilayer. Bilayer preferring lipids like DGDG, PC and PG stabilize a bilayer structure of the MGDG. Membranes very rich in MGDG like the inner chloroplast envelope membrane and the thylakoid membrane system will form relatively unstable structures because they are close to a phase transition (bilayer › non-bilayer). For thylakoid systems it has been described that a slight shift in conditions to promote the formation of non-bilayer structures (for instance temperature increase) results in induction of the H_{II} phase.

There have been many suggestions on the functional role of non-bilayer lipids in membranes (Cullis and De Kruijff, 1979) including their involvement in membrane passage of proteins. Recently, the first experimental evidence is obtained that non-bilayer lipids are important for efficient protein transport across the *E.coli* inner membrane (Rietveld et al., 1995). The general believe is that the lipid packing properties arising from the presence of non-bilayer lipids in membranes are important for proper integration of proteins into membranes.

^{2}H NMR studies on mixed model membranes of ^{2}H acyl chain labeled synthetic MGDG and PC revealed another interesting property of MGDG (Chupin et al., 1994). ^{2}H NMR can provide information on the order of specific segments of a lipid molecule in an aqueous dispersion of that lipid. MGDG and PC were synthesized with oleoyl chains in which on the 11-position the protons were replaced by deuterons. The peak separation in the characteristic doublet type of the ^{2}H NMR spectrum is the residual quadrupolar splitting which is a measure of the local order. For pure $^2H_4$-PC bilayers a quadrupolar splitting of ~ 6.2 kHz is observed (see Fig. 19). Pure $^2H_4$-MGDG in water does not form bilayers but is in a H_{II} phase. The quadrupolar splitting of the ^{2}H NMR spectrum of the lipid in that phase is ~ 2kHz (Chupin et al., 1994) which small value reflects both the rapid diffusion of the MGDG molecules around the tubes of the H_{II} phase and the geometry of the phase allowing more disordered acyl chain conformations. MGDG can be incorporated in PC in bilayers up to 60 mol%.

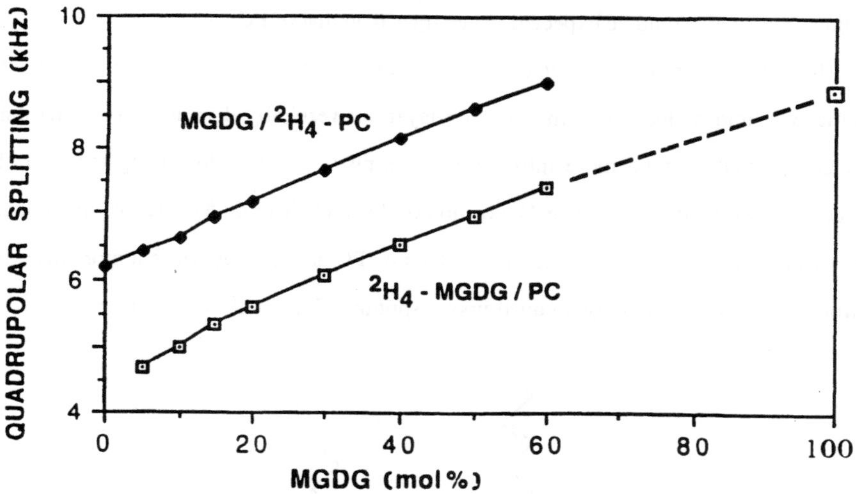

Fig. 19: Effect of the MGDG concentration on the quadrupolar splitting of mixed $^2H_4$-MGDG/PC (■) and MGDG/$^2H_4$-PC (u) dispersions. The values were obtained from 46.1 MHz ^2H NMR spectra of aqeous dispersions of these lipids. Reproduced from (Chupin et al., 1994).

By preparing two samples in which each lipid is alternatively labeled, insight into the individual behavior of each lipid can be obtained (Fig. 19). The quadrupolar splitting of $^2H_4$-MGDG increases with the MGDG concentration. Extrapolation predicts that the quadrupolar splitting of the acyl chains in a hypothetical pure MGDG bilayer would be around 8 kHz which is higher than the quadrupolar splitting of pure PC bilayer and reflects a tight acyl chain packing in a MGDG bilayer. This is exactly according to expectation. Because of the strong inter head group interactions and the unsaturated acyl chains the MGDG/water interface will try to acquire a convex (inverted) curvature but this is counteracted by the opposed monolayer resulting in a "frustrated" bilayer in which the acyl chains are more squeezed towards each other resulting in a more ordered system with larger quadrupolar splittings. The surprise lies in the fact that the quadrupolar splitting of $^2H_4$-MGDG in the mixture with PC is <u>smaller</u> than the value observed for PC in the mirror sample. If the mixture would behave in an ideal way then an <u>identical</u> quadrupolar splitting would be expected which would lineairy increase with the molar fraction of MGDG. Alternatively, if the two molecules in the mixture would maintain more their individual behavior the quadrupolar splitting of MGDG would be expected to be larger than that of PC.

How can this unexpected behavior be explained? It was proposed (Chupin et al., 1994) that the two lipid molecules in the mixed bilayer are vertically displaced such that the PC head group would extend more into the aqueous phase thereby displacing the acyl chain between the two lipid classes. The 11-position of the acyl chain on MGDG would be located more towards the methyl end of the oleoyl chain on PC and thereby sense a more disordered environment causing a decrease in quadrupolar splitting. Fig. 20 illustrates this model.

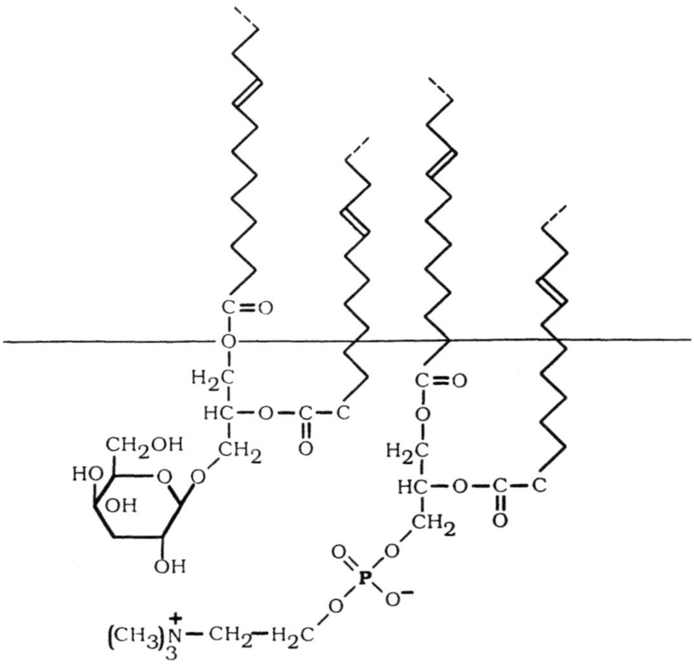

Fig. 20: A model for the organization of mixed PC-MGDG bilayers. The line is to indicate the direction of the surface of the bilayer. Both lipids are depicted with part of two oleoyl chains including the double bondbetween the 9-10 carbon position. The figure merely illustrates the vertical displacement of the two molecules. The precize conformation of the molecules is not known.

An intriguing possibility provided by the irregular hydrocarbon-polar interface is that it might provide potential insertion sites for the transit sequence. When the precursor of ferredoxin interacts with such mixed MGDG/PC bilayers a transit sequence dependent reorganization of both lipid molecules was revealed by ^{31}P and ^{2}H NMR spectroscopy (Chupin et al., 1994). The precursor induces a bilayer › isotropic transition for part of the lipid molecules. The isotropic structure was interpreted to be of an inverted nature. Isotropic structures are commonly encountered as intermediates between the bilayer and the H_{II} phase. This observation demonstrates that the transit sequence promotes the formation of type II non-bilayer lipid structures in this system. Extrapolation of this observation to chloroplasts protein import leads to the view that upon interaction of the transit sequence with the membrane lipids the lipid packing equilibrium is shifted towards a non-bilayer situation which could be important for the translocation process.

5. A model for import

The knowledge of transit sequence-lipid interactions summarized in this overview can be integrated in a model for import of precursor proteins into chloroplasts. The model pictured in Fig. 21 solely concentrates on transit sequence-lipid interactions in the outer membrane. Also, in the inner membrane such interactions might occur which similary could contribute to translocation and processing.

The model proposes that the newly synthesized precursor initially interacts via its transit sequence with cytosolically exposed lipid domains in the outer membrane. The interaction is two fold. The C-terminus of the transit sequence is assumed to interact with anionic lipids whereas the N-terminus interacts preferentially with chloroplast specific MGDG. The interaction is followed by insertion of both domains into the lipid layer. This is accompanied by the induction of a specific folding pattern and membrane topology of the transit sequence possibly involving an N-terminal helix. The local lipid structure is changed towards a non-bilayer organization. The membrane inserted transit sequence encounters via 2-dimensional diffusion the proteins of the transport apparatus. Here the transit sequence docks on the translocator complex with high affinity resulting in a productive interaction. The transit sequence translocates across the outer membrane in an ATP dependent manner. It is proposed

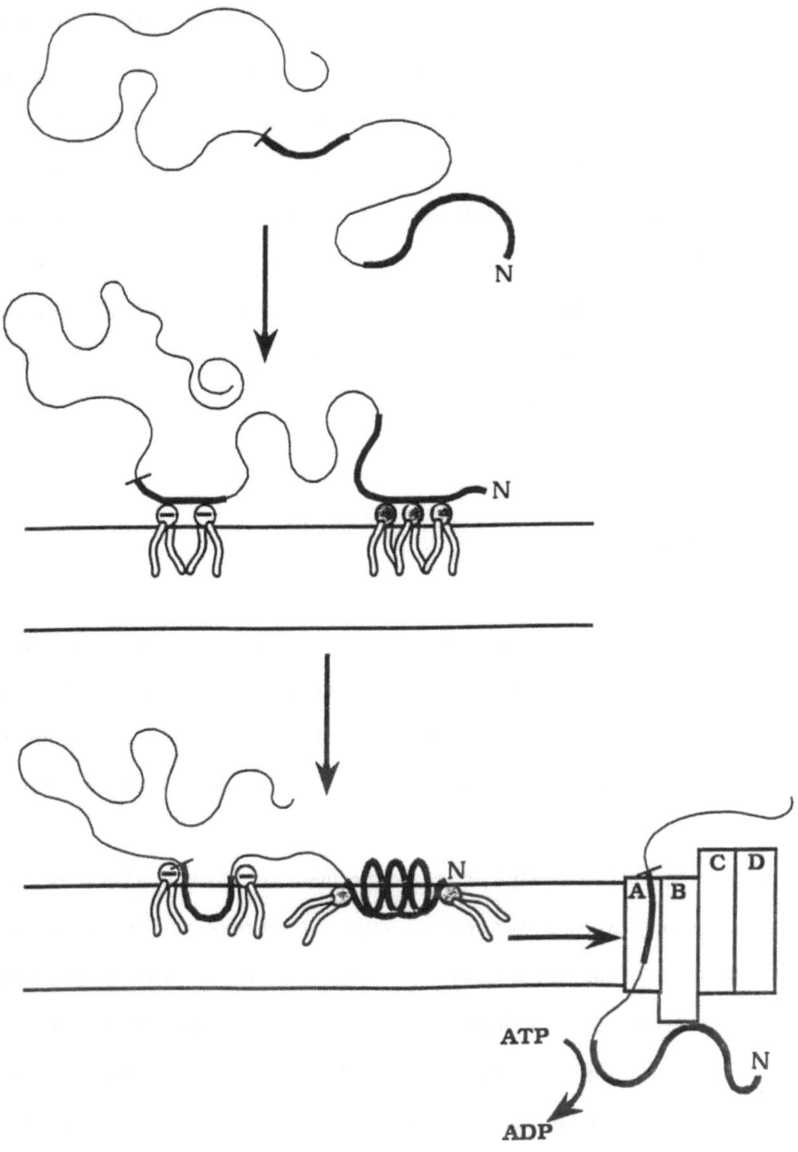

Fig. 21: A model of the initial stage of import of precursor proteins into chloroplasts.

that for these steps two 75 kDa protein (A or B) are responsible. These proteins were identified via cross linking and affinity purification approaches (Perry and Keegstra, 1994; Schnell et al., 1994) and do not expose domains to the cytosol. The B component is an integral protein and a member of the heatshock protein 70 family (Schnell et al., 1994) it exposes a domain to the inter membrane space and might be a protein responsible for the ATP dependency of protein translocation. The precursor is shown to interact with two other components of the translocation complex i.e. the 34 and 86 kDa proteins (Kessler et al., 1994). These are both GTP binding proteins and expose domains to the cytosol. They might act as unfoldases and reductases. The translocation complex might act as a pore to allow passage of the precursor across the outer membrane. In the inner membrane a separate translocation complex might be present which facilitates translocation of the precursor across that membrane. The main attractions of the model are that it gives some insight into the way transit sequences function in the early steps of the path way. The model explains organelle specific targeting of the precursor as achieved by a combination of specific protein-lipid and protein-protein interactions and provides a mechanism for efficient delivery of the incoming precursor to the translocation complex. Two dimensional diffusion over the membrane surface can deliver the precursor 3 orders of magnitude faster to a receptor then 3 dimensional diffusion through the aqceous phase.

6. Concluding remarks

Much attention has been paid in recent years to the lipid composition of chloroplasts in relation to chloroplast functioning and in particular to photosynthesis. It has been clearly established and is described in other sections of this book that a specific and well controlled composition is vital to optimal functioning of the chloroplasts.

Research from our group over the past years has identified another possible important role for chloroplast lipids. They interact specifically with the targeting sequences of proteins to be imported into chloroplasts. This suggests that they play an important role in the biogenesis of the organelle. The present view on the translocation apparatus and the transit sequence structure and function is still very limited. However, recent advances such as isolation of translocation complexes (Perry and Keegstra, 1994; Schnell et al., 1994; Kessler et al., 1994), introduction of multi-dimensional NMR techniques to determine the structure of lipid associated amphipathic peptides (Lancelin et al., 1994), the further developments in plant genetics and the possibility to analyze the chloroplast envelope directly via electrophysiological methods (Bulychev et al., 1994) offer the promise that in the near future a much more detailed picture of the process and the lipid-protein interactions involved will emerge.

Acknowledgements

Most of the research described in this review is the result of the collaboration between the groups of Peter Weisbeek and Ben de Kruijff on chloroplast protein import. The research was supported by grants from the University of Utrecht, the Institute of Biomembranes, the Life Science Foundation and the Netherlands Organization for Research. Monique Regter is thanked for preparing the manuscript and Nico van Galen for several of the drawings

References

Billecocq, A (1974) Structures of biological membranes: Localization of galactosyldiglycerides in chloroplasts by means of specific antibodies. Biochim Biophys Acta 352: 245--251

Bulychev, A, Pilon, M, Dassen, H, Van 't Hof, R, Vredenberg, W & De Kruijff, B (1994) Precursor-mediated opening of translocation pores in chloroplast envelopes. FEBS Lett 356: 204--206

Chupin, V, Van 't Hof, R & De Kruijff, B (1994) The transit sequence-dependent binding of the chloroplast precursor protein ferredoxin to lipid vesicles and its implications for membrane stability. FEBS Lett 350: 104--108

Cornwell, KL & Keegstra, K (1987) Evidence that a chloroplast surface protein is associated with a specific binding site for the precursor to the small subunit of ribulose-1,5-biphosphate carboxylase. Plant Physiol 85: 780--785

Cullis, PR & De Kruijff, B (1979) Lipid polymorphism and the functional roles of lipids in biological membranes. Biochim Biophys Acta 559: 399--420

De Boer, AD & Weisbeek, PJ (1991) Chloroplast topogenesis: Protein import, sorting and assembly.Biochim Biophys Acta 1071: 221--253

Demel, RA, De Swaaf, ME, Mannock, D, Van 't Hof, R & De Kruijff, B (1995) The specificity of glycolipid-preferredoxin interaction. Mol Membr Biol 12: 255--261

Dorne, A-J, Joyard, J, Block, MA & Douce, R (1985) Do thylakoids really contain phosphatidylcholine. Proc Natl Acad Sci USA 87: 71--74

Douce, R, Block, MA, Dorne, A-J & Joyard, J (1984) The plastid envelope membranes: their structure, composition and role in chloroplast biogenesis. Subcell Biochem 10: 1--84

Douce, R & Joyard, J (1990) Biochemistry and function of the platid envelope. Annu Rev Cell Biol 6: 173--216

Douglas, MG, McCammon, MT & Vassaroti, A (1986) Targeting proteins into mitochondria. Microbiol Rev 50: 166--178

Eilers, M & Schatz, G (1986) Binding of a specific ligand inhibits import of a purified precursor protein into mitochondria. Nature 322: 228--232

Ellis, RJ (1981) Chloroplast proteins: synthesis, transport and assembly. Annu Rev Plant Physiol 32: 111--137

Endo, T, Kawamura, M & Nakai, M (1992) The chloroplast-targeting domain of a plastocyanin transit peptide can form a helical structure but does not have a high affinity for lipid bilayers. Eur J Biochem 207: 671--675

Friedman, AL & Keegstra, K (1989) Quantitative analysis of precursor binding. Plant Physiol 89: 993--999

Gasser, SM, Daum, G & Schatz, G (1982) Import of proteins into mitochondria. Energy-dependent uptake of precursors by isolated mitochondria. J Biol Chem 257, 13034-13041

Gierash, LM (1989) Signal sequences. Biochemistry 28, 923--930

Hageman, J, Baecke, C, Ebskamp, M, Pilon, M Smeekens, S & Weisbeek, PJ (1990) Protein import into and sorting inside the chloroplast are independent processes. Plant Cell 2: 479--494

Hall, PO & Rao, KK (1977) in: Encyclopedia of Plant Physiology 5: Pirson A and Zimmerman MH, Eds 206--215

Hall, DQ & Rao, KK (1987) Photosynthesis 4th Ed Edward Arnold Ltd, London

Hay, R, Bahni, P & Gasser, S (1984) How mitochondria import proteins. Biochim Biophys Acta 779: 65--87

Hirsch S, Michael E, Heemeyer F, Von Heijne G and Soll J (1994) A receptor component of the chloroplast protein translocation machinery. Science 266: 1989--1992

Horniak, L, Pilon, M, Van 't Hof, R & De Kruijff, B (1993) The secondary structure of the ferredoxin transit sequence is modulated by its interaction with negatively charged lipids. FEBS Lett 334: 241--246

Hovius, R, Lambrechts, H, Nicolay, K & De Kruijff, B (1990) Improved methods to isolate and subfractionate rat liver mitochondria. Lipid composition of the inner and outer membrane. BBA 1021: 217--226

Howard, KP & Prestigard, JH (1995) Membrane and solution conformations of monogalactosyldiacylglycerol using NMR/molecular modeling methods. J Am Chem Soc 117: 5031--5040

Keegstra, K (1989) Transport and routing of proteins into chloroplasts. Cell 56, 247--253

Kessler, F, Blobel, G, Patel, HA & Schnell, DJ (1994) Identification of two GTP-binding proteins in the chloroplast protein import machinery. Science 266: 1035--1039

Lancelin, J-M, Bally, I, Arland, GJ, Blackedge, M, Gans, P, Stein, M & Jacquot, J-P (1994) NMR structure of ferredoxin chloroplastic transit peptide from Chlamydomonas reinhardtii promoted by trifluoroethanol in aqueous solution. FEBS Lett 350: 104--108

Muller, G & Zimmermann, R (1988) Import of honeybee prepromelittin into the endoplasmic reticulum: energy requirements for membrane insertion. EMBO J 7: 639--648

Neupert, W & Lill, R (1992) Elsevier, Amsterdam. Membrane biogenesis and protein targeting

Olsen, LJ, Theg, SM, Selman, BR & Keegstra, K (1989) ATP is required for the binding of precursor proteins to chloroplasts. J Biol Chem 264: 6724--672

Perry, SE & Keegstra, K (1994) Envelope membrane proteins that interact with chloroplast precursor proteins. The plant cell 6: 93--105

Pilon, M, de Kruijff, B & Weisbeek, PJ (1992) New insights into the import mechanism of the ferredoxin precursor into chloroplasts. J Biol Chem 267: 2548--2556

Pilon, M, Rietveld, AG, Weisbeek, PJ & de Kruijff, B (1992) Secondary structure and folding of a functional chloroplast precursor protein. J Biol Chem 267: 19407--19413

Pilon, M, Weisbeek, PJ & De Kruijff, B (1992) Kinetic analysis of translocation into isolated chloroplasts of the purified ferredoxin precursor. FEBS Lett 302: 65--68

Pilon, M, Wienk, H, Sips, W, De Swaaf, ME, Talboom, F, Van 't Hof, R, De Korte-Kool, G, Weisbeek, PJ & De Kruijff, B (1995) Functional domains of the ferredoxin transit sequence involved in chloroplast import. J Biol Chem 270: 3882--3893

Biochem Rapoport, TA (1992) Transport of proteins across the endoplasmic reticulum membrane. Science 258: 931--936

Rietveld, AG, Koorengevel, MC & De Kruijff, B (1995) Non-bilayer lipids are required for efficient protein transport across the plasma membrane of *Escherichia coli*. EMBO J: in press

Robinson, G & Ellis, RJ (1984) Transport of proteins into chloroplasts. Partial purification of a chloroplast protease involved in the processing of imported precursor polypeptides. Eur J 142: 337--342

Roise, D & Schatz, G (1988) Mitochondrial presequences. J Biol Chem 263: 4509--4511

Sanders, SL & Schekman, R (1992) Polypeptide translocation across the endoplasmic reticulum membrane. J Biol Chem 267, 13791-13794

Schleyer, M, Schmidt, B & Neupert, W (1982) Requirement of a membrane potential for the posttranslational transfer of proteins into mitochondria. Eur J Biochem 125: 109--116

Schnell, DJ, Blobel, G & Pain, D (1990) The chloroplast import receptor is an integral membrane protein of chloroplast envelope contact sites. J Cell Biol 111: 1825--1838

Schnell, DJ, Kessler, F & Blobel, G (1994) Isolation of components of the chloroplast protein import machinery. Science 266: 1007--1012

Seedorf, M & Soll, J (1995) Copper chloride, an inhibitor of protein import into chloroplasts. FEBS Lett 367: 19--22

Shipley, GG, Green, JP & Nichols, BW (1973) The phase behavior of monogalactosyl, digalactosyl and sulphoquinovosyl. Biochem Biophys Acta 311: 531--544

Smeekens, S, van Binsbergen, J & Weisbeek, PJ (1985) The plant ferredoxin precursor: nucleotide sequence of a full length cDNA done. Nucleic Acid Res 13: 3179--3194

Smeekens, S, van Steeg, H, Bauerle, C, Bettenbroek, H, Keegstra, K & Weisbeek, PJ (1987) Import into chloroplasts of a yeast mitochondrial protein directed by ferredoxin and plastocyanin transit peptides. Plant Mol Biol 9: 377--388

Sugiura, M (1992) The chloroplast genome. Plant Mol Biol 19: 149--168

Theg, SM, Bauerle, C, Olsen, LJ, Selman, BR & Keegstra, K (1989) Internal ATP is the only energy requirement for the translocation of precursor proteins across chloroplastic membranes. J BiolChem 264: 6730--6736

Umesono, K & Ozeki, H (1987) Chloroplast gene organization in plants. Trends Genet 3: 281--287

Van 't Hof, R, Demel, RA, Keegstra, K & De Kruijff, B (1991) Lipid-peptide interactions between fragments of the transit peptide of ribulose-1,5-bisphosphate carboxylase/oxygenase and chloroplast membrane lipids. FEBS Lett 291: 350--354

Van 't Hof, R, Van Klompenburg, W, Pilon, M, Kozubek, A, De Korte-Kool, G, Demel, RA, Weisbeek, PJ & De Kruijff, B (1994) The transit sequence mediates the specific interaction of the precursor of ferredoxin with chloroplast envelope membrane lipids. J Biol Chem 268: 4037--4042

Van 't Hof, R & De Kruijff, B (1995) Transit sequence-dependent binding of the chloroplast precursor protein ferredoxin to lipid vesicles and its implications for membrane stability. FEBS Lett 356: 204--206

Van 't Hof, R & De Kruijff, B (1995) Characterization of the import process of a transit peptide into chloroplasts. J Biol Chem (in press)

Von Heijne, G, Steppuhn, J & Herrmann, RG (1989) Domain structure of mitochondrial and chloroplast targeting peptides. Eur J Biochem 180: 535-545

Von Heijne, G & Nishikawa, K (1991) Chloroplast transit peptides. The perfect random coil? FEBS Lett 278: 1--3

Waegemann, K, Paulsen, H & Soll, J (1990) Translocation of proteins into isolated chloroplasts requires cytosolic factors to obtain import competence. FEBS Lett 261: 89--92

PROTEIN TRANSPORT INTO AND ACROSS THE MITOCHONDRIAL OUTER MEMBRANE: RECOGNITION, INSERTION AND TRANSLOCATION OF PREPROTEINS

Roland Lill*, Gyula Kispal, Klaus-Peter Künkele, Andreas Mayer, Bernd Risse, Harald Steiner, Petra Heckmeyer, Ida Van der Klei#, and Deborah A. Court

Institut für Physiologische Chemie der Universität München,
Goethestrasse 33, 80336 München, Federal Republic of Germany

Biological Sciences, University of Groningen
Laboratory for Electron Microscopy, Kerklaan 30, 9571 NN Haren,
The Netherlands

* Corresponding author : Phone +49-89-5996 304
Fax +49-89-5996 270
E-mail: Lill@bio.med.uni.muenchen.de

Introduction

The compartmentation of cells requires the accurate subcellular sorting of proteins after their synthesis on cytoplasmic ribosomes. Any cellular membrane possesses a distinct transport machinery which specifically recognizes, inserts and translocates precursor proteins destined for this particular membrane or organelle (for reviews see articles in Neupert and Lill, 1992). For instance, preproteins are inserted into the membrane of the endoplasmic reticulum and become translocated into the lumen by the action of the Sec61 and Sec63 complexes (see Rapoport et al., this issue). For mitochondria, the transport process is particularly complex, since these organelles are bounded by two

NATO ASI Series, Vol. H 96
Molecular Dynamics of Biomembranes
Edited by Jos A. F. Op den Kamp

membranes, the outer and inner membrane which enclose the intermembrane space and the matrix, and separate the mitochondrion from the cytoplasm. The sub-compartmentation requires distinct sub-organellar sorting pathways which ensure the accurate distribution of the proteins. Despite the complexity of the sorting process, our knowledge about the mechanisms and the machinery of mitochondrial protein translocation has well advanced during the last decade (for recent reviews see Pfanner and Neupert, 1990; Glick and Schatz, 1991; Segui-Real et al., 1993b; Schwarz and Neupert, 1994). In this contribution, we will focus mainly on our investigations on mechanistic aspects underlying protein transport across the mitochondrial outer membrane. The identification and properties of the components of the receptor complex have been reviewed in depth elsewhere (Kiebler et al., 1993a; Lill et al., 1994). For a detailed description of the events occurring at the mitochondrial inner membrane, the reader is referred to another article in this issue (Schneider et al.).

Protein translocation into and across the mitochondrial outer membrane is facilitated by a large complex (the "MOM complex" or "receptor complex"; Söllner et al., 1992) composed of at least 6 proteins (the MOM proteins; Fig. 1).
According to their apparent sensitivity or resistance to attack by external proteases, these components have been classified as "receptors", exposing large domains at the mitochondrial surface, or as constituents of the "general insertion pore (GIP)". Both these terms should be considered operational, and only recently, we have begun to learn more about the detailed function these proteins perform in the translocation process. The topographical arrangement of the surface-exposed proteins is known. MOM19 and MOM72 (the numbers indicate the apparent molecular mass of the MOM proteins) are anchored via an N-terminal hydrophobic segment to the membrane (Söllner et al., 1989; Söllner et al., 1990). MOM22 is attached to the membrane through a central hydrophobic segment with the N-terminus facing into the cytosol and the C-terminal portion exposed to the intermembrane space (Kiebler et al., 1993b). Information about the structure of the membrane-embedded components of the MOM complex is scarce. MOM8 (Isp6p; Kassenbrock et al., 1993) seems to be anchored to the membrane by one trans-membrane segment leaving a short N-terminal segment (about 30 amino acid residues) exposed to the cytosol. As MOM38 is lacking obvious membrane-spanning regions (Kiebler et al., 1990), a prediction of its topography is rather difficult. As a possible structural arrangement, a ß-barrel conformation was suggested (Court et al., 1995). However, this idea has not been proven experimentally.

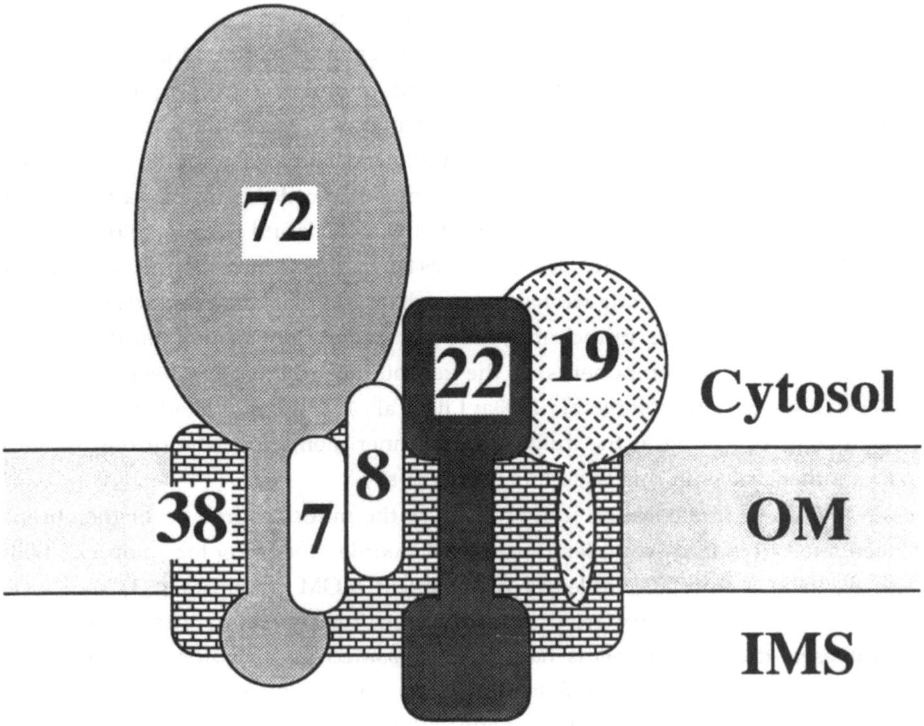

Figure 1: The MOM complex involved in preprotein insertion and translocation at the mitochondrial outer membrane. The numbers define the various MOM proteins according to their apparent molecular mass. The model accounts for the topographical orientation of the individual proteins. However, with a few exceptions the exact protein-protein interactions within the complex are unknown. For details see text. The homologous proteins of the yeast mitochondrial receptor complex are also referred to as Isp6 (=MOM8; Kassenbrock et al., 1993), Mas20p (=MOM19; Ramage et al., 1993; Moczko et al., 1994), Mas22p (=MOM22; Lithgow et al., 1994), Isp42p (=MOM38; Baker et al., 1990), and Mas70 (=MOM72; Steger et al., 1990; Hines et al., 1990). Abbreviations used: OM outer membrane; IMS, intermembrane space.

Previous perceptions of protein transport into mitochondria assumed that the outer membrane translocation machinery could only act in cooperation with the inner membrane. This was based on the observations that protein import across the outer membrane requires the presence of a membrane potential, $\Delta\Psi$, at the inner membrane, and that translocation across the two membranes is strictly coordinated in time and space (see Glick et al., 1991; Pfanner et al., 1992; Segui-Real et al., 1993a for detailed discussion). Studies during the past few years, however, have demonstrated the existence of individual translocation machineries in the two mitochondrial membranes (Hwang et al., 1989; Mayer et al., 1993; Horst et al., 1995; Berthold et al., 1995) which can act independently of each other (Jascur et al., 1992; Segui-Real et al., 1993a). Crucial for the description of the independent action of the outer membrane translocation machinery were detailed investigations of the import pathway taken by a protein of the intermembrane space, cytochrome *c* heme lyase (CCHL; Lill et al., 1992). We therefore will first outline aspects of the biogenesis of this protein and the homologous cytochrome c_1 heme lyase (CC_1HL; Steiner et al., 1995). Then, we will describe studies on protein insertion and translocation using highly purified outer membrane vesicles (OMV; Mayer et al., 1993). These investigations served to initially identify the endogenous translocation activity of isolated outer membranes, and allowed further characterization of its substrate specificity. Finally, we will summarize our current functional knowledge on the mechanism of the protein transport across this membrane (Mayer et al., 1995b) and on the particular role of individual constituents of the MOM complex (Mayer et al., 1995a).

Biogenesis of mitochondrial heme lyases

Heme lyases are enzymes which catalyze the covalent attachment of heme to the apoforms of *c*-type cytochromes. The proteins are located in the intermembrane space, where they are bound to the outer face of the inner membrane (Fig. 2). The majority of the protein is associated with the cristae membranes. At present, structural information for three enzymes is available, namely for CCHL from *Neurospora crassa* (Drygas et al., 1989) and *Saccharomyces cerevisiae* (Dumont et al., 1988), and for CC_1HL from *S. cerevisiae* (Zollner et al., 1992). The enzymes share 35% sequence identity (50% similarity); yet they are distinct in their substrate specificity, i.e. each heme lyase can only catalyze the covalent heme attachment to its own substrate. A characteristic feature of heme lyases is the lack of typical mitochondrial presequences at their N-termini. Most other enzymes of the intermembrane space like cytochromes b_2 and c_1 or nucleoside diphosphate kinase are synthesized with such targeting signals. The lack of an N-terminal

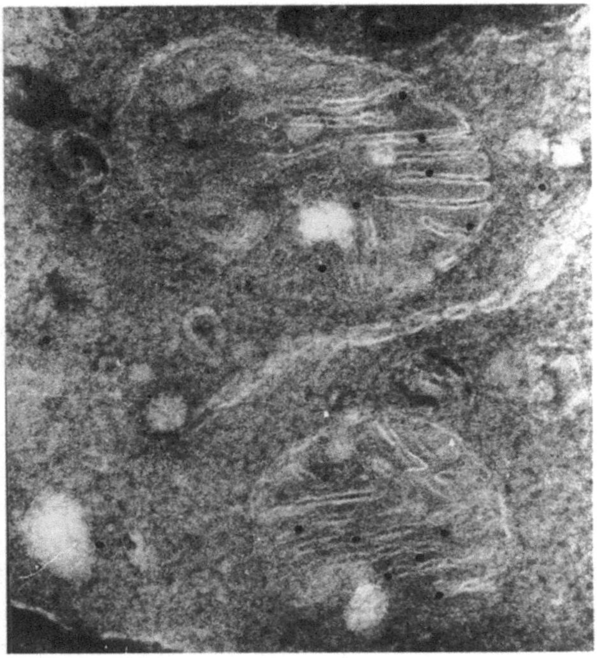

Figure 2: Submitochondrial localization of cytochrome *c* heme lyase (CCHL) from *Neurospora crassa*. Immuno-gold labelling of CCHL (Van der Klei et al., 1994) indicates the predominant localization at the cristae membranes. Additional biochemical data suggest CCHL to be associated with the outer face of the mitochondrial inner membrane (Segui-Real et al., 1993a).

targeting signal is shared by a few constituents of the inner membrane, such as the ADP/ATP translocator (AAC), the phosphate translocator and a few subunits of the electron transport chain complexes. All these proteins require a membrane potential across the inner membrane, indicating the participation of the mitochondrial inner membrane translocation machinery in guiding these preproteins to their final submitochondrial location. A hallmark distinguishing heme lyases from these preproteins is the independence of the heme lyase import into the intermembrane space on a membrane potential.

Biochemical studies have provided us with a detailed picture of the import pathway of heme lyases (Fig. 3; Lill et al., 1992; Steiner et al., 1995). The proteins are synthesized on cytoplasmic ribosomes. The chaperones binding to these

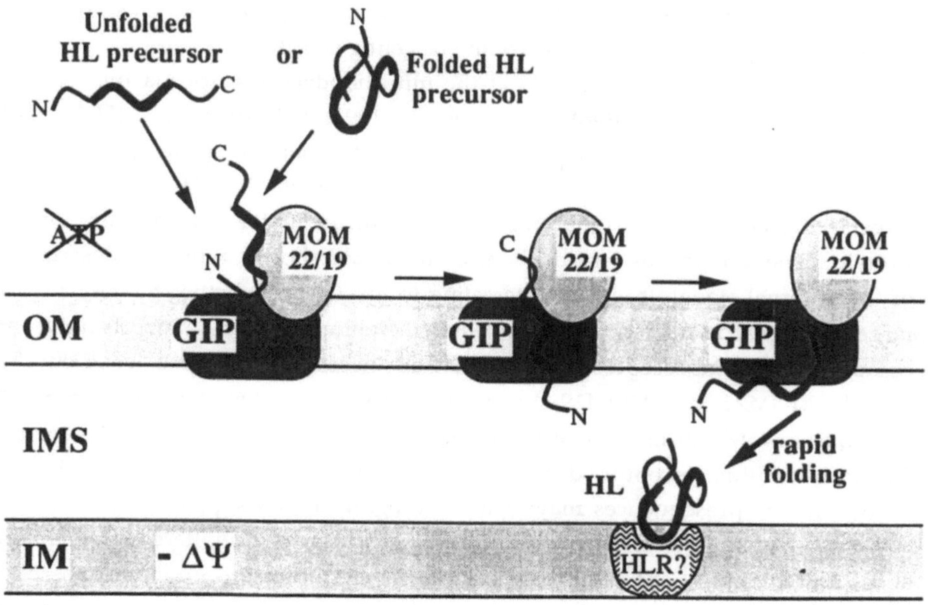

Figure 3: Biogenesis of mitochondrial heme lyases (HL). The model depicts the events leading to recognition, import and folding of these proteins. For further explanations see text. Abbreviations used in addition to those given in Fig. 1: GIP, general insertion pore; IM, inner membrane; HLR, putative heme lyase receptor at the inner membrane.

precursors are still unknown. The independence of the import process on the presence of nucleoside triphosphates makes it unlikely that Hsp70 (Deshaies et al., 1988) or MSF (Hachiya et al., 1994) are involved in targeting, since both proteins use ATP for their action. In the case of heme lyases, the function of a chaperone might not be required for maintaining an "open" conformation, since the folded proteins can also cross the membrane (Steiner et al., 1995). Rather, a chaperone may potentially suppress the high tendency of heme lyases to unspecifically associate with membranes and even liposomes. Furthermore, a chaperone may contribute to increase the targeting specificity.

Heme lyases are recognized at the mitochondrial surface via the receptor proteins MOM22 and MOM19 (Lill et al., 1992; Kiebler et al., 1993b; Lill, unpublished). The exact nature of the targeting signal in heme lyases is still unknown. Preliminary studies using truncated versions of CCHL from *N. crassa* have indicated that the signal should be localized in the central third of the protein. Deletions in this part of the protein should shortly allow the description of this internal signal. It will be interesting to investigate whether the heme lyase targeting signal is related to that of outer membrane proteins. Surprisingly, the components of the receptor complex that are necessary for heme lyase import, namely MOM22 and MOM19, have been shown to interact with N-terminal presequences (Mayer et al., 1995a; see below). This opens two possibilities for the function of this receptor during heme lyase import. Features similar to mitochondrial presequences may also be recognized as a signal in heme lyases. Alternatively, another feature within the heme lyases is deciphered, most likely at a different site of the receptor. Direct binding studies are underway to solve these important questions. From the receptors, the heme lyases are transferred across the membrane (Fig. 3). The N-terminus becomes translocated first creating a transport intermediate in which the C-terminus is still outside the translocation site and can be prevented from undergoing membrane passage by the binding of specific antibodies against this region of the protein (R. Lill et al., in preparation).

Translocation of heme lyases into the intermembrane space does not require external energy sources such as ATP or a membrane potential, $\Delta\Psi$, across the inner membrane (Lill et al., 1992; Steiner et al., 1995). This raises the important question of what drives the translocation process. Previously, we had suggested as an attractive possibility the energy derived from folding of the protein after its entry into the intermembrane space (Lill et al., 1992). To directly test this proposal, we initiated an investigation on the folding of heme lyases in the intermembrane space. Up to now, for this compartment there exists no information as to how endogenous proteins become folded. The generation of a large, folded domain of CC_1HL was used to study the folding process (Steiner et al., 1995). Folding occurs at the same rate as import indicating that heme lyases become folded either during or immediately after their transfer across the outer membrane. Folding is not affected by depletion of ATP and $\Delta\Psi$, or by inhibitors of peptidylprolyl *cis-trans* isomerases, i.e. it does not involve homologs of known folding factors. We conclude that protein folding in the intermembrane space obeys principles different from those established for other subcellular compartments such as the mitochondrial matrix or the cytoplasm (for reviews see Stuart et al., 1994; Hendrick and Hartl, 1993). The folded domain of CC_1HL was present in small amounts after synthesis in reticulocyte lysate. This folded protein could be imported into mitochondria with the same efficiency as the unfolded protein (Fig. 3). Thus, the net energy gained from folding of the polypeptide chain is zero, ruling out folding as a potential driving force for import.

Which forces may then promote the accumulation of heme lyases in the intermembrane space? We propose that heme lyases, during the targeting process, undergo a series of interactions with increasing affinity. After the initial interaction with the outer surface of the mitochondria, heme lyases may be transferred to the inner face of the membrane (Fig. 3). This reaction may be the energetic basis for the import of CCHL observed with isolated OMVs (see below). Then, a specific interaction with the inner membrane may direct the protein to the outer face of the inner membrane, the functional location of heme lyases. As a specific binding partner, one might envisage a component of the inner membrane (a putative "heme lyase receptor"; Fig. 3). Alternatively, the unique lipid composition of the inner membrane may be responsible for assuring the final localization of the enzyme.

In summary, mitochondrial heme lyases are passed into the intermembrane space along a direct sorting pathway using the general protein transport machinery of the outer membrane. Unlike other proteins of the intermembrane space like cytochromes b_2 and c_1, import occurs independently of the inner membrane. Thus, heme lyases define a distinct sorting pathway which most likely is used by other components of the intermembrane space.

Protein transport into and across the isolated mitochondrial outer membrane

For the mechanistic study of protein transport across the mitochondrial outer membrane, we set up a translocation system with isolated outer membrane vesicles (OMV). These vesicles were purified from mitochondria after detaching the outer membrane from the organelles by swelling under hypotonic conditions (Mayer et al., 1993). The outer membrane was purified more than 1000-fold over the inner membrane by sucrose density gradient centrifugation. The isolated membranes form vesicles which are tightly sealed and in a right-side-out orientation. The known components of the protein import complex are major constituents of the purified outer membrane suggesting that protein transport is a central function of this membrane.

For the investigation of protein insertion of outer membrane proteins, the classical protein import assays cannot generally be adopted. First, proteins of this membrane do not contain cleavable targeting signals. Second, many of these proteins expose domains to the cytosol and therefore are accessible to added protease. Thus, alternative ways to monitor the insertion into the membrane had to be developed. In many cases, the acquired resistance of imported material to extraction under alkaline conditions can be used. This method was applied successfully for proteins such as MOM72 and Bcl-2 (Nargang et al., 1995; Nguyen et al., 1993). A powerful tool to follow the insertion into the membrane is the generation of proteolytic fragments upon import. The unassembled form of the precursor usually does not give rise to characteristic fragments (or at least at much lower protease concentrations). Therefore, the assay yields additional information about the proper folding after integration into the membrane. Such an assay has been used to describe the import mechanism for MOM19 (Schneider et al., 1991) and MOM22 (Keil and Pfanner, 1993). The major component of the outer membrane, porin, on the contrary, attains a conformation in the membrane which is highly resistant to proteases. This property makes it feasible to employ the protease protection assay allowing the convenient distinction of folded porin from uninserted, i.e. unfolded precursor which is highly sensitive to proteolytic attack (Kleene et al., 1987).

We used these assays to study the competence of isolated OMV for protein insertion (Mayer et al., 1993). Import of a number of precursors destined for the mitochondrial outer membrane occurred efficiently and fulfilled all the criteria established in import studies with intact mitochondria. For instance, porin insertion was fully dependent on the presence of ATP and required MOM22 and MOM19 function. Similarly, MOM22 insertion was abolished after inactivating both MOM19 and MOM72 receptors. These studies demonstrated that the outer membrane is fully competent for the insertion of endogenous proteins. Thus, the outer membrane contains a protein transport apparatus which can insert membrane proteins independently of the inner membrane.

Can the outer membrane also promote translocation of precursor proteins across the membrane and accumulate preproteins in the lumen of the OMV? Since heme lyases reach the intermembrane space without the need for $\Delta\Psi$, this question was tested with *N. crassa* CCHL. Efficient import was observed at rates comparable to those seen with intact mitochondria (Mayer et al., 1993). Transport required the function of MOM19 and MOM22. OMV extracted under alkaline conditions are fully active in translocation showing that soluble proteins, e.g., of the intermembrane space are not involved in the transport reaction (A. Mayer, unpublished). Thus, the translocation apparatus of the mitochondrial outer membrane is competent to translocate certain preproteins without a requirement of components of the inner membrane and in the absence of a direct contact to the inner membrane. For CCHL, the results unequivocally demonstrate that membrane passage does not involve components of the inner membrane, even if such components would act in the absence of a membrane potential. Where was CCHL bound after the import reaction? We would like to speculate that binding specifically occurs to a binding site on the inner face of the translocation machinery (see Fig. 3). The higher affinity of this site as compared to the surface interaction may be the reason for passage of CCHL across the translocation pore. The internal binding site may be identical or overlapping to the presequence-specific *trans*-site discussed in the following chapter.

The molecular mechanism of protein transport across the mitochondrial outer membrane

How do preproteins carrying an N-terminal presequence become translocated across the outer membrane? Since, in intact mitochondria, there is an absolute requirement of $\Delta\Psi$ for import of these preproteins, even for translocation across the outer membrane, it was not surprising that these preproteins did not become imported into purified OMV (as judged from their protease sensitivity, Mayer et al., 1993). We observed, however, that the presequence portion became exposed to the intermembrane space (Mayer et al., 1995b), a prerequisite for further recognition and transport across the inner membrane which is facilitated by the endogenous translocation apparatus of this membrane. To monitor such a partial translocation of the N-terminal part of the preprotein, an analytical trick was employed. Purified matrix processing peptidase (MPP) was enclosed in the lumen of the OMV and used to test the appearance of the cleavage site of the preprotein in the lumen of the vesicles (corresponding to the intermembrane space of intact mitochondria). These and other related biochemical experiments yielded a number of detailed mechanistic insights into how preproteins carrying N-terminal presequences become translocated across the membrane and are presented to the inner membrane for further transport. These observations are summarized in the model depicted in Fig. 4 and will be explained in the following.

Recognition of the presequence at the mitochondrial surface

Binding of preproteins to the mitochondrial surface is difficult to determine directly. With isolated mitochondria, there is a high degree of unspecific binding of preproteins. Furthermore, as outlined below preproteins cannot be prevented from transferring at least part of their presequence across the membrane. We circumvented these obstacles i) by using OMV which show low levels of unspecific binding and ii) by employing preproteins which can be prevented from membrane insertion by stabilizing the folded state of their mature part (Mayer et al., 1995a). For the latter purpose, fusion proteins between mitochondrial presequences and dihydrofolate reductase (DHFR) were used which can be stabilized by folate antagonists like methotrexate. The initial interaction of preproteins at the mitochondrial surface studied this way occurs specifically with the presequence and is mediated by protease-sensitive components ("receptors"). Binding to this so-called *cis*-site (Mayer et al., 1995b) was abolished by blocking either MOM19 or MOM22 by prebinding specific antibodies (Fig. 4). Likewise, only residual non-specific binding was seen using OMV isolated from MOM19 and MOM22 null mutants lacking the respective protein (Harkness et al., 1994; Nargang et al., 1995). Thus, only the cooperative action of MOM22 and MOM19 leads to detectable specific binding of preproteins at the mitochondrial surface. MOM22 and MOM19 can be crosslinked and therefore may be viewed as a two-subunit receptor harbouring the *cis*-recognition site for mitochondrial presequences. Binding to the *cis*-site is readily reversible and at higher salt concentrations becomes very labile precluding detection by a sedimentation binding assay. This behaviour may reflect the involvement of electrostatic forces in binding preproteins to the mitochondrial surface. Most likely, the salt bridges are formed between the acidic cytosolic domain of MOM22 and the positively charged presequence. The labile character of presequence binding may be important for the rapid release of non-cognate substrates from the mitochondrial surface. The *cis*-site is specific for the presequence, since presequence peptides compete for binding equally well as purified preproteins (in the micromolar concentration range). Therefore, at least a large fraction of the binding energy at the *cis*-site is provided by interaction with the presequence. Interestingly, the import of preproteins lacking an N-terminal presequence (such as porin and CCHL) is also supported by MOM22/MOM19. According to direct binding studies using OMV these preproteins specifically associate with these receptor proteins raising the challenging question of how the targeting signals are deciphered in such preproteins.

Binding of the presequence on the trans side of the outer membrane
After the interaction of the presequence with the *cis*-site, the polypeptide chain becomes inserted into the (putative) translocation pore and finally reaches the intermembrane space side of the outer membrane (Mayer et al., 1995b; Fig. 4). How this reaction occurs in molecular terms is completely unknown. A plausible scenario might be that the presequence slides along a furrow from the MOM22/MOM19 binding site into the pore. It is well possible that the *cis*-site itself forms this furrow, and thus plays an important function not only in presequence recognition, but also in its transfer into the translocation pore. In this view, MOM22/MOM19 could be regarded as a device for threading the polypeptide chain into the pore and may act like a funnel. It should be mentioned in this context, that in all likelyhood the MOM22/MOM19 receptor provides a large binding area for the presequence rather than a well-defined binding site. This would allow for the dynamic formation of multiple salt bridges between the receptor and presequences which are known to be variable in their charge distribution and in length. During insertion into the translocation pore, multiple unspecific interactions of the presequence with the walls of the pore could counterbalance part of the energy necessary for the release from the *cis*-site, thereby lowering the energy of the transition state. Detailed investigations are necessary to elucidate this interesting mechanistic aspect.

At the *trans* side of the outer membrane, the presequence associates with a specific binding site termed *trans*-site. Thus, a translocation intermediate (termed "outer membrane translocation intermediate") is created in which the presequence is translocated across the outer membrane, the N-terminal portion of the mature part spans the membrane, and the C-terminus is still exposed to the cytosol, giving rise to its accessibility for external proteases (Fig. 4). In OMV, binding to this *trans*-site is comparatively stable and accounts for the net translocation of the N-terminal part of the preprotein. What is the evidence for translocation of the N-terminal region of a preprotein across the mitochondrial outer membrane? MPP introduced into the lumen of the OMV by a freeze-thaw procedure (Mayer et al., 1995) can cleave the presequence of various preproteins. In contrast, MPP added from outside is unable to process the presequence of this translocation intermediate, demonstrating that the cleavage site of the preproteins had undergone complete transfer into the lumen of the OMV. After removal of the presequence, the mature part dissociated at a very fast rate from the OMV (Fig. 4). Three major conclusions can be drawn from this observation. First, the initial translocation of the mature part of the preprotein is fully reversible. Second, the translocation channel serves as a passive pore which does not undergo a stable interaction with the mature part of the preprotein. Third, binding to the *trans*-site occurs specifically via the presequence.

According to these data, translocation of preproteins across the outer membrane is initiated by the sequential interaction of the mitochondrial presequence with two specific binding sites on either side of a translocation pore which allows the reversible sliding of the polypeptide chain. Reversibility appears to be a general feature of all membrane transport reactions, as retrograde translocation has been reported for the membrane of the endoplasmic reticulum (Ooi and Weiss, 1992), the mitochondrial inner membrane (Ungermann et al., 1994) and the bacterial plasma membrane (Schiebel et al., 1991). The molecular basis of this property of the translocation machineries may be the existence of (hydrophilic) translocation channels which permit the movement of the translocating chain in both directions (Simon and Blobel, 1991; Crowley et al., 1993). Reversibility of translocation across the mitochondrial outer membrane is also observed for the transfer of the intact preprotein (Fig. 4), although the reverse reaction was considerably slower than the dissociation of the mature part alone. Furthermore, significant dissociation was only seen, when the equilibrium of the reaction was shifted by stabilizing the conformation of the mature part of the preprotein (see above), thus precluding its reentry into the translocation channel.

This brings us to the discussion of another important aspect of protein transfer across the mitochondrial outer membrane. Upon entry into the translocation pore, folded parts of the preprotein immediately adjacent to the presequence become unfolded (Mayer et al., 1995b; Fig. 4). Unfolding does not require the presence of ATP, indicating that chaperones like Hsp70 proteins may not be involved in this reaction. Obviously, the binding of the presequence to the *trans*-site is strong enough to shift the equilibrium to the side of translocation. Thus, the function of the *trans*-site is not only restricted to driving the transport of the presequence across the outer membrane. It also provides the energy for unfolding the polypeptide chain upon its entry into the translocation pore. At least for the initial stages of the import reaction, an unfolding of the translocating chain does not require a "pulling" force from the internal face of the membrane. Such a function has been proposed for mitochondrial Hsp70 (Neupert et al., 1990;

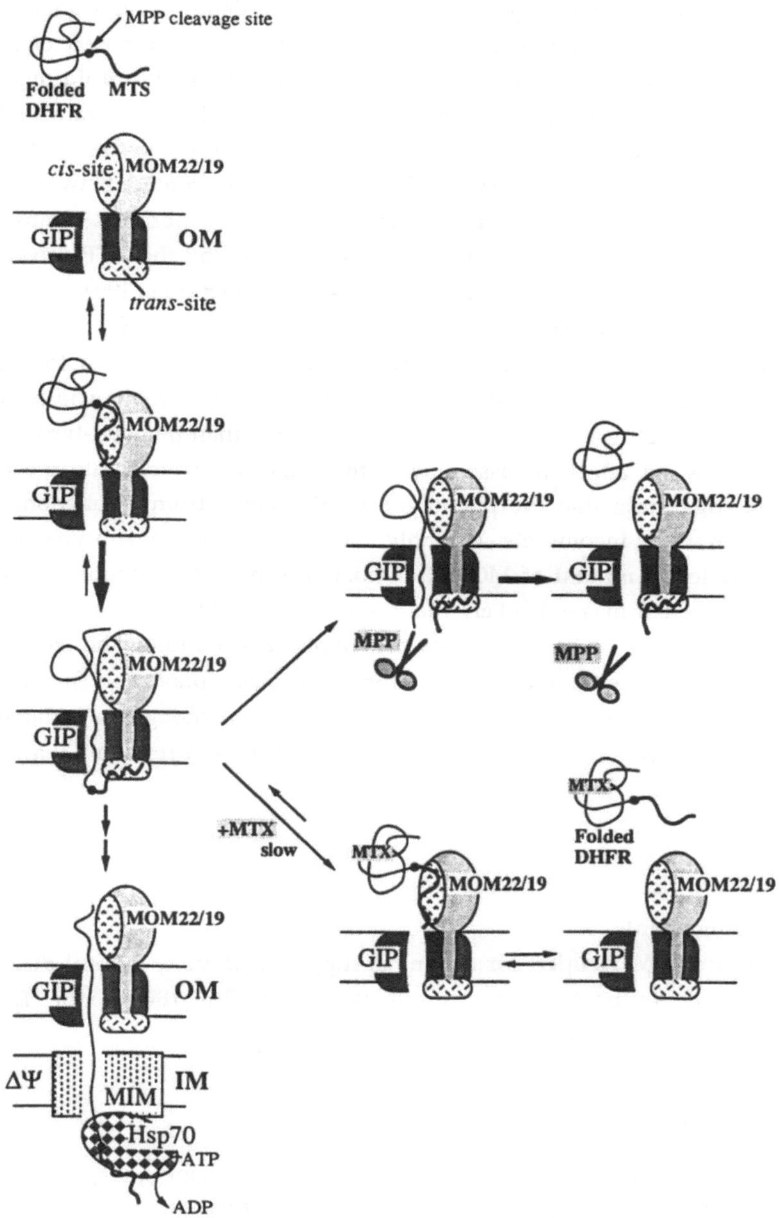

Figure 4: Molecular mechanism of preprotein translocation across the mitochondrial outer membrane. A detailed discussion of the various aspects of the model is found in the text. Abbreviations used in addition to those given in Figs. 1 and 3: MPP, matrix processing peptidase; MTS, matrix targeting signal; DHFR, dihydrofolate reductase; MTX, methotrexate; MIM, translocation machinery of the mitochondrial inner membrane.

Glick, 1995; Pfanner and Meijer, 1995) but may be needed only during later stages of the transport process. Interaction with the translocation machinery of the outer membrane is apparently sufficient to result in efficient unfolding. In this unfolded state, the polypeptide chain represents in ideal substrate to become further passed into and across the inner membrane. How this transfer occurs is still enigmatic, even though one might expect a direct contact between the two translocation machineries of the outer and inner membranes. Translocation across the latter membrane is rendered unidirectional by the ATP-driven action of the MIM44/Hsp70 chaperone system (Fig. 4; Schneider et al., 1994).

What is the molecular basis of the *trans*-site? Preliminary results suggest that the C-terminal segment of MOM22 may play some role in forming this binding site. Deletion of this short C-terminal portion (49 amino acid residues in *N. crassa* MOM22) results in cells which grow slower than the wild-type cells. The deletion does not affect the assembly of the truncated protein into the receptor complex suggesting that the phenotype results solely from a functional defect rather than from incomplete assembly. The rather mild phenotype of the C-terminal deletion mutant of MOM22 in comparison to the essential character of the whole protein makes it likely that the intermembrane space regions of other components of the receptor complex are involved in building the *trans*-site. As found for the *cis*-site, ionic interactions may also play a role during the recognition of the presequence at the *trans*-site, even though in isolated OMV binding of the presequence was found to be more salt-resistant than *cis*-site binding.

Acknowledgements
We thank Prof. W. Neupert for continous support and many helpful discussions. The work is supported by a grant from the SFB 184 of the Deutsche Forschungsgemeinschaft.

REFERENCES

Baker, K.P., Schaniel, A., Vestweber, D., and Schatz, G. (1990). A yeast mitochondrial outer membrane protein is essential for protein import and cell viability. Nature *348*, 605-609.

Berthold, J., Bauer, M.F., Schneider, H., Klaus, C., Dietmeier, K., Neupert, W., and Brunner, M. (1995). The MIM complex mediates preprotein translocation across the mitochondrial inner membrane and couples it to the mt-Hsp70/ATP driving system. Cell *81*, 1085-1093.

Court, D.A., Lill, R., and Neupert, W. (1995). The protein import apparatus of the mitochondrial outer membrane. Can. J. Bot. *in press.*,

Crowley, K.S., Reinhart, G.D., and Johnson, A.E. (1993). The signal sequence moves through a ribosomal tunnel into an noncytoplasmic aqueous environment at the ER membrane early in translocation. Cell *73*, 1101-1115.

Deshaies, R.J., Koch, B.D., and Schekman, R. (1988). The role of stress proteins in membrane biogenesis. Trends Biochem. Sci. *13*, 384-388.

Drygas, M.E., Lambowitz, A.M., and Nargang, F.E. (1989). Cloning and analysis of the *Neurospora crassa* gene for cytochrome c heme lyase. J. Biol. Chem. *264*, 17897-17907.

Dumont, M.E., Ernst, J.F., and Sherman, F. (1988). Coupling of heme attachment to import of cytochrome c into yeast mitochondria. J. Biol. Chem. *263*, 15928-15937.

Glick, B.G. and Schatz, G. (1991). Import of proteins into mitochondria. Annu. Rev. Genet. *25*, 21-44.

Glick, B.G., Wachter, C., and Schatz, G. (1991). Protein import into mitochondria: two systems acting in tandem? Trends Cell Biol. *1*, 99-103.

Glick, B.S. (1995). Can Hsp70 proteins act as force-generating motors? Cell *80*, 11-14.

Hachiya, N., Komiya, T., Alam, R., Iwahasi, J., Sakaguchi, M., Omura, T., and Mihara, N. (1994). MSF, a novel cytoplasmic chaperone which functions in precursor targeting to mitochondria. EMBO J. *13*, 5146-5154.

Harkness, T.A.A., Nargang, F.E., Van der Klei, I., Neupert, W., and Lill, R. (1994). A crucial role of the mitochondrial protein import receptor MOM19 for the biogenesis of mitochondria. J. Cell Biol. *124*, 637-648.

Hendrick, J.P. and Hartl, F.U. (1993). Molecular chaperone function of heat shock proteins. Ann. Rev. Biochem. *62*, 349-384.

Hines, V., Brandt, A., Griffiths, G., Horstmann, H., Brütsch, H., and Schatz, G. (1990). Protein import into yeast mitochondria is accelerated by the outer membrane protein *MAS70*. EMBO J. *9*, 3191-3200.

Horst, M., Hilfiker-Rothenfluh, S., Oppliger, W. and Schatz, G. (1995). Dynamic interaction of the protein translocation systems in the inner and outer membranes of yeast mitochondria. EMBO J. *14*, 2293-2297.

Hwang, S.T., Jascur, T., Vestweber, D., Pon, L., and Schatz, G. (1989). Disrupted yeast mitochondria can import precursor proteins directly through their inner membrane. J. Cell Biol. *109*, 487-493.

Jascur, T., Goldenberg, D.P., Vestweber, D., and Schatz, G. (1992). Sequential translocation of an artificial precursor protein across the two mitochondrial membranes. J. Biol. Chem. *267*, 13636-13641.

Kassenbrock, C.K., Cao, W., and Douglas, M.G. (1993). Genetic and biochemical characterization of *ISP6*, a small mitochondrial outer membrane protein associated with the protein translocation complex. EMBO J. *12*, 3023-3034.

Keil, P. and Pfanner, N. (1993). Insertion of MOM22 into the mitochondrial outer membrane strictly depends on surface receptors. FEBS Lett. *321*, 197-200.

Kiebler, M., Becker, K., Pfanner, N., and Neupert, W. (1993a). Mitochondrial protein import: Specific recognition and membrane translocation of preproteins. J. Membrane Biol. *135*, 191-207.

Kiebler, M., Keil, P., Schneider, H., van der Klei, I., Pfanner, N., and Neupert, W. (1993b). The mitochondrial receptor complex: A central role of MOM22 in mediating transfer of preproteins from receptors to the general insertion pore. Cell *74*, 483-492.

Kiebler, M., Pfaller, R., Söllner, T., Griffiths, G., Horstmann, H., Pfanner, N., and Neupert, W. (1990). Identification of a mitochondrial receptor complex required for recognition and membrane insertion of precursor proteins. Nature *348*, 610-616.

Kleene, R., Pfanner, N., Pfaller, R., Link, T.A., Sebald, W., Neupert, W., and Tropschug, M. (1987). Mitochondrial porin of *Neurospora crassa*: cDNA cloning, in vitro expression and import into mitochondria. EMBO J. *6*, 2627-2633.

Lill, R., Hergersberg, C., Schneider, H., Söllner, T., Stuart, R.A., and Neupert, W. (1992). General and exceptional pathways of protein import into the sub-mitochondrial compartments. In Neupert, W. and Lill, R. (Eds.), Membrane biogenesis and protein targeting pp. 265-276. Elsevier Science Publishers, Amsterdam.

Lill, R., Mayer, A., Steiner, H., Kispal, G., and Neupert, W. (1994). Molecular mechanisms of protein translocation across the mitochondrial outer membrane. In Hartl, F. (Eds.), Advances in Molecular and Cell Biology pp. in press. Academic Press, New York.

Lill, R., Stuart, R.A., Drygas, M.E., Nargang, F.E., and Neupert, W. (1992). Import of cytochrome *c* heme lyase into mitochondria: a novel pathway into the intermembrane space. EMBO J. *11*, 449-456.

Lithgow, T., Junne, T., Suda, K., Gratzer, S., and Schatz, G. (1994). The mitochondrial outer membrane protein Mas22p is essential for protein import and viability of yeast. Proc. Natl. Acad. Sci. U.S.A. *91*, 11973-11977.

Mayer, A., Driessen, A.J.M., Neupert, W., and Lill, R. (1995). The use of purified and protein-loaded mitochondrial outer membrane vesicles for the functional analysis of preprotein transport. Methods Enzymol. *in press.*

Mayer, A., Lill, R., and Neupert, W. (1993). Translocation and insertion of precursor proteins into isolated outer membranes of mitochondria. J. Cell Biol. *121*, 1233-1243.

Mayer, A., Nargang, F.E., Neupert, W., and Lill, R. (1995a). MOM22 is a receptor for mitochondrial presequences and cooperates with MOM19. EMBO J. *in press.*

Mayer, A., Neupert, W., and Lill, R. (1995b). Mitochondrial protein import: Reversible binding of the presequence at the *trans* side of the outer membrane drives partial translocation and unfolding. Cell *80*, 127-137.

Moczko, M., Ehmann, B., Gärtner, F., Hönlinger, A., Schäfer, E., and Pfanner, N. (1994). Deletion of the receptor MOM19 strongly impairs import of cleavable preproteins into *Saccharomyces cerevisiae* mitochondria. J. Biol. Chem. *269*, 9045-9051.

Nargang, F.E., Künkele, K., Mayer, A., Ritzel, R.G., Neupert, W., and Lill, R. (1995). "Sheltered disruption" of *Neurospora crassa* MOM22, an essential component of the mitochondrial protein import complex. EMBO J. *14*, 1099-1108.

Neupert, W., Hartl, F., Craig, E.A., and Pfanner, N. (1990). How do polypeptides cross the mitochondrial membranes? Cell *63*, 447-450.

Neupert, W. and Lill, R. (Ed.). (1992). Membrane biogenesis and protein targeting. In New Comprehensive Biochemistry, Vol. 22, Neuberger, A. and Van Deenen, L. L. M. (Series Eds.). Amsterdam, Elsevier Science Publishers.

Nguyen, M., Millar, D.G., Yong, V.W., Korsmeyer, S.J., and Shore, G.C. (1993). Targeting of Bcl-2 to the mitochondrial outer membrane by a carboxy-terminal signal anchor sequence. J. Biol. Chem. *268*, 25265-25268.

Ooi, C.E. and Weiss, J. (1992). Bidirectional movement of the nascent polypeptide across microsomal membranes reveals requirements for vectorial translocation of proteins. Cell *71*, 87-96.

Pfanner, N. and Meijer, M. (1995). Pulling in the proteins. Current Biology *5*, 132-135.

Pfanner, N. and Neupert, W. (1990). The mitochondrial protein import apparatus. Annu. Rev. Biochem. *59*, 331-353.

Pfanner, N., Rassow, J., van der Klei, I.J., and Neupert, W. (1992). A dynamic model of the mitochondrial protein import machinery. Cell *68*, 999-1002.

Ramage, L., Junne, T., Hahne, K., Lithgow, T., and Schatz, G. (1993).Functional cooperation of mitochondrial protein import receptors in yeast. EMBO J. *12*, 4115-4123.

Schiebel, E., Driessen, A.J.M., Hartl, F., and Wickner, W. (1991). Dm_{H}+ and ATP function at different steps of the catalytic cycle of preprotein translocase. Cell *64*, 927-939.

Schneider, H., Berthold, J., Bauer, M.F., Dietmeier, K., Guiard, B., Brunner, M., and Neupert, W. (1994). Mitochondrial Hsp70/MIM44 complex facilitates protein import. Nature *371*, 768-774.

Schneider, H., Söllner, T., Dietmeier, K., Eckerskorn, C., Lottspeich, F., Trülzsch, K., Neupert, W., and Pfanner, N. (1991). Targeting of the master receptor MOM19 to mitochondria. Science *254*, 1659-1662.

Schwarz, E. and Neupert, W. (1994). Mitochondrial protein import: Mechanisms, components, and energetics. Biochim. Biophys. Acta *1187*, 270-274.

Segui-Real, B., Kispal, G., Lill, R., and Neupert, W. (1993a). Functional independence of the protein translocation machineries in mitochondrial outer and inner membranes: passage of preproteins through the intermembrane space. EMBO J. *12*, 2211-2218.

Segui-Real, B., Stuart, R.A., and Neupert, W. (1993b). Transport of proteins into the various subcompartments of mitochondria. FEBS Lett. *313*, 2-7.

Simon, S.M. and Blobel, G. (1991). A protein-conducting channel in the endoplasmatic reticulum. Cell *65*, 371-380.

Söllner, T., Griffiths, G., Pfaller, R., Pfanner, N., and Neupert, W. (1989). MOM19, an import receptor for mitochondrial precursor proteins. Cell *59*, 1061-1070.

Söllner, T., Pfaller, R., Griffiths, G., Pfanner, N., and Neupert, W. (1990). A mitochondrial import receptor for the ADP/ATP carrier. Cell *62*, 107-115.

Söllner, T., Rassow, J., Wiedmann, M., Schlossmann, J., Keil, P., Neupert, W., and Pfanner, N. (1992). Mapping of the protein import machinery in the mitochondrial outer membrane by crosslinking of translocation intermediates. Nature *355*, 84-87.

Steger, H.F., Söllner, T., Kiebler, M., Dietmeier, K.A., Trülzsch, K.S., Tropschug, M., Neupert, W., and Pfanner, N. (1990). Import of ADP/ATP carrier into mitochondria: two receptors act in parallel. J. Cell Biol. *111*, 2353-2363.

Steiner, H., Zollner, A., Haid, A., Neupert, W., and Lill, R. (1995). Biogenesis of mitochondrial heme lyases in yeast. Import and folding in the intermembrane space. J. Biol. Chem. *in press.*

Stuart, R.A., Cyr, D.M., Craig, E.A., and Neupert, W. (1994). Mitochondrial molecular chaperones: their role in protein translocation. Trends Biochem. Sci. *19*, 87-92.

Ungermann, C., Neupert, W., and Cyr, D.M. (1994). The role of Hsp70 in conferring unidirectionality on protein translocation into mitochondria. Science *266*, 1250-1253.

Van der Klei, I.J., Veenhuis, M., and Neupert, W. (1994). A morphological view on mitochondrial protein targeting. Microscopy Res. Technique *27*, 284-293.

Zollner, A., Rödel, G., and Haid, A. (1992). Molecular cloning and characterization of the *Saccharomyces cerevisiae CYT2* gene encoding cytochrome c_1 heme lyase. Eur. J. Biochem. *207*, 1093-1100.

PROTEIN IMPORT ACROSS THE INNER MITOCHONDRIAL MEMBRANE

Hans-Christoph Schneider, Jutta Berthold, Matthias F. Bauer, Christian Klaus, Walter Neupert and Michael Brunner,
Institut für Physiologische Chemie der Universität München,
Goethestraße 33,
80336 München,
Germany

Introduction

Most mitochondrial proteins are synthesized in the cytosol and must then be transported across the organelle membranes to reach their functional destination. Mitochondria contain two translocation systems, one in the outer membrane and one in the inner membrane (Glick *et al.*, 1991; Pfanner *et al.*, 1992). The system in the mitochondrial outer membrane (MOM complex) consists of at least six protein components most of which have been isolated, cloned and sequenced (for review see Kiebler *et al.*, 1993).

In *Saccharomyces cerevisiae* three essential proteins of the inner membrane have been described which are required for protein translocation. MIM17 (SMS1) (Dekker *et al.*, 1993; Kübrich *et al.*, 1994; Maarse *et al.*, 1994; Ryan *et al.*, 1994) and MIM23 (MAS6) (Dekker *et al.*, 1993; Emtage and Jensen, 1993; Ryan and Jensen, 1993; Kübrich *et al.*, 1994) are integral membrane proteins with three or four predicted membrane spanning segments.

The third component, MIM44 (Isp45), is a hydrophilic protein which is peripherally associated with the inner face of the inner membrane (Maarse *et al.*, 1992; Scherer *et al.*, 1992; Blom *et al.*, 1993; Horst *et al.*, 1993). No membrane spanning region is predicted from the primary sequence. However, it is not clear whether its carboxy terminus is exposed to the intermembrane space (Blom *et al.*, 1993; Horst *et al.*, 1993).

The mitochondrial heat shock protein of 70 kDa (mt-Hsp70) (Craig *et al.*, 1989; Morishima *et al.*, 1990) also plays a critical role in protein import. It was found to be in contact with preproteins arrested in transit across both mitochondrial membranes (Ostermann *et al.*, 1990; Scherer *et al.*, 1990; Manning-Krieg *et al.*, 1991). In addition, mitochondria which harbour temperature sensitive mt-Hsp70s are defective in protein import under non-permissive

NATO ASI Series, Vol. H 96
Molecular Dynamics of Biomembranes
Edited by Jos A. F. Op den Kamp

conditions (Kang *et al.*, 1991; Gambill *et al.*, 1993; Voos *et al.*, 1993). Mt-Hsp70 is proposed to interact in an ATP-dependent fashion with the incoming unfolded polypeptide chains and thereby to energetically drive the translocation into the matrix space (*Neupert et al.*, 1990; Stuart *et al.*, 1994a; Stuart *et al.*, 1994b). Recently MIM44 was found to recruit mt-Hsp70 to the sites, where preproteins enter the matrix space (Kronidou *et al.*, 1994; Rassow *et al.*, 1994; Schneider *et al.*, 1994).

In our studies (Schneider *et al.*, 1994; Berthold *et al.*, 1995) we investigated the role of the mt-Hsp70/MIM44 complex in protein import. We could show that the formation of this complex is regulated by ATP hydrolysis and binding of unfolded polypeptide. Our results suggest a mechanism of mt-Hsp70/MIM44 function by which MIM44 first binds the unfolded polypeptide chain emerging from the putative translocation channel, then transfers it to mt-Hsp70 concomitant with ATP hydrolysis and complex dissociation. This allows the polypeptide chain to move further into the mitochondria and to undergo another cycle of binding to the complex. The mt-Hsp70/MIM44 system interacts with a complex in the mitochondrial inner membrane (MIM complex) which contains MIM17 and MIM23 as well as two novel components, MIM14 and MIM33.

MIM44 forms a high molecular weight complex with mt-Hsp70

To address if MIM44 is associated with other proteins we determined the apparent molecular mass of MIM44 in solubilized mitochondria. Analysis on a gel filtration column revealed a molecular weight of 250 kDa. The same apparent size was found by sedimentation studies on sucrose and glycerol gradients.

To characterize the proteins in contact with MIM44 we performed co-immunoprecipitation with affinity purified MIM44 specific antibodies under native conditions. In addition to MIM44 a protein with a molecular mass of 70 kDa and one of 22 kDa were found in the immunoprecipitate. Specific antibodies identified the 70 kDa protein as mt-Hsp70 and the 22 kDa protein as MGE, the mitochondrial homologue of bacterial GrpE (Bolliger *et al.*, 1994; Ikeda *et al.*.). The complex described here contains MIM44 and mt-Hsp70 in about 1:1 stoichiometry. MGE was found in lower amounts. The apparent molecular weight of 250 kDa suggests that the complex may consist of two molecules of each, MIM44 and mt-Hsp70.

On treatment of intact mitochondria with chemical crosslinker specific MIM44 containing adducts were generated. Low concentrations of crosslinker were sufficient to form a major adduct with MIM44 which displayed an apparent molecular mass of approximately 116 kDa. In addition a cross-linked species with 160 kDa was generated. When the adducts were immunoprecipitated with anti MIM44 IgG, both could be decorated with antibodies against mt-Hsp70 on a Western Blot. This indicates that in intact mitochondria MIM44 is in physical contact with mt-Hsp70. The 116 kDa adduct is apparently composed of one MIM44 and one mt-Hsp70, whereas the 160 kDa adduct could contain two MIM44 and one mt-Hsp70.

The interaction of mt-Hsp70 with MIM44 is regulated by ATP and by binding of unfolded polypeptide

Molecular chaperones of the Hsp70 class possess ATPase activity (Rothman, 1989). We therefore investigated the effect of adenine nucleotides on the mt-Hsp70/MIM44 complex. In the presence of EDTA MIM44 sedimented in a high molecular weight form in glycerol gradient centrifugation. However, when Mg-ATP was present, MIM44 sedimented corresponding to a lower molecular weight. When mitochondria were solubilized in the presence or absence of Mg-ATP followed by an immunoprecipitation with MIM44 specific antibodies, mt-Hsp70 was only coprecipitated in the absence of Mg-ATP. Thus, lysis of mitochondria in the presence of Mg-ATP resulted in complete dissociation of the mt-Hsp70/MIM44 complex. However, the complex was stable when the non-hydrolysable ATP analogue AMP-PNP was present. Likewise EDTA preserved the complex. Thus, hydrolysis of ATP seems to be required for the dissociation of mt-Hsp70 from MIM44.

As preproteins cross the mitochondrial membranes in an unfolded conformation (Rassow *et al*, 1990) and as chaperones of the Hsp70 class interact with unfolded polypeptides (Palleros *et al.*, 1991; Langer *et al.*, 1992) we studied the influence of unfolded proteins on the stability of the mt-Hsp70/MIM44 complex. With low concentrations of Mg-ATP (5 μM) only partial dissociation of the mt-Hsp70/MIM44 complex could be observed. When the permanently unfolded protein RCMLA (reduced carboxymethylated lactalbumin) or synthetic peptides were included under these conditions the amount of mt-Hsp70 complexed with MIM44 was strongly reduced while native proteins had no effect.

After ATP-dependent dissociation of the complex MIM44 could be crosslinked to RCMLA but not to native lactalbumin. Furthermore, MIM44 binds to RCMLA covalently linked to an affinity resin, suggesting that MIM44 can bind unfolded proteins in the absence of mt-Hsp70. MIM44 appears to have a lower affinity for RCMLA than mt-Hsp70 since RCMLA could be co-immunoprecipitated with mt-Hsp70 but not with MIM44.

In summary, the mt-Hsp70/MIM44 complex dissociates in an ATP dependent manner. The complex is stabilized when ATP-hydrolysis is inhibited by EDTA or competed by AMP-PNP. Unfolded proteins and synthetic peptides stimulated the ATP-dependent disintegration of the complex. MIM44 in the absence of mt-Hsp70 binds unfolded polypeptide chains with low affinity.

Mt-Hsp70/MIM44 complex formation is correlated with protein import

Two temperature sensitive yeast mutants of mt-Hsp70 (ssc1-2 and ssc1-3) have been described (Gambill *et al.*, 1993). Mitochondria isolated from these strains are defective in protein import under non-permissive conditions. When the phenotype was induced in both mutants only minor amounts of mt-Hsp70 were found in complex with MIM44. Thus, the loss of the mt-Hsp70/MIM44 complex was paralleled by a defect in import competence.
A similar result was obtained when the ATP levels of the mitochondrial matrix were manipulated. Depletion of matrix ATP is known to block protein import into the matrix (Stuart *et al.*, 1994a). When high concentrations of matrix ATP were present, mt-Hsp70 was recovered in complex with MIM44, whereas with low matrix ATP levels only traces were detected. This result shows that high matrix ATP levels promote complex formation whereas low ATP levels shift the equilibrium towards complex dissociation.
Chimeric preproteins with a DHFR (mouse dihydrofolate reductase) domain fused to the carboxy terminus of a mitochondrial import signal were arrested during translocation by addition of methotrexate (MTX) which stabilizes the folded DHFR moiety (Rassow *et al.*, 1989). These membrane spanning import intermediates were specifically crosslinked to MIM44 and mt-Hsp70 and must therefore be in close proximity to these proteins. The crosslinks were diminished when the mitochondrial matrix was depleted of ATP prior to crosslinking. Under these conditions preproteins which expose only short segments into the matrix fall completely out of the mitochondria while longer translocation intermediates slide back but might be held by mt-Hsp70 molecules that are bound to more N-terminal segments of the translocating chain or by partially folded structures in the N-terminal region of the preprotein (Ungermann *et al.*, 1994).
All these experiments suggest that the Hsp70/MIM44 complex is required to prevent retrograde movement of the preprotein in transit. When the mitochondrial matrix is depleted of ATP the Hsp70/MIM44 complex disintegrates and is no longer in contact with the translocating chain. The import intermediate is now able to slide back in the import channel. Thus, the mt-Hsp70/MIM44 complex has an important role in locking the chain in transit to prevent retrograde movement, thereby allowing only forward movement of a preprotein into the matrix.

MIM17 and MIM23 are components of a protein complex in the inner membrane which interacts with MIM44

MIM17 and MIM23 were identified as integral proteins in the inner mitochondrial membrane with a function in protein import. When mitochondria were solubilized in digitonin both proteins coeluted from a gel filtration column in a fraction corresponding to a molecular mass of 280 kDa. Upon solubilization in n-octy-β,D-glucopyranoside (octylglucoside) MIM23 eluted corrsponding to a molecular mass of 110 kDa and MIM17 corresponding to 90 kDa. When immunoprecipitation was performed in octylglucoside anti-MIM23 IgG precipitated only MIM23, and anti-MIM17 IgG precipitated only MIM17. In digitonin both antibodies precipitated the same set of four proteins. This indicates that MIM17 and MIM 23 are in a complex in the mitochondrial inner membrane (MIM complex). In addition the complex contains a protein of 33 kDa (MIM33) and a protein of 14 kDa (MIM14).

MIM44 is also associated with the MIM complex but its interaction is very sensitive to salt concentration and detergent. To detect an interaction between MIM44 and the MIM complex, very mild solubilization conditions had to be used. When the mitochondria were lysed with 0.5% digitonin in the presence of 50 mM NaCl MIM44 was specifically detected in immunoprecipitates with anti-MIM23 IgG. Even under these conditions only a fraction of MIM44 was recovered with the MIM complex and the amount decreased with increasing salt concentration in the lysis buffer.

MIM17 and MIM23 could be crosslinked to an arrested preprotein in transit indicating that the the MIM complex is in contact with the translocating chain during import. When a preprotein was arrested in a membrane spanning conformation and the mitochondria were subsequently solubilized with digitonin to preserve the MIM complex, the translocation intermediate was co-precipitated with IgGs against MIM17, MIM23, MIM44, and mt-Hsp70. When octylglucoside was used for solubilization of the mitochondria the arrested preprotein was precipitated with antibodies against MIM44 and mt-Hsp70 but not with antibodies against MIM23 or MIM17. As the MIM complex disintegrates in the presence of octylglucoside, the intactness of the MIM complex is required for co-immunoprecipitation of a preprotein in transit. This suggests that the individual components MIM23 and MIM17 do not bind to the translocating chain with high affinity. An explanation of these results might be that the MIM complex forms a channel which surrounds the translocating chain. The components of this channel do not tightly interact with the translocating chain and the arrested intermediate is held in this channel by a folded DHFR moiety on one side and by a tightly associated mt-Hsp70 molecule on the other side.

Model for the protein import across the inner mitochondrial membrane

On the basis of the data described and of previous studies we propose a model of how MIM44 and mt-Hsp70 cooperate in the translocation of preproteins. After translocation across the outer membrane the presequence of the preprotein is passed on to the MIM complex and inserted into the inner membrane import machinery. This process requires an

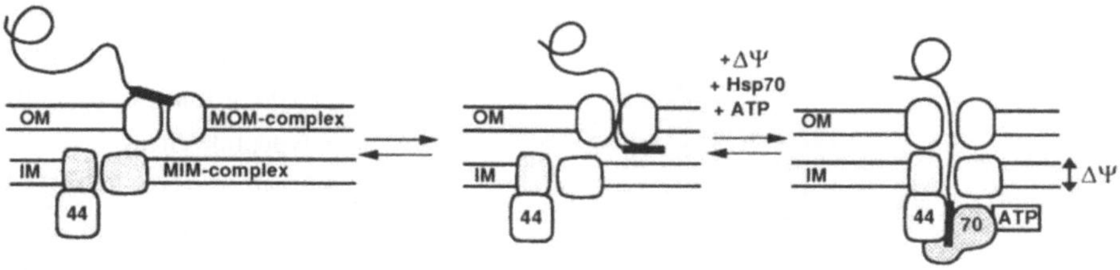

Figure 1: Initiation of protein translocation across the mitochondrial membranes. In the first step the presequence of the preprotein is translocated from the cis-site to the trans-site of the outer membrane import machinery (Mayer *et al.*, 1995). Subsequently, the presequence is inserted into the inner membrane import machinery in a membrane potential dependent manner. At the inner face of the inner membrane the preprotein is passed over to the mt-Hsp70/MIM44 machinery which drives the vectorial translocation in an ATP-dependent fashion. OM = outer membrane, IM = inner membrane, 44 = MIM44, 70 = mt-Hsp70, ΔΨ = membrane potential.

electrical membrane potential ΔΨ (Figure 1). The mt-Hsp70/MIM44 complex awaits the incoming polypeptide chain on the inner side of the inner membrane. Interaction of the translocating polypeptide with this complex stimulates the ATPase of mt-Hsp70. Upon ATP hydrolysis the complex dissociates and mt-Hsp70 is transferred to the incoming polypeptide chain. Mt-Hsp70 in the ADP form binds the preprotein with high affinity (Palleros *et al.*, 1993) and locks the translocation intermediate in the matrix. Now the polypeptide chain can slide further through the import channel and MIM44 rebinds to a more C-terminal segment and thereby another cycle of mt-Hsp70 binding is initiated (Figure 2).

In this reaction pathway MIM44 binds to the incoming polypeptide chain and transfers it to mt-Hsp70. The direct regulated interaction of MIM44 with mt-Hsp70 would provide a molecular machinery which confers unidirectionality to the import process by preventing the translocating chain from retrograde movement. Thus the mt-Hsp70/MIM44 complex could be viewed as a molecular ratchet. Interestingly, a 'brownian ratchet' has been suggested previously on theoretical considerations as a possible principle of membrane transfer of proteins (Simon *et al.*, 1992). It was recently discussed that a nucleotide dependent conformational change of mt-Hsp70 could facilitate the translocation process by exerting a 'pulling force' on the translocating polypeptide chain (Glick, 1995). However, ligand induced

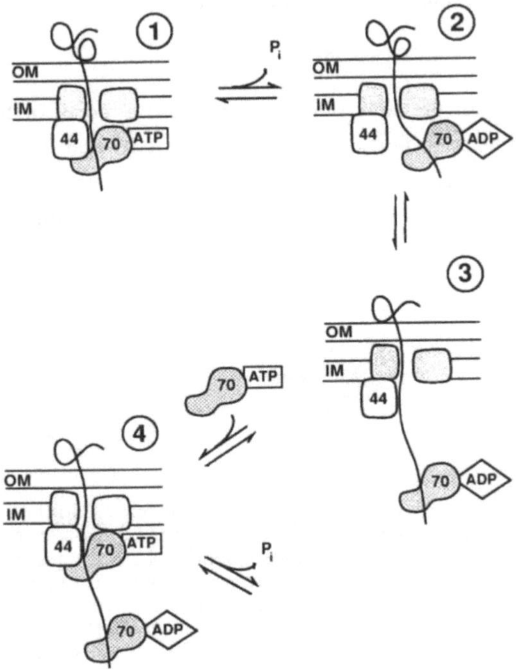

Figure 2: Model for the cooperation of mt-Hsp70 and MIM44 in protein translocation; 44 = MIM44, 70 = mt-Hsp70, OM = outer membrane, IM = inner membrane.

conformational changes of a protein of the size of mt-Hsp70 would be relatively small (Vale, 1994). It may be speculated that such a translocation motor could assist unfolding of domains on the outside of the mitochondria at critical stages rather than drive translocation of an entire polypeptide chain.

As the mt-Hsp70/MIM44 complex would facilitate mt-Hsp70 binding to the translocating protein as soon as it enters the matrix, aggregation or unproductive folding of the unfolded polypeptide chains would be prevented. Some mt-Hsp70 may remain bound until folding to the native state occurs or until the polypeptide is transferred to Hsp60. Thus, the mt-Hsp70/MIM44 complex would have a role in driving the import reaction and in keeping the incoming polypeptide chain in a state competent for folding.

Acknowledgements

We would like to thank Margarete Moser and Colette Zobywalski for excellent technical assistance. This work was supported by grants from the Deutsche Forschungsgemeinschaft Sonderforschungsbereich 184 (Teilprojekt B12), the human Frontiers of Science Program, the Fonds der Chemischen Industrie and the Münchener Medizinische Wochenschrift (to M.B.).

References

Berthold J, Bauer MF, Schneider HC, Klaus C, Dietmeier K, Neupert W and Brunner M (1995) The MIM complex mediates preprotein translocation across the mitochondrial inner membrane and couples it to the mt-Hsp70/ATP driving system. Cell in press

Blom J, Kübrich M, Rassow J, Voos W, Dekker PJT, Maarse AC, Meijer M and Pfanner N (1993) The essential yeast protein MIM44 (encoded by MPI1) is involved in an early step of preprotein translocation across the mitochondrial inner membrane. Mol. Cell. Biol. 13: 7364-7371

Bolliger L, Deloche O, Glick BS, Georgopulos C, Jenö P, Kronidou N, Horst M, Morishima N and Schatz G (1994) A mitochondrial homolog of bacterial GrpE interacts with mitochondrial hsp70 and is essential for viability. EMBO J. 13: 1998-2006

Craig EA, Kramer J, Shilling J, Werner-Washburne M, Holmes S, Kosic-Smither J and Nicolet CM (1989) SSC1, an essential member of the S. cerevisiae HSP70 multigene family, encodes a mitochondrial protein. Mol. Cell. Biol. 9: 3000-3008

Dekker PJT, Keil P, Rassow J, Maarse AC, Pfanner N and Meijer M (1993) Identification of MIM23, a putative component of the protein import machinery of the mitochondrial inner membrane. FEBS Lett. 331: 66-70

Emtage JLT and Jensen RE (1993) MAS6 encodes an essential inner membrane component of the yeast mitochondrial import pathway. J. Cell Biol. 122: 1003-1012

Gambill BD, Voos W, Kang PJ, Miao B, Langer T, Craig EA and Pfanner N (1993) A dual role for mitochondrial heat shock protein 70 in membrane translocation of preproteins. J. Cell. Biol. 123: 109-117

Glick BS, Wachter C and Schatz G (1991) Protein import into mitochondria: two systems acting in tandem? Trends Cell Biol. 1: 99-103

Glick BS (1995) Can Hsp70 proteins act as force generating motors? Cell 80: 11-14

Horst M, Jenö P, Kronidou NG, Bollinger L, Opplinger W, Scherer P, Manning-Krieg U, Jascur T and Schatz G (1993) Protein import into yeast mitochondria: the inner membrane import site protein ISP45 is the MPI1 gene product. EMBO J. 12: 3035-3041

Ikeda E, Yoshida S, Mitsuzawa H, Uno I and Toh-e A (1994) YGE1 is a yeast homologue of *Escherichia coli* grpE and is required for maintenance of mitochondrial functions. FEBS Lett. 339: 265-268

Kang PJ, Ostermann J, Shilling J, Neupert W, Craig EA and Pfanner N (1990) Requirement for hsp70 in the mitochondrial matrix for translocation and folding of precursor proteins. Nature 348: 137-143

Kiebler M, Becker K, Pfanner N and Neupert W (1993) Mitochondrial protein import: Specific recognition and membrane translocation of preproteins. J. Membrane Biol. 135: 191-207

Kronidou NG, Opplinger W, Bollinger L, Hannavy K, Glick B, Schatz G and Horst M (1994) Dynamic interaction between Isp45 and mitochondrial hsp70 in the protein import system of yeast mitochondrial inner membrane. Proc. Natl. Acad. Sci. USA 91: 12818-12822

Kübrich M, Keil P, Rassow J, Dekker PJT, Blom J, Meijer M and Pfanner N (1994) The polytopic mitochondrial inner membrane proteins MIM17 and MIM23 operate at the same preprotein import site. FEBS Lett. 349: 222-228

Langer T, Lu C, Echols H, Flanagan J, Hayer MK and Hartl FU (1992) Successive action of DnaK, DnaJ and GroEL along the pathway of chaperone mediated folding. Nature 356: 683-689

Maarse AC, Blom J, Grivell LA and Meijer M (1992) MPI1, an essential gene encoding a mitochondrial membrane protein, is possibly involved in protein import into yeast mitochondria. EMBO J. 11: 3619-3628

Maarse AC, Blom J, Keil P, Pfanner N and Meijer M (1994) Identification of the essential yeast protein MIM17, an integral mitochondrial inner membrane protein involved in protein import. FEBS Lett. 349: 215-221

Manning-Krieg UC, Scherer PE and Schatz G (1991) Sequential action of mitochondrial chaperones in protein import into the matrix. EMBO J. 10: 3273-3280

Mayer A, Neupert W and Lill R (1995) Mitochondrial protein import: reversible binding of the presequence at the trans side of the outer membrane drives partial translocation and unfolding. Cell 80: 127-137

Morishima N, Nakagawa K, Yamamoto E and Shibata T (1990) A subunit of yeast site-specific endonuclease SceI is a mitochondrial version of the 70-kDa heat shock protein. J. Biol. Chem. 265: 15189-15197

Neupert W, Hartl F-U, Craig EA and Pfanner N (1990) How do polypeptides cross the mitochondrial membranes? Cell 63: 447-450

Ostermann J, Voos W, Kang PJ, Craig EA, Neupert W and Pfanner N (1990) Precursor proteins in transit through mitochondrial contact sites interact with hsp70 in the matrix. FEBS Lett. 277: 281-284

Palleros DR, Welch WJ and Fink AL (1991) Interaction of hsp70 with unfolded proteins: Effects of temperature and nucleotides on the kinetics of binding. Proc. Natl. Acad. Sci. USA 88: 5719-5723

Palleros DR, Reid KL, Shi L, Welch WJ and Fink AL (1993) ATP-induced protein -Hsp70 complex dissociation requires K^+ but not ATP hydrolysis. Nature 365: 664-666

Pfanner N, Rassow J, van der Klei I and Neupert W (1992) A dynamic model of the mitochondrial import machinery. Cell 68: 999-1002

Rassow J, Guiard B, Winhues U, Herzog V, Hartl FU and Neupert W (1989) Translocation arrest by reversible folding of a precursor protein imported into mitochondria: A means to quantitate translocation contact sites. J. Cell Biol 109: 1421-1428

Rassow J, Hartl FU, Guiard B, Pfanner N and Neupert W (1990) Polypeptides traverse the mitochondrial envelope in an extended state. FEBS Lett. 275: 190-194

Rassow J, Maarse AC, Krainer E, Kübrich M, Müller H, Meijer M, Craig EA and Pfanner N (1994) Mitochondrial protein import: biochemical and genetic evidence for interaction of matrix hsp70 and the inner membrane protein MIM44. J. Cell Biol. 127: 1547-1556

Rothman JE (1989) Polypeptide chain binding proteins: catalysts of protein folding and related processes in cells. Cell 59: 591-601

Ryan KR and Jensen RE (1993) Mas6p can be cross-linked to an arrested precursor and interacts with other proteins during mitochondrial protein import. J. Biol. Chem. 268: 23743-23746

Ryan KR, Menold MM, Garrett S and Jensen RE (1994) SMS1, a high-copy supressor of the yeast mas6 mutant, encodes an essential inner membrane protein required for mitochondrial protein import. Mol. Biol. Cell 5: 529-538

Scherer PE, Krieg UC, Hwang ST, Vestweber D and Schatz G (1990) A precursor protein partly translocated into yeast mitochondria is bound to a 70 kd mitochondrial stress protein. EMBO J. 9: 4315-4322

Scherer PE, Manning-Krieg UC, Schatz G and Horst M (1992) Identification of a 45-kDa protein at the import site of the yeast mitochondrial inner membrane. Proc. Natl. Acad. Sci. USA 89: 11930-11934

Schneider HC, Berthold J, Bauer MF, Dietmeier K, Guiard B, Brunner M and Neupert W (1994) Mitochondrial Hsp70/MIM44 complex facilitates protein import. Nature 371: 768-774

Simon SM, Peskin CS and Oster GF (1992) What drives the translocation of proteins? Proc. Natl. Acad Sci. USA 89: 3770-3774

Stuart RA, Gruhler A, van der Klei I, Guiard B, Koll H and Neupert W (1994a) The requirement of matrix ATP for the import of precursor proteins into the mitochondrial matrix and intermembrane space. Eur. J. Biochem. 220: 9-18

Stuart RA, Cyr DM and Neupert W (1994b) Mitochondrial molecular chaperones: their role in protein translocation. Trends Biochem. Sci. 19: 87-92

Ungermann C, Neupert W and Cyr DM (1994) The role of Hsp70 in conferring unidirectionality on protein translocation into mitochondria. Science 266: 1250-1253

Vale RD (1994) Getting a grip on myosin. Cell 78: 733-737

Voos W, Gambill BD, Guiard B, Pfanner N and Craig EA (1993) Presequence and mature part of preproteins strongly influence the dependence of mitochondrial import on heat shock protein 70 in the matrix. J. Cell Biol. 123: 119-126

HOW MITOCHONDRIA RECOGNIZE AND BIND PRECURSOR PROTEINS AT THE SURFACE

Volker Haucke
Biozentrum, University of Basel
Klingelbergstrasse 70
CH-4056 Basel
Switzerland

The mitochondrial protein import machinery

Mitochondria are essential organelles which perform many important functions, such as respiration, oxidative phosphorylation and the synthesis of heme, lipids and amino acids. Because the mitochondrial genome encodes only a few proteins, most mitochondrial precursor proteins are synthesized in the cytosol and are subsequently imported into mitochondria. Almost all mitochondrial proteins are thought to be imported by the same general machinery (Fig. 1) (Attardi and Schatz, 1988).
Most mitochondrial proteins are synthesized as larger precursors which contain a basic and amphipathic targeting signal of 15-35 amino acids which can adopt an α-helical conformation (Fig. 2). These targeting signals are always found at or close to the amino-terminus and are sufficient to direct an attached non-mitochondrial passenger protein into the mitochondrial matrix (Roise and Schatz, 1988).
Precursors first bind to receptors on the mitochondrial surface and then insert into the translocation channel in the outer mitochondrial membrane. Precursors must be in an unfolded conformation in order to be translocated across the mitochondrial membranes (Eilers and Schatz, 1986); targeting to and insertion into the outer membrane is in many cases aided by cytosolic, ATP-dependent chaperones, such as mitochondrial import stimulating factor (MSF) (Hachiya et al., 1993). The mitochondrial outer membrane contains at least four different proteins that function as receptors for protein import: Mas20p, Mas22p, Mas37p and Mas70p (Lithgow et al., 1995). The receptor-bound precursor is then transferred into a proteinaceous channel in the outer membrane, which is composed of several subunits (i.e. Isp42p, Isp6p) (Fig. 1). Outer membrane proteins,

NATO ASI Series, Vol. H 96
Molecular Dynamics of Biomembranes
Edited by Jos A. F. Op den Kamp

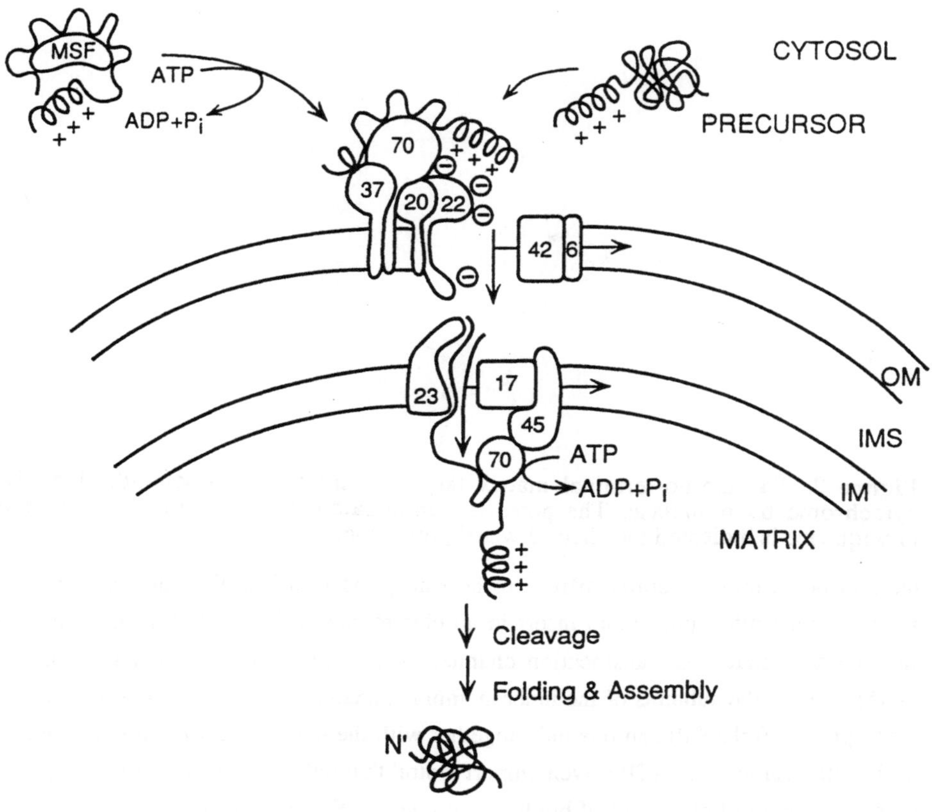

Figure 1. The mitochondrial protein import machinery. Precursor proteins can interact with ATP-dependent cytosolic chaperones, such as MSF (mitochondrial import stimulating factor). They are then recognized by receptors (Mas20p, Mas22p, Mas37p and Mas70p) on the mitochondrial surface and subsequently inserted into the outer membrane (OM) translocation channel. Outer membrane proteins may diffuse laterally to become firmly anchored in the outer membrane, whereas most other proteins interact with the inner membrane (IM) translocation machinery. This step requires an electric potential ($\Delta\varphi$) across the inner membrane. Translocation across the inner membrane requires in addition the ATP-dependent action of mitochondrial hsp70 on the matrix side of the inner membrane. Cleavable matrix targeting sequences are removed by the general matrix processing peptidase (MPP). Finally, the imported proteins undergo folding and, in some cases, assembly.

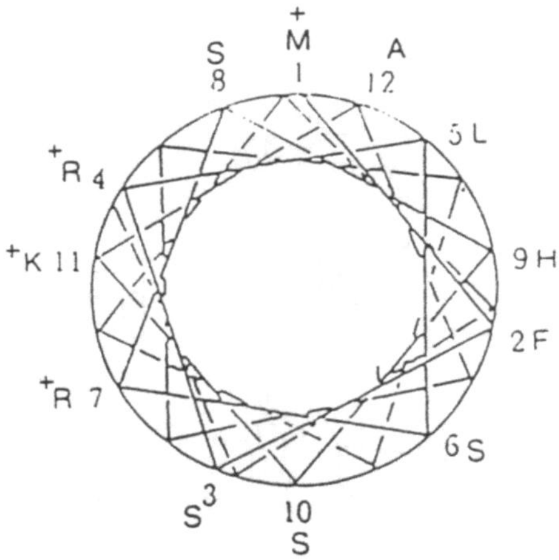

Figure 2. The amino-terminal matrix targeting signal of mitochondrial NADH-cytochrome b5 reductase. The potential amphipathic helical conformation of the presequence is indicated by a helical wheel projection.

such as porin may laterally diffuse to be stably inserted into the outer membrane, whereas most other precursors insert in an electrochemical potential-dependent step into the proteinaceous translocation channel in the mitochondrial inner membrane. Isp45p, one of the subunits of the inner membrane channel, is bound to a 70 kDa heat-shock protein (mhsp70) and this unit, together with the matrix-localized homologue of GrpEp may serve as an ATP-driven 'import motor' that pulls precursors across the inner membrane by multiple cycles of binding and release (Kronidou et al., 1994; Schneider et al., 1994; Glick, 1995) (Fig. 1). The matrix-targeting signals of most precursor molecules are then removed by an endoproteinase (Fig. 1, MPP), which consists of two non-identical, but homologous subunits. Finally, the imported protein undergoes folding and assembly, which is often aided by the chaperonins hsp60 and cpn10.

Precursor recognition at the mitochondrial surface

Protein import into mitochondria is initiated by the binding and recognition of cytoplasmically-synthesized precursor proteins by receptor proteins in the mitochondrial outer membrane. Early studies with *Neurospora crassa* and *Saccharomyces cerevisiae* mitochondria identified two import-receptor subunits: a protein of about 20 kDa, termed MOM19 in *N. crassa* and Mas20p in yeast (Ramage et al., 1993), and a protein of about 70 kDa, termed MOM72 in *N. crassa* and Mas70p in yeast (Hines et al., 1990; Steger et al., 1990). Recently, two other receptor components have been identified: a 22 kDa protein, termed MOM22 in *N. crassa* and Mas22p in yeast (Kiebler et al., 1993; Lithgow et al., 1994) and a 37 kDa protein, termed Mas37p, which so far has been identified only in yeast (Gratzer et al., 1995). Mas70p and Mas37p form a heterodimeric subcomplex that co-purifies from solubilized outer membranes (Gratzer et al., 1995). Similarly, Mas20p and Mas22p appear to be part of another subcomplex in the mitochondrial outer membrane. A complex containing at least three of the four known protein import receptor subunits has been isolated from mitochondria of *N. crassa* (Kiebler et al., 1990). Import studies with isolated yeast mitochondria have shown that different subsets of precursors differ in their dependence on extramitochondrial ATP and the presence of Mas37p/ Mas70p (Wachter et al., 1994; Gratzer et al., 1995). The import of one class of precursors, such as the ADP/ ATP carrier (AAC), cytochrome c_1, the F_1-ATPase β-subunit ($F_1\beta$), and the mitochondrial isozyme III of alcohol dehydrogenase (ADH III), is strongly dependent on extramitochondrial ATP and the Mas37p/ Mas70p subcomplex. These precursors appear to be delivered to the Mas37p/ Mas70p complex at the mitochondrial surface while being attached to cytosolic chaperones, such as MSF (Hachiya et al., submitted). In contrast, the import of the second class of precursors, exemplified by cytochrome b_2, hsp60 and DHFR-containing fusion proteins, occurs independently of extramitochondrial ATP and Mas37p/ Mas70p.

Recently, we have attempted to determine which mitochondrial receptors specifically recognize amino-terminal mitochondrial presequences (Haucke et al., 1995). To this aim we have studied the productive binding of several mitochondrial precursors to the surface of yeast mitochondria. We first incubated the *in vitro* translated, radiolabeled precursors with isolated yeast mitochondria that had been uncoupled with the protonophore carbonyl-cyanide m-chlorophenylhydrazone (CCCP). We then reenergized the mitochondria by inactivating CCCP and adding an energy source and chased the prebound precursor into the organelles. Finally, we assessed the effectiveness of the chase by determining the amount of the prebound precursor that had become inaccessible to added protease. When Su9-DHFR (a fusion protein consisting of the first 69 amino acids of the *N. crassa* F_0F_1-ATPase subunit 9 precursor fused to mouse DHFR) was allowed to bind to uncoupled mitochondria isolated either from the wild-type or a mas20 deletion strain, only the wild-type mitochondria were able to productively bind the Su9-DHFR precursor (Fig. 3). Mitochondria lacking both Mas37p and Mas70p were as active as wild-type mitochondria in the binding assay (not shown). In contrast, productive binding of the ADP/ATP carrier (AAC, a protein which lacks a positively charged mitochondrial targeting sequence) was unaffected by the absence of Mas20p, but was completely dependent on the Mas37p/ Mas70p complex (Fig. 3 & data not shown).

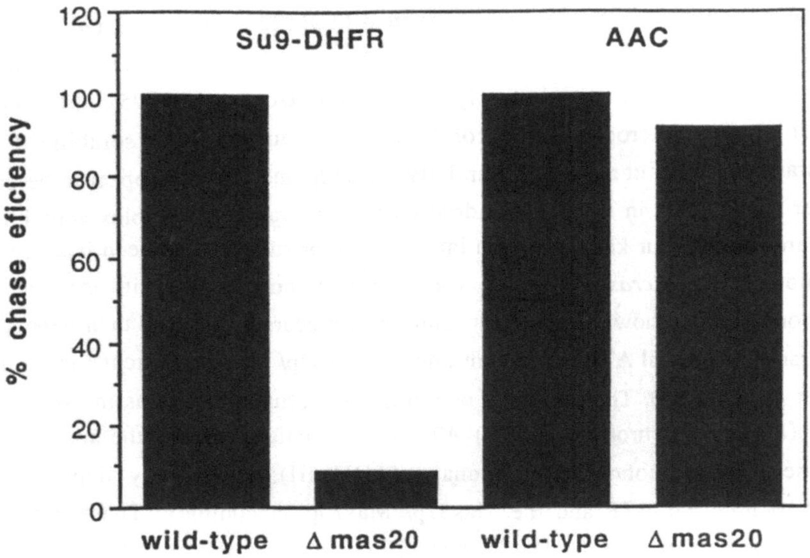

Figure 3. Productive binding of Su9-DHFR, but not AAC requires the presence of the import receptor Mas20p. The amount of precursor chased into wild-type mitochondria was taken as 100%.

Productive binding of Su9-DHFR could also be strongly inhibited by low concentrations of a mitochondrial presequence peptide (added during the initial binding step), whereas productive binding of the AAC was much less affected (not shown). Thus, productive binding of Su9-DHFR to yeast mitochondria is predominantly mediated by Mas20p and probably occurs via the precursor's targeting sequence. To investigate the mechanism by which different precursors are bound to the mitochondrial surface, we measured productive binding of the precursors of Su9-DHFR, ADHIII, hsp60, the Rieske Fe/S protein, and AAC at different salt concentrations. Su9-DHFR and hsp60 belong to the class of precursors whose import is preferentially mediated by Mas20p and does not require extramitochondrial ATP (and, hence, cytosolic chaperones). The import of ADHIII, the Rieske Fe/S protein and AAC is dependent on Mas37p/ Mas70p and requires ATP in the cytosol for efficient import (Wachter et al., 1994). The precursors were allowed to bind to deenergized mitochondria in the presence of varying salt concentrations, the mitochondria were reisolated, washed, and chased in low-salt buffer. The addition of 50-200 mM NaCl to the binding buffer inhibited productive binding of Su9-DHFR and hsp60 almost completely, but had little effect on productive binding of AAC (Fig. 4). Productive binding of ADHIII and the Rieske Fe/S protein (both are synthesized as precursors with a basic mitochondrial targeting sequence) showed an intermediate salt-sensitivity.

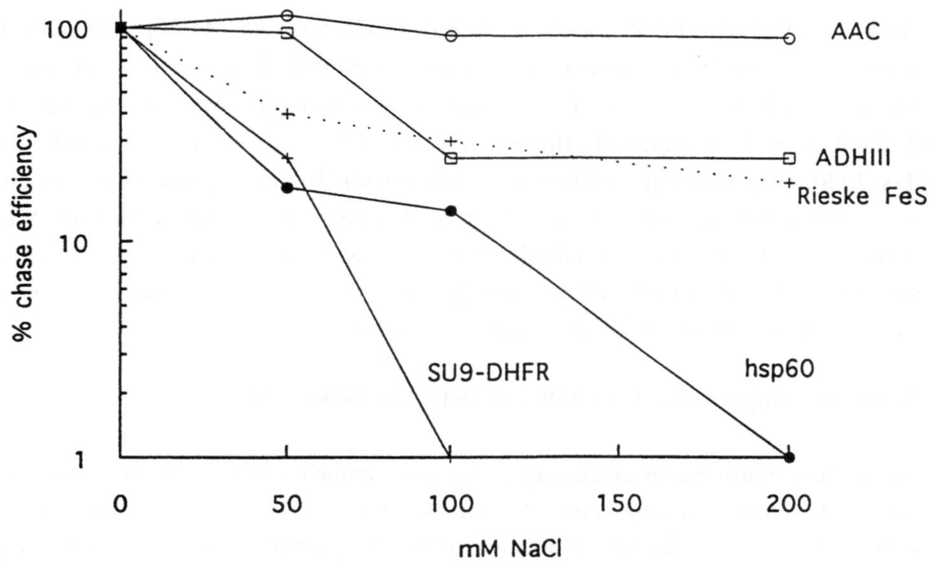

Figure 4. Productive binding of the precursors of Su9-DHFR and hsp60 is salt-sensitive. The precursors of Su9-DHFR, hsp60, ADHII, the Rieske Fe/S protein and the AAC were translated *in vitro* and incubated with deenergized mitochondria in import buffer containing the indicated concentrations of NaCl. The prebound precursors were chased into the mitochondria, surface-bound precursor molecules were removed by protease treatment and the imported material was analyzed by SDS-PAGE, fluorography and densitometric quantification of the bands. The amount of precursor chased in the absence of added NaCl was taken as 100% (adapted from Haucke et al., 1995).

The experiments outlined above and described elsewhere (Haucke et al., 1995; Lithgow et al., submitted) indicate that the import receptor Mas20p binds mitochondrial precursor proteins through electrostatic interactions with the positively charged presequence and may, together with its partner subunit Mas22p, function as a 'presequence receptor'. Recent experiments from our laboratory (Haucke et al., submitted) have shown that Mas20p and Mas70p interact with one another via the tetratricopeptide repeat (TPR) motifs in their cytosolic domains. Mas20p containing a mutation in its single TPR domain fails to associate with Mas70p and the mitochondria are defective in the import of those precursors which require Mas37p/Mas70p for efficient import. In contrast, import of precursors imported independently of Mas37p/Mas70p occurs at wild-type rates (not shown). The phenoptype of the mutant mitochondria thus resembles that of mitochondria lacking Mas70p or Mas37p. We postulate that the two receptor subcomplexes associate with one another via the TPR domains of Mas20p and Mas70p, resulting in a multifunctional receptor that binds simultaneously to different regions of a precursor molecule.

Protein sorting events in the mitochondrial outer membrane

The protein translocation machinery in the outer mitochondrial membrane not only participates in protein import into the mitochondrial matrix, but is also involved in sorting of proteins to the outer membrane or the intermembrane space. A particularly interesting example is exemplified by mitochondrial NADH-cytochrome b_5 reductase (MCR1). The primary translation product has an amino-terminal matrix-targeting signal, followed by a stretch of 21 uncharged amino acids, which serve as a stop-transfer signal. This precursor protein is inserted into the outer membrane, but only about one third of the molecules become firmly anchored to the outer face of that membrane. The remaining molecules are further transported to the inner membrane, are cleaved by inner membrane protease 1, and are released into the intermembrane space (Hahne et al., 1994). This 'leaky stop-transfer' in the outer membrane represents a novel mechanism for targeting the product of a single gene to different compartments of the same organelle.

Conclusions

The past few years have witnessed considerable progress in understanding how mitochondria are able to bind and import cytosolically made precursor proteins. We have been able to identify many of the components involved in the early recognition and binding steps in the import process. However, the mechanisms by which this is achieved are still poorly understood. Future experiments will be designed toward reconstituting the interactions between precursor proteins and the purified receptor components and to reproduce the translocation process in proteoliposomes.

Acknowledgements

The work described here has been carried out in the laboratory of Dr. G. Schatz. I thank members of the Schatz' lab for discussion, Dr. T. Lithgow for numerous ideas and suggestions and Margrit Jäggi for help with the artwork. This work was supported by grants to G.S. from the Swiss National Science Foundation and the European Economic Community, and a fellowship to V.H. from the Boehringer Ingelheim Fonds.

References

Attardi, G. and Schatz, G. (1988). The biogenesis of mitochondria. Annu. Rev. Cell Biol. 4, 289-333.

Eilers, M. and Schatz, G. (1986). Binding of a specific ligand inhibits import of a purified precursor protein into mitochondria. Nature 322, 228-232.

Glick, B. S. (1995). Can hsp70 proteins act as force-generating motors? Cell 80, 11-14.

Gratzer, S., Lithgow, T., Bauer, R. E., Lamping, E., Paltauf, F., Kohlwein, S. D., Haucke, V., Junne, T., Schatz, G. and Horst, M. (1995). Mas37p, a novel receptor subunit for protein import into mitochondria. J. Cell Biol. 129, 25-34.

Hachiya, N., Alam, R., Sakasegawa, Y., Sakaguchi, M., Mihara, K. and Omura, T. (1993). A mitochondrial import factor purified from rat liver cytosol is an ATP-dependent conformational modulator for precursor proteins. EMBO J. 12, 1579-1586.

Hahne, K., Haucke, V., Ramage, L. and Schatz, G. (1994). Incomplete arrest in the outer membrane sorts NADH-cytochrome b_5 reductase to two different submitochondrial compartments. Cell 79, 829-839.

Haucke, V., Lithgow, T., Rospert, S., Hahne, K. and Schatz, G. (1995). The yeast mitochondrial import receptor Mas20p binds precursor proteins through electrostatic interaction with the positively charged presequence. J. Biol. Chem. 270, 5565-5570.

Hines, V., Brandt, A., Griffiths, G., Horstmann, H., Brutsch, H. and Schatz, G. (1990). Protein Import into Yeast Mitochondria Is Accelerated by the Outer Membrane Protein Mas70. EMBO J. 9, 3191-3200.

Kiebler, M., Keil, P., Schneider, H., van der Klei, I., Pfanner, N. and Neupert, W. (1993). The mitochondrial receptor complex: a central role of MOM22 in mediating preprotein transfer from receptors to the general insertion pore. Cell 74, 483-492.

Kiebler, M., Pfaller, R., Soellner, T., Griffiths, G., Horstmann, H., Pfanner, N. and Neupert, W. (1990). Identification of a mitochondrial receptor complex required for recognition and membrane insertion of precursor proteins. Nature 348, 610-616.

Kronidou, N. G., Oppliger, W., Bolliger, L., Hannavy, K., Glick, B. S., Schatz, G. and Horst, M. (1994). Dynamic interaction between Isp45 and mitochondrial hsp70 in the protein import system of yeast mitochondrial inner membrane. Proc. Natl. Acad. Sci. USA 91, 12818-12822.

Lithgow, T., Glick, B. S. and Schatz, G. (1995). The protein import receptor of mitochondria. Trends Biochem. Sci. 20, 98-101.

Lithgow, T., Junne, T., Suda, K., Gratzer, S. and Schatz, G. (1994). The mitochondrial outer membrane protein Mas22p is essential for protein import and viability of yeast. Proc. Natl. Acad. Sci. USA 91, 11973-11977.

Ramage, L., Junne, T., Hahne, K., Lithgow, T. and Schatz, G. (1993). Functional cooperation of mitochondrial protein import receptors in yeast. EMBO J. 12, 4115-4123.

Roise, D. and Schatz, G. (1988). Mitochondrial presequences. J. Biol. Chem. 263, 4509-4511.

Schneider, H. C., Berthold, J., Bauer, M. F., Dietmeier, K., Guiard, B., Brunner, M. and Neupert, W. (1994). Mitochondrial Hsp70/ MIM44 complex facilitates protein import. Nature 371, 768-773.

Steger, H. F., Söllner, T., Kiebler, M., Dietmeier, K. A., Pfaller, R., Trulzsch, K. S., Tropschug, M., Neupert, W. and Pfanner, N. (1990). Import of ADP/ATP carrier into mitochondria - two receptors act in parallel. J. Cell Biol. 111, 2353-2363.

Wachter, C., Schatz, G. and Glick, B. S. (1994). Protein import into mitochondria: the requirement for external ATP is precursor-specific whereas intramitochondrial ATP is universally needed for translocation into the matrix. Mol. Biol. Cell 5, 465-474.

The General Features of Membrane Traffic During Endocytosis in Polarized and Non-Polarized Cells

Ira Mellman
Department of Cell Biology
Yale University School of Medicine
333 Cedar Street, PO Box 208002
New Haven, Connecticut 06520-8002
US

One of the most remarkable features of cellular membranes in their ability to engage in dynamic interactions across organelle boundaries without compromising either the biochemical specificity or the function of the interacting organelles. This feature is particularly well illustrated in the case of endocytosis, where the continuous formation of endocytic vesicles can result in the interiorization of membrane vesicles accounting for over 100-200% of the plasma membrane's entire surface area every hour (Steinman et al., 1983). Although this membrane is delivered to and recycled from an array of endocytic organelles, the individual properties of these organelles and of the plasma membrane itself are preserved. The mechanisms responsible for governing these events lie at the foundation of our understanding of intracellular membrane transport and include the ability of cells to selectively sort membrane and soluble proteins into specific transport vesicles as well as the ability to target those transport vesicles to their appropriate destinations. This chapter will review the salient features of the endocytic pathway in animal cells and discuss our understanding of mechanisms of protein sorting. Although it is usually more advantageous to consider such matters in the simplest available cell type, in this case, it turns out that our consideration of membrane transport in polarized epithelial cells over the past several years has yielded fundamental principles that would not have otherwise been immediately obvious.

NATO ASI Series, Vol. H 96
Molecular Dynamics of Biomembranes
Edited by Jos A. F. Op den Kamp

Types of endocytosis in animal cells.

Classically, two basic forms of endocytosis are distinguished, phagocytosis and pinocytosis. Phagocytosis refers to the uptake of large particles (>1 μM in diameter) such as bacteria, yeast, or other microorganisms. It is most typical of professional phagocytes of the immune system (macrophages, neutrophils) although most cell types can be induced to mediate phagocytosis at least inefficiently. Phagocytosis is a regulated process, and only occurs following the binding of ligand-coated particles to receptors capable of triggering uptake (Greenberg, 1995). In leukocytes, receptors for the Fc domain of IgG (Fc receptors) are among the most common "phagocytic receptors". They appear to work by generating a cascade of events beginning with inductive tyrosine phosphorylation of a variety of cytoplasmic substrates and ending with the ordered assembly of actin-containing microfilaments (Stossel, 1989). It is the actin polymerization that is thought to provide the motile force resulting in the ingestion of the bound particle essentially by pseudopod extension. The internalized particle resides within a "phagosome" within the cytoplasm which then fuses with hydrolase-rich lysosomes. Exposure to these lysosomal hydrolases generally results in the digestion of the particle. Interestingly, however, much of the plasma membrane used during particle internalization escapes degradation and is returned or recycled back out to the plasma membrane.

The second major form of endocytosis is pinocytosis, or "cell drinking". This appears to be a constitutive process in which there is a continuous formation of small (0.1 - 0.2 μM) vesicles at the plasma membrane (Steinman et al., 1983). These vesicles contain solutes dissolved in the extracellular medium (fluid-phase pinocytosis) or macromolecules bound to any of a wide variety of specific cell surface receptors (receptor-mediated endocytosis). In general, macromolecules internalized either in the fluid phase or after binding to receptors are ultimately delivered to lysosomes and degraded. Enormous quantities of fluid and membrane are internalized as a result of the ongoing endocytic activity, exceeding by many-fold the capacity of cells to synthesize membrane proteins and lipids. Accordingly, it has long been appreciated that, unlike internalized cargo, most of the membrane that is internalized as endocytic vesicles must avoid lysosomal degradation and be reutilized or recycled. Over the past several years, the basic features of how cells accomplish the continuous accumulation of internalized solutes in the face of continuous bidirectional membrane transport have become well established. So much so, that understanding the how the endocytic pathway works provides the clearest and most instructive example for understanding the logic underlying all types of intracellular vesicular transport.

The basic pathway of receptor-mediated endocytosis.

Receptor-ligand complexes that form at the plasma membrane are selectively accumulated at regions of the cell surface that are highly differentiated for forming nascent endocytic vesicles (Pearse and Robinson, 1990; Trowbridge et al., 1993). These domains are referred to as "clathrin-coated pits" due to the presence of a characteristic cytoplasmic coat consisting largely of the 180 kD protein clathrin as well as several associated proteins (see below). These coated pits bud off to yield free coated vesicles which rapidly uncoat, perhaps due to the activity of a cytosolic HSP70. After uncoating, the vesicles then fuse with early endosomes, a heterogeneous array of tubular-vesicular structures found throughout the cytoplasm (Kornfeld and Mellman, 1989). This fusion, which appears to be under control of the monomeric GTPase rab5 presumably in conjunction with yet to be identified v- and t-SNARE proteins (Bucci et al., 1992; Gorvel et al., 1991), results in the delivery of the receptor-ligand complex to an environment which is slightly acidic (pH 6.3-6.8) but generally hydrolase-poor. The early endosome lumen is, however, sufficiently acidic to favor the dissociation of many ligands from their receptors. The vacant receptors then selectively accumulate in the early endosome's tubular extensions which, in turn, bud off to yield a vesicles that migrate on microtubule tracks to the perinuclear region and accumulates as a distinct population of recycling vesicles near the microtubule organizing center (Kornfeld and Mellman, 1989; van der Sluijs et al., 1995; Yamashiro et al., 1984). From this population, receptors recycle back to the plasma membrane for re-use.

The discharged ligands have a decidedly different fate (Kornfeld and Mellman, 1989). Free in the endosome lumen, the ligands are effectively sorted away from the receptors they used to enter the cell largely for reasons of simple geometry. Since the tubular extensions of the early endosome that represent sites of nascent recycling vesicle formation (Geuze et al., 1983, 1984) have, by definition, a higher surface to volume ratio relative to the spherical clathrin-coated vesicles that mediated initial uptake, a relatively small amount of fluid-dissolved ligands can follow the ligands into the tubules (Marsh, 1986). Instead, most of the ligands accumulate in the more vesicular portions of the early endosome complex. After all the tubules have pinched off, these vesicular elements (termed carrier vesicles) also bind to microtubules and travel to the perinuclear region (Gruenberg et al., 1989). However, rather than joining the recycling vesicle population, the vesicles fuse with a third element of the endosomal system termed late endosomes (Kornfeld and Mellman, 1989). In contrast to early endosomes, late endosomes are enriched in lysosomal hydrolases and begin the degradation of the ligands and other macromolecules they receive. Late endosomes also have a lower internal pH than do early endosomes, being as acidic as pH 5.3. Finally, the ligands are transferred to lysosomes, the last stop for internalized molecules on the endocytic pathway. Although the relationship between late endosomes and lysosomes remains somewhat obscure, it seems most likely that the two structures are in something of a dynamic equilibrium with each other such that when all membrane and content available for degradation is digested, late endosomes by definition become lysosomes (Kornfeld and Mellman, 1989).

Thus, even in simple non-polarized cells, early endosomes serve as a critical sorting station in which incoming receptor ligands are dissociated and targeted to two distinct destinations: recycling vesicles vs. late endosomes and lysosomes. To some extent, sorting occurs as a result of the characteristic geometry of these organelles, but it is apparent that Euclidean logic alone cannot completely explain their mode of action. Compelling evidence exists, particularly in polarized epithelial cells (see below), that early endosomes actively sort recycling receptors by selectively accumulating them in nascent recycling vesicles in a signal-dependent fashion directly analogous to the initial localization of receptors in clathrin-coated pits on the plasma membrane. It is of fundamental and general importance that such a sorting station exists. First, early endosomes provide a site for receptors to discharge their ligands without risking exposure to potentially destructive lysosomal hydrolases. In addition, early endosomes clearly exhibit the principles that must underlie the mechanism of molecular sorting between any two organelles that engage in continuous bidirectional transport. An example is the poorly understood "intermediate compartment" or "cis-Golgi network" between the ER and the Golgi complex, thought to be the site from which resident ER proteins are returned from the "Golgi" back to the ER (Mellman and Simons, 1992).

Biochemistry of molecular sorting on the endocytic pathway: clathrin-coated pits.

While investigation of the secretory pathway has led to our greatest understanding of how vesicles recognize and fuse with each other, analysis of the endocytic pathway has revealed more insights into the mechanisms of molecular sorting. First among these is the paradigm established by receptor localization at clathrin-coated pits. From the work of a large number of groups, it is now clear that most receptors accumulate at clathrin-coated pits due to the presence of characteristic, if degenerate, signals found on the cytoplasmic domains of all coated pit membrane proteins (Trowbridge et al., 1993). These signals most often involve a critical tyrosine (less commonly phenylalanine) residue surrounded by a quasi-predictable array of other amino acids. In the human LDL receptor, for example, the tetrapeptide NPXY is necessary and sufficient to allow efficient receptor accumulation at clathrin-coated pits. More recently, a second coated pit localization domain has been identified which involves a critical dileucine motif. Less well characterized, the dileucine-based signal seems to be more common among receptors expressed in leukocytes; for example, it is the internalization motif used in the case of the IgG Fc receptors mentioned above. Why this is the case is unknown, but may reflect the fact that tyrosine-containing motifs similar to the coated pit motif in leukocyte receptors are also involved in the recruitment of cytoplasmic tyrosine kinases in response to receptor activation. The binding of such a kinase might in such cases block an endocytosis signal that would otherwise be required for mediating receptor internalization and down regulation.

To understand how these coated pit localization motifs are likely to work, it is necessary to explore what we know concerning the structure and function of clathrin coated pits. As mentioned above, coated pits consist principally of the protein clathrin which exists as a trimer of three 180 kD heavy chains each complexed with a 30-35 kD light chain (Pearse and Robinson, 1990). These clathrin trimers form characteristic three-armed structures termed triskelions, similar in shape to the fertility symbols of many ancient cultures. In vitro, these triskelions will self-assemble into icosahedral cages of pentagonal and hexagonal arrays similar in shape to the clathrin coats themselves.

In addition to clathrin, coated pits and coated vesicles contain a second protein complex known as adaptors (Robinson, 1994). Two types of adaptor complexes are known. HA-1 adaptors (so named because they were the first to elute from a hydroxylapatite column) and HA-2 adaptors each consist of four subunits: two 100 kD subunits (α, β in HA-2; β', γ in HA-1), one ~50 kD subunit (AP47 in HA-1, AP50 in HA-2), and one ~20 kD subunit (AP20 vs. AP17, respectively). The adaptor complexes serve at least three critical functions. First, they bind to the membrane. Second, they provide the membrane binding sites for clathrin assembly. Third, they appear to be the machinery that decodes the coated pit localization signals in receptor cytoplasmic domains. Precisely which subunit is responsible for these functions is not yet certain, although it seems that the α and γ adaptins have conserved clathrin binding sites as well as domains that may be involved in membrane attachment. The β and β' subunits are almost identical to each other.

The two adaptor complexes differ functionally in one important respect, namely that each binds to different organelles. Whereas the HA-2 complex binds to the plasma membrane and nucleates the formation of cell surface coated pits, the HA-

1 complex binds to elements of the trans-Golgi network (TGN), another major site of coated pit formation in the cell. Like their plasma membrane-derived counterparts, TGN-derived coated vesicles also appear to be targeted to endosomes (perhaps even early endosomes) and are involved in the delivery of newly synthesized lysosomal components to the endocytic pathway (Kornfeld and Mellman, 1989). These components, including receptors such as the mannose-6-phosphate receptor and perhaps lysosomal membrane glycoproteins such as the lgp/lamp family, appear to have cytoplasmic domain signals for localization at clathrin-coated buds of the TGN that are closely related by sequence to those involved in coated pit localization at the plasma membrane. Indeed, both tyrosine and dileucine-based motifs, presumably modified by sequence context to permit interaction with HA-1 adaptors rather than (or in addition to) HA-2 adaptors have been associated with targeting from the Golgi complex directly to endosomes. Whether early or late endosomes represent the destination for TGN-derived coated pits remains uncertain, although there is limited evidence that early endosomes can at least in part serve this function (Ludwig et al., 1991).

Finally, although both adaptor complexes recruit from the same pool of clathrin, it is not yet clear why the two complexes themselves are recruited to different organelles. In addition, the binding of HA-1 to the TGN is clearly mediated in association with a soluble small GTP binding protein ARF (ADP ribosylation factor) (Stenbeck et al., 1992), while the binding of HA-2 to the plasma membrane appears not to involve ARF, at least not a soluble form of ARF.

There are two basic possible mechanisms as to how adaptors bind to the plasma membrane. Since it is apparent that they can interact with the cytoplasmic tails of coated pit receptors, one possibility is that receptor tails form the adaptor binding sites. Thus, receptors would essentially nucleate the formation of their own coated pits, presumably recruiting more receptors and more adaptors to the site. There is some indirect evidence in support of this mechanism. The other possibility is that a generalized non-receptor protein serves as a site for adaptor recruitment; thus, the coated pit would form first, recruiting receptors subsequently. One recent suggestion is that the neuronal protein synaptotagmin may serve such a function, since it has been shown to interact with HA-2 adaptors with high affinity in vitro. However, synaptotagmin is not normally a plasma membrane protein (it is found only in synaptic vesicles) and synaptotagmin knock-outs in mice and Drosophila do not appear to interfere with coated pit formation even in neurons. Nevertheless, this remains a possibility since there is an increasing number of synaptotagmin-related proteins expressed in both neuronal and non-neuronal cells, any one of which might serve as the authentic adaptor binding site (Zhang et al., 1994).

Mechanism of endosome acidification.

Sorting of ligands and receptors in endosomes in non-polarized cells involves two distinct activities: low pH-mediated dissociation of the receptor-ligand complex and physical segregation of recycling receptors away from the bulk of the dissociated ligand pool. A fair amount of detailed has been learned concerning the events associated with the ligand dissociation step; we are only now beginning to unravel how receptor segregation occurs.

Acidification of endosomes, as well as of all other organelles of the vacuolar pathway that maintain acidic internal pH's (i.e., secretory granules, lysosomes, TGN), occurs via the activity of a newly appreciated class of proton pump (Mellman et al., 1986). These pumps, designated vacuolar ATPases (V-ATPase), are similar in overall structure to the mitochondrial/prokaryotic F_1F_0 class of ATPase and accomplish the transport of protons from the cytosol into the endosome lumen in exchange for ATP hydrolysis (Nelson, 1992). The pumps are electrogenic, meaning that proton transport results in the net accumulation of positive charges in the endosome, creating an interior-positive membrane potential in addition to a pH gradient. Although proton transport is not coupled directly to the co-transport of another ion, the development of the membrane potential is controlled by the permeability characteristics of the endosomal membrane to alkali cations and to anions. The potential gradient is also influenced, in early endosomes, by the presence of at least one additional ion transport ATPase, namely the Na^+,K^+-ATPase whose topology also results in the accumulation of net positive charge by pumping 3 Na^+ into the endosome for every 2 K^+ removed. These are important considerations since the V-ATPase is exceedingly potential sensitive. In the presence of an interior-positive membrane potential, the V-ATPase will no longer pump protons. This provides the most likely mechanism by which internal pH is regulated in endocytic organelles (Fuchs et al., 1989; Schmid et al., 1989). Early endosomes, which have Na^+, K^+-ATPase activity as well as limited K^+ permeability, are similarly limited in the extent to which they can generate a pH gradient, achieving values of pH 6.3 - 6.8. Late endosomes and lysosomes, in contrast, typically achieve internal pH's of 4.7 - 5.7, appropriate for creating an optimal environment for lysosomal hydrolase activity as well as for protein denaturation.

Molecular sorting of receptors in endosomes and the TGN.

After discharging their ligands, itinerant plasma membrane receptors must then return to the cell surface. That they do this has long been thought to occur in the absence of any specific signal, i.e., "by default". Evidence in favor of default recycling from endosomes includes the fact that NBD-labeled ceramide-containing glycolipid analogs, when used to label endosomes in cultured cells, rapidly return to the cell surface at kinetics indistinguishable from those of authentic recycling receptors (e.g., transferrin receptor). For many receptors, there is also usually a substantial intracellular receptor pool found in early endosomes and/or recycling vesicles. The existence of such a pool may reflect an equilibrium which reflects the selective and concentrative internalization of receptors via clathrin-coated pits and the non-selective, non-concentrative recycling of receptors from endosomes back to the cell surface. Finally, many receptors can be prevented from recycling and instead directed towards lysosomes simply by cross-linking following interaction with a non-dissociating polyvalent ligand. This observation suggested that something out of the ordinary needed to occur to access the lysosomal pathway. Indeed, transport of some receptors to and degradation in lysosomes is an important function known as "down regulation" that regulates the surface expression of many receptors involved in signal transduction.

More recent evidence, however, has demonstrated that receptor recycling cannot occur strictly by default. It has become increasingly clear that endosomes serve as a complex sorting station from which various membrane proteins are directed to very specific destinations. For example, in neurons or neuroendocrine cells, endosomes serve not only as intermediates in the pathway of transferrin receptor recycling but also in the formation of synaptic vesicles. Thus, endosomes must have the capacity of distinguishing between synaptic vesicle components and transferrin receptors in order to sort them into separate vesicles that form from a continuous membrane. Similar situations exist for the targeting of the insulin-regulated glucose transporter (GLUT4) in adipocytes, polymeric IgA receptors into transcytotic vesicles in epithelial cells, as well as for newly synthesized MHC class II molecules in lymphocytes. However, it cannot be inferred from these examples whether it is the recycling receptors, the specialized membrane protein (e.g., GLUT4), or both that contain specific signals governing their transport.

Recent work on epithelial cells appears now to have solved this fundamental problem and, in the process, also solved how these and perhaps other polarized cells achieve their characteristic asymmetric distribution of plasma membrane components (Matter and Mellman, 1994).

It is well known that epithelial cells are differentiated into two biochemically and functionally distinct plasma membrane domains (Mellman et al., 1993; Rodriguez-Boulan and Nelson, 1989). The apical domain, facing the lumen of any given organ, is often characterized by multiple actin-containing microvilli that increase the cell's effective surface area, facilitating the absorptive function associated with many epithelial cell types. The apical membrane in such cells also contains a wide array of highly specialized hydrolases and transporters for amino acids, sugars, and other compounds that also serve absorptive activity. The apical plasma membrane is also highly enriched in complex glycolipids as well as GPI-anchored proteins. On the other hand, the basolateral domain contains receptors, transporters, and channels that are more closely associated with the "house-keeping" functions common to all cells. Thus, early on the concept emerged that the apical surface, being epithelial cell-specific, was also the specialized surface requiring that apical components possess specific targeting determinants or other features that allowed their selective apical targeting. While not completely wrong, this concept has now turned out to be fundamentally incorrect.

With the advent of cell culture systems allowing the growth of model epithelial cell lines, such as Madin-Darby canine kidney (MDCK) cells, on permeable supports, it became possible to search for targeting signals responsible for polarized transport. We first noted that IgG Fc receptors, when transfected into MDCK cells, were expressed largely at the basolateral surface unless specific splice isoforms were used that inactivated the receptor's dileucine-based coated pit localization signal (Hunziker and Mellman, 1989). In this case, the receptor was transported from the TGN to the apical, rather than the basolateral, surface. These studies were subsequently confirmed for several other examples, both by us and by other groups leading to our current understanding that targeting of simple membrane proteins in epithelial cells is controlled by a distinct basolateral targeting signal found in the cytoplasmic domain. The absence or inactivation of this signal typically led to selective appearance on the apical surface rather than random distribution across both surfaces. Accordingly, it appears that sorting in epithelial cells involves the activity of two hierarchically arranges signals: a dominant basolateral signal and an apical signal that is only revealed upon inactivation of the basolateral signal (Hunziker et al., 1991; Matter et al., 1992; Matter et al., 1993; Matter et al., 1994).

These findings had two very important general implications (Matter and

Mellman, 1994). First, the fact that virtually any basolateral protein could be turned into an apical protein simply by inactivating its basolateral targeting determinant indicated that there was nothing particularly special about apical proteins, and that the apical "signal" must be a generally shared feature of most if not all membrane glycoproteins. Second, the fact that basolateral targeting signals were found on proteins that were expressed in both polarized and non-polarized cells suggested that their function may not be limited to epithelial cells. Indeed, there is evidence that a wide array of epithelia make use of the same signals. Even more interestingly, even neurons -- a highly polarized non-epithelial cell type -- may also rely on the same logic: for some proteins, apical expression in epithelial cells correlates with axonal expression in cultured neurons. Perhaps even non-polarized cells are continuously sorting, albeit non-productively, apical from basolateral proteins via cognate pathways. Such a situation would imply that even lymphocytes may be polarized cells waiting to happen; which, of course, they are (in the case of cytotoxic T-cells recognizing their targets).

One further important finding is that these basolateral and apical targeting determinants appear to be decoded not only in the TGN, but also in endosomes (Matter and Mellman, 1994). This is evident from the fact that some epithelia, e.g., intestinal and hepatic epithelial cells, do not rely entirely on the TGN to sort apical from basolateral proteins on the biosynthetic pathway. Instead, both apical and basolateral proteins may be transported to the basolateral surface, internalized, and separated in endosomes. Proteins destined for the apical surface are then sequestered into specific vesicles that are targeted to the apical surface (transcytosis) while basolateral proteins are returned via recycling back to their surface of origin. This can also be demonstrated by transfection in MDCK cells that normally sort completely in the TGN. Accordingly, endosomes and the TGN use the same logic to distinguish between apical and basolateral proteins. Thus, endosomes must have the ability to decode a cytoplasmic domain sorting signal on the recycling pathway: i.e., recycling does not occur strictly by default.

The nature of the basolateral targeting determinants has been characterized and, surprisingly, they appear to resemble determinants used for coated pit localization. Perhaps the most extensive analysis has been performed using LDL receptor, whose cytoplasmic tail has been shown to contain two distinct determinants (Matter et al., 1994). The first overlaps with the receptor's tyrosine-dependent coated pit signal. Although tyrosine-dependent itself, other residues absolutely required for endocytosis are not required for basolateral sorting, and vice versa. The second signal occurs in the distal COOH-terminal region of the receptor's cytoplasmic domain. It, too, is tyrosine-dependent, but this tyrosine is not at all involved in coated pit localization. Detailed analysis of the two signals by mutagenesis has revealed a crude consensus motif consisting of a tyrosine and one critical adjacent residue followed by a variable number of "unimportant" residues ending in a conserved stretch of three acidic amino acids. Although this motif will clearly not apply to all examples, some have already been predicted on this basis. In addition, it is also clear that dileucine motifs (as in Fc receptor), in addition to the tyrosine-containing motif found in LDL receptor, can accomplish basolateral targeting from both endosomes and the TGN. This again stresses the relationship, still understood, between basolateral targeting and coated pit localization.

Prospects for future discoveries.

The next major challenge, of course, will be to find the cytoplasmic proteins that are responsible for decoding and sorting basolateral proteins. Given the mystical relationship between basolateral targeting determinants and coated pit localization signals, it seems conceivable that such proteins will be analogous, in function and perhaps in structure, to clathrin adapator proteins or other known coat proteins. Work to identify these mechanisms is proceeding using a wide range of approaches including genetic screening, biochemistry, and functional reconstitution of receptor sorting in vitro. Also of importance will be to identify the mechanism of apical sorting. The fact that the ability to selectively reach the apical surface is widely distributed and does not require the presence of a cytoplasmic domain, apical targeting information is likely to be found in a proteins lumenal or membrane anchoring domain (the latter rendered less likely by the fact that many secretory glycoproteins which lack membrane spanning segments are also directed to the apical domain). One property common to all these proteins would appear to be the presence of N-linked sugar. Indeed, there is now evidence form K. Simons and colleagues that apically-targeted vesicles contain a lectin-like molecule which may, at low affinity, collect glycoproteins which are unable to interact with a higher affinity basolateral targeting system (Fiedler and Simons, 1994). Interaction with such a putative lectin might then lead to sequestration in apically-targeted vesicles, which also may represent a differentiated lipid domain.

Although there is yet much to do, the basic rules of the game have now been defined and, accordingly, the conceptual framework for proceeding has been established. Despite the difficulty of the experiments ahead, proceeding on the basis of a sound set of assumptions will make the wait bearable if not necessarily guarantee success.

REFERENCES

Bucci, C., Parton, R. G., Mather, I. H., Stunnenberg, H., Simons, K., Hoflack, B., and Zerial, M. (1992) The small GTPase rab5 functions as a regulatory factor in the early endocytic pathway. Cell 70, 715-728.

Fiedler, K., and Simons, K. (1994) A putative novel class of animal lectins in the secretory pathway homologous to leguminous lectins. Cell 77, 625-626.

Fuchs, R., Schmid, S., and Mellman, I. (1989) A possible role for Na+,K+-ATPase in regulating ATP-dependent endosome acidification. Proc. Natl. Acad. Sci. (USA) 86, 539-543.

Geuze, H. J., J. W. Slot and G. J.A.M. Strous (1983) Intracellular site of asialoglycoprotein receptor-ligand uncoupling: Double-label immunoelectron microscopy during receptor-mediated endocytosis. Cell 32, 277-287.

Geuze, H. J., Slot, J. W., Strous, G. J., Peppard, K., von Figura, K., Hasilik, A., and Schwartz, A. L. (1984) Intracellular receptor sorting during endocytosis: Comparative immunoelectron microscopy of multiple receptors in rat liver. Cell 37, 195-204.

Gorvel, J. P., Chavrier, P., Zerial, M., and Gruenberg, J. (1991) rab5 controls early endosome fusion in vitro. Cell 64, 915-25.

Greenberg, S. (1995) Signal transduction of phagocytosis. Trends in Cell Biol. 5, 93-99.

Gruenberg, J., Griffiths, G., and Howell, K. E. (1989) Characterization of the early endosome and putative carrier vesicles in vivo and with an assay of vesicle fusion in vitro. J. Cell Biol. 108, 1301-1316.

Hunziker, W., Harter, C., Matter, K., and Mellman, I. (1991) Basolateral sorting in MDCK cells involves a distinct cytoplasmic domain determinant. Cell 66, 907-920.

Hunziker, W., and Mellman, I. (1989) Expression of macrophage-lymphocyte Fc receptors in MDCK cells: polarity and transcytosis differ for isoforms with or without coated pit localization domains. J. Cell Biol. 109, 3291-3302.

Kornfeld, S., and Mellman, I. (1989) The biogenesis of lysosomes. Annu. Rev. Cell Biol. 5, 483-525.

Ludwig, T., Griffiths, G., and Hoflack, B. (1991) Distribution of newly synthesized lysosomal enzymes in the endocytic pathway of normal rat kidney cells. J. Cell Biol. 115, 1561-72.

Marsh, M., G. Griffiths, G. E. Dean, I. Mellman, and A. Helenius (1986) Three dimensional structure of endosomes in BHK-21 cells. Proc. Natl. Acad. Sci. (USA) 83, 2899 - 2903.

Matter, K., Hunziker, W., and Mellman, I. (1992) Basolateral sorting of LDL receptor in MDCK cells: The cytoplasmic domain contains two tyrosine-dependent determinants. Cell 71, 741-753.

Matter, K., and Mellman, I. (1994) Mechanisms of cell polarity: sorting and transport in epithelial cells. Curr. Opinion in Cell Biol. 6, 545-554.

Matter, K., Whitney, J. A., Yamamoto, E. M., and Mellman, I. (1993) Common signals control low density lipoprotein receptor sorting in endosomes and the Golgi complex of MDCK cells. Cell 74, 1053-64.

Matter, K., Yamamoto, E. M., and Mellman, I. (1994) Structural requirements and sequence motifs for polarized sorting and endocytosis of LDL and Fc receptors in MDCK cells. Journal of Cell Biology 126, 991-1004.

Mellman, I., Fuchs, R., and Helenius, A. (1986) Acidification of the endocytic and exocytic pathways. Ann. Rev. Biochem. 55, 663-700.

Mellman, I., and Simons, K. (1992) The Golgi complex: in vitro veritas? Cell 68, 829-40.

Mellman, I., Yamamoto, E., Whitney, J. A., Kim, M., Hunziker, W., and Matter, K. (1993) Molecular sorting in polarized and non-polarized cells: common problems, common solutions. Journal of Cell Science Supplement 17, 1-7.

Nelson, N. (1992) Organellar proton-ATPases. Curr. Opinion in Cell Biol. 4, 654-60.

Pearse, B. M. F., and Robinson, M. S. (1990) Clathrin, adaptors, and sorting. Ann. Rev. Cell Biol. 6, 151-171.

Robinson, M. S. (1994) The role of clathrin, adaptors and dynamin in endocytosis. Current Opinion in Cell Biology 6, 538-44.

Rodriguez-Boulan, E., and Nelson, W. J. (1989) Morphogenesis of the polarized epithelial cell phenotype. Science 245, 718-725.

Schmid, S., Fuchs, R., Kielian, M., Helenius, A., and Mellman, I. (1989) Acidification of endosome subpopulations in wild-type Chinese hamster ovary cells and temperature-sensitive acidification-defective mutants. J. Cell Biol. 108, 1291-1300.

Steinman, R. M., I.S. Mellman, W.A. Muller, and Z.A. Cohn (1983) Endocytosis and the recycling of plasma membrane. J. Cell Biol. 96, 1-27.

Stenbeck, G., Schreiner, R., Herrmann, D., Auerbach, S., Lottspeich, F., Rothman, J. E., and Wieland, F. T. (1992) Gamma-COP, a coat subunit of non-clathrin-coated vesicles with homology to Sec21p. Febs Letters 314, 195-8.

Stossel, T. P. (1989) From signal to pseudopod. How cells control actin assembly. J. Biol. Chem. 264, 18261-18264.

Trowbridge, I. S., Collawn, J. F., and Hopkins, C. R. (1993) Signal-dependent membrane protein trafficking in the endocytic pathway. [Review]. Annual Review of Cell Biology 9, 129-61.

van der Sluijs, P., Daro, E., Lewin, D. A., Kim, M., Peter, P., Bachmann, A., Galli, T., and Mellman, I. (1995) Rab4, rab5, and cellubrevin define different early endosome populations on the pathway of transferrin receptor recycling. Submitted.

Yamashiro, D., Tycko, B., Fluss, S., and Maxfield, F. R. (1984) Segregation of transferrin to a mildly acidic (pH 6.5) para-golgi compartment in the recycling pathway. Cell 37, 789-800.

Zhang, J. Z., Davletov, B. A., Sudhof, T. C., and Anderson, R. G. (1994) Synaptotagmin I is a high affinity receptor for clathrin AP-2: implications for membrane recycling. Cell 78, 751-60.

Mitotic fragmentation of the Golgi apparatus

Graham Warren
Imperial Cancer Research Fund
PO Box 123
44 Lincoln's Inn Fields
London WC2A 3PX

Introduction

Ten years ago I put forward a simple model to explain the fragmentation of the Golgi apparatus during mitosis in animal cells (Warren, 1985). The single copy that characterises the interphase organelle is converted into thousands of small vesicles (Lucocq, et al., 1989), a process that is thought to aid the partitioning of this organelle between daughter cells (Birky, 1983).

The model was based on the two observations, the first being the general inhibition of membrane traffic that occurs at the onset of mitosis and persists through to telophase (Warren, 1993). All vesicle-mediated steps that have been studied are inhibited, some up to 30-fold when compared to interphase cells. The second observation was that the products of Golgi fragmentation appeared to be small vesicles, similar in size to Golgi transport vesicles (Palade, 1975; Zeligs and Wollman, 1979).

I suggested that vesicle-mediated transport was inhibited at the level of vesicle fusion (Warren, 1985). Golgi transport vesicles would continue to bud at the onset of mitosis but they would be unable to fuse with their target membrane. Since the membrane lost during the budding of a vesicle is normally replenished by the fusion of an incoming vesicle, the consequence of continued budding in the absence of fusion would be the conversion of cisternae into small vesicles.

Many lines of evidence attest to the validity of this model though our most recent work suggests that there may be an additional fragmentation pathway that does not require the mechanism that generates budding transport vesicles. Nevertheless, the underlying principles for both pathways appear to be the same and lead to a model which has wider applicability.

NATO ASI Series, Vol. H 96
Molecular Dynamics of Biomembranes
Edited by Jos A. F. Op den Kamp

COP I-mediated fragmentation

COP I-coated vesicles are involved in both anterograde and retrograde transport through the Golgi stack (Letourneur, et al., 1994; Rothman, 1994). Using a cell-free system that mimics the mitotic fragmentation of the Golgi apparatus, Tom Misteli was able to show that the budding rate of COP I-coated vesicles was the same under both interphase and mitotic conditions. The same was also true for the uncoating of these vesicles suggesting that the entire mechanism responsible for generating Golgi transport vesicles continues unabated during mitosis (Misteli and Warren, 1994).

Mitotic cytosol, immuno-depleted of coatomer (one of the subunits of the COP I coat), was unable to generate Golgi transport vesicles confirming the role of COP I-coated vesicles in the fragmentation process. There was, however, an unexpected conversion of Golgi stacks into extensive tubular networks which fragmented slowly over time (Misteli and Warren, 1994).

A COP I-independent pathway

The possibility of a COP I-independent pathway was confirmed by Tom Misteli when he measured the amount of Golgi membrane that could be consumed by the COP I-mediated budding mechanism. Irrespective of the amount of cytosol present, only 50-60% of the membrane was consumed suggesting that the other 40-50% was fragmented by another mechanism. This mechanism was shown to involve tubular networks which were formed from Golgi cisternae and then fragmented eventually yielding a heterogeneous collection of vesicles larger in size than the COP I-coated vesicles (Misteli and Warren, 1995).

Periplasmic fusion

Despite the apparent difference between these two fragmentation pathways it seems likely that they have a common underlying mechanism (Rothman and Warren, 1994). The release of COP I-coated vesicles follows the fusion of the periplasmic (or non-cytoplasmic) surfaces of the cisternal membrane. This fusion differs from cytoplasmic fusions in that no specificity is required: the only membranes that can fuse are within the same compartment. Furthermore, it is conceivable that the fusion is triggered simply by bringing together the periplasmic membrane surfaces. Many different devices need to penetrate or sever membranes so this fusion trigger would permit many different devices to utilise a common fusion process. Such a device would, however, be susceptible to random triggering. The turbulent nature of the cell cytoplasm might inadvertently bring the periplasmic surfaces together and so trigger the fusion process. This could be prevented by membrane scaffolds (both lumenal and cytoplasmic) which would serve to keep the membrane surfaces apart. If periplasmic fusion happened, a fenestration would appear in a cisternal membrane. This could be repaired by fusion of the membranes surrounding the hole. This type of cytoplasmic fusion is referred to as homotypic fusion since it involves the fusion of equivalent membranes. The other type of cytoplasmic fusion, heterotypic fusion, is exemplified by the fusion of transport vesicles with their target membrane.

During mitosis, the membrane scaffolds break down and cytoplasmic fusions (both homotypic and heterotypic) would be inhibited. Periplasmic fusion would continue in the form of COP I-coated vesicle budding. Spontaneous periplasmic fusion would be enhanced by the breakdown of the membrane scaffolds. In the absence of a repair mechanism (homotypic fusion) cisternae would be converted into tubular networks which would then fragment yielding a heterogeneous population of remnants and vesicles. This model for Golgi fragmentation is summarised in Fig. 1.

Inhibition of vesicle fusion

The mechanism which inhibits homotypic fusion during mitosis is still obscure but some progress has been made in identifying the target for mitotic inhibition of heterotypic fusion.

The SNARE hypothesis provides a simple mechanism for the specific fusion of vesicles with their target membranes (Rothman, 1994)(Fig. 2). Each vesicle contains a specific vesicle, or v-SNARE which recognises the cognate target, or t-SNARE, in the acceptor membrane. The assembly of this SNARE complex requires a number of accessory proteins, including the rab family of small GTP-binding proteins, which are thought to ensure the fidelity of the pairing process. Once assembled, the SNARE complex is acted upon by the NSF ATPase in association with SNAPs. Hydrolysis of ATP triggers the break-up of the SNARE complex and subsequent fusion of the two membranes.

Given the general inhibition of membrane traffic during mitosis, it was reasonable to assume that the target would either be a common component of all vesicle-mediated transport steps (such as NSF and SNAPs) or a common site on a protein family (such as the rab proteins), members of which were involved in each of the steps of vesicle fusion in the cell.

Tim Levine focused on p115, a homodimer which has a superficial resemblance to myosin. This protein is involved in ER to Golgi (Nakajima, et al., 1991; Seog, et al., 1994), intra-Golgi (Sapperstein, et al., 1995)and transcytotic transport (Barroso, et al., 1995) so is likely to be a common component of all vesicle-mediated transport steps. It appears to function in the docking of vesicles (Barroso, et al., 1995), tethering them to the membrane surface so that they can sample the t-SNARE that is there (Fig. 2).

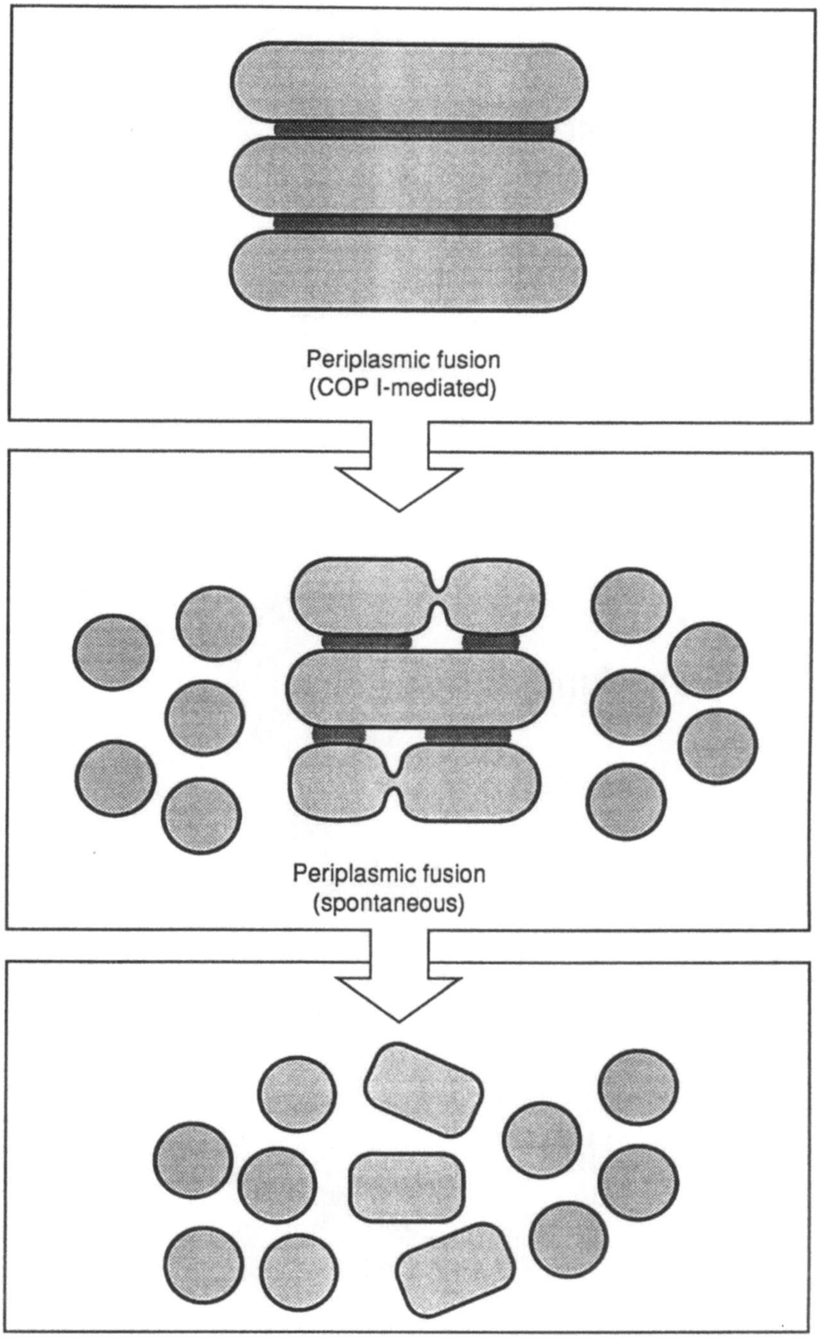

Fig. 1 Working model for mitotic fragmentation of the Golgi apparatus

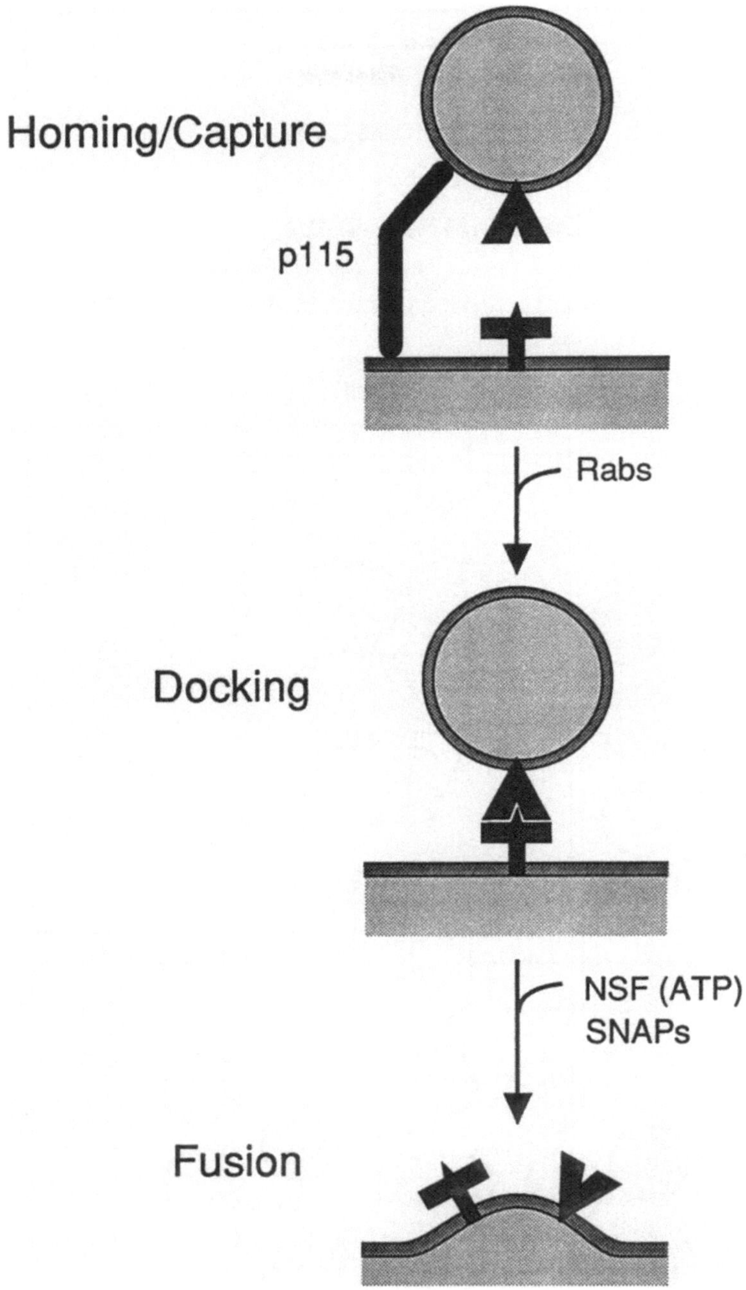

Fig. 2 The SNARE hypothesis and p115

Tim was able to show that the receptor for p115 was inactivated during mitosis so that p115 could no longer bind to membranes (Levine, et al., 1995). If p115 cannot bind, vesicles cannot dock. If vesicles cannot dock, they cannot fuse. This provides a neat explanation for the inhibition of vesicle-mediated traffic in mitotic cells.

Summary

A working model for the fragmentation of the Golgi apparatus suggests that it is triggered by enhanced periplasmic fusion in the absence of cytoplasmic fusions (Fig. 1). Heterotypic fusion is likely inhibited by the inactivation of the receptor for the vesicle-docking protein, p115.

Such a model is applicable to organelles other than the Golgi apparatus. The ER, including the nuclear envelope, fragments during mitosis but there is only one example where this might result from the continued budding of transport vesicles (Warren, 1993). In general it would seem that the breakdown of the membrane scaffolds, which includes the loss of ribosomes (Zeligs and Wollman, 1979), would permit spontaneous periplasmic fusion, and an inhibition of homotypic fusion would prevent repair. Other cellular organelles such as endosomes (Hopkins, et al., 1990) might also succumb to what may well be a general fragmentation mechanism.

Acknowledgements

I would like to thank all those members of my laboratory who, over the years, have contributed to the work described in this review.

References

Barroso, M., D.S. Nelson and E. Sztul (1995). Transcytosis-associated protein (TAP)/p115 is a general fusion factor required for binding of vesicles to acceptor membranes. Proc Natl Acad Sci U S A. 92:527-531.

Birky, C.W. (1983). The partitioning of cytoplasmic organelles at cell division. Int. Rev. Cytol. 15:49-89.

Hopkins, C.R., A. Gibson, M. Shipman and K. Miller (1990). Movement of internalized ligand-receptor complexes along a continuous endosomal reticulum. Nature. 346:335-9.

Letourneur, F., E.C. Gaynor, S. Hennecke, C. Demolliere, R. Duden, S.D. Emr, H. Riezman and P. Cosson (1994). Coatomer is essential for retrieval of dilysine-tagged proteins to the endoplasmic-reticulum. Cell. 79:1199-1207.

Levine, T.P., R.H. Kieckbusch and G. Warren (1995). The binding of the vesicle docking protein p115 is reduced under mitotic conditions: a mechanism for mitotic fragmentation of the Golgi apparatus. Embo J. in press

Lucocq, J.M., E.G. Berger and G. Warren (1989). Mitotic Golgi fragments in HeLa cells and their role in the reassembly pathway. J Cell Biol. 109:463-74.

Misteli, T. and G. Warren (1994). COP-coated vesicles are involved in the mitotic fragmentation of Golgi stacks in a cell-free system. J Cell Biol. 125:269-282.

Misteli, T. and G. Warren (1995). The role of tubular networks and a COP I-independent pathway in the mitotic fragmentation of Golgi stacks in a cell-free system. J.Cell Biol. in press:
Nakajima, H., A. Hirata, Y. Ogawa, T. Yonehara, K. Yoda and M. Yamasaki (1991). A cytoskeleton-related gene, usol, is required for intracellular protein transport in Saccharomyces cerevisiae. J Cell Biol. 113:245-60.
Palade, G. (1975). Intracellular aspects of the process of protein synthesis. Science. 189:347-58.
Rothman, J.E. (1994). Mechanisms of intracellular protein transport. Nature. 372:55-63.
Rothman, J.E. and G. Warren (1994). Implications of the SNARE hypothesis for intracellular membrane topology and dynamics. Curr. Biol. 4:220-233.
Sapperstein, S.K., D.M. Walter, A.R. Grosvenor, J.E. Heuser and M.G. Waters (1995). p115 is a general vesicular transport factor-related to the yeast endoplasmic-reticulum to Golgi transport factor uso1p. Proc Natl Acad Sci U S A. 92:522-526.
Seog, D.H., M. Kito, K. Yoda and M. Yamasaki (1994). Uso1 protein contains a coiled-coil rod region essential for protein-transport from the ER to the Golgi-apparatus in Saccharomyces-cerevisiae. J Biochem. 116:1341-1345.
Warren, G. (1985). Membrane traffic and organelle division. Trends Biochem. Sci. 10:439-443.
Warren, G. (1993). Membrane partitioning during cell division. Annu. Rev. Biochem. 62: 323-348.
Zeligs, J.D. and S.H. Wollman (1979). Mitosis in rat thyroid epithelial cells in vivo. I. Ultrastructural changes in cytoplasmic organelles during the mitotic cycle. J. Ultrastruct. Res. 66:53-77.

KINETIC MEASUREMENTS OF FUSION BETWEEN VESICLES DERIVED FROM THE ENDOPLASMIC RETICULUM

Joke G. Orsel, Ingrid Bartoldus and Toon Stegmann
Department of Biophysical Chemistry
Biozentrum of the University of Basel
Klingelbergstrasse 70
CH 4056 Basel
Switzerland

Membrane fusion plays a crucial role in biological processes such as the intracellular vesicular transport of proteins, fertilization, the re-formation of organelles after mitosis, and the entry of enveloped animal viruses into their host cells. Specialized proteins are involved in inducing, regulating and timing the fusion of biological membranes, resulting in a merger of the lipids of the two membrane bilayers. The proteins involved in endoplasmic fusion events, which take place after initial contact between cytoplasmically oriented surfaces of cellular membranes, have properties different from those involved in exoplasmic events, where initial contact is between externally oriented surfaces (Stegmann *et al.*, 1989). Endoplasmic fusion takes place in the cytosol, a tightly controlled and uniform environment with many factors, like ATP or Ca^{2+}, that could be used to drive or regulate a protein-mediated fusion reaction. Many different endoplasmic fusion events occur simultaneously in the cytosol of one cell, which places stringent demands on fusion specificity. In contrast, exoplasmic fusion takes place in an environment which is not as tightly regulated, and soluble factors are not available. Endoplasmic fusion involves complexes of cytosolic and integral membrane proteins, some of which are common to many intracellular fusion events, while others provide for specificity (Rothman, 1994). Exoplasmic fusion events, like virus-cell fusion, are mostly caused by a single protein present on one of the two membranes (Stegmann *et al.*, 1989).

The mechanism of membrane fusion is best understood for exoplasmic fusion mediated by the membrane proteins of enveloped animal viruses (reviewed in Bentz, 1993), particularly the hemagglutinin (HA) of influenza virus. Studies on virus-induced fusion have benefited greatly from the

NATO ASI Series, Vol. H 96
Molecular Dynamics of Biomembranes
Edited by Jos A. F. Op den Kamp

availability of quantitative and kinetic fluorescence-based methods for measuring fusion. Much less is known about the mechanism whereby endoplasmic fusion proteins achieve the merger of membranes, and in part, this is due to the lack of suitable kinetic fusion assays for these systems. Viral fusion is mostly measured using liposomes as target membranes. Since these artificial membranes are made *de novo* from purified lipids, fluorescent phospholipid analogues can be incorporated in the liposomal membrane. Fusion assays based on the relief of quenching of the fluorescence of phospholipid analogues (Struck *et al.*, 1981; Silvius *et al.*, 1987) are dependable, because molecular exchange of these probes between labeled and unlabeled membranes, which would lead to a false-positive fusion signal (see below), is rare. However, already formed membranes, such as biological membranes, do not take up significant quantities of exogenously added phospholipids. Therefore, non-kinetic biochemical assays are mostly used to assess the fusion of these membranes.

We are trying to develop a kinetic, fluorescence-based assay for the measurement of endoplasmic fusion. Two methods could be used in principle. The most commonly used assay at the moment is the one based on the dequenching of the small hydrophobic molecule octadecylrhodamine (R18) (Hoekstra *et al.*, 1984). R18 spontaneously incorporates into membranes if added from an ethanolic solution, and thus biological membranes can be labeled with it. Although this method is reliable for quantifying fusion in some cases, there is spontaneous exchange of the probe between membranes in others (Wunderli-Allenspach and Ott, 1990; Stegmann *et al.*, 1993; Stegmann *et al.*, 1995). Alternatively, cells can be grown in the presence of pyrene-labeled fatty acids, which are then metabolically incorporated into the phospholipids of cellular membranes (Morand *et al.*, 1982; Pal *et al.*, 1988; Stegmann *et al.*, 1993), thus labeling biological membranes with fluorescent phospholipid analogues. Here, we present measurements of fusion between vesicles derived from the endoplasmic reticulum (ER) by either assay.

In vivo, fusion of fragments derived from the ER with each other occurs during mitosis. Fragmentation of the ER begins at prometaphase, and later, during telophase, ER fragments fuse to re-form the ERs of the daughter cells (Warren, 1993). Fusion between microsomes or vesicles derived from the ER has been reconstituted *in vitro* in a variety of systems (Paiement *et al.*, 1980; Comerford and Dawson, 1991; Watkins *et al.*, 1993; Latterich and Schekman,

1994). Fusion between tubules or reticular elements of the ER takes place continuously in cells (Lee and Chen, 1988), and this process has also been reconstituted *in vitro* (Dabora and Sheetz, 1988).

In order to measure the fusion between ER-derived vesicles with the R18 assay, vesicles were produced from CHO K1 cells logarithmically growing in suspension culture by means of a ball-bearing "cell cracker" (Balch and Rothman, 1985). A microsomal fraction enriched in ER marker enzymes was then purified by differential and sucrose density gradient centrifugation in the presence of protease inhibitors at 0°C. Half the vesicle preparation was labeled by the addition of a small quantity of an ethanolic solution of R18, while vortexing the suspension, and the vesicles were then further incubated with the probe at 0°C for 20 min. Next, free dye was separated from membrane-associated R18 by chromatography (Hoekstra *et al.*, 1984). Upon injection of a 1:4 mixture of labeled and unlabeled vesicles into the cuvette of a fluorimeter at 37°C, an increase in fluorescence was seen (Figure 1a). An increase in fluorescence means that the

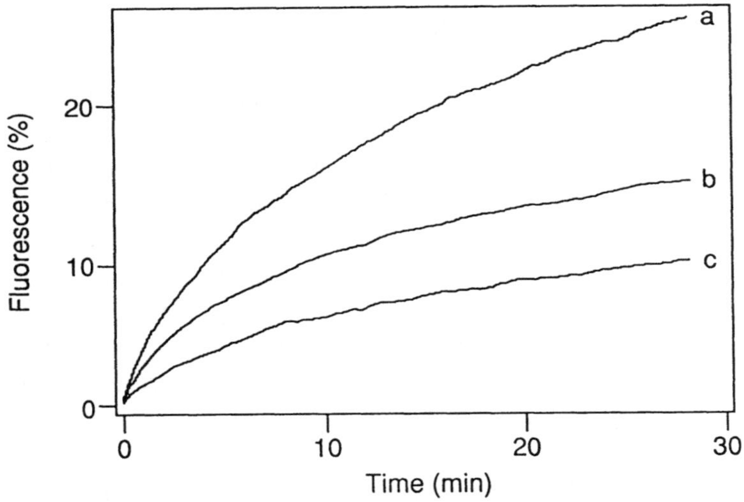

Figure 1. Fluorescence increases upon injection of a 1:4 mixture of R18-labeled and unlabeled vesicles at 37°C (a), or labeled vesicles alone, incubated with R18 either at 0°C for 20 min. (b), or at 37°C for 2 min. (c), into the fluorimeter cuvette.

average distance between the R18 molecules increased, and it could be the result of the dilution of the probe upon fusion of a labeled with an unlabeled membrane, or it could reflect the molecular transfer of probe to unlabeled membranes. However, surprisingly, we found that if labeled vesicles were introduced into the cuvette of the fluorimeter in the absence of unlabeled membranes, there also was an increase in fluorescence (Figure 1b). The increase depended on the time and temperature of incubation of the vesicles with R18; with increasing temperature or time the increase in fluorescence was reduced. For comparison, Figure 1c shows the increase in fluorescence observed for vesicles that were labeled for 2 minutes at 37°C.

An explanation for this phenomenon, which we have also observed with some other biological membranes (Stegmann *et al.*, 1995), could be that after addition of an ethanolic solution of R18 to membranes there are not only randomly dispersed monomers of the probe in the membranes, but there is a second population of R18 in micelles or clusters that remains associated with the vesicles. In these structures, the probe is completely quenched. Upon warming, some of the clustered R18 partitions into membranes, forming monomers, which results in dequenching. Thus, when the labeling of liposomes or some viral membranes with different concentrations of an ethanolic solution of R18 was studied, it was found that within a certain range of concentrations, the fluorescence quenching decreased with decreasing concentration, as expected. However, at very low concentrations of R18 there still was significant quenching, and this did not decrease at even lower concentrations, suggesting the presence of structures with very high concentrations of R18 (Stegmann *et al.*, 1993). To determine the contribution of partitioning or molecular transfer of R18 vs. fusion for R18-labeled ER-derived vesicles, labeled vesicles were mixed with unlabeled liposomes at 37°C. An increase in fluorescence was observed (not shown). However, a fluorescence increase was not seen when unlabeled ER-derived vesicles were incubated with liposomes, labeled with a pair of fluorescent phospholipid analogues that are commonly used to measure fusion (Struck *et al.*, 1981). These data indicate that most of the increase in fluorescence seen with the R18 assay in this system did not reflect membrane fusion.

In order to be able to measure fusion with an assay based on fluorescent phospholipid analogues, CHO K1 cells were grown on dishes in the presence of the fatty acid analogue 1-pyrenehexadecanoic acid (P16), added from a stock solution in DMSO. With an optimized labeling protocol, more than 90% of the cell-associated P16 was incorporated into phospholipids. Pyrene-labeled phospholipids can be used to measure fusion because excited pyrene molecules can form excimers (dimers) with pyrene molecules in the ground state in a concentration-dependent fashion. Since excimers emit at a higher wavelength than monomers, the ratio of excimer to monomer fluorescence is a measure of the concentration of the probe. Dilution of the probe upon fusion of a labeled with an unlabeled membrane causes an increase in monomer fluorescence, and a decrease in excimer fluorescence (Pal *et al.*, 1988; Stegmann *et al.*, 1993).

Upon injection of P16-labeled and unlabeled ER-derived vesicles (1: 10 ratio) into a cuvette at 37 °C, in the presence of GTP, ATP, an ATP-regenerating system and cytosol, there was an increase in monomer fluorescence (Figure 2a) and a corresponding decrease in excimer fluorescence (not shown). If labeled

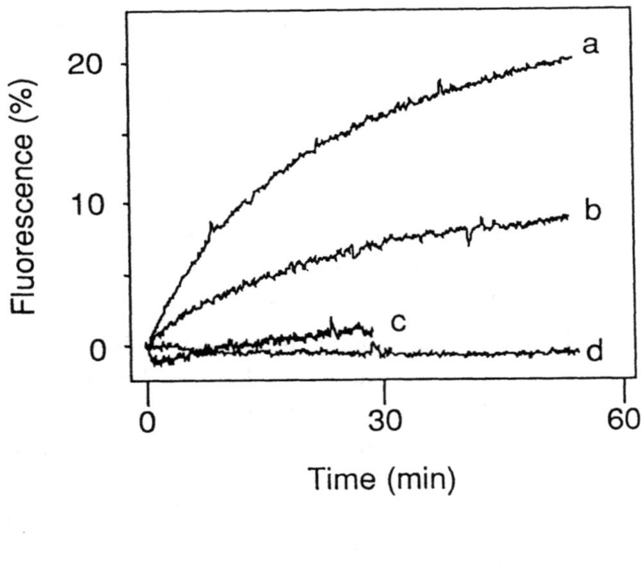

Figure 2. Fluorescence increases upon injection of pyrene-phospholipid-labeled ER-derived vesicles, mixed with a 10-fold excess of unlabeled vesicles in the presence (a) or absence (b) of ATP (400 μM), GTP (70 μm) and cytosol (150 μg of protein) into the cuvette. In the presence of these factors, labeled vesicles were also injected together with liposomes (c) or alone (d). Tracing b is representative of the results obtained under a variety of conditions (see text).

vesicles were injected into the cuvette in the absence or presence of nucleotides and cytosol, but without unlabeled vesicles, the fluorescence remained stable (Figure 2d), at least at vesicle protein concentrations at and above 57 μg/mL. Below that concentration, the fluorescence decreased continuously. It was found that the decrease was not due to photobleaching or oxygen quenching of pyrene. However, the lower the concentration of vesicles, the more pronounced was the decrease (results not shown), suggesting that labeled vesicles were being adsorbed to the cuvette wall or the stirrer. Therefore, all experiments were performed at 57 μg/mL.

The increase in fluorescence shown in Figure 2a could be caused by fusion or by the molecular exchange of probe between membranes. To distinguish between these possibilities, we first tested the effect of the addition of liposomes to labeled vesicles. Since specific proteins are involved in biological fusion, no fusion is expected with liposomes. In the presence of liposomes, the fluorescence first decreased slightly over the course of 5-10 minutes, and then increased slightly (fig 2c). Although we do not completely understand this behavior, these data indicate that there was not much fusion with, or probe transfer to, liposomes. Likewise, if labeled and unlabeled membranes were treated with proteases, the increase in fluorescence was much reduced (not shown).

ATP, frequently GTP, and soluble proteins from the cytosol have been implicated in endoplasmic fusion. For example, fusion between yeast ER elements in vivo requires ATP (Latterich and Schekman, 1994), fusion between mammalian microsomes requires GTP (Paiement *et al.*, 1980; Comerford and Dawson, 1991), and the latter can be inhibited by a non-hydrolysable analogue of GTP, GTPγS (Watkins *et al.*, 1993). In our system, if GTP was omitted, ATP depleted by apyrase, or if fusion was measured in buffer without any added nucleotides or cytosol, the increase in fluorescence was much reduced (Figure 2b). If GTPγS (1 mM) was added to a mixture of labeled and unlabeled membranes in the presence of nucleotides and cytosol, the increase was affected to a similar extent (not shown), and resembled the curve shown in Figure 2b. Endoplasmic fusion in mammalian cells has been shown to almost universally involve a molecule called NSF· which can be inactivated with N-ethylmaleimide (NEM) (Rothman, 1994). If both populations of ER-derived membranes were pretreated with NEM for 20 min. at 0°C, the increase in fluorescence was reduced to the extent seen in the absence of nucleotides and cytosol (cf. Figure 2b).

These results indicate that the fluorescence increase shown in Figure 2a is protein dependent, affected by agents such as NEM or GTPγS that are known to inhibit fusion in many cases, and is stimulated by factors which are often required for fusion, such as ATP and GTP. Therefore, the increase in fluorescence could be due to membrane fusion. However, it is not clear why the omission of nucleotides and cytosol, treatment of membranes with NEM, or incubation in the presence of GTPγS all resulted in such a remarkably similar, partial reduction of the fluorescence increase. If the reduced fluorescence increase seen in these instances is due to fusion, then it must mean that there is a population of vesicles that has already acquired all the proteins and nucleotides that are necessary for fusion, and is "primed" to fuse with other vesicles as soon as it is warmed up. It could also indicate that the nucleotide and protein requirements for this type of fusion are less absolute than for other endoplasmic fusion events.

Using a biochemical assay to study the fusion between ER-derived vesicles from mammalian cells, Watkins *et al.* (1993) found that fusion could be inhibited completely by micromolar concentrations of GTPγS, or the depletion of ATP. Therefore, in their hands, it seems that the presence of ATP is a necessary prerequisite for fusion, and the activity of a GTPase is required for fusion. The extent of inhibition expected for NEM treatment is less clear. Fusion between ER membranes in yeast does not require the yeast homologue of NSF (Latterich and Schekman, 1994), and would therefore not be inhibited by NEM. On the basis of these results, it has been suggested that homotypic fusion, between vesicles derived from the same organelle, as opposed to heterotypic fusion, would not require NSF (Latterich and Schekman, 1994). On the other hand, homotypic fusion between Golgi derived vesicles from mammalian cells did require NSF (Acharya *et al.*, 1995), and could be inhibited by NEM (Rabouille *et al.*, 1995), so that this may reflect a difference between yeast and mammalian cells. Since the latter results were obtained by morphological techniques (Rabouille *et al.*, 1995), a precise dose-response relationship is difficult to establish for NEM.

Partial inhibition by GTPγS, ATP depletion and NEM could also indicate that there is a second process which, in addition to fusion, contributes to the fluorescence increase. Since there is little transfer of probe to liposomes (Figure 2d), unspecific molecular exchange of material between labeled and unlabeled lipid bilayers of the vesicles should not contribute significantly. But at this point, we cannot completely exclude a form of molecular exchange of pyrene-labeled lipids which would be specific for biological membranes. Perhaps residual pyrene-labeled fatty acids, that were not incorporated into phospholipids, contributed to the signal.

The final level of fluorescence increase in the presence of nucleotides and cytosol reached 40% after several hours (not shown). Assuming the increase is due to fusion, and that labeled and unlabeled vesicles are of the same size and fuse only once, this means that 80% of the labeled vesicles participated in fusion; if multiple rounds of fusion took place this figure would be closer to 40% (Nir *et al.*, 1986). Fusion between ER-derived vesicles from mammalian cells measured with a biochemical assay (Watkins *et al.*, 1993), was found to have an efficiency of 25% for purified vesicles and 50% if crude postnuclear supernatant was used. These extents were reached after one hour, and the half-time for fusion was about 40 min., which would be quite comparable to our data (Figure 2a).

In conclusion, we have developed a fluorescence-based assay for endoplasmic membrane fusion, based on the metabolic incorporation of pyrene-labeled fatty acids into cellular phospholipids which should enable kinetic, quantitative measurements of fusion. Since the fluorescence signal was only partially reduced by factors known to inhibit fusion, the results of the assay remain to be confirmed by measurements that are independent of pyrene phospholipids.

References

Acharya U, McCaffery JM, Jacobs R and Malhotra V (1995) Reconstitution of vesiculated Golgi membranes into stacks of cisternae: requirement of NSF in stack formation. J. Cell Biol. 129: 577-589

Balch WE and Rothman JE (1985) Characterization of protein transport between succesive compartments of the Golgi apparatus: asymmetric properties of donor and acceptor activities in a cell-free system. Arch. Biochem. Biophys. 240: 413-425

Bentz J (eds) (1993) Viral fusion mechanisms. CRC Press Boca Raton

Comerford JG and Dawson AP (1991) Fluoroaluminate treatment of rat liver inhibits GTP-dependent vesicle fusion. Biochem. J. 280: 335-340

Dabora S and Sheetz MP (1988) The microtubule dependent formation of a tubulovesicular network with characteristics of the ER from cultured cell extracts. Cell 54: 27-35

Hoekstra D, de Boer T, Klappe K and Wilschut J (1984) Fluorescence method for measuring the kinetics of fusion between biological membranes. Biochemistry 23: 5675-5681

Latterich M and Schekman R (1994) The karyogamy gene KAR2 and novel proteins are required for ER-Membrane fusion. Cell 78: 87-98

Lee C and Chen LB (1988) Dynamic behavior of endoplasmic reticulum in living cells. Cell 54: 37-46

Morand O, Fibach E, Dagan A and Gatt S (1982) Transport of fluorescent derivatives of fatty acids into cultured human leukemic myeloid cells and their subsequent metabolic utilization. Biochim. Biophys. Acta 711: 539-550

Nir S, Stegmann T and Wilschut J (1986) Fusion of influenza virus with cardiolipin liposomes at low pH: mass action analysis of kinetics and extent. Biochemistry 25: 257-266

Paiement J, Beaufay H and Godelaine D (1980) Coalescence of microsomal vesicles from rat liver: a phenomenon occuring in parallel with enhancement of the glycosylation activity during incubation of stripped rough microsomes with GTP. J. Cell Biol. 86: 29-37

Pal R, Barenholz Y and Wagner RR (1988) Pyrene phospholipid as a biological fluorescent probe for studying fusion of virus membrane with liposomes. Biochemistry 27: 30-36

Rabouille C, Misteli T, Watson R and Warren G (1995) Reassembly of Golgi stacks from mitotic fragments in a cell-free system. J. Cell Biol. 129: 605-618

Rothman JE (1994) Mechanisms of intracellular protein transport. Nature 372: 55-63

Silvius JR, Leventis R, Brown PM and Zuckermann M (1987) Novel fluorescent assays of lipid mixing between membranes. Biochemistry 26: 4679-4287

Stegmann T, Doms RW and Helenius A (1989) Protein-mediated membrane fusion. Annu. Rev. Biophys. Biophys. Chem. 18: 187-211

Stegmann T, Orsel J, Jamieson JD and Padfield PJ (1995) Limitations of the octadecylrhodamine dequenching assay for membrane fusion. Biochem. J. 307: 875-878

Stegmann T, Schoen P, Bron R, Wey J, Bartoldus I, Ortiz A, Nieva JL and Wilschut J (1993) Evaluation of viral membrane fusion assays. Comparison of the octadecylrhodamine dequenching assay with the pyrene excimer assay. Biochemistry 32: 11330-11337

Struck DK, Hoekstra D and Pagano RE (1981) Use of resonance energy transfer to monitor membrane fusion. Biochemistry 20: 4093-4099

Warren G (1993) Membrane partitioning during cell division. Ann. rev. Biochem. 323-348

Watkins JD, Hermanowski AL and Balch WE (1993) Oligomerization of Immunoglobulin G heavy and light chains in vitro. J. Biol. Chem. 268: 5182-5192

Wunderli-Allenspach H and Ott S (1990) Kinetics of fusion and lipid transfer between virus receptor containing liposomes and influenza viruses as measured with the octadecylrhodamine B chloride assay. Biochemistry 29: 1990-1997

THE SORTING OF MEMBRANE PROTEINS DURING THE FORMATION OF ER-DERIVED TRANSPORT VESICLES

Joseph L. Campbell and Randy Schekman
Department of Molecular and Cellular Biology
Howard Hughes Medical Institute
401 Barker Hall
University of California, Berkeley
Berkeley, CA 94720, USA.

SUMMARY

Fusion proteins between Sec22p and Sec12p were constructed to identify the sorting determinants present on ER membrane proteins. Both proteins are type II ER-membrane proteins involved in ER to Golgi traffic. Sec22p is a transmembrane protein that has been shown to be efficiently packaged into ER derived transport vesicles(Rexach et al., 1994; Barlowe et al., 1994). It is a member of a family of proteins known as v-SNAREs which are thought to be responsible for the targeting of transport vesicles(Reviewed in Rothman and Warren, 1994). Sec12p is an ER-resident transmembrane protein that is not efficiently packaged into ER-derived transport vesicles, but is required for the formation of vesicles. The first fusion we made has the cytoplasmic domain of Sec22p fused to the membrane and lumenal domains of Sec12p. This fusion protein (called 22-12p) behaves like Sec22p in two ways. First, it complements a temperature sensitive (ts) allele of *sec22 in vivo*. Second, it is efficiently packaged into ER-derived transport vesicles *in vitro*. The second fusion has the cytoplasmic domain Sec12p fused to the transmembrane and lumenal domains of Sec22p. This fusion protein (called 12-22p) behaves like Sec12p *in vivo* because it complements both a ts allele of *sec12* and a deletion of SEC12. Additionally, 12-22p is not efficiently packaged *in vitro* into ER-derived transport vesicles. Taken together these results indicate that sorting information is present on the cytoplasmic domain of one or both of these proteins. We currently favor the hypothesis that the cytoplasmic domain of Sec22p is a positive signal for entry into transport vesicles because immuno EM analysis indicates that Sec22p is concentrated into vesicles as they are formed.

NATO ASI Series, Vol. H 96
Molecular Dynamics of Biomembranes
Edited by Jos A. F. Op den Kamp

Additionally, overproduction of wild type Sec22p leads to a reduction in the efficiency of release of Sec22p *in vitro*. This suggests that a factor necessary for the inclusion of Sec22p in ER-derived vesicles becomes limiting when Sec22p is overproduced.

INTRODUCTION

Proteins enter the transport pathway of eukaryotic cells when they are translocated into or across the membrane of the ER. Once in the transport pathway, proteins can be secreted or targeted to the Golgi cisternae, lysosomes, or the plasma membrane. Repeated cycles of vesicle budding and fusion move proteins through the transport pathway. In the yeast *Saccharomyces cerevisiae*, a combined genetic and biochemical approach has identified many of the components required for the vesicular transport of proteins from the ER to the Golgi (Salama et al., 1993; Barlowe et al., 1994). The formation of ER-derived vesicles from ER membranes *in vitro* has now been observed, and this assay requires the addition of cytosol or a specific set of proteins (Sar1p, Sec23p complex, and Sec13p complex). The vesicles formed in this assay will fuse with Golgi membranes *in vitro*. Additionally, this reaction retains the selectivity observed *in vivo*. Specifically, ER resident proteins such as Sec61p and Sec12p are not included in these vesicles. Despite this progress, little is known about how protein sorting occurs during the process of vesicular budding from ER membranes. Several amino acid sequences have been found that are responsible for the ER-localization of proteins. For example, the C-terminal addition of the sequence HDEL to soluble proteins or KKXX to membrane-bound proteins leads to ER localization. These sequences do not however, appear to lead to the exclusion of proteins from ER-derived transport vesicles. Instead, proteins containing such sequences are retrieved from early Golgi compartments after they have been released from the ER. To date, no sequences that affect the sorting of proteins into ER derived transport vesicles have been identified. As a first step towards identifying such signals, we have constructed

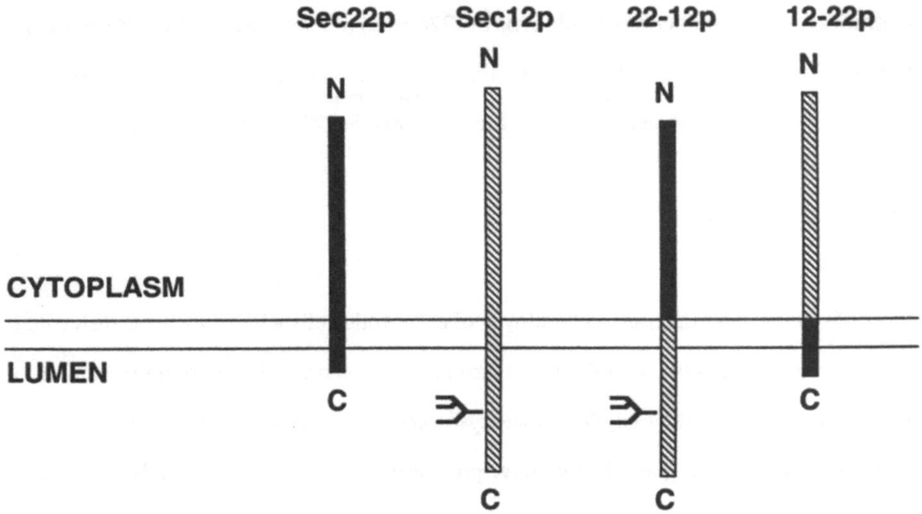

Figure 1

SCHEMATIC VIEW OF FUSION PROTEINS IN THE ER MEMBRANE

Sec22p, Sec12p, and the two fusion proteins are shown in the ER membrane. As indicated the N-terminal portions of the proteins are in the cytoplasm, and the C-terminal portions of the proteins are in the lumen. The glycosylation site in the lumenal domain of Sec12p and the fusion protein 22-12p is indicated.

fusions between two type II membrane proteins that display different sorting properties. One of these proteins, Sec22p is efficiently packaged into ER-derived transport vesicles and the other, Sec12p, is not. Our results with these fusions indicate that sorting information is present on the cytoplasmic domain of one or both of these proteins.

RESULTS

Two fusion proteins have been constructed. The first fusion protein, called 22-12, contains the cytoplasmic domain of Sec22p fused to the transmembrane and lumenal domains of Sec12p. The second fusion protein, called 12-22, contains the cytoplasmic domain of Sec12p fused to the transmembrane and lumenal domains of Sec22p. A schematic view of Sec22p, Sec12p, and the two fusion proteins is shown in figure 1. As shown in figure 1, the lumenal domain of Sec12p contains a glycosylation site.

First these fusions were tested for their ability to allow growth of ts alleles of *sec22* and *sec12* at the restrictive temperature of 37°C. 22-12p was found to complement a ts allele of *sec22*, and 12-22p was found to complement a ts allele of *sec12*. The glycosylation site on the lumenal domain of Sec12p allows one to monitor the appearance of Sec12p or 22-12p in the Golgi *in vivo*. A pulse chase analysis of 22-12p performed in a wild type strain demonstrated that this fusion is rapidly converted to a Golgi modified form *in vivo* (figure 2). In these pulse chase experiments, we also examined the maturation of carboxypeptidase Y (CPY) and found that 22-12p is delivered to the Golgi at a rate comparable to CPY. This is in contrast to previous studies that have shown that wild type Sec12p is delivered very slowly to Golgi (d'Enfert et al., 1991). This *in vivo* analysis suggests that 22-12p is efficiently packaged into ER derived vesicles.

We next wished to test if these fusion proteins were efficiently packaged into ER derived transport vesicles *in vitro*. The *in vitro* budding reaction developed in this laboratory was employed. Briefly, an ER membrane fraction is isolated from yeast cells.

Figure 2

Pulse Chase Analysis of 22-12 Fusion Protein

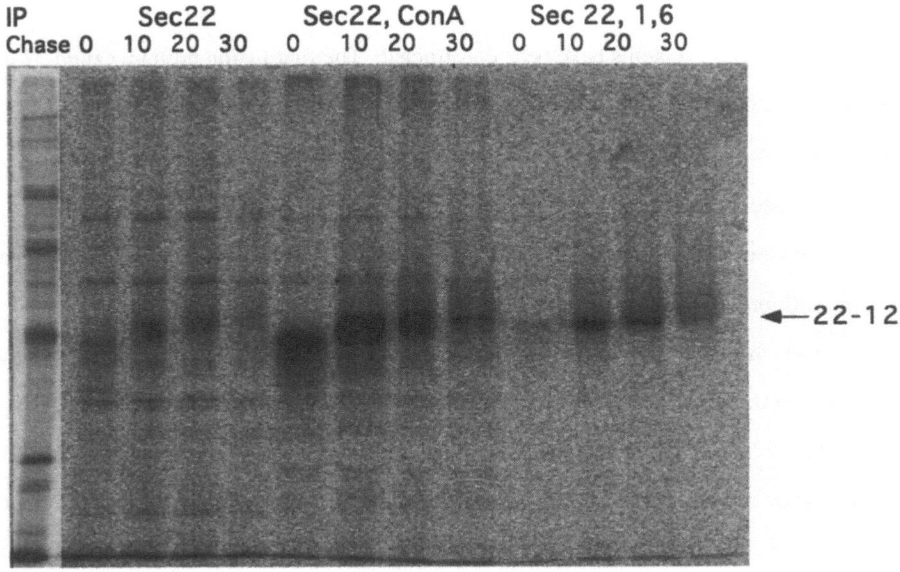

Cells containing a plasmid encoding 22-12p were radio labeled for 5 minutes and chased for the time (in minutes) indicated in the above the samples in the figure. The first lane contains molecular weight standards. The samples in the next four lanes were immunoprecipitated with Sec22 antiserum. The samples in the four lanes after that were immunoprecipitated with Sec22 antiserum and then subjected to precipitation with ConA (ConA precipitation is indicative of ER specific modifications). The samples in the last four lanes were sequentially immunoprecipitated; first with Sec22 antiserum and then with α 1,6 antiserum (Precipitation with α 1,6 antiserum is indicative of Golgi specific modifications.).

When incubated with cytosol or three proteins fractions (Sar1p, Sec23p complex, and Sec13p complex) and the appropriate nucleotides (GTP and ATP), this membrane fraction will produce ER-derived transport vesicles. These vesicles can be separated from the ER membrane by differential centrifugation (The donor ER-membranes are in the medium speed pellet (MSP) and the released vesicles are in the medium speed supernatant (MSS)). When this assay was first developed the release of ^{35}S labeled pre-pro-α factor (ppαF) that had been translocated into ER membranes was monitored. When these reactions are performed on membranes from a wild type strain, ppαF is efficiently packaged into vesicles (25-35% of the translocated ppαF is found in the MSS) (Table 1). As assayed by western blotting, Sec22p is also efficiently released from wild type membranes (20-25% of the Sec22p is found in the MSS). This is in contrast to Sec12p, which remains in the ER membrane (less than 3% of the Sec12p is found in the MSS). In order to test how the fusions performed in this assay, we prepared membranes from strains transformed with a plasmid encoding either the 22-12 fusion or the 12-22 fusion. We found that membranes containing either fusion still efficiently packaged ppαF and Sec22p and that Sec12p was still excluded from the vesicles (Table 1). Interestingly, we found that the 22-12 fusion protein was packaged into vesicles (about 10% of the 22-12p is found in the MSS). By contrast, 12-22p is not found in the MSS (Less than 3%). Since 22-12p contains the cytoplasmic domain of Sec22p and 12-22p contains the cytoplasmic domain of Sec12p, we conclude that there is a sorting domain on the cytoplasmic domain of one or both of these proteins (either a positive domain on Sec22p or a negative domain on Sec12p).

We wished to investigate the possibility that the cytoplasmic domain of Sec22p contained a positive signal that led to Sec22p inclusion in ER-derived vesicles. In order to do this, membranes were prepared from cells that were overproducing Sec22p. By comparing the ratios of Sec61p and Sec22p in these membranes to the same ratios in control membranes, we estimated that these membranes contained 20 fold more Sec22p. Release of Sec22p from the overproducer membranes was 3-fold less efficient *in vitro* than

Table 1

Release of proteins from ER membranes containing fusion proteins.

	Sec22p	ppαF	Sec12p	22-12p	12-22p
WILD TYPE	24%	25%	<3%	NA	NA
22-12p	21%	36%	<3%	10%	NA
12-22p	25%	31%	<3%	NA	<3%

Release of ppαF into the medium speed supernatant was assayed by measuring trypsin resistant conA precipitable counts. Release of the other proteins into the medium speed supernatant was monitored by quantitative western blotting.

Table 2

Release of proteins from ER membranes that harbor excess Sec22p.

	Sec22p	ppαF
WILD TYPE	24%	26%
Sec22-OP	8%	30%

Release of ppαF into the medium speed supernatant was assayed by measuring trypsin resistant conA precipitable counts. Release of Sec22p into the medium speed supernatant was monitored by quantitative western blotting.

Table 3

Release of 22-12Tp from ER membranes.

	Sec22p	ppαF	22-12p	22-12Tp
WILD TYPE	24%	26%	NA	NA
22-12p	21%	36%	10%	NA
22-12Tp	23%	35%	NA	18%

Release of ppαF into the medium speed supernatant was assayed by measuring trypsin resistant conA precipitable counts. Release of the other proteins into the medium speed supernatant was monitored by quantitative western blotting.

the release of Sec22p from control membranes (Table 2). In contrast to Sec22p, we found that the membranes prepared from the Sec22 overproducer released ppαF as efficiently as wild type membrane do (Table 2). Therefore, membranes prepared from strains overproducing Sec22p are not defective for budding in general. We believe that when Sec22p is overproduced a factor necessary for its packaging becomes limiting. Other cargo or v-SNARE molecules that also require this limiting factor may experience a similar reduction in packaging efficiency in membranes that harbor excess Sec22p.

The observation that 22-12p is not packaged as efficiently as the wild type Sec22p is also consistent with the existence of a negative signal on the lumenal domain of Sec12p. In order to test this hypothesis we made a truncated version of this fusion protein that lacks the last 81 residues of the 108 residue C-terminal lumenal domain of Sec12p. We find that this truncated fusion, called 22-12Tp is packaged almost as efficiently as wild type Sec22p (15-20% of the 22-12Tp is in the MSS) (Table 3). This observation indicates that the lumenal domain of Sec12p hinders the packaging of 22-12p into vesicles. One might speculate that the lumenal domain positions proteins into a region of the ER that is not easily packaged into vesicles. This positioning could hinder the packaging of the full length 22-12p into vesicles. Consistent with this hypothesis, when the analogous truncation is made in the wild type Sec12p, the truncated version of Sec12p can only complement a SEC12 deletion when it is overproduced. This observation, can explained if the overproduction allows the truncated version of Sec12p to reach its proper destination without containing targeting information. Thus, 22-12p may display an intermediate phenotype in our vesicle formation assay because it has a positive signal from Sec22p and a negative signal from Sec12p.

REFERENCES

Barlowe, C., Orci, L., Yeung, T., Hosobuchi, M., Hamamoto, S., Salama, N., Rexach, M.F., Ravazzola, M., Amherdt, M., and Schekman, R. (1994). COPII: a membrane coat formed by Sec proteins that drive vesicle budding from the Endoplasmic Reticulum. Cell *77*, 895-907.

d'Enfert, C., Barlowe, C., Nishikawa, S., Nakano, A., and Schekman, R. (1991) Structural and functional dissection of a membrane glycoprotein required for vesicle budding from the Endoplasmic Reticulum. MCB *11*, 5727-5734.

Rexach, M.F., Latterich, M., and Schekman, R.W. (1994). Characteristics of Endoplasmic Reticulum-derived transport vesicles. JCB *126*, 1133-1148.

Rothman, J.E., and Warren, G. (1994). Implications of the SNARE hypothesis for intracellular membrane topology and dynamics. Current Biology *4*, 220-232.

Salama, N.R., Yeung, T., and Schekman, R.W. (1993). The Sec13p complex and reconstitution of vesicle budding from the ER with purified cytosolic proteins. EMBO J. *12*, 4073-4082.

ISOLATION AND CHARACTERIZATION OF YEAST MUTANTS DEFECTIVE IN THE DOLICHOL PATHWAY FOR N-GLYCOSYLATION

Jack Roos, Jin Xu, Sandra Centoducati, José Luz, Natarajan Ramani, Qi Yan, Rolf Sternglanz and William J. Lennarz
Department of Biochemistry and Cell Biology
State University of New York at Stony Brook
Stony Brook, New York 1794-5215
USA

SUMMARY

A number of important issues about the assembly of the oligosaccharide chains linked to Dol-PP and their translocation and transfer to nascent polypeptide chains have not been resolved. We have decided that a new approach to studying these issues is necessary. Accordingly, we have developed a novel screen for isolating temperature-sensitive (ts) yeast mutants that may have defects in the assembly, translocation or transfer of oligosaccharide-lipid to N-glycosylation sites on proteins. This screen uses a convenient, highly sensitive *in vitro* assay that utilizes the endogenous end product of the dolichylphosphate synthetic pathway, oligosaccharyl PP-Dol, and a exogenous ^{125}I-labeled peptide as the substrates for N-linked glycosylation. Defects in synthesis of Dol-P, assembly of oligosaccharyl-PP-Dol or in transfer of the oligosaccharide chain to the ^{125}I-labeled -Asn-x-Ser/Thr- peptide are expected to result in a decrease in formation of ^{125}I-glycopeptide. Peptide glycosylation is assayed initially in crude cell lysates prepared from small cultures of strains from a collection of temperature-sensitive mutants. Subsequently, strains that exhibit a decrease in the amount of ^{125}I-labeled glycopeptide formed are further tested using isolated microsomes. Strains that exhibit decreased peptide glycosylation could have a defect in oligosaccharyl transferase. This would be likely if analysis revealed that endogenous substrate, oligosaccharyl-PP-Dol, is present in equal amounts in the mutant and in parental, wild-type strain. Alternatively, the decrease in peptide glycosylation could be due either to a defect in assembly (or translocation) of the oligosaccharide chain linked to Dol-PP or to

NATO ASI Series, Vol. H 96
Molecular Dynamics of Biomembranes
Edited by Jos A. F. Op den Kamp

a defect in synthesis of Dol-P. We have established that strains defective in Dol-P availability can be identified by the fact that addition of exogenous Dol-P plus sugar nucleotides to microsomes corrects the defect in glycosylation of ^{125}I-peptide. Moreover, we have established that in some strains that are defective in peptide glycosylation, but not corrected by exogenous Dol-P, the defect is in assembly of the oligosaccharide chain. This is done by [^{3}H]Man labeling of the oligosaccharide-PP-Dol of that strain *in vivo*, and subsequent analysis of the amount and the size of the oligosaccharide chain. Using these techniques we have examined 440 ts mutants and identified several candidates in each of three classes of mutants. Two mutants with apparent defects in new subunits of oligosaccharyl transferase are being characterized.

INTRODUCTION

Since the discovery of the involvement of the long chain polyprenyl phosphate, dolichyl phosphate, in the pathway of assembly of the oligosaccharide chain of N-linked glycoproteins a great deal of progress has been made in understanding the individual steps of this complex process. As shown in Figure 1, the overall process can be considered to occur in three phases.

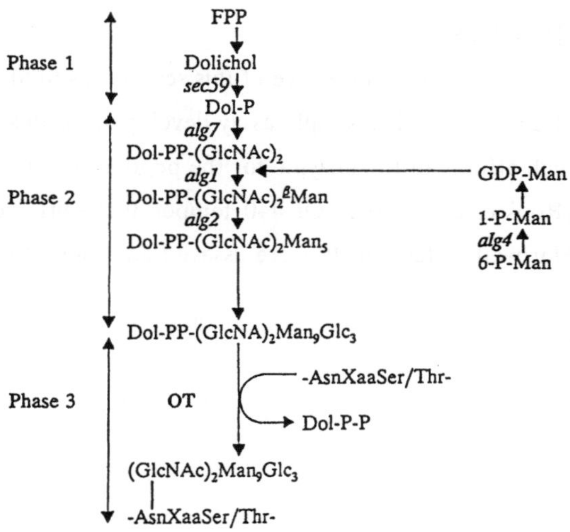

Figure 1. Potential sites of defects in the glycoprotein synthetic pathway. Yeast mutants defective in specific steps of the pathway are indicated.

A number of the early steps in Phase 1, the synthesis of dolichol phosphate have been elucidated. However, the details of the late steps in formation of dolichol phosphate, as well as the regulation of its synthesis, are not well understood. Phase 2 is the multi-step, lipid-linked oligosaccharide assembly process, whereby dolichol phosphate serves as a membrane anchor for the assembly of oligosaccharylpyrophosphoryldolichol. As shown, numerous steps in this process have been delineated in yeast as the result of the availability of *alg* mutants, although at this stage, too, the details of some steps, and their topology and the regulation of their assembly, remain to be elucidated. Characterization of the enzyme involved in the final step (Phase 3), in which the oligosaccharide chain is transferred from the lipid anchor to -Asn-X-Ser/Thr- in protein has, for a number of years, eluded investigators. Insight into the topology and subunit composition of the oligosaccharyl transferase was first revealed upon purification of the canine and avian OT as a complex of ribophorin I (66kD), ribophorin II (63kD) and OST48 (48kD) (Kelleher *et al.*, 1992; Kumar *et al.*, 1994). Subsequently, the OT of yeast was purified as a hetero-oligomer consisting of multiple subunits (Kelleher and Gilmore, 1994); te Heesen *et al.*, 1992 and 1993; Silberstein *et al.*, 1995). At present six protein subunits, Wbp1, Swp1, as well as Ost1, 2, 3 and 4 have been identified as components of OT.

RESULTS

The initial objective of this screen was to identify yeast mutants defective in peptide glycosylation. The simple assay developed for this screen is extremely sensitive due to the high affinity of the enzyme for the peptide substrate and its very high specific radioactivity (Roos *et al.*, 1994). Of 440 temperature-sensitive (ts) strains available to us from the Hartwell collection, 19 were assayed and found to have defects in glycosylation (Table 1).

Table 1. **Peptide glycosylation activity assay**

Class	Strain	*In vitro* activity in cell lysates (% of wt at 25^0 C)	
		25^0 C	36^0 C
Wild-type	A364A	100	95
I	JRY64	37	71
	JRY149	38	49
	JRY279	41	35
	JRY283	64	19
	JXY358	20	35
II	JRY47	51	38
	JRY59	24	43
	JRY125	67	57
	JRY128	99	71
	JRY148	46	61
	JRY157	53	41
	JXY374	40	40
	JXY436	45	40
III	JRY163	9	9
	JRY249	90	62
	JRY265	19	29
	JXY361	100	50
	*MA7-B	13	9

*from Markus Aebi

A second step screen used to classify these mutants was to label the mutant cells with [^{14}C]mannose *in vivo*, isolate the oligosaccharyl-PP-dolichol, hydrolyze it to produce [Man-^{14}C]-oligosaccharide, and then determine its size by gel filtration. The results of this analysis indicates that the mutants either lacked an oligosaccharide chain (Class 1), had a truncated chain (Class 2) or had a normal oligosaccharide (Class 3). This analysis is shown in Figure 2.

CLASS 1

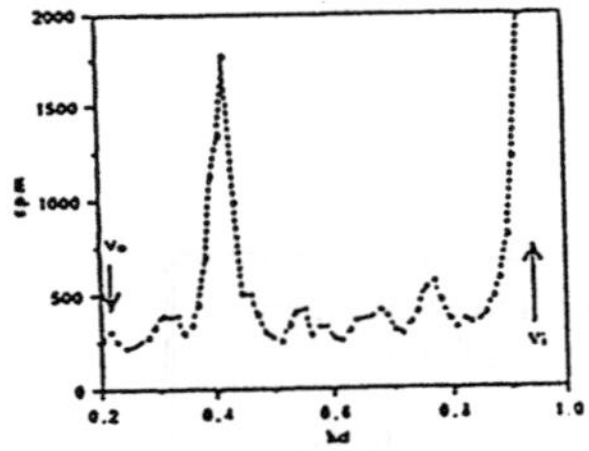

A364A:wild type

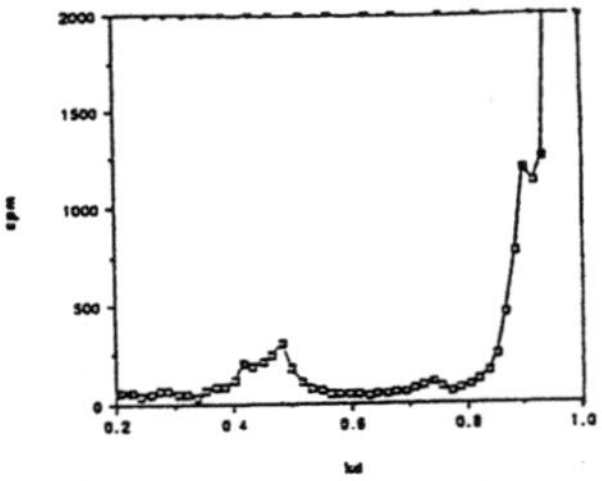

JRY149:Class 1

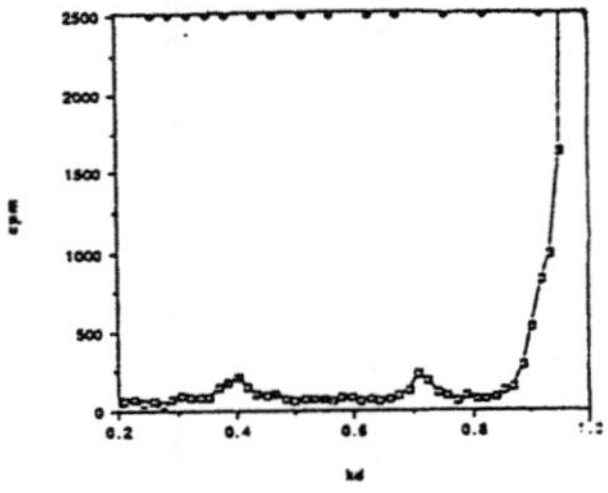

JRY64:Class 1

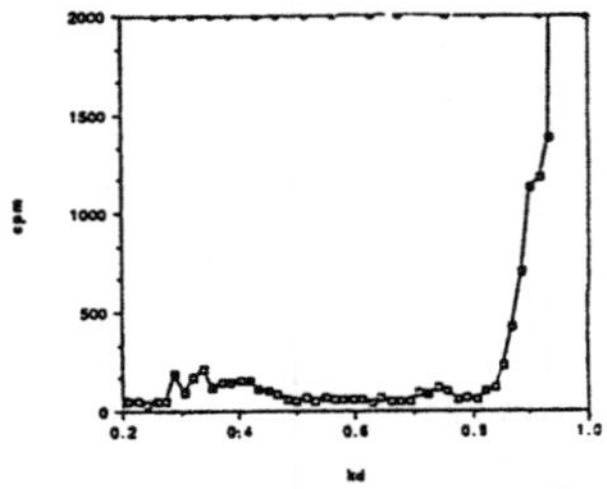

JRY279:Class 1

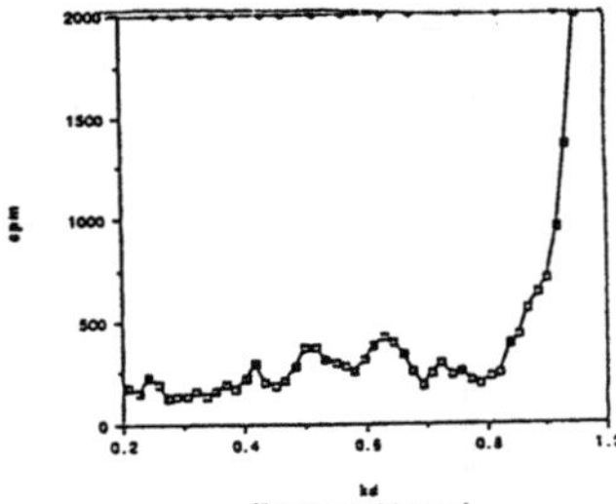

JRY283:Class 1

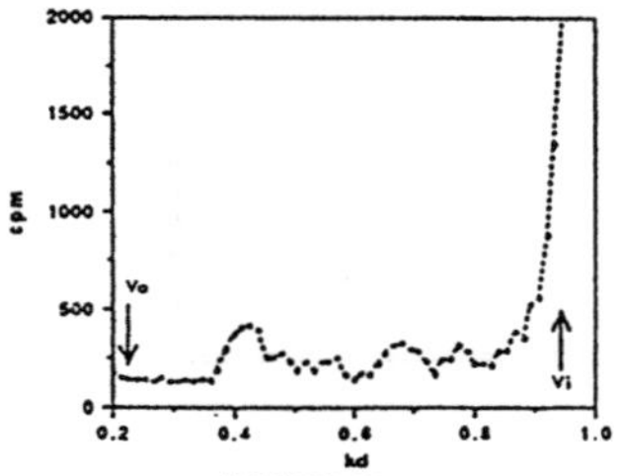

JXY358:Class 1

CLASS 2

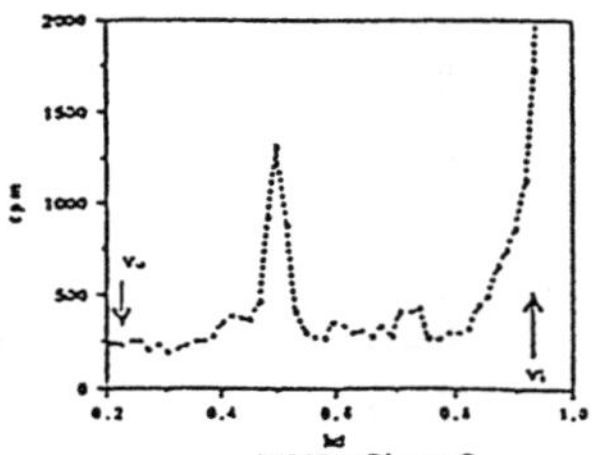

JRY59:Class 2

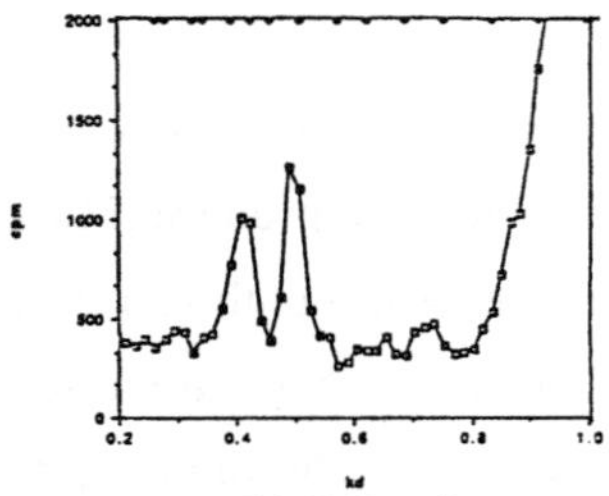

JRY125:Class 2

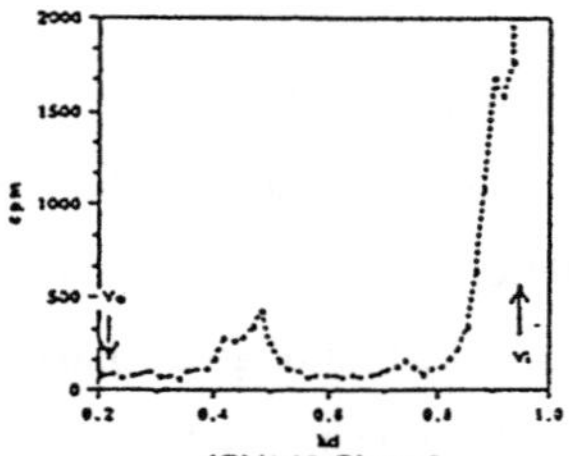

JRY148:Class 2

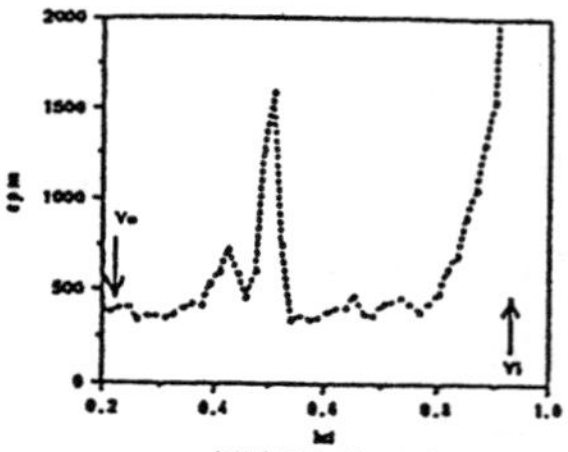

JRY157:Class 2

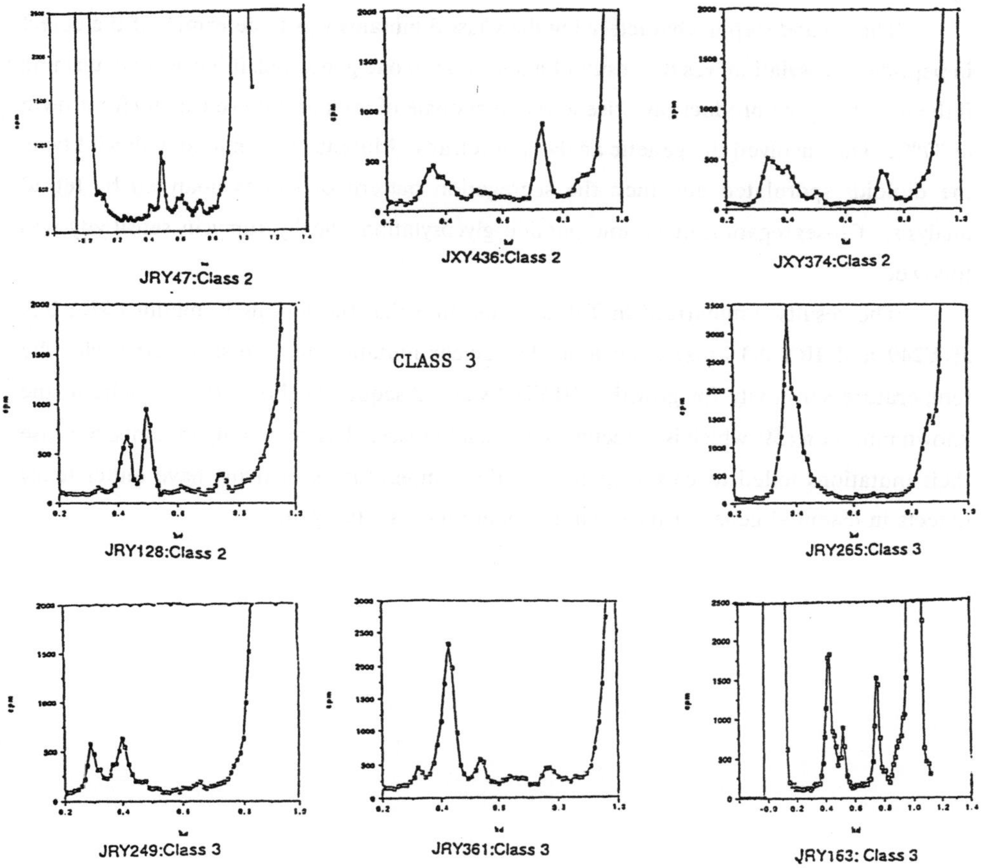

Figure 2. Size analysis of oligosaccharide chains from peptide glycosylation defective mutants. The full length oligosaccharide ($Glc_3Man_9GlcNAc_2$) elutes at 0.4kd.

The second step in characterizing the Class 3 mutants was to determine if the defect in peptide glycosylation was the result of a mutation in one gene, and, if it was, to determine if the mutated gene product gave rise to the phenotype of temperature sensitivity for growth at 37°C. This involved the genetic analysis of tetrads. Mutants were mated with wild type, the diploids sporulated and then the segregation pattern of spores analyzed by tetrad analysis. Co-segregation of ts, low peptide glycosylation and hygromycin sensitivity was followed.

The results summarized in Table 2 indicated that the lesions in mutants JRY163, JRY249 and JRY283 arose from a single nuclear mutation that co-segregated with the temperature sensitivity for growth. JRY283 was subsequently shown to be allelic to the known mutant *sec59*, which is defective in dolichol kinase. The remaining mutants, because their mutations failed to co-segregate with the temperature sensitivity, have either leaky defects in essential genes or have mutations in non-essential genes.

Table 2. **Summary of tetrad analysis in Class 1, 2, and 3 mutants***

Mutant Class	Strain	Impaired Peptide Glycosylation	Temperature Sensitivity	Hygromycin Sensitivity
I	JRY64	2:2	2:2	2:2
	JRY149	2:2	2:2	2:2
	JRY279	2:2	2:2	2:2
	JRY283	2:2	2:2	R
	JXY358	2:2	2:2	R
II	JRY47	2:2	R	R
	JRY59	2:2	R	2:2
	JRY125	1:3	2:2	2:2
	JRY128	2:2	R	R
	JRY148	2:2	2:2	R
	JRY157	2:2	R	R
	JXY374	2:2	2:2	2:2
	JXY436	R	R	R
III	JRY163	2:2	2:2	2:2
	JRY249	2:2	2:2	2:2
	JRY265	2:2	2:2	2:2
	JXY361	1:3	2:2	R
	Ma7-B	2:2	2:2	2:2

*Lines connecting the different methods of analysis indicate co-segregation. Only in the case of 163, 249, as well as MA7-B (carrying the *Wpb1* subunit) was co-segregation of all three characteristics is observed.

Current and Future Studies. The current focus has been on the temperature sensitive mutants, JRY 163 and JRY249. The approach utilized has been to transform these with a wild type yeast gamete library and screen for recovery of growth at reasonable temperature, and for expression of wild type levels of peptide glycosylation. Genes thus isolated were recovered and cloned and sequenced.

The gene that rescued JRY163 was found to encode for a small, 3.9kD protein of extreme hydrophobicity that we have named *OST4*. A second gene, *MEG1*, isolated following transformation acted as a multi- copy suppressor of the temperature sensitivity of JRY163, but did not restore peptide glycosylation activity. Currently we are studying how this gene suppresses the temperature sensitivity. Current work on JRY249 suggests that it may encode for another subunit of OT; the sequencing of this gene, tentatively called *OST5*, is in progress.

ACKNOWLEDGEMENTS

This work was supported by grants GM33184 and GM33185 to WJL and GM28220 to RS from the National Institutes of Health.

REFERENCES

Kelleher, D. J., Kreibich, G. and Gilmore, R. (1992) Oligosaccharyl transferase activity is associated with a protein complex composed of ribophorins I and II and a 48kd protein. Cell 69, 55-65.

Kelleher, D. J. and Gilmore, R. (1994) The *Saccharomyces cerevisiae* oligosaccharyl transferase is a protein complex composed of wbp1p, swp1p and four additional polypeptides. J. Biol. Chem. 269, 12908-12917.

Roos, J., Sternglanz, R. and Lennarz, W. J. (1994) A screen for yeast mutants with defects in the dolichol-mediated pathway for N-glycosylation. Proc. Natl. Acad. Sci., USA 91, 1485-1489.

Silberstein, S., Collins, P, G., Kelleher, D. J. Rapiejko, P. J. and Gilmore, R. (1995) The α subunit of the *Saccharomyces cerevisiae* oligosaccharyl transferase complex is essential for vegetative growth of yeast and is homologous to mammalian ribophorin I. J. Cell Biol. 128, 525-536.

te Heesen, S., Knauer, R., Lehle, L. and Aebi, M. (1993) Yeast wbp1p and swp1p form a protein complex essential for oligosaccharyl transferase activity. EMBO J. 12, 279-284.

te Heesen, S., Janetzky, B., Lehle, L. and Aebi, M. (1992) The yeast *WBP1* is essential for oligosaccharyl transferase activity *in vivo* and *in vitro*. EMBO J. 11, 2071-2075.

te Heesen, S. and Aebi, M. (1993) *wbp1* and *kar2* genetically interact N-linked glycosylation and BiP are essential factors in the protein processing pathway. Abstracts of the Yeast Molecular Biology Meeting, p. XIX, Cold Spring Harbor, New York.

The Importance of Lipid-Protein Interactions in Signal Transduction Through the Calcium-Phospholipid Second Messenger System

Stuart McLaughlin, Carolyn Buser, Gennady Denisov, Michael Glaser†,
W.Todd Miller, Andrew Morris, Mario Rebecchi, and Suzanne Scarlata
Basic Health Sciences, HSC
SUNY, Stony Brook
NY 11794-8661 USA

Introduction. As illustrated in Fig. 1, the binding of a hormone (H), neurotransmitter or growth factor to a receptor (R) in a membrane can activate a phosphoinositide-specific phospholipase C (PLC) that hydrolyzes phosphatidylinositol 4,5-bisphosphate (PIP_2) into the two second messengers inositol 1,4,5-trisphosphate (IP_3) and diacylglycerol (DAG). IP_3 diffuses through the cytoplasm and releases calcium ions from the endoplasmic reticulum (ER). The increase in the cytoplasmic concentration of Ca^{++} produces translocation of protein kinase C (PKC) to the plasma membrane and concomitant activation of this enzyme. Maximal activation of PKC requires DAG, which remains in the membrane, and an acidic lipid such as phosphatidylserine (PS). The requirement for PS suggests that basic residues on the protein (+ signs) interact with acidic lipids in the membrane. The membrane-bound, activated form of PKC then phosphorylates its membrane-bound substrates, which include the myristoylated alanine-rich C kinase substrate (MARCKS) and $pp60^{src}$ (Src).

The translocation of PKC from the cytoplasm to the membrane facilitates its interaction with membrane-bound substrates by a mechanism that is often referred to as "reduction of dimensionality" (Adam and Delbrück, 1968). In its simplest form, reduction of dimensionality implies that when PKC moves from the cytoplasm to the membrane, the enzyme encounters about a 1000-fold higher effective concentration of membrane-bound substrates. Specifically, for a spherical cell of radius $r \approx 10\ \mu m$, translocation moves the enzyme from a volume $4\pi r^3/3$ to a volume $4\pi r^2 d$, where d, the thickness of the spherical shell that contains the membrane-bound PKC, Src and MARCKS, is of order 1 nm (see Fig. 1). The advantage of membrane binding is apparent for all the proteins illustrated in Fig. 1. Specifically, it facilitates interaction of the G protein subunits α_q and $\beta\gamma$ with both the activated receptor, R, and membrane-bound effector, PLC-β; it enhances the ability of PLC to hydrolyze its membrane-bound substrate, PIP_2; and it

†Dept. Biochemistry, University of Illinois, Urbana IL 61801

NATO ASI Series, Vol. H 96
Molecular Dynamics of Biomembranes
Edited by Jos A. F. Op den Kamp

increases the probability that PKC will phosphorylate its membrane-bound substrates Src and MARCKS.

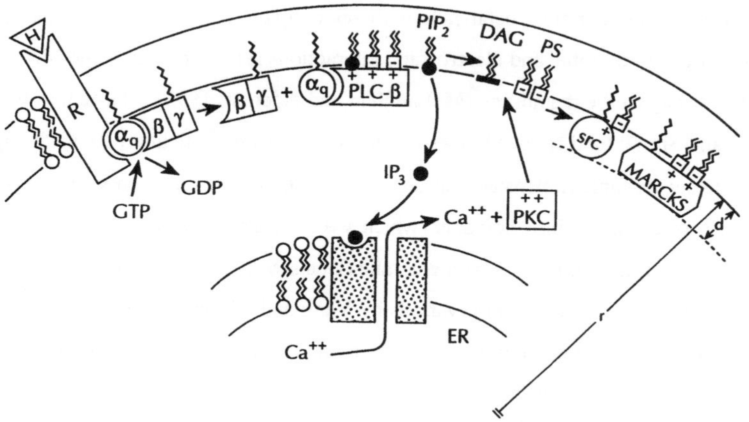

Fig. 1. Sketch of the calcium/phospholipid second messenger system. See text for details

In this chapter we review the different mechanisms the major components of the calcium/phospholipid second messenger system (Berridge, 1993) use to attach themselves to membranes. Briefly, the α subunits of heterotrimeric guanine nucleotide binding (G) proteins typically contain both a myristate (14 carbon saturated acyl chain) linked by a stable amide bond to the N-terminal glycine residue and a palmitate (16 carbon saturated acyl chain) attached through a labile thioester bond to a cysteine near the N terminus (Wedegaertner et al., 1995). The γ component of the βγ subunit typically contains a geranylgeranyl (20 carbon isoprenoid) attached via a stable thioether bond to a Cys residue located in the C-terminal CAAX box (Wedegaertner et al., 1995). The available evidence suggests these acyl and isoprenyl lipid modifications attach the G protein subunits to the membrane by inserting into the interior of the bilayer, as illustrated in Fig. 1; the insertion is driven by the hydrophobic force, which is discussed in detail elsewhere (e.g. Tanford, 1991). The δ (and probably also the β) isoforms of PLC bind specifically to the highly acidic lipid PIP$_2$ via a spatially juxtaposed cluster of basic residues in a region termed the pleckstrin homology (PH) domain (Harlan et al., 1994, 1995; Garcia et al., 1995; James et al., 1995). PKC binds to monovalent acidic lipids such as PS, probably through a cluster of basic residues in its C2 domain, which is known to bind Ca^{++} and phospholipids

(Newton, 1993; Sutton et al., 1995). MARCKS and Src contain two different membrane attachment sites: a myristate attached covalently to the N-terminal glycine residue and a cluster of basic residues. The myristate inserts into the hydrocarbon interior of the lipid membrane and the basic residues interact electrostatically with acidic lipids (e.g. PS) in the membrane. This combination of hydrophobic and electrostatic interactions serves to anchor the MARCKS and Src proteins firmly to the membrane (McLaughlin and Aderem, 1995). Note that PLC, PKC, MARCKS and Src all bind to membranes by interacting electrostatically with acidic lipids. As we discuss below, a simple electrostatic calculation suggests these components may self-assemble laterally with acidic lipids into domains (which we shall also refer to as signal transduction ensembles) on the surface of the plasma membrane (Denisov, Glaser, McLaughlin, unpublished). Direct fluorescence microscope observations by Yang and Glaser (1995) show that a peptide corresponding to the basic region of MARCKS will self assemble into domains with acidic lipids such as PS in phospholipid vesicles.

Binding of PLCs to membranes. There are three classes of phosphoinositide-specific phospholipase Cs (PLCs). The β class of PLCs is activated by both the α and the βγ subunits of the G_q class of G proteins (Sternweis and Smrcka, 1992). The γ class of PLCs contains Src-homology 2 (SH2) domains that bind to phosphorylated tyrosine residues on activated receptor tyrosine kinases such as the epidermal growth factor (EGF) receptor (Carpenter, 1992), which then phosphorylate and activate PLC-γ (Rhee and Choi, 1992; Lee and Rhee, 1995). The factors that activate the δ class of PLCs are not known, but the mechanism by which PLC-δ attaches itself to membranes is now well understood. Rebecchi et al. (1992) and Pawelczyk and Lowenstein (1993) showed that PLC-δ from bovine brain binds with high specificity and affinity (μM dissociation constant) to PIP_2 in phospholipid vesicles. Subsequent experiments showed that the PIP_2-binding site is located in the N-terminal region of the molecule and is distinct from the catalytic site that hydrolyzes PIP_2 (Cifuentes et al., 1993).

The N-terminal region of PLC-δ contains a pleckstrin homology or PH domain (Haslam et al., 1993; Parker et al, 1994; Gibson et al., 1994; Cohen et al., 1995; Pawson, 1995). In an important series of experiments, Fesik and coworkers (Harlan et al., 1994, 1995) showed that PH domains from several different proteins bind to PIP_2 with high affinity. Recent experiments showed that the PH domain from PLC-δ binds to PIP_2 in membranes with an affinity and specificity comparable to that of the native enzyme (Garcia et al, 1995).

The structures of three PH domains are known: the overall fold of the PH domain from human dynamin (Ferguson et al., 1994) is similar to the structures of PH domains from spectrin (Macias et al., 1994) and pleckstrin (Yoon et al., 1994), which suggests that PH domains are well defined protein modules with a distinct structure even though they share less sequence identity than other structural modules such as SH2 and SH3 domains. The PH domain consists of two nearly orthogonal antiparallel β sheets, one with four and one with three strands, and a C-terminal α helix. Harlan et al. (1994) showed that binding of PIP_2 to the PH domain of pleckstrin perturbs the NMR signals from a cluster of basic residues (K13, K14, K22) located in the loop that joins the first and second β strands; these positively charged residues interact electrostatically with the negatively charged phosphate moieties of PIP_2. The binding pocket for PIP_2 is located at the lip of the β sheet in the N-terminal half of the PH domain. As expected, mutations in any of these three basic residues reduce the affinity of the PH domain for PIP_2 (Harlan et al., 1995). These basic residues are conserved in many PH domains, which also often contain an excess of other basic residues proximal to the PIP_2-binding site. This presumably accounts for the observation that the apparent association constants of the native PLC-δ and PLC-β enzymes and of isolated PH domains from PLC-δ for membranes containing PIP_2 are higher if the membrane also contains monovalent acidic lipids (Rebecchi et al., 1992; Garcia et al., 1995; James et al., 1995). (It should be noted that the C-terminal region of several PH domains is also capable of binding βγ subunits of G proteins (Pitcher et al., 1995).)

The PH domain of PLC-δ anchors the enzyme to the membrane, placing PLC-δ in close proximity to its membrane-bound substrate PIP_2 and allowing it to act in a precessive manner. That is, the enzyme can "scoot" over the surface, hydrolyzing many PIP_2 molecules before desorbing from the membrane (Jain and Berg, 1989). The PH domain should also direct PLC-δ to specific membranes enriched in PIP_2, such as the plasma membrane, and may direct the protein laterally to signal transduction ensembles enriched in acidic lipids such as PS and PIP_2 (see below).

Binding of PKC to membranes. All protein kinases share a structurally similar catalytic core and many, including PKC, contain a pseudosubstrate region in the regulatory domain. Activation of PKC presumably involves removal of the pseudosubstrate region from the substrate binding site of the catalytic domain (Newton, 1993). The pseudosubstrate region contains a cluster of five basic residues that may bind to acidic lipids (Mosior and McLaughlin, 1991). The forms of PKC activated by Ca^{++} also have C2 domains, which contain a cluster of basic residues that could bind to acidic lipids in the membrane. The structure of these domains, which are also found in PLC isoforms, has recently been deduced (Sutton et al., 1995). Although it is not clear which basic residues on PKC bind to acidic phospholipids, it is well established that the binding of PKC to membranes is a steep sigmoidal function of the mole fraction of PS in the membrane (Newton, 1993). In terms of the simple model we discuss below, this sigmoidal dependence of binding on the fraction of PS in the membrane suggests PKC will form domains or ensembles with acidic lipids and its substrates such as MARCKS.

Binding of PKC substrates MARCKS and Src to membranes. MARCKS, a major PKC substrate, contains two domains involved in membrane binding: an N-terminal myristoylation domain and a basic effector domain that contains the PKC phosphorylation sites as well as the calmodulin- and actin-binding sites (Aderem, 1992; Blackshear, 1993). While the precise function of MARCKS is not yet clear, it has been implicated in motility, secretion, membrane traffic and mitogenesis; it may act by regulating actin structure at the membrane and/or the level of free calmodulin. Src also contains an N-terminal myristoylation domain and a cluster of basic residues that are important for membrane binding. In addition, it contains SH2 and SH3 domains, which mediate protein-protein interactions, and a tyrosine kinase domain (Resh, 1993). The physiologically important substrates of the Src tyrosine kinase have not yet been determined.

Src and MARCKS, like about 100 other proteins, are myristoylated cotranslationally on their N-terminal Gly residues by the enzyme N-myristoyl transferase (Towler et al., 1988). Many, but not all, of these myristoylated proteins bind to membranes. (The myristate chain on protein kinase A (PKA), for example, bends back into a hydrophobic groove on the protein and presumably serves a structural function (Zheng et al., 1993); PKA binds only weakly to membranes.) Src and MARCKS do bind to membranes and myristate is necessary for both this membrane binding and for proper functioning of the proteins (McLaughlin and Aderem, 1995). For example, mutating the N-terminal Gly residues of Src (Resh, 1993) or MARCKS (Aderem, 1992; Blackshear, 1993) prevents acylation by N-myristoyl transferase. The nonacylated proteins do not bind to membranes and the non-acylated v-Src, for example, does not transform cells (Resh, 1993).

The available evidence suggests the acyl chain attaches a myristoylated protein to a membrane by inserting hydrophobically into the phospholipid bilayer. Direct evidence for this paradigm was obtained by Vergères et al. (1995), who showed that a radioactive probe developed to photolabel the apolar core of membrane proteins labeled MARCKS on the myristoyl chain. How strong is this attachment? The binding energy of acylated peptides to phospholipid vesicles increases linearly with the number of carbons in the chain (Peitzsch and McLaughlin, 1993; Silvius and l'Heureux , 1994): the slope of 0.8 kcal/mol per CH_2 group is similar to that observed by Tanford (1991) for the partitioning of the neutral form of a fatty acid from water into a bulk alkane phase. The simplest interpretation is that the membrane binding energy is due to the classical hydrophobic effect (Tanford, 1991). The observations that there is little dependence of membrane binding on temperature, phospholipid composition of the vesicles, or chemical composition of the peptides are all consistent with this interpretation (Peitzsch and McLaughlin, 1993). Specifically, myristoylated glycine (myr-G), myr-GA, myr-GAA and a myrisotylated peptide corresponding to the first 15 aa of Src all bind to electrically neutral vesicles with a unitary Gibbs free energy of ΔG_u = -8 kcal/mol (Peitzsch and McLaughlin, 1993; Buser et al., 1994). This implies that about 8/0.8 = 10 CH_2 groups are removed from water and embedded into the hydrocarbon interior of the membrane. This ΔG_u = -8 kcal/mol corresponds to a partition coefficient, or effective association constant with lipids, of $K_A = 10^4\ M^{-1}$. (As discussed in detail elsewhere (e.g. Peitzsch and McLaughlin, 1993), the change in the unitary Gibbs free energy differs from the change in the standard Gibbs free energy by the cratic contribution, about 2.4 kcal/mol.) The binding of intact Src and MARCKS proteins to electrically neutral vesicles is 3-10 fold weaker than that of myristoylated peptides, with $K_A \sim 10^3\ M^{-1}$ (Kim et al., 1994b; Sigal et al., 1994). To calculate the fraction of myristoylated proteins bound to membranes, we note the effective concentration of lipids in a spherical cell of radius 10 μm is 10^{-3} M. Thus only half the myristoylated protein would be bound, indicating that myristate will not firmly anchor these myristoylated proteins to the plasma membrane.

If myristate cannot anchor proteins like Src and MARCKS to membranes, what other factors are important? Several observations suggest that electrostatic interactions provide additional binding energy for these proteins. For example, adding 33% acidic lipids to electrically neutral vesicles increases the binding of both a myristoylated peptide corresponding to the first 15 residues of Src and the intact c-Src protein about 1000-fold (Buser et al., 1994; Sigal et al., 1994). The membrane binding of N-terminal mutants of c-Src decreases as the number of basic residues decreases, which confirms that it is these basic residues in Src that bind to the acidic lipids (Sigal et al., 1994). Similar results have been obtained with the MARCKS protein: adding 20% acidic lipid to electrically neutral vesicles increases the binding of MARCKS 100-fold (Kim et al., 1994b).

Experiments with simple basic peptides (e.g. Lys_5) demonstrate their binding to membranes containing acidic lipids is essentially independent of the chemical nature of both the basic residues (Lys vs Arg) and the acidic phospholipids (PS vs PG). These peptides (and peptides that correspond to the basic regions of Src and MARCKS) do not bind to electrically neutral (PC) vesicles, and increasing the ionic strength reduces their binding to vesicles containing acidic lipids. These observations suggest the binding is driven mainly by electrostatic interactions. NMR experiments indicate hydrophilic basic peptides like Lys_5 do not penetrate the polar head group region (Roux et al., 1988), and high pressure fluorescence measurements suggest a layer of water molecules probably remains between the bound basic peptide and the bilayer (Montich et al., 1993). Theoretical electrostatic calculations based on molecular models of bilayers (Peitzsch et al., 1995) and basic peptides are consistent with all these results (N. Ben Tal, B. Honig, G. Denisov, R. Peitzsch, and S. McLaughlin, unpublished calculations). The electrostatic calculations predict the long range attraction between the basic residues and the acidic lipids is opposed by a shorter range repulsion of the charged peptide from the low dielectric membrane, which is due to the "image charge" or Born effect (Parsegian, 1969). The calculations thus predict basic peptides will have an energy well about 3 Å from the surface, which agrees with the experimental observations. In summary, the interactions of myristate with the membrane interior and of the basic domains of Src and MARCKS with acidic lipids can be understood in terms of the fundamental physical principles of hydrophobic (e.g. Tanford, 1991) and electrostatic (McLaughlin, 1989; Davis and McCammon, 1990; Sharp and Honig, 1990) interactions.

How do electrostatic and hydrophobic interactions combine to anchor a myristoylated peptide or protein to a membrane? The magnitude of the electrostatic interactions can be determined by measuring the membrane binding of the nonmyristoylated forms of peptides. For example, a nonmyristoylated peptide that mimics the N-terminal region of Src, nonmyr-Src(2-16), binds to vesicles comprised of 33% acidic lipid with an apparent association constant, K_B, of 10^3 M^{-1} (Buser et al., 1994); vesicles of this composition mimic the acidic lipid content of the inner surface of a typical plasma membrane (Op den Kamp, 1979). Thus neither the electrostatic ($K_B = 10^3$ M^{-1}) nor the hydrophobic ($K_A = 10^4$ M^{-1}) association constant is strong enough to anchor Src to membranes. The myristoylated form of the basic peptide, however, binds to 2:1 PC:PS vesicles with a $K = 10^7$ M^{-1}, which is sufficiently strong to anchor both this peptide to vesicles and the intact Src protein to the plasma membrane. This is because the hydrophobic and electrostatic energies add, or the binding constants ($K_A = 10^4$ M^{-1}, $K_B = 10^3$ M^{-1}) multiply. This can be understood in terms of a simple model in which the peptide is depicted by 2 balls joined by a string of length r (Buser et al., 1994; McLaughlin & Aderem, 1995): when the ball representing the acyl chain binds to the membrane, the ball representing the cluster of basic residues is confined to a hemisphere of radius r, which greatly increases the chance that it will associate electrstatically with acidic lipids. The model predicts that the overall apparent association constant, K, is:

$$K \approx K_A(1 + \alpha K_B/r) \approx K_A K_B(\alpha/r) \quad (1)$$

where $\alpha = 4$ nm/M^{-1} is a constant that depends only on the area/phospholipid (Buser et al., 1994). For the myr-Src peptide, r ~ 1–3 nm and $\alpha/r \approx 1$. Thus Eq. 1 predicts correctly the 1000-fold increase in the binding of both the myristoylated peptide (Buser et al., 1994) and intact Src protein (Sigal et al., 1994) observed when 33% acidic lipid is incorporated into the membrane.

In terms of the simple ball and string model represented by Eq. 1, MARCKS differs from Src because the region connecting the N-terminal myristate and the cluster of basic residues is longer (150 aa) and contains many acidic amino acids, which will tend to reduce the electrostatic interaction. However, the basic domain also contains more basic residues (13), which will increase the electrostatic interaction with lipids. Peptides that correspond to the basic region of MARCKS bind strongly to membranes containing acidic lipids, and PKC phosphorylation of the peptides causes them to desorb from the membranes (Taniguchi and Manenti, 1993; Kim et al., 1994a). More importantly, phosphorylation also causes the intact MARCKS protein to move off phospholipid vesicles containing acidic lipids (Taniguchi and Manenti, 1993; Kim et al., 1994b). These data are consistent with a simple "electrostatic switch" mechanism discussed in detail elsewhere (McLaughlin & Aderem, 1995): phosphorylation of serines within the basic cluster reduces the electrostatic attraction of MARCKS for the membrane, and the hydrophobic interaction of the myristate is not sufficient to anchor the protein. As expected, addition of calcium-calmodulin (which also binds to the basic domain) or an increase in ionic strength (which weakens the electrostatic interaction) causes MARCKS to desorb from phospholipid vesicles, mimicking the effect of phosphorylation (Kim et al., 1994b). The results obtained with phospholipid vesicles parallel the results obtained in living cells: activation of PKC causes MARCKS to translocate from the plasma membrane to the cytoplasm in intact synaptosomes (Wang et al., 1989), macrophages (Rosen et al., 1990) and neutrophils (Thelen et al., 1991), but not in all cell types (e.g. James and Olson, 1989).

Lateral organization of proteins. MARCKS has a punctate distribution in the plasma membranes of macrophages (Rosen et al., 1990), which demonstrates directly that it forms domains or ensembles in intact cells. (The word "domain" can be confusing because it is used widely to describe both a region of a protein with a defined structure (e.g. the SH2, SH3 and PH domains) and a region of a membrane where proteins or lipids are assembled laterally. We use the term "ensemble" as a synonym for a membrane domain.) The cluster of basic residues on MARCKS (residues 151-175), which contains 13 positively charged amino acids and is responsible in part for anchoring the protein to membranes (Kim et al., 1994b), apparently is also involved in this ensemble/domain formation. Specifically, a peptide corresponding to this cluster self assembles spontaneously into lateral domains with acidic lipids in phospholipid vesicles (Yang and Glaser, 1995). We propose here a simple biophysical mechanism that can account for the formation of these ensembles/domains. This mechanism implies that PLC, PKC, Src and other molecules that bind to membranes through electrostatic interactions with acidic lipids also will be located in these signal transduction ensembles.

We focus first on the problem of how the MARCKS (150-175) peptide self-assembles into lateral domains enriched in acidic phospholipids when it binds to phospholipid vesicles (Yang and Glaser, 1995). In our model, this basic peptide causes lateral domain formation by decreasing the electrostatic free energy of the system (Denisov, Glaser & McLaughlin, unpublished). The negatively charged phospholipids in a membrane and the counterions (e.g. K^+) in the aqueous diffuse double layer adjacent to the surface may be regarded as a parallel plate capacitor with the counterions located a Debye length ($1/\kappa \approx 1$ nm in a 0.1 M monovalent salt solution) from the surface. (e.g. McLaughlin, 1989); electrical energy is stored in this capacitor. As shown in Fig. 2 A, the acidic lipids (stippled circles) in a phospholipid vesicle are distributed randomly in the absence of peptides. There are two reasons why the acidic lipids do not spontaneously form a domain (Fig. 2B) in the absence of basic peptides: domain formation would increase both the free energy due to a mixing entropy term (e.g.Träuble, 1977), ΔG_{mix}, and the electrostatic free energy, ΔG_{el}, which can be calculated within the framework of Gouy-Chapman theory by a formula derived by Jahnig (1976) and Träuble et al. (1976). In our simple model we assume the change in the Gibbs free energy of the system upon domain formation is the sum of these two terms: $\Delta G = \Delta G_{mix} + \Delta G_{el}$.

When basic peptides (bars in Fig. 2C and D) bind to the membrane surface, domain formation is favored because the bound peptides reduce the charge on the membrane, decreasing the electrical free energy stored in the diffuse double layer "electrical capacitor". For domain formation to occur ($\Delta G < 0$), the reduction in electrical free energy produced by peptides binding

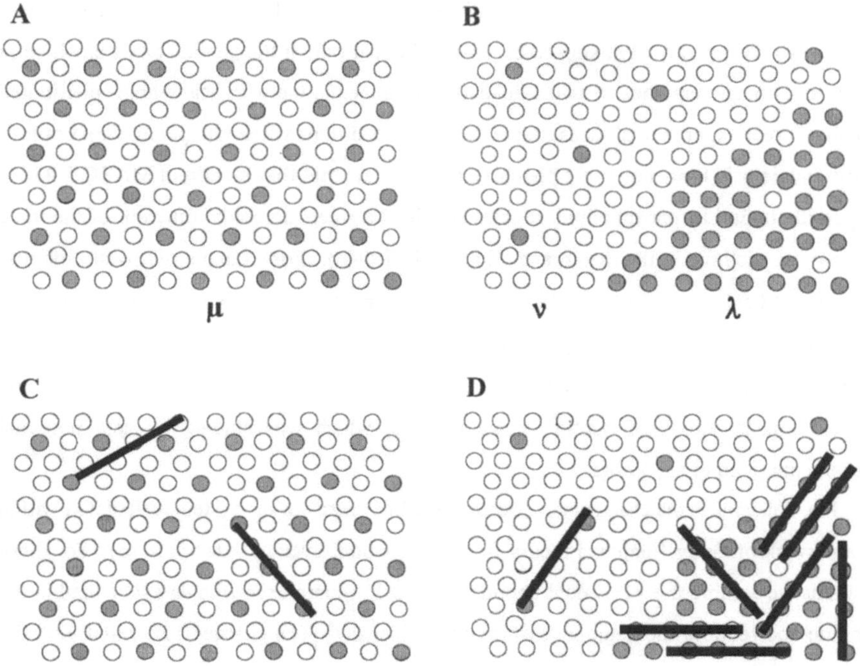

Fig. 2. Sketch of domain formation. The open circles represent zwitterionic lipids (e.g. PC, PE) and the stippled circles represent acidic phospholipids (e.g. PS, PIP_2). In the absence of basic peptides (Fig. 2A), acidic lipids like PS are distributed uniformly in the plane of the membrane (Yang and Glaser, 1995). Acidic lipids will not spontaneously form domains (Fig. 2B) because the change in the Gibbs free energy due to both entropy of mixing and electrostatics is unfavorable. When basic peptides (bars, Fig. 2C) are added to the membrane they stabilize domains (Fig.2D) because they bind much more strongly to membranes (or regions of membranes) that contain high fractions of acidic lipids (Kim et al., 1994a). Yang and Glaser (1995) showed that fluorescently labeled MARCKS(151-175) does form domains enriched in acidic lipids such as that shown in Fig. 2D. μ is the mole fraction of acidic lipids in the membrane, λ is the fraction of acidic lipids in the domain and ν is the fraction of acidic lipids in the non-domain region of the membrane

to domains enriched in acidic lipids must be sufficient to overcome the unfavorable change in free energy due to the entropy of mixing term. This will occur only if the peptide binds much more strongly to a domain enriched in acidic lipids than to a membrane with acidic lipids distributed randomly. Direct experimental measurements (under conditions where the peptide/lipid ratio is very low and domain formation is not possible), demonstrate that the fraction of MARCKS(151-175) peptide bound to phospholipid vesicles is indeed a steep sigmoidal function of the mole fraction of acidic lipid in the membrane (Kim et al., 1994a). For example, when the solution contains 1 mM lipid, , < 5% of the MARCKS (151-175) peptide binds to membranes containing 5% acidic lipid but > 95% of the peptide binds to membranes containing 17% acidic lipid. This binding can be described by a simple Gouy-Chapman/mass action theory, as can the adsorption of higher concentrations of peptide to membranes, which was studied by zeta potential measurements (Kim et al., 1994a).

Fig. 3 illustrates the change in the electrostatic free energy of the system, calculated within the framework of Gouy-Chapman theory from an equation given by Jahnig (1976). For these calculations we assume the +13 valent peptide forms a 1:1 complexes with an acidic lipid in the membrane and reduces the negative surface charge density. The charges associated with the acidic lipids and bound peptides are assumed to be smeared uniformly over the surfaces of each of the three regions (μ, ν and λ in Fig. 2). Fig. 3 also illustrates the change in free energy due to the entropy of mixing the zwitterionic and acidic lipids, as calculated from a conventional thermodynamic relation (e.g. Träuble, 1977). These curves are plotted as a function of the mole fraction of acidic lipids in the membrane, μ. We also plot the sum of these two curves, the net change in free energy of the system upon domain formation. For these calculations we assume that the salt concentration is 10 mM, the concentration employed by Yang and Glaser (1995) in most of their experiments with the MARCKS peptide, and that the mole fraction of acidic lipids in the domain (λ in Fig. 2D) is 0.95. We also assume the peptide is present in the aqueous phase at a concentration where it binds significantly only to membranes (or domains) with a high fraction of acidic lipids. Fig. 3 illustrates that ΔG has two maxima; this is characteristic of systems where domain formation occurs (Raudino, 1995). Our calculations also indicate that domain formation will be favored by low ionic strength because the magnitude of the electrostatic free energy term increases as the ionic strength decreases, but the mixing entropy term is independent of salt concentration. This prediction agrees with the results of Yang and Glaser (1995) who noted that increasing the ionic strength to 100 mM reduces the stability of the

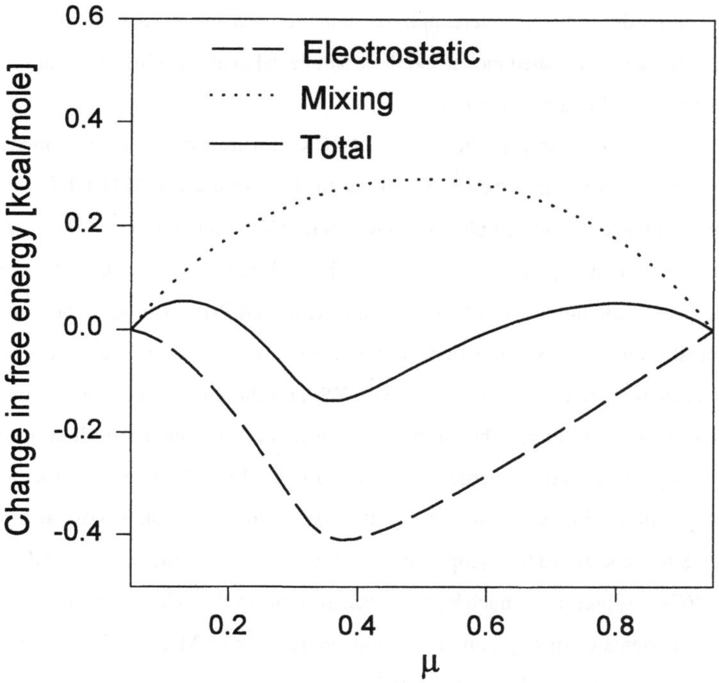

Fig. 3. Change in the Gibbs free energy upon domain formation vs mole fraction of acidic lipids in the membrane, μ, as calculated from the model. Domain formation is induced by the binding of a peptide of a valence +13 (which corresponds to the MARCKS (151-175) peptide) to an acidic lipid in 1:1 complex. The aqueous bulk solution contains 0.01M monovalent electrolyte. The fractions of acidic lipids in the domain and non-domain portions of the membrane are λ=0.95 and ν=0.05, and the value of μ varies in the range $\nu < \mu < \lambda$. The change in the free energy is a sum of two terms, an electrostatic term that favors domain formation and a mixing term that opposes it. The total ΔG has two maxima and one minimum, which is characteristic of systems where lateral phase separation occurs (e.g., Träuble, 1977; Raudino, 1995). Our calculations show that a decrease in the ionic strength produces a deeper minimum in ΔG and therefore favors domain formation.

domains. The model is highly oversimplified, but the electrostatic and mixing terms will be present in any more sophisticated model and these simple calculations illustrate that electrostatics is sufficient to cause domain formation.

Theory aside, experiments show that MARCKS forms domains in macrophages (Rosen et al., 1990) and that a peptide corresponding to the the basic domain of MARCKS self assembles into domains with acidic lipids on phospholipid veicles (Yang and Glaser, 1995). The domains MARCKS forms in cells, like those on phospholipid vesicles, are likely to contain high concentrations of acidic lipids, including the multivalent lipid PIP_2. All our experimental results and theoretical calculations suggest that the acidic lipids are associated only loosely with basic peptides, or basic regions of proteins like MARCKS, and should be accessible to other proteins with clusters of basic residues. Thus, other components of the signal transduction system that utilize acidic lipids to bind to membranes (e.g. PLC-δ_1, PLC-β_1, PKC, Src) should also be localized preferentially in these domains. Furthermore, when PKC phosphorylates MARCKS, and MARCKS translocates to the cytoplasm (Rosen et al., 1990), the domains will disintegrate because MARCKS is present at much higher concentrations in the cell (e.g. 10 μM in brain) than the other components of this system. Thus, phosphorylation of MARCKS could be a negative feedback mechanism to provide temporal and spatial localization of the signal passing through this calcium/phosphospholipid second messenger system.

This hypothesis is obviously speculative, but it can be tested experimentally. For example, Yang and Glaser (unpublished) have shown that fluorescently labeled PKC colocalizes with the MARCKS peptide and acidic lipids in phospholipid vesicles. We are labelling PLC-δ, PLC-β, and Src with fluorescent probes and will test whether these proteins also colocalize with the MARCKS peptide and acidic lipids in phospholipid vesicles.

References

Adam, G. and Delbrück, M. (1968) Reduction of dimensionality in biological diffusion processes. In: *Structural Chemistry and Molecular Biology.* edited by A. Rich and N. Davidson. San Francisco: W.H. Freeman and Company, p. 198-215.

Aderem, A. (1992) The MARCKS brothers: a family of protein kinase C substrates. *Cell* 71:713-716.

Berridge, M.J. (1993) Inositol trisphosphate and calcium signalling. *Nature* 361:315-325.

Blackshear, P. J. (1993) The MARCKS family of cellular protein kinase C substrates. *J. Biol. Chem.* 268:1501-1504.

Buser, C. A., Sigal, C. T., Resh, M. D., and McLaughlin, S. (1994) Membrane binding of myristylated peptides corresponding to the NH_2-terminus of Src. *Biochemistry* 33:13093-13101.

Carpenter, G. (1992) Receptor tyrosine kinase substrates: src homology domains and signal transduction. *FASEB J.* 6:3283-3289.

Cifuentes, M. E., Honkanen, L., and Rebecchi, M. J. (1993) Proteolytic fragments of phosphoinositide-specific phospholipase C-δ_1. Catalytic and membrane binding properties. *J. Biol. Chem.* 268:11586-11593.

Cohen, G. B., Ren, R., and Baltimore, D. (1995) Modular binding domains in signal transduction proteins. *Cell* 80:237-248.

Davis, M. E. and McCammon, J. A. (1990) Electrostatics in biomolecular structure and dynamics. *Chem. Rev.* 90:509-521.

Ferguson, K. M., Lemmon, M. A., Schlessinger, J., and Sigler, P. B. (1994) Crystal structure at 2.2 Å resolution of the pleckstrin homology domain from human dynamin. *Cell* 79:199-209.

Garcia, P., Gupta, R., Shah, S., Morris, A.J., Rudge, S.A., Scarlata, S., Petrova, V., McLaughlin, S., and Rebecchi, M. J. (1995) Binding of the pleckstrin homology domain of phospholipase C-δ_1 to membrane bilayers and inositol 1,4,5-trisphosphate. *Biochemistry* submitted.

Gibson, T. J., Hyvonen, M., Musacchio, A., Saraste, M., and Birney, E. (1994) PH domain: the first anniversary. *Trends Biochem. Sci.* 19:349-353.

Harlan, J. E., Hajduk, P. J., Yoon, H. S., and Fesik, S. W. (1994) Pleckstrin homology domains bind to phosphatidylinositol-4,5-bisphosphate. *Nature* 371:168-170.

Harlan, J.E., Yoon, H.S., Hajduk, P.J., and Fesik, S.W. (1995) Structural characterization of the interaction between a pleckstrin homology domain and phosphatidylinositol 4,5-bisphosphate. *Biochemistry* 34:9859-9864.

Haslam, R. J., Kolde, H. B., and Hemmings, B. A. (1993) Pleckstrin domain homology. *Nature* 363:309-310.

Jahnig, F. (1976) Electrostatic free energy and shift of the phase transition for charged lipid membranes. *Biophys. Chem.* 4:309-318.

Jain, M. K. and Berg, O. G. (1989) The kinetics of interfacial catalysis by phospholipase A2 and regulation of interfacial activation: hopping versus scooting. *Biochim. Biophys. Acta* 1002:127-156.

James, G. and Olson, E. N. (1989) Myristoylation, phosphorylation, and subcellular distribution of the 80-kDa protein kinase C substrate in BC3H1 myocytes. *J. Biol. Chem.* 264: 20928 -20933.

James, S.R., Paterson, A., Harden, T.K., and Downes, C.P. (1995) Kinetic analysis of phospholipase Cβ isoforms using phospholipid-detergent mixed micelles. Evidence for interfacial catalysis involving distinct micelle binding and catalytic steps. *J. Biol. Chem.* 270:11872-11881.

Kim, J., Blackshear, P.J., Johnson, J.D. and McLaughlin, S. (1994a) Phosphorylation reverses the membrane association of peptides that correspond to the basic domains of MARCKS and neuromodulin. *Biophys. J.* 67:227-237.

Kim, J., Shishodo, T., Jiang, X., Aderem, A., and McLaughlin, S. (1994b) Phosphorylation, high ionic strength, and calmodulin reverse the binding of MARCKS to phospholipid vesicles. *J. Biol. Chem.* 269:28214-28219.

Lee, S.B., and Rhee, S.G. (1995) Significance of PIP_2 hydrolysis and regulation of phospholipase C isozymes. *Cur. Opin. Cell Biol.* 7:183-189.

Macias, M. J., Musacchio, A., Ponstingl, H., Nilges, M., Saraste, M., and Oschkinat H. (1994) Structure of the pleckstrin homology domain from beta-spectrin. *Nature* 369:675-677.

McLaughlin, S. (1989) The electrostatic properties of membranes. *Annu. Rev. Biophys. Biophys. Chem.* 18:113-136.

McLaughlin, S. and Aderem, A. (1995) The myristoyl-electrostatic switch: A modulator of reversible protein-membrane interactions. *Trends Biochem. Sci.* 20:272-276.

Montich, G., Scarlata, S., McLaughlin, S., Lehrmann, R., and Seelig, J. (1993) Thermodynamic characterization of the association of small basic peptides with membranes containing acidic lipids. *Biochim. Biophys. Acta* 1146:17-24.

Mosior, M. and McLaughlin, S. (1991) Peptides that mimic the pseudosubstrate region of protein kinase C bind to acidic lipids in membranes. *Biophys J.* 60:149-159.

Newton, A. C. (1993) Interaction of proteins with lipid headgroups: lessons from protein kinase C. *Annu. Rev. Biophys. Biomol. Structure* 22:1-25.

Op den Kamp, J. A. (1979) Lipid asymmetry in membranes. *Annu. Rev. Biochem.* 48:47-71.

Parker, P. J., Hemmings, B. A., and Gierschik, P. (1994) PH domains and phospholipases — a meaningful relationship?. *Trends Biochem. Sci.* 19:54-55.

Parsegian, A. (1969) Energy of an ion crossing a low dielectric membrane; solutions to four relevant electrostatic problems. *Nature* 221:844-846.

Pawelczyk, T., and Lowenstein, J.M. (1993) Binding of phosholipase C-δ_1 to phospholipid vesicles. *Biochem. J.* 291:693-696.

Pawson, T. (1995) Protein modules and signalling networks. *Nature* 373:573-580.

Peitzsch, R. M., Eisenberg, M., Sharp, K. A., and McLaughlin, S. (1995) Calculations of the electrostatic potential adjacent to model phospholipid bilayers. *Biophys. J.* 68:729-738.

Peitzsch, R. M. and McLaughlin, S. (1993) Binding of acylated peptides and fatty acids to phospholipid vesicles: pertinence to myristoylated proteins. *Biochemistry* 32:10436-10443.

Pitcher,J.A., Touhara,K., Payne,E.S., and Lefkowitz,R.J. (1995) Pleckstrin homology domain-mediated membrane association and activation of the β-adrenergic receptor kinase requires coodinate interaction with G βγ subunits and lipid. *J. Biol. Chem.* 270:11707-11710.

Raudino, A. (1995) Lateral inhomogeneous lipid membranes: theoretical aspects. *Adv. Colloid Interfac. Sci.* 57:229-285.

Rebecchi, M.,Peterson,A., and McLaughlin, S. (1992) Phosphoinositide-specific phospholipase C-δ_1 binds with high affinity to phospholipid vesicles containing phosphatidylinositol 4,5-bisphosphate. *Biochemistry* 31:12742-12747.

Resh, M. D. (1993) Interaction of tyrosine kinase oncoproteins with cellular membranes. *Biochim. Biophys. Acta* 1155:307-322.

Rhee, S. G. and Choi, K. D. (1992) Regulation of inositol phospholipid-specific phospholipase C isozymes. *J. Biol. Chem.* 267:12393-12396.

Rosen, A., Keenan, K. F., Thelen, M., Nairn, A. C., and Aderem, A. (1990) Activation of protein kinase C results in the displacement of its myristoylated, alanine-rich substrate from punctate structures in macrophage filopodia. *J. Exp. Med.* 172:1211-1215.

Roux, M., Neumann, J. M., Bloom, M., and Devaux, P. F. (1988) ^{2}H and ^{31}P NMR study of pentalysine interaction with headgroup deuterated phosphatidylcholine and phosphatidylserine. *Eur. Biophys. J.* 16:267-273.

Sharp, K. A. and Honig, B. H. (1990) Electrostatic interactions in macromolecules: theory and applications. *Annu. Rev. Biophys. Biophys. Chem.* 19:301-332.

Sigal, C. T., Zhou, W., Buser, C. A., McLaughlin, S., and Resh, M. D. (1994) The amino terminal basic residues of Src mediate membrane binding through electrostatic interaction with acidic phospholipids. *Proc. Natl. Acad. Sci. U. S. A.* 91:12253-12257.

Silvius, J. R. and l'Heureux, F. (1994) Fluorimetric evaluation of the affinities of isoprenylated peptides for lipid bilayers. *Biochemistry* 33:3014-3022.

Sternweis, P. C. and Smrcka, A. V. (1992) Regulation of phospholipase C by G proteins. *Trends Biochem. Sci.* 17:502-506.

Sutton, R. B., Davletov, B. A., Berghuis, A. M., Sudhof, T. C., and Sprang, S. R. (1995) Structure of the first C_2 domain of synaptotagmin I: A novel Ca^{2+}/phospholipid-binding fold. *Cell* 80:929-938.

Tanford, C. (1991) The Hydrophobic Effect: Formation of Micelles and Biological Membranes. Malabar, FL: Krieger Publishing Co.

Taniguchi, H. and Manenti, S. (1993) Interaction of myristoylated alanine-rich protein kinase C substrate (MARCKS) with membrane phospholipids. *J. Biol. Chem.* 268:9960-9963.

Thelen, M., Rosen, A., Nairn, A. C., and Aderem, A. (1991) Regulation by phosphorylation of reversible association of a myristoylated protein kinase C substrate with the plasma membrane. *Nature* 351:320-322.

Towler, D. A., Gordon, J. I., Adams, S. P., and Glaser, L. (1988) The biology and enzymology of eukaryotic protein acylation. *Annu. Rev. Biochem.* 57:69-99.

Träuble, H. (1977) Membrane electrostatics. In: *Structure of Biological Membranes.* Edited by S. Abrahamsson and I. Pascher. New York: Plenum Press, p. 509-550.

Träuble, H., Teubner, M., Woolley, P., and Eibl, H. (1976) Electrostatic interactions at charged lipid membranes. I. Effects of pH and univalent cations on membrane structure. *Biophys. Chem.* 4:319-342.

Vergères, G., Manenti, S., and Weber, T. (1995) Interaction of MARCKS, a major protein kinase C substrate, with the membrane. In: *Signalling Mechanisms: From Transcription Factors to Oxidative Stress.* edited by L. Packer and K. Wirtz. NATO ASI Series. Berlin: Springer-Verlag.

Wang, J. K., Walaas, S. I., Sihra, T. S., Aderem, A., and Greengard, P. (1989) Phosphorylation and associated translocation of the 87-kDa protein, a major protein kinase C substrate, in isolated nerve terminals. *Proc. Natl. Acad. Sci. U. S. A.* 86:2253-2256.

Wedegaertner, P. B., Wilson, P. T., and Bourne, H. R. (1995) Lipid modifications of trimeric G proteins. *J. Biol. Chem.* 270:503-506.

Yang, L. and Glaser, M. (1995) Membrane domains containing phosphatidylserine and substrate can be important for the activation of protein kinase C. *Biochemistry* 34:1500-1506.

Yoon, H. S., Hajduk, P. J., Petros, A. M., Olejniczak, E. T., Meadows, R. P., and Fesik, S. W. (1994) Solution structure of a pleckstrin-homology domain. *Nature* 369:672-675.

Zheng, J., Knighton, D. R., Xuong, N. H., Taylor, S. S., Sowadski, J. M., and Ten Eyck, L.F. (1993) Crystal structures of the myristylated catalytic subunit of cAMP-dependent protein kinase reveal open and closed conformations. *Protein Science* 2:1559-1573.

COVALENTLY ATTACHED LIPID BILAYERS ON PLANAR WAVEGUIDES FOR MEASURING PROTEIN BINDING TO FUNCTIONALIZED MEMBRANES

Stephan Heyse [1], Michael Sänger [2], Hans Sigrist [2], Günther Jung[3], Karl-Heinz Wiesmüller[3] and Horst Vogel [1]
[1] Institute of Physical Chemistry 4, Chemistry Department
Swiss Federal Institute of Technology, CH-1015 Lausanne,
Switzerland

OUTLINE

Many biologically important signal transduction processes occur at the level of cell membranes. Specialized membrane receptors, such as ligand-gated ion channels (Unwin, 1993) or G-protein coupled receptors (Strader, 1994), specifically recognize ligands and transform this binding event into a corresponding signal. While in the former case extremely sensitive techniques have been developed to observe the opening and closing of single channels upon ligand binding (Rudy and Iverson, 1992), in the latter case the electrical properties of the membrane remain unchanged upon receptor activation. In order to elucidate the molecular events involved for example in G protein signal cascades, radioactive or fluorescent labels have been introduced by complicated and time-consuming experimental procedures (Feder et al., 1986; Heithier et al., 1992).

Here we present a new method to investigate the interactions between membrane receptors and their ligands in an artificial, reconstituted membrane system. Lipid bilayers are anchored via covalently attached thiolipids to TiO_2/SiO_2 waveguide surfaces to produce mechanically stable, long-lasting and reusable membranes. The coupling scheme is designed to keep the membrane at a distance from the surface, thus maintaining a certain membrane flexibility and allowing for an aqueous phase between surface and bilayer. The method yields imperfect, covalently bound thiolipid layers which can be converted to lipid bilayers by self-assembly of phospholipids.

---

[2] Institute of Biochemistry, University of Berne, Freiestrasse 3, CH-3012 Berne, Switzerland
[3] Institute of Organic Chemistry, University of Tübingen, Germany

NATO ASI Series, Vol. H 96
Molecular Dynamics of Biomembranes
Edited by Jos A. F. Op den Kamp

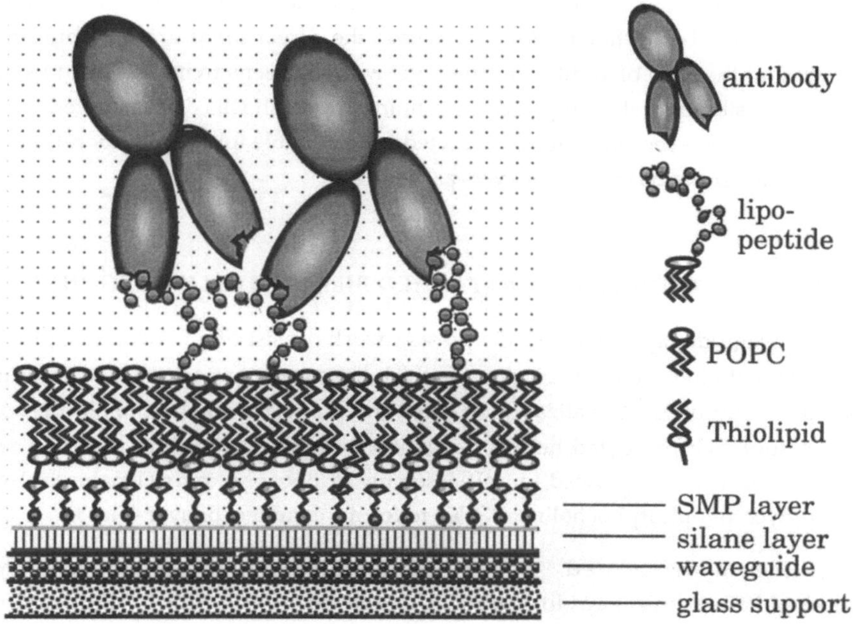

Fig. 1: Schematic representation of a supported, lipopeptide-containing bilayer on a waveguide with antibodies bound to it. It is formed in the following way: Waveguides are functionalized with an aminosilane and reacted with the bifunctional crosslinker SMP to produce a surface exposing maleimide groups. Binding of thiolipids from mixed micelles to the maleimide-modified waveguide surface and removal of detergent by washing with buffer yields an anchoring thiolipid layer. Subsequent self-assembly of a phospholipid / $(NANP)_3$-lipopeptide mixture from an octyl glucoside mixed micellar solution completes this layer to a bilayer containing lipid-anchored peptide antigens. Specific and nonspecific binding of a monoclonal anti-NANP antibody to this membrane were investigated.

Receptor molecules are easily incorporated into these membranes. In our study, they comprised varied amounts of lipid-anchored peptide antigens. The peptide part of the lipopeptides (a threefold repeat of the sequence Asn-Ala-Asn-Pro = NANP) was chosen because it is the major antigenic determinant of the cell surface protein of the malaria parasite *Plasmodium falciparum*, (Godson, 1985) and therefore is relevant to vaccine development. The $(NANP)_3$-lipopeptides served as a model of a membrane receptor to detect the binding of monoclonal antibodies raised against NANP sequences, thus mimicking recognition reactions occurring at cell membrane surfaces.

The layer formation and the binding of ligands were monitored with an integrated optics technique (Tiefenthaler, 1992). It allows the direct monitoring of binding events without the need of additional labels, yet with high sensitivity and in aqueous media. The feasibility of the supported membrane approach was demonstrated with membrane surface receptors, but it is expected to be suitable for the incorporation of transmembrane proteins into the bilayers as well.

FORMATION OF COVALENTLY ATTACHED BILAYERS ON WAVEGUIDES

The thin waveguide layers used in the present work consist of a hydrophilic glass-like material with high refractive index, to which the covalent attachment of lipids is not possible without functionalization of either the support or the lipid molecules or both. The approach presented here consists of a three-step procedure to set up an anchoring lipid leaflet, followed by self-assembly of phospholipids (1-palmitoyl-2-oleoyl-*sn*-glycero-3-phosphocholine, POPC) to form a complete bilayer (Fig. 1).

Firstly, the hydroxylated waveguide surface is modified with γ-aminopropyltriethoxysilane following a standard liquid-phase silanization procedure, similarly to the protocol used by Kallury et al. (1989). Secondly, the surface amino functionalities are reacted with the heterobifunctional crosslinker N-succinimidyl-3-maleimidopropionate (SMP), thus leading to a surface exposing maleimide functional groups ("maleimide surface"). Thirdly, synthetic thiolipids presented in small octyl glucoside (OG) mixed micelles are immobilized by the spontaneous formation of thioether bonds between thiol functions at the lipid head groups and the thiol-reactive maleimido groups on the waveguide surface (Fig. 1). This highly selective reaction is performed in aqueous media, thus enabling optical monitoring of layer formation *in situ*. The self-assembly of thiolipid on modified waveguides was complete (>95%) after 30 - 60 minutes.

Surface coverages of 50-60% were calculated for thiolipid layers composed of 1,2-dioleoyl-*sn*-glycero-3-phosphothioethanol (DOPSH) or 1,2-dimyristoyl-*sn*-glycero-3-phosphothioethanol (DMPSH). This result was obtained by comparing the mean thickness values of self-assembled thiolipid layers on waveguides (12 Å) with those of thiolipids on gold surfaces (21 Å, as determined by surface plasmon resonance), where the hydrocarbon chains of gold-attached alkylthiols and thiolipids pack as closely as possible in a nearly hexagonal manner (Fenter et al., 1994; Lang et al., 1994). This experimentally obtained value of 21 Å corresponds to a monolayer of DOPSH on gold, as inferred from the transmembrane distance of 44 Å between the choline groups of a DOPC bilayer (Wiener and White, 1992) and the monolayer thickness of 22-27 Å (n=1.45) determined by Lang et al. (1994) for a different thiolipid with a longer spacer than DOPSH and with saturated palmitoyl chains. So we conclude that thiolipid layers formed on waveguides are incompletely packed monolayers, with on the average only half of the geometric waveguide surface covered by thiolipids. In principle the incompleteness of the anchoring layer is an advantage for future incorporation of membrane proteins.

The covalent attachment of the thiolipid layer was demonstrated (1) by its resistance to detergent washing, as opposed to thiolipid adsorbed to waveguide surfaces where the maleimide groups had been previously inactivated by mercaptoethanol, (2) by competition experiments with radiolabeled cysteine, and (3) by the increased hydrophobicity of the maleimide surface after thiolipid binding, which persisted after washing with methanol (contact angle measurements).

In a final step, the surface-attached thiolipids serve as templates and anchors for the formation of a complete lipid bilayer by vesicle fusion or by self-assembly from mixed lipid-detergent micelles, which contain a certain fraction of receptor molecules, where needed (Fig. 2). This procedure is simple, fast, reproducible and most importantly, very well suited for the reconstitution of membrane-bound receptors into supported lipid membranes. Optical measurements indicated that the supported bilayers were stable under buffer and also did not desorb upon alteration of the ionic strength. By washing with OG, the second (phospholipid) layer could be entirely removed and afterwards reconstructed by vesicle fusion or micelle dilution.

Evidence for the formation of a lipid bilayer comes from the observation that the resulting layer composed of thiolipids and phospholipids had always an identical thickness of about 47 Å, despite variations in thiolipid coverage and in method of phospholipid self-assembly (either vesicle fusion or micelle dilution). Therefore it is assumed that adsorbing phospholipids fill up "holes" in the first thiolipid leaflet. Interestingly, DOPSH / POPC bilayers on gold surfaces show essentially the same thickness of about 46 Å (data not shown). It was shown by capacitance measurements (unpublished) that in this case, too, the adsorbing phospholipids seemed to fill up "holes", as the layer capacitance approached a bilayer value during adsorption. The lipid film thickness of 47 Å on modified waveguides is in the same range as published data on DOPC bilayers (44 Å, Wiener and White, 1992), fluid DMPC bilayers adsorbed on quartz (43 Å, Johnson et al., 1991) and mixed POPC / thiolipid layers on gold (47 Å calculated from the angle shift with n=1.45, Lang et al., 1994).

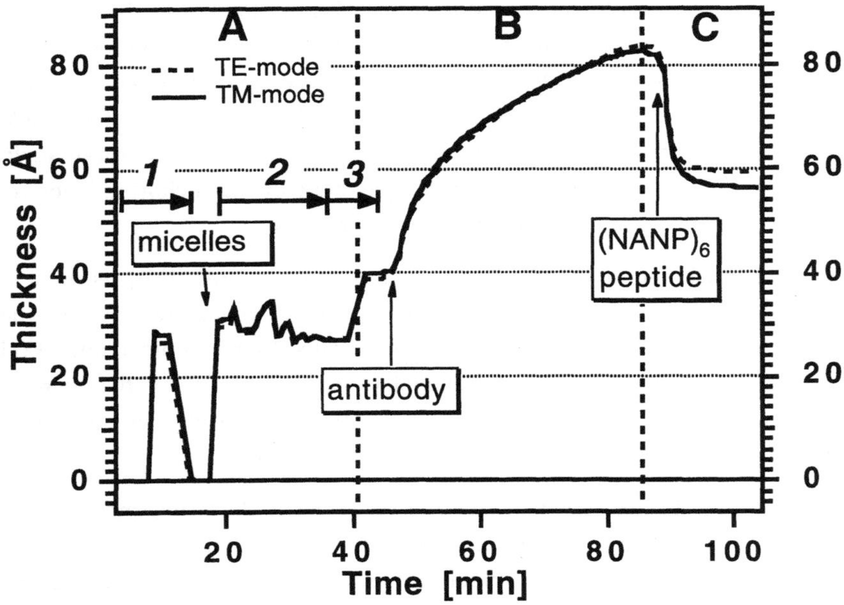

Fig. 2: Antibody binding to a lipid membrane containing $(NANP)_3$-lipopeptide. Shown are the apparent thickness values of adsorbed lipid and protein layers on the waveguide versus time, based on measurement of the effective waveguide indices for the two light polarizations (TE and TM) during the different binding reactions. Experimental protocol: The waveguide was incubated with DOPSH overnight. **(A)** After one washing step with OG *(1)*, a second lipid layer was formed by applying a 50 mM OG mixed micellar solution containing 1.24 mM POPC and 0.027 mM lipopeptide (= 2 mol%) and subsequent dilution 10 times 1:1 (v:v) with buffer *(2)*. **(B)** The buffer was exchanged for antibody buffer *(3)*, causing a change in the index of refraction of the medium, which manifested itself as a jump in the measured signal. Then a solution of 200 nM antibody was added (t=47 min), resulting in a continous binding reaction. **(C)** After washing with buffer, part of this antibody was finally displaced upon addition of 75 µM free antigen $(NANP)_6$. Thickness values were calculated by transforming the effective waveguide indices first into an optical thickness $\Delta n * d_A$ of the lipid and protein adlayer, which is directly related to the adsorbed mass on the waveguide (Spinke et al., 1993) and thereby to the number of adsorbed molecules (Terrettaz et al., 1993). Δn is the difference in refractive index between the organic film and the surrounding medium, d_A is the geometrical thickness of the organic film. By assuming further a refractive index of n=1.45 for lipids and proteins (Terrettaz et al., 1993; Lang et al., 1994, and references therein), the optical thickness was converted into a mean geometrical thickness of the adlayer, which is a measure of molecule density rather than molecular dimensions.

The lipopeptide-containing layers produced by the micelle dilution technique were in the same thickness range as pure POPC layers. It seems therefore justified to conclude that the membranes obtained were stable bilayers in this case, too.

PROTEIN BINDING TO FUNCTIONALIZED MODEL MEMBRANES

The utility of waveguide-anchored membranes for studying reactions at membrane surfaces was demonstrated by incorporating lipid-anchored peptides into the supported membranes in order to optically detect the specific binding of an antibody to its lipid-anchored peptide antigen. Fig. 2 shows a standard optical measurement.

On a thiolipid template, a mixed POPC / lipopeptide layer is formed by dilution of mixed POPC/lipopeptide/OG micelles (Fig. 2, part A). After exchange of the buffer solution, antibody solution is added, resulting in a high mass adsorption on the membrane (Fig. 2, part B). The total mass adsorption depended on the lipopeptide content of the membrane-forming mixed micelle solution (Fig. 3). In the absence of lipopeptide (pure POPC second layer), the amount of nonspecifically bound antibody corresponded to a protein layer with an average thickness of 5 - 7 Å. With increasing lipopeptide content, more antibody adsorbed. At 0.5% lipopeptide / 99.5% POPC the adsorbed antibody layer was approximately 10 Å thick, at 2% lipopeptide/ 98% POPC it was 40 Å thick.

After antibody binding, free $(NANP)_6$ antigen in high concentration was added. This led to a partial desorption of the protein from the surface (Fig. 2, part C). The $(NANP)_6$ concentration had been chosen such that it assured maximal antibody desorption. Bound antibodies which could thus be desorbed by addition of free $(NANP)_6$ were considered specifically bound, whereas the remaining fraction was considered nonspecifically bound. This evaluation method was validated by control experiments on gold surfaces using surface plasmon resonance detection: (1) Antibody bound to pure POPC membranes on DMPSH thiolipid templates could not be desorbed by $(NANP)_6$. (2) On highly ordered mixed POPC / 2 mol% lipopeptide layers formed on alkylthiols all bound antibody could be desorbed by $(NANP)_6$. (3) Exactly the same amount of desorption was observed on mixed POPC / 2 mol% lipopeptide layers on waveguides, but there remained additionally some antibody unspecifically bound.

Nonspecific antibody binding to supported membranes on waveguides became more prominent with increasing lipopeptide content of the membrane, indicating a disturbing effect of the lipopeptide on the membrane structure.

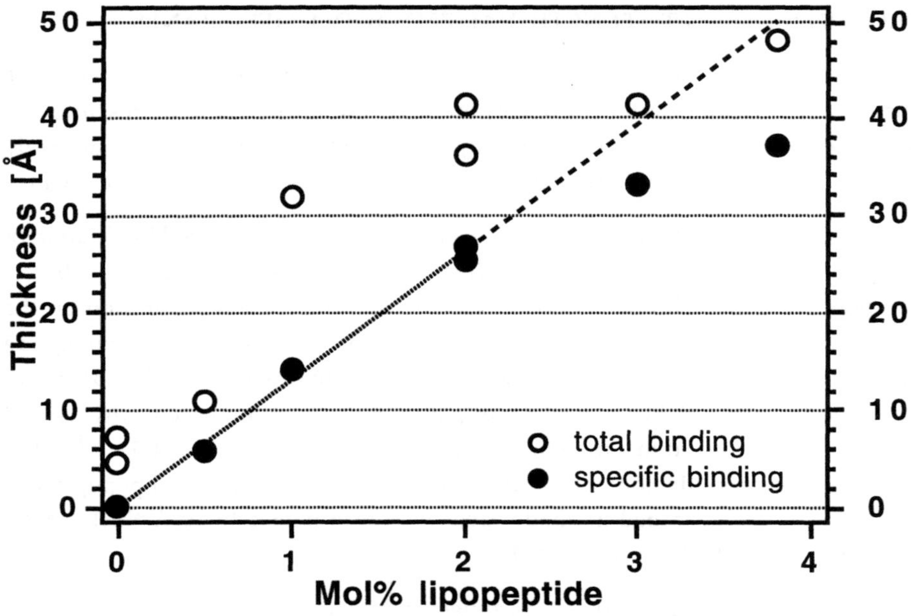

Fig. 3: Binding of anti-NANP antibodies to $(NANP)_3$-lipopeptide-doped supported membranes. Shown are the thickness values of the antibodies bound versus the ratio lipopeptide / total lipid in the mixed micelle solution used for the formation of the supported bilayer. Total binding (O) corresponds to the increase in the apparent thickness of the adlayer observed upon addition of the 200 nM antibody solution. Specific binding (●) designates the amount of protein which is desorbed after addition of 75 µM $(NANP)_6$ peptide. The dotted line is the linear regression of the values for specific binding up to 2 mol% lipopeptide.

Specific antibody binding to the membrane correlates linearly up to 2 mol% with the lipopeptide concentration in the membrane-forming micelle solution (Fig. 3). Modeling the antibodies as spheres, a densely packed protein layer with an average thickness of 47 Å is obtained at 2.8 mol% lipopeptide, assuming that every antibody binds two lipopeptides and all lipopeptides are involved. At 2 mol% lipopeptide the layer of bound antibodies would be 33 Å thick. The actual lipopeptide concentration in the supported membrane may, of course, not be 2 mol%. However, the linear dependence of the specific antibody binding on the lipopeptide content of the mixed micelle solution (up to 2 mol%) shows that it is possible to control the amount of membrane-incorporated lipopeptide.

On raising further the lipopeptide : POPC ratio, this specific binding increased more slowly. The amount of nonspecifically bound antibodies stayed now constant, corresponding to a 10 Å thick protein layer. Control experiments showed that free $(NANP)_6$ peptide bound only slightly to a membrane doped with 2% lipopeptide, producing a layer of about 1Å mean thickness. The binding of antibody and subsequent desorption by free $(NANP)_6$ can be repeated on the same supported membrane, yielding in the second cycle almost the same specific binding (>90%) as in the first cycle.

The results obtained were, within experimental error, identical for both DMPSH and DOPSH anchoring lipids.

CONCLUSION

Functionalized lipid bilayers anchored by covalently bound thiolipids have been established on waveguides by simple self-assembly processes in aqueous media. Membrane formation and recognition reactions at the membrane surface are monitored *in situ* optically and without the need of labels. The lipid bilayer serves not only as a matrix for the incorporation of membrane-anchored receptors but simultaneously reduces the nonspecific binding of proteins, such as antibodies.

The coupling scheme is applicable to hydroxyl-exposing surfaces in general. It could for example be extended to supported bilayers on glass or glass-like surfaces, for the application of different optical techniques.

The presented concept is designed to allow the incorporation of transmembrane proteins in their active state into supported lipid bilayers. These proteins require aqueous phases on both sides of a fluid membrane.

Immobilized membranes have great potential for the development of new biosensors and the study of reconstituted biological signal cascades, which mostly involve transmembrane proteins and protein-protein interactions at the membrane surface.

REFERENCES

Feder D, Im M-J, Klein HW, Hekman M, Holzhöfer A, Dees C, Levitzki A, Helmreich EJM and Pfeuffer T (1986) Reconstitution of β_1-adrenoceptor-dependent adenylate cyclase from purified components. EMBO J. 5: 1509-1514

Fenter P, Eberhardt A and Eisenberger P (1994) Self-assembly of n-alkyl thiols as disulfides on Au(111). Science 266: 1216-1218

Godson GN (1985) Molecular approaches to malaria vaccines. Scientific American 32-39

Heithier H, Fröhlich M, Dees C, Baumann M, Haring M, Gierschik P, Schiltz E, Vaz VL, Hekman M and Helmreich EJ (1992) Subunit interactions of GTP-binding proteins. Eur. J. Biochem. 204: 1169-1181

Johnson SJ, Bayerl TM, McDermott DC, Adam GW, Rennie AR, Thomas RK and Sackmann E (1991) Structure of an adsorbed dimyristoylphosphatidylcholine bilayer measured with specular reflection of neutrons. Biophys. J. 59: 289-294

Kallury KMR, Ghaemmaghami V, Krull UJ and Thompson M (1989) Immobilization of phospholipids on silicon, platinum, indium/tin oxide and gold surfaces with characterization by X-ray photoelectron spectroscopy and time-of-flight secondary ion mass spectroscopy. Anal. Chim. Acta 225: 369-389

Lang H, Duschl C and Vogel H (1994) A new class of thiolipids for the attachment of lipid bilayers on gold surfaces. Langmuir 10: 197-210

Rudy B and Iverson LE, J. N. Abelson and M. I. Simon (1992) Ion channels. Meth. Enzymol. 207, Academic Press, London

Spinke J, Liley M, Guder H-J, Angermaier L and Knoll W (1993) Molecular recognition of self-assembled monolayers: the construction of multicomponent multilayers. Langmuir 9: 1821-1825

Strader CD (1994) Structure and function of G-protein-coupled receptors. Annu. Rev. Biochem. 63: 101-132

Terrettaz S, Stora T, Duschl C and Vogel H (1993) Protein binding to supported lipid membranes: Investigation of the cholera toxin-ganglioside interaction by simultaneous impedance spectroscopy and surface plasmon resonance. Langmuir 9: 1361-1369

Tiefenthaler K (1992) Integrated Optical Couplers as Chemical Waveguide Sensors. Advances in Biosensors 2: 261-289

Unwin N (1993) Neurotransmitter action: opening of ligand-gated ion channels. Cell 72: 31-41

Wiener MC and White SH (1992) Structure of a fluid dioleoylphosphatidylcholine bilayer determined by joint refinement of x-ray and neutron diffraction data. Biophys. J. 61: 434-447

THE EFFECT OF STEROL SIDE CHAIN CONFORMATION ON LATERAL LIPID DOMAIN FORMATION IN MONOLAYER MEMBRANES

Peter Mattjus and J.Peter Slotte
Department of Biochemistry and Pharmacy,
Åbo Akademi University, FIN-20520 Turku, Finland

Catherine Vilchèze and Robert Bittman
Department of Chemistry and Biochemistry,
Queens College of the City University of New York,
Flushing, New York 11367-1597, USA

Introduction

The unique structure of cholesterol appears to have evolved to give the molecule important functions and properties as a membrane component. The amphiphilic properties of cholesterol are provided by the hydrophilic 3β-hydroxy group, the hydrophobic tetracyclic ring structure, and the isooctyl side chain at position C-17. The side chain seems to be important for proper interactions with phospholipids in membranes [Demel et al., 1972a; Demel et al., 1972b]. 5-Androsten-3β-ol, which lacks the isooctyl side chain, can neither condense dipalmitoylphosphatidylcholine (DPPC) monolayers [Demel et al. 1972a; Slotte et al., 1994], nor reduce the solute permeability of phosphatidylcholine liposomes [Nakamura et al., 1980]. Sterols having shorter or longer side chains, as compared to the isooctyl chain of cholesterol, are also less effective as rigidifiers in phospholipid bilayer membranes than cholesterol [Suckling & Boyd, 1976; Craig et al., 1978; Suckling et al., 1979]. A study from our laboratory using cholesterol oxidase as a probe for the strength of sterol-phospholipid interaction also reported a significant difference in sterol/phospholipid interaction in both small unilamellar vesicles and monolayers as a function of the sterol side chain composition [Slotte et al., 1994].

The difference in molecular interactions between cholesterol and phosphatidylcholine on one hand and sphingomyelin on the other, has over the years been examined by several research groups using a variety of different techniques [Mattjus & Slotte, 1994; Bittman et al., 1994; McLean & Phillips, 1982; Lund-Katz et al., 1988; McIntosh et al., 1992]. The most striking difference between phosphatidylcholine and sphingomyelin concerning their interaction with cholesterol appears to be their different ability to solubilize cholesterol [Slotte, 1992; McIntosh, et al., 1992]. Further, our resent monolayer studies have suggested that the amide function in sphingomyelin is of special importance and is at least in part responsible for the increased affinity between cholesterol and sphingomyelin [Bittman et al., 1994]. The

NATO ASI Series, Vol. H 96
Molecular Dynamics of Biomembranes
Edited by Jos A. F. Op den Kamp

advantage of the monolayer technique is that lipid films can be directly viewed and observed using fluorescence microscopy [for a review, see Weis, 1991]. Laterally condensed domains in a "sea" of a laterally expanded phase can be visualized, since a suitable fluorophore present in the membrane partition differently into these two lateral phases [von Tscharner & McConnell, 1981]. The fluorescent probe NBD-cholesterol used in the present study has been shown to partition preferentially into a expanded phases, and is consequently excluded from more condensed phases [Slotte & Máttjus, 1994].

In this work we have examined the formation of lateral sterol-rich domains in mixed sterol/phospholipid monolayers using monolayer fluorescence microscopy (Mattjus et al., 1995). The objective was to examine the effect of the sterol side chain composition on the formation of sterol-rich domains. We have used a number of sterol analogues which have either an unbranched (*n*-series) or a single methyl-branched (*iso*) side chain of different length (from 3 to 10 carbons). A further comparison was made between DPPC and *N*-palmitoylsphingomyelin (*N*-PSPM), since the affinity of interaction between a sterol and these two phospholipids differ.

Results

Monolayers containing 33 mole % cholesterol in DPPC were slowly compressed at the gas/water interface (less than 4 Å^2/molecule, min) at constant temperature (22 °C). The mixed monolayers also contained 0.5 mol % NBD-cholesterol as a fluorescent probe which selectively partitions into loosely packed (expanded) domains. When a cholesterol/DPPC mixed monolayer was compressed to a surface pressure of 2.0 mN/m, cholesterol-rich condensed domains (dark areas) could be seen to coexist with an expanded phase (bright

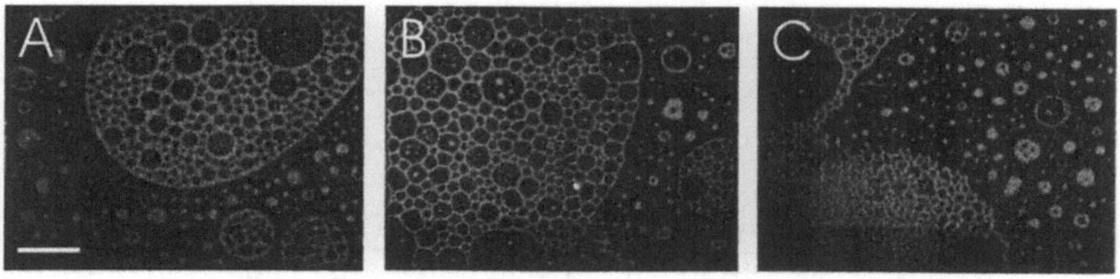

Figure 1. Cholesterol-poor liquid-expanded domains (light) in a cholesterol-rich liquid-condensed (dark) background. A monolayer was prepared containing 33 mol % cholesterol, 66.5 mol % DPPC and 0.5 mol % NBD-cholesterol (fluorescent probe) and subjected to symmetrical compression at a speed not exceeding 4 Å^2/molecule, min. The micrograph sequence shows the phase transformation where the domain boundary line between liquid-condensed and liquid-expanded phases dissipates, A at 2.0 mN/m, B at 2.5 mN/m and C at the onset pressure of 3.0 mN/m. The scale bar is 150 µm.

areas; Fig. 1A). If the monolayer was compressed further, one reached a characteristic surface pressure at which the line boundary between condensed and expanded phases became unstable (Fig. 1B), and eventually dissolved (Fig. 1C). This characteristic surface pressure is termed the phase transformation pressure. The point of phase transformation pressure can be approached both from a low (below) or a high surface pressure (above the phase transformation pressure). If one approaches the phase transformation pressure from the apparent one phase region (at a surface pressure above the phase transformation pressure), the monolayer will again enter a multiple phase coexistence state. If the monolayers have experienced a compression/expansion cycle, the morphology of the condensed domains are not necessarily similar before and after the compression/expansion cycle. It is assumed that the domain morphology after a compression/expansion cycle is closer to equilibrium structure than it is during initial compression.

Phase transformation pressures of sterol/DPPC and sterol/N-SPM mixed monolayers.

The onset phase transformation pressure at which the domain boundary line between condensed and expanded phases dissipated was dependent on the sterol side chain composition, and was determined by visual observation. Sterols with an unbranched side chain being 5 to 7 carbons in length were observed to have a phase transformation pressure ranging between 1.3 and 1.5 mN/m in the DPPC mixed monolayer (Fig. 2A, solid line). On the other

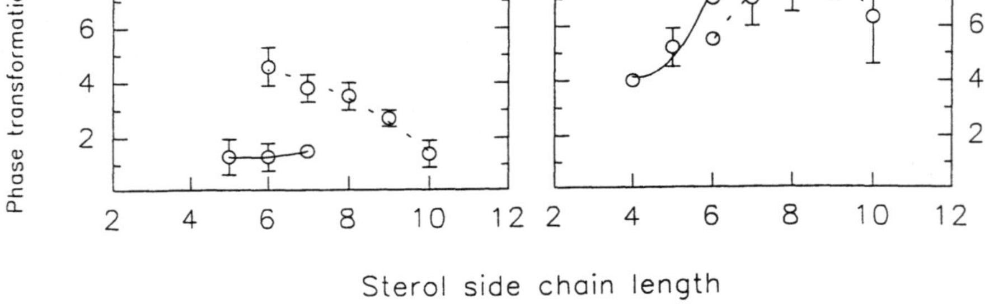

Figure 2. Onset phase transformation pressure as a function of sterol side chain composition. Sterol/phospholipid mixed monolayers were prepared to contain either 66 mol % DPPC (panel A) or *N*-PSPM (panel B), together with 1 mol % NBD-cholesterol. The solid line represents sterols from the *n*-series, whereas dotted lines represent *iso*-sterols. Values are averages ± S.E.M. from three different experiments.

hand, the phase transformation pressure decreased from 4.6 to 1.4 mN/m with increasing the side chain length for the branched side chain sterols (Fig. 2A, dotted line). This difference in phase transformation pressure as a function of side chain length and conformation clearly imply that the mode of the sterol/DPPC interaction was very sensitive to alterations in the sterol side chain.

The onset phase transformation pressure in monolayers containing sterol analogues and *N*-P-SPM increased markedly with increasing the side chain length for unbranched sterols (Fig. 2B, solid line), whereas the function was bell-shaped with *iso*-sterol analogues, so that *i*-C8, and *i*-C9 sterols displayed the highest phase transformation pressure, whereas other side chain lengths gave lower phase transformation pressures (Fig. 2B, dotted line).

Lateral domain formation in mixed sterol/phospholipid monolayers.

The surface pressure at which the documentation was performed was chosen to be 0.5 mN/m, since this surface pressure was below the phase transformation pressure of all domain-forming sterol analogues. At 33 mol % sterol in the mixed monolayers, *n*-C3 and *n*-C4 failed to form discrete, microscopically observable condensed domains with DPPC (during initial compression, at 0.5 mN/m; Fig. 3a & b). With longer chain unbranched analogues (*n*-C5, *n*-C6 and *n*-C7), the monolayer surface texture was dark with occasional (*n*-C5 and *n*-C7) or numerous (*n*-C6) brightly fluorescent inclusions (Fig. 3c-e). Of the single methyl branched analogues, the *i*-C5 failed to form lateral domains (Fig 3f), whereas the other sterol formed a characteristic surface texture. The surface texture was foam-like with the *i*-C6 to *i*-C9 analogues (Figs. 3g-j), while the monolayer with *i*-C10 had a surface texture which was characterized by small brightly fluorescent domains against a darker background (Fig. 3k). The monolayer surface texture after a compression/expansion cycle did not markedly change for the unbranched sterols, whereas the foamy texture of the *iso*-analogues became more homogenous (laterally expanded droplets against a condensed phase). The surface texture of the *i*-C9 and *i*-C10 monolayers was only marginally altered by a compression/expansion cycle.

At a sterol concentration of 33 mol % in *N*-SPM monolayers, the *n*-C3 mixture was uniformly fluorescent with no domains (Fig. 4a), but all other unbranched sterol analogues formed laterally condensed domains (Fig. 4b-e). While the *n*-C4 mixed monolayer had small circular-shape liquid-expanded domains of heterogeneous size (Fig. 4b), the *n*-C5, *n*-C6 and *n*-C7 sterol monolayers formed a surface texture with coalesced liquid-expanded and liquid-condensed domains (Fig. 4c-e). Of the single methyl-branched sterol analogues, only *i*-C5 failed to form lateral domains (Fig. 4f). All other sterols in the *iso*-series had both liquid-expanded and coalesced liquid-condensed domains which were non-uniformly distributed (Fig. 4g-k). The shapes of the domains changed surprisingly little when the sterol/*N*-PSPM monolayers were passed through a compression/expansion cycle (micrographs not shown).

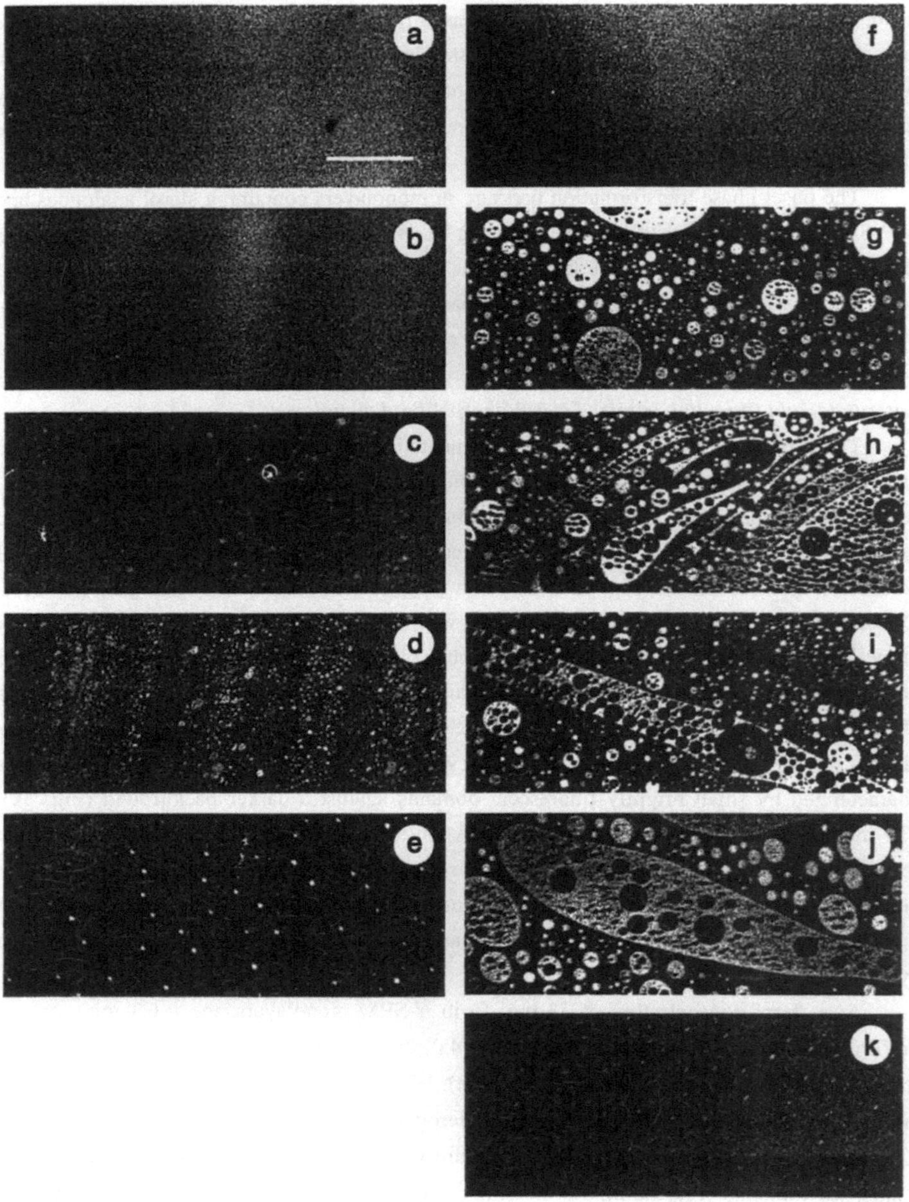

Figure 3. Lateral domain formation in sterol/DPPC monolayers. The monolayers contained 33 mol % sterol, 66 mol % DPPC, and 1 mol % NBD-cholesterol. Micrographs were documented at 0.5 mN/m during the initial compression of the monolayer. Panel a is *n*-C3, b is *n*-C4, c is *n*-C5, d is *n*-C6, e is *n*-C7, f is *i*-C5, g is *i*-C6, h is *i*-C7, i is *i*-C8 (cholesterol), j is *i*-C9, and k is *i*-C10. The scale bar is 100 μm.

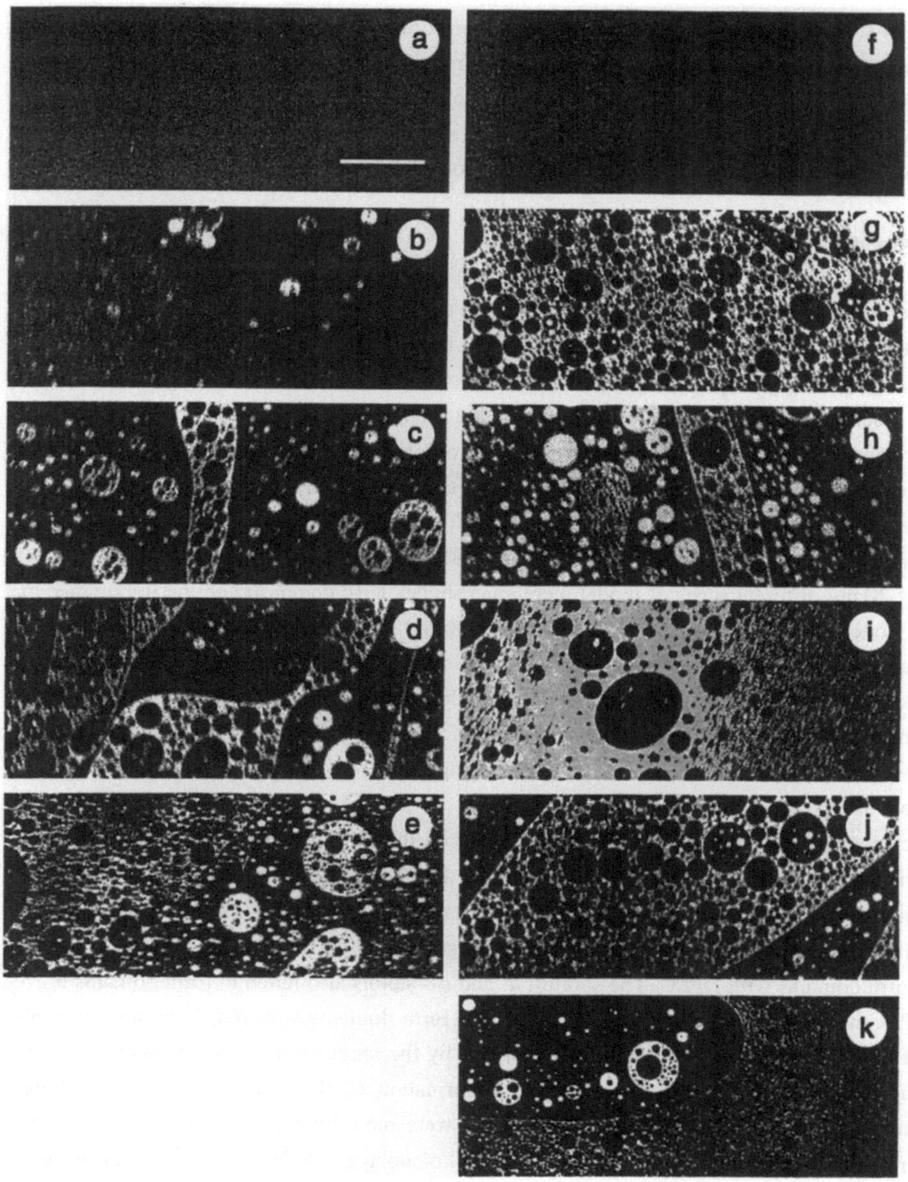

Figure 4. Lateral domain formation in sterol/*N*-PSPM monolayers. The monolayers contained 33 mol % sterol, 66 mol % *N*-PSPM, and 1 mol % NBD-cholesterol. Micrographs were documented at 0.5 mN/m during the initial compression of the monolayer. Panel a is *n*-C3, b is *n*-C4, c is *n*-C5, d is *n*-C6, e is *n*-C7, f is *i*-C5, g is *i*-C6, h is *i*-C7, i is *i*-C8 (cholesterol), j is *i*-C9, and k is *i*-C10. The scale bar is 100 μm.

The *n*-C4 monolayer had smaller and more uniformly distributed fluorescent inclusions, whereas the other unbranched sterol monolayers remained unchanged. With the branched side chain analogues the compression/expansion cycle triggered the fusion of condensed domains in *i*-C8, *i*-C9 and *i*-C10 containing monolayers, whereas the shorter chain sterol monolayers remained unchanged by the compression/expansion cycle.

Discussion

The formation of specific lateral domains in mixed sterol/phospholipid monolayers must arise from molecular interactions between sterols and phospholipids. Therefore, the formation of domains can be interpreted to demonstrate the occurrence of a specific interaction between these two molecular species. However, due to the limited resolution of the light microscope, the absence of visible lateral domains in a mixed monolayer does not rule out the possibility that domains still exist which are too small to be observed by this microscopic technique. The laterally condensed domains formed in a mixed cholesterol/DPPC monolayers are known to be sterol-rich (as opposed to being a pure DPPC domain), because the fractional area of the condensed domains increases with increasing cholesterol concentration [Slotte, 1995].

Many different types of model membrane studies have demonstrated the importance of the isooctyl side chain of cholesterol for the optimal interaction of this sterol with various phospholipids. In this respect, it is not surprising to find that the side chain composition of the sterols of this study had marked effects on the ability of the sterols to form sterol/phospholipid domains in mixed monolayers. Although all of the sterols used in this study have been previously shown to condense the lateral packing density of both DPPC and 1-stearoyl-2-oleoyl-*sn*-glycero-3-phosphocholine [Slotte et al., 1994], not all of them were able to form macroscopic sterol/phospholipid domains. Based on these results, one can conclude that the ability of a sterol to condense the lateral packing density of a phospholipid is not related to its ability to form lateral sterol/phospholipid domains. Sterol analogues with short side chains of both the unbranched (*n*-C3 and *n*-C4) and the single methyl branched type (*i*-C5) were unable to form domains with DPPC. The shortest *n*- and *iso*-sterols also failed to form domains in *N*-PSPM monolayers, whereas *n*-C4 did appear to form domains with *N*-P-SPM but not with DPPC. The morphology of the domains formed by the longer side chain analogues differed somewhat, depending on the length and conformation of the side chain. In general the domains formed in *N*-PSPM mixed monolayers were more homogeneous in size, and more evenly distributed than was the situation in DPPC monolayers. This finding is consistent with our recent data showing a more homogeneous size and lateral distribution of cholesterol/sphingomyelin domains as compared to cholesterol/DPPC domains, when examined as a function of cholesterol concentration [Slotte, 1995].

The formation of lateral domains is the result of the interplay of different forces, some of which are attractive (e.g., hydrophobic and van der Waals forces), while other are disruptive (e.g., electrostatic and dipole-induced repulsion; [Seul & Sammon, 1990; Keller et al., 1987]). Since all sterols of this study are 3β-hydroxy sterols, the permanent dipole was similar within the sterol series, although the orientation of the overall sterol dipole may differ in a way which possibly depends on the side chain composition, e.g., due to the difference in the angle of the dynamic cone of the rotating sterols [Gallay & de Kruijff, 1982]. However, it is clear that the strength of sterol/phospholipid association must in part depend on the possibility for attractive van der Waals forces to form. These forces are expected to be more numerous if the sterol side chain is longer rather than shorter. The observation that sterol/phospholipid domains were formed with intermediate and long side chain sterol analogues, but not with short side chain analogues is consistent with the idea that the formation of lateral domains in some way relates to the formation of an apparently critical number of attractive van der Waals forces. This in turn may relate to a phenomenon of hydrophobic mismatch which has been described for bilayer membrane containing DPPC and sterols with different length side chains [McMullen, et al., 1995]. In the study of McMullen et al. (1995) using some of the same sterol analogues that we have used in this study, it was shown that whereas the *n-* or *iso-*conformation of the side chain did not alter the thermotropic phase behavior of DPPC, even small changes in the length of the C17 linked side chain had marked effects on the phase behavior of the bilayers. This behavior of the sterols was described to arise from the hydrophobic mismatch between different sterols and DPPC.

Although monolayer membranes turn into an apparent one-phase state when the surface pressure is raised above the phase transformation pressure, it is still likely that a heterogeneous distribution of sterols in the plane of the membrane remains. Different experimental approaches using bilayer membranes as models have shown that the lateral distribution of cholesterol in phospholipid membranes is laterally heterogeneous [Schroeder et al., 1991; Rubinstein et al., 1980; Ben-Yashar & Barenholz, 1989]. These findings suggest that cholesterol-rich domains (although not similar to those we observed in monolayers at low surface pressures) can exist at the higher surface pressures of bilayer membranes.

In conclusion, we have been able to further demonstrate the importance of the sterol side chain conformation on sterol/phospholipid interaction, as evidenced by the formation (or lack thereof) of lateral domains in mixed sterol/phospholipid monolayers. There also appeared to be a difference in how sterols interacted with *N*-P-SPM as compared with DPPC.

Acknowledgments

This study was supported by generous grants from the Academy of Finland, the Sigrid Juselius Foundation, and the Åbo Akademi University (PM and JPS), and NIH HL-16660 (RB).

References

Ben-Yashar, V. and Barenholtz, Y. (1989) The interaction of cholesterol and cholest-4-en-3-one with dipalmitoylphosphatidylcholine. Comparison based on the use of three fluorophores. Biochim. Biophys. Acta 985, 271-278.

Bittman, R., Kasireddy, C.R., Mattjus, P. and Slotte, J.P. (1994) Interaction of cholesterol with sphingomyelin in monolayers and vesicles. Biochemistry 33, 11776-11781.

Craig, I.F., Boyd, G.S. and Suckling, K.E. (1978) Optimum interaction of sterol side chains with phosphatidylcholine. Biochim. Biophys. Acta 508, 418-421.

Demel, R.A., Bruckdorfer, K.R. and van Deenen, L.L.M. (1972a) The effect of sterol structure on the permeability of liposomes to glucose, glycerol and Rb^+. Biochim. Biophys. Acta 255, 321-330.

Demel, R.A., Geurts van Kessel, W.S.M. and van Deenen, L.L.M. (1972b) The properties of polyunsaturated lecithins in monolayers and liposomes and the interaction of these lecithins with cholesterol. Biochim. Biophys. Acta 266, 26-40.

Gallay, J. and de Kruijff, B. (1982) Correlation between molecular shape and hexagonal H_{II} phase promoting ability of sterols. FEBS Lett. 143, 133-136.

Keller, D., Korb, J.P. and McConnell, H.M. (1987) Theory of shape transitions in two-dimensional phospholipid domains. J. Phys. Chem. 91, 6417-6422.

Lund-Katz, S., Laboda, H.R., McLean, L.R. and Phillips, M.C. (1988) Influence of molecular packing and phospholipid type on rates of cholesterol exchange. Biochemistry 27, 3416-3423.

Mattjus, P., Bittman, R., Vilchèze, C. and Slotte, J.P. (1995), submitted to Biochim. Biophys. Acta.

Mattjus, P. and Slotte, J.P. (1994) Availability for enzyme-catalyzed oxidation of cholesterol in mixed monolayers containing both phosphatidylcholine and sphingomyelin. Chem. Phys. Lipids 71, 73-81.

McIntosh, T. J., Simon S.A., Needham D. and Huang C-h. (1992) Structure and cohesive properties of sphingomyelin/cholesterol bilayers. Biochemistry 31, 2012-2020.

McLean, L.R. and Phillips, M.C. (1982) Cholesterol desorption from clusters of phosphatidylcholine and cholesterol in unilamellar vesicle bilayers during lipid transfer exchange. Biochemistry 21, 4053-4059.

McMullen, T.P.W., Vilchèze, C., McElhaney, R.N. and Bittman, R. (1995) Differential scanning calorimetric study of the effects of sterol side-chain length and structure on dipalmitoylphosphatidylcholine thermotropic phase behavior. Biophys. J. 69, 169-176.

Nakamura, T., Nishikawa, M., Inoue, K., Nojima, S., Akiyama, T. and Sankawa, U. (1980) Phosphatidylcholine liposomes containing cholesterol analogues with side chains of various lengths. Chem. Phys. Lipids 26, 101-110.

Rubinstein, J.L.R., Owicki, J.C. and McConnell, H.M. (1980) Dynamic properties of binary mixtures of phosphatidylcholines and cholesterol. Biochemistry 19, 569-573.

Schroeder, F., Jefferson, J.R., Kier, A.B., Knittel, J., Scallen, T.J., Wood, W.G. and Hapala, I. (1991) Membrane cholesterol dynamics: Cholesterol domains and kinetic pools. Proc. Soc. Exp. Biol. Med. 196, 235-252.

Seul, M. and Sammon, M.J. (1990) Competing interactions and domain-shape instabilities in a monomolecular film at an air-water interface. Phys. Rev. Lett. 64, 1903-1906.

Slotte, J.P. (1992) Enzyme-catalyzed oxidation of cholesterol in mixed phospholipid monolayers reveals the stoichiometry at which free cholesterol clusters disappears. Biochemistry 31, 5472-5477.

Slotte, J.P. (1995) Lateral domain formation in mixed monolayers containing cholesterol and dipalmitoyl phosphatidylcholine or *N*-palmitoyl sphingomyelin. Biochim. Biophys. Acta (in press).

Slotte, J.P. and Mattjus, P. (1994) Visualization of lateral phases in cholesterol and phosphatidylcholine monolayers at the air/water interface - a comparative study with two different reporter molecules. Biochim. Biophys. Acta 1254, 22-29.

Slotte, J.P., Jungner, M., Vilchéze, C. and Bittman, R. (1994) Effects of sterol side-chain structure on sterol-phosphatidylcholine interactions in monolayers and small unilamellar vesicles. Biochim. Biophys. Acta 1190, 435-443.

Subramaniam, S. and McConnell, H.M. (1987) Critical Mixing in Monolayer Mixtures of Phospholipid and Cholesterol. J. Phys. Chem. 91, 1715-1718.

Suckling, K.E. and Boyd, G.S.(1976) Interaction of the cholesterol side-chain with egg lecithin. Biochim. Biophys. Acta 436, 295-300.

Suckling, K.E., Blair, H.A.F, Boyd, G.S., Craig, I.F. and Malcolm, B.R. (1979) The importance of the phospholipid bilayer and the length of the cholesterol molecule in membrane structure. Biochim. Biophys. Acta 551, 10-21.

von Tscharner, V. and McConnell, H.M. (1981) An alternative view of phospholipid phase behavior at the air-water interface: Microscope and film balance studies. Biophys. J. 36, 409-419.

Weis, R.M. (1991) Fluorescence microscopy of phospholipid monolayer phase transitions. Chem. Phys. Lipids 57, 227-239.

Yeagle, P.L. (1993) The biophysics and cell biology of cholesterol: An hypothesis for the essential role of cholesterol in mammalian cells. In Cholesterol in Model Membranes (Finegold, L.X., ed.), pp.1-12, CRC Press, Boca Raton, FL.

Yu, H. and Hui, S.W. (1992) Methylation effects on the microdomain structures of phosphatidylethanolamine monolayers. Chem. Phys. Lipids 62, 69-78.

The Kinetics, Specificities and Structural Features of Lipases.

Stéphane Ransac, Frédéric Carrière, Ewa Rogalska and Robert Verger

Laboratoire de Lipolyse Enzymatique, UPR 9025, IFRC1 du CNRS, 31 chemin Joseph Aiguier, 13402 Marseille cedex 20, France.

Frank Marguet and Gérard Buono

Réactivité et Catalyse, ENSSPICAM, URA 1410-CNRS, Faculté des Sciences de l'Université d'Aix-Marseille III, Av. Escadrille Normandie-Niémen, 13397 Marseille cedex 13, France.

Eduardo Pinho Melo, Joaquim M. S. Cabral

Laboratório de Engenharia Bioquímica, Instituto Superior Técnico, av. Rovisco Pais, 1000 Lisboa, Portugal.

Marie-Pierre E. Egloff, Herman van Tilbeurgh and Christian Cambillau

Laboratoire de Cristallisation et de Cristallographie des Macromolécules Biologiques, URA 1296, IFRC1 du CNRS, 31 chemin Joseph Aiguier, 13402 Marseille cedex 20, France.

Introduction

The four main classes of biological substances are carbohydrates, proteins, nucleic acids and lipids. The first three of these substances have been clearly defined on the basis of their structural features, whereas the property which is common to all lipids is a physicochemical one. Lipids are in fact a group of structurally heterogeneous molecules which are all insoluble in water but soluble in apolar and slightly polar solvents such as ether, chloroform and benzene.

Lipids have been classified by Small [1] depending on how they behave in the presence of water (Figure 1). This makes it possible to distinguish between polar and apolar lipids (*e.g.* hydrocarbon, carotene). The polar lipids can be further subdivided into three classes. Class I consists of those lipids which do not swell in contact with water and form stable monomolecular films (these include triacylglycerols, diacylglycerols, phytols, retinols, vitamin A, K and E, waxes and numerous sterols). The class II lipids (which include

NATO ASI Series, Vol. H 96
Molecular Dynamics of Biomembranes
Edited by Jos A. F. Op den Kamp

phospholipids, monoacylglycerols and fatty acids) spread evenly on the surface of water, but since they become hydrated, they swell up and form well-defined lyotropic (liquid crystalline) phases such as liposomes. The class III lipids (such as lysophospholipds and bile salts) are partly soluble in water and form unstable monomolecular films, and beyond the critical micellar concentration level, micellar solutions.

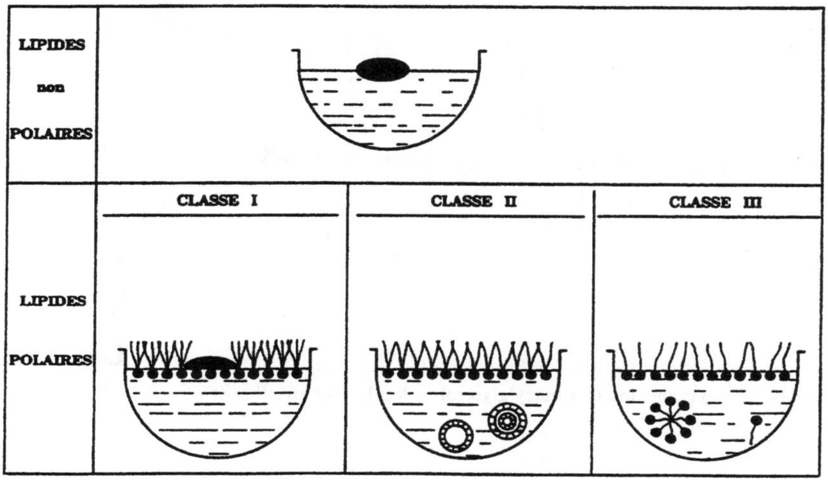

Figure 1: A classification of biological lipids based upon their interaction in aqueous systems; (from Small [1]).

Lipases are carboxylic ester hydrolases and have been termed glycerol-ester-hydrolase (EC 3.1.1.3) in the international system of classification. They differ greatly as regards both their origins (which can be bacterial, fungal, mammalian, etc..) and their properties, and they can catalyze the hydrolysis, or synthesis, of a wide range of different carboxylic esters. They all show highly specific activity however towards glyceridic substrates. Under physiological conditions, since natural triacylglycerols are water-insoluble, lipases, which are generally soluble in water, catalyze the hydrolysis of carboxylic ester bonds at lipid/water interfaces.

Since the plane of symmetry of the glycerol molecule is a prochiral plane, the two primary hydroxyl groups are stereochemically distinct (*i.e.*, they are enantiotopic groups). A Fischer projection of a glycerol molecule was drawn up with the secondary alcohol chain branching off to the left of the main hydrocarbon chain (Figure 2), and the carbon atoms were numbered 1, 2 and 3 working downwards. With this system of numbering, glycerol becomes *sn*-glycerol (*i.e.* stereospecifically numbered glycerol). This makes it possible to obtain unambiguous expressions for the stereoisomeric forms of (phospho)glycerides. Natural phospholipids, for example, all belong to the *sn*-glycero-3-phosphate series. In the case of natural triacylglycerols, the fatty acids which esterify positions *sn*-1, *sn*-2 and *sn*-3 are often different, which results in various chiral substrates.

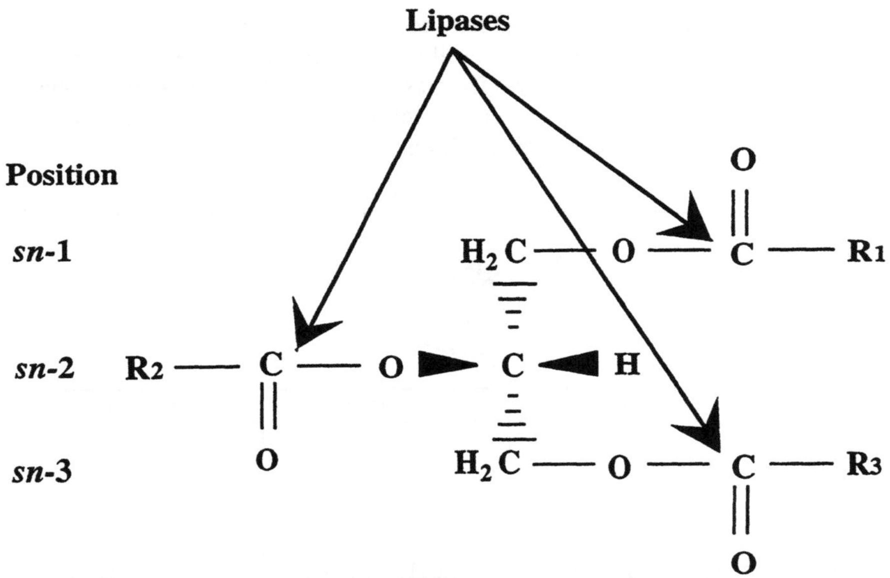

Figure 2: Fisher representation of a triacylglycerol molecule. Identification of potentially hydrolysable ester bonds.

I. Kinetics

1.1. Definition of lipases and "interfacial activation".

What exactly is a lipase ? Is it enough to say that it is a carboxyl esterase which specifically hydrolyzes triacylglycerols ? In 1958, Sarda and Desnuelle [2] defined lipases in kinetic terms, based on the "interfacial activation" phenomenon. This property was not to be found for example among the enzymes which have been classified as esterases, *i.e.*, those acting only on carboxylic ester molecules which are soluble in water. The "interfacial activation" phenomenon was in fact first observed as far back as 1936 by Holwerda *et al* [3] and then in 1945 by Schønheyder and Volqvartz [4]. It amounts to the fact that the activity of lipases is enhanced on insoluble substrates (such as emulsions) than on the same substrates in true monomeric solutions. It therefore emerged from the studies mentioned above that lipases might constitute a special category of esterases which are highly efficient at hydrolyzing molecules having a carboxylic ester group and are aggregated in water.

This property was used for a long time to distinguish between lipases and esterases. A conceptual shift has unfortunately occurred however, as the result of which "interfacial activation" has been taken to mean a hypothetical conformational change occurring as the result of interfacial adsorption [5] . Worse still, "interfacial activation" has sometimes been wrongly taken to refer to the increase in the catalytic activity of an enzyme on a triacylglycerol substrate occurring in the presence of a tensioactive agent.

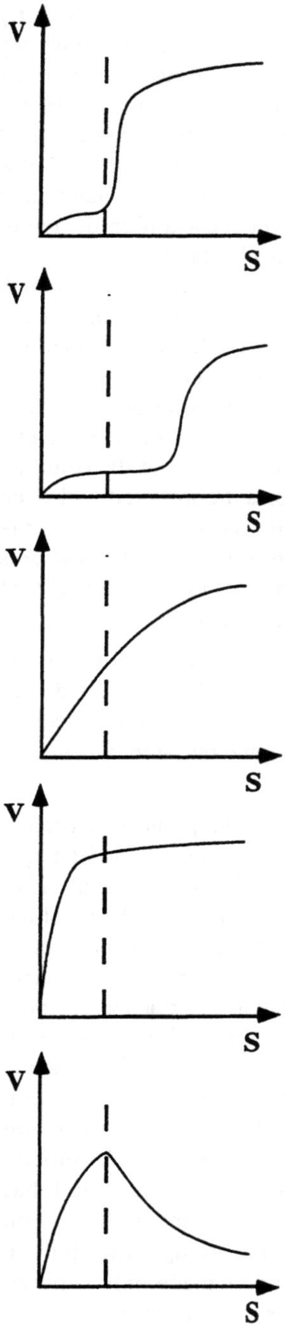

Figure 3: Hydrolysis rates (V) as a function of the amounts (S) of a partly water soluble ester. Dotted vertical lines represent the limit of solubility or the critical micellar concentration of the ester used. These profiles, observed with different lipases, were based on published results.

The results of recent lipase research have nevertheless shown how careful one has to be about extrapolating any kinetic and/or structural characteristics observed to all lipases in general. The catalytic activity of many lipolytic enzymes has been measured using carboxylic esters which are partly soluble in water, and many differences have been found to exist between the resulting profiles (Figure 3). The first three-dimensional structures to be elucidated [6, 7] suggested that the "interfacial activation" phenomenon might be due to the presence of an amphiphilic peptidic loop covering the active site of the enzyme in solution, just like a lid. When contact occurs with a lipid/water interface, this lid might undergo a conformational rearrangement as the result of which the active site becomes accessible. It is worth noting however that the hydrolysis of a substrate having the form of a truly monomeric solution might well also require the lid to be open without any "interfacial activation" being involved. "Interfacial activation" seems in fact to have to do with the respective lifetimes of the open and closed forms of lipases.

In the framework of the European Bridge-T Lipase project (1990 to 1994), some new three-dimensional structures and numerous biochemical data provided new insights into lipases at the molecular and atomic levels (see § III). It emerged from these studies that lipases do not all subscribe to the phenomenon of "interfacial activation". The main exceptions noted so far are the lipases from *Pseudomonas glumae* [8,9], *Pseudomonas aeruginosa* [10, 11] and *Candida antarctica* B [12], all of which nevertheless have an amphiphilic lid covering the active site. On the other hand, some new pancreatic lipases have been recently identified [13]. Comparisons between their primary amino acid sequences have shown that they have a fairly high degree of homology, but they can nevertheless be divided into three sub-groups: (i) the "classical" pancreatic lipases; (ii) the type 1 (RP1) pancreatic lipases, and (iii) the type 2 (RP2) pancreatic lipases.

Although the kinetic properties of the "classical" pancreatic lipases, particularly as regards "interfacial activation", have been quite fully documented, it was only fairly recently that the lipases of the coypu and the guinea-pig were found not to show any "interfacial activation" [14, 15]. Paradoxically, coypu lipase, while not showing any "interfacial activation", has a 23 amino-acid lid which is homologous to that of the "classical" pancreatic lipases, whereas the guinea-pig lipase has a mini-lid consisting of only five amino-acid residues. Other lipases, such as *Staphylococcus hyicus*, show "interfacial activation" only with some substrates [16]. Lastly, the pancreatic phospholipases A_2 have no identified amphiphilic lid, and yet show a high degree of "interfacial activation" [17] Another structurally very similar phospholipase A_2 (*Naja melanoleuca*) on the contrary shows no "interfacial activation" with medium chain-length phospholipid substrates [18]. The three-dimensional structure of porcine pancreatic phospholipase A_2, present in a 40 kDa ternary complex with micelles and competitive inhibitor, has been determined recently by van den Berg *et al* [19] using multidimensional heteronuclear NMR spectroscopy. Whereas free in solution Ala^1, Leu^2 and Trp^3 are disordered, with the α-amino group of Ala^1 pointing out into the solvent, in the ternary complex these residues have an α-helical conformation with the α-amino group buried inside the protein. As a consequence, the important conserved hydrogen bonding network which is also seen in the crystal structures is present only in the ternary complex, but not in free phospholipase A_2. Comparison of the NMR structures of the free enzyme and the enzyme in the ternary complex indicates that conformational changes play a role in the "interfacial activation" of phospholipase A_2.

"Interfacial activation" as well as the presence of a lid domain are therefore not in the least appropriate criteria on the basis of which to determine whether or not such and such an esterase belongs to the lipase sub-family. The safest experimental criterion nowadays for identifying a lipase seems to be, as in the early days of enzyme research, whether or not it hydrolyzes long-chain acylglycerols.

The greatest caution requires to be exercised both when performing and interpreting kinetic measurements with lipids. Since the medium is heterogeneous, adding any amphilic compound to the system is liable to have both quantitative and qualitative effects on the interface for physico-chemical reasons. The presence of a phospholipid substrate in the micellar state will lead for example to the aggregation of pancreatic phospholipase A_2 into multimolecular lipoprotein complexes [20]. On the other hand, some lipases, such as gastric lipases, rapidly become denatured at an interface with a pure tributyrin emulsion. Consequently, it is impossible to experimentally assess what "interfacial activation" may have occurred with substrates of this kind. In addition, some esters which are partly soluble in water sometimes form monomolecular adsorption films on the surface of the air-bubbles which are produced by stirring the reaction mixture. This artefact is responsible for great disparity between initial velocity measurements, depending on whether or not mechanical stirring methods are used. It has in fact been established that the lipid/water "interfacial quality", in terms of the tension at the interface, is one of the most decisive parameters when working with lipolytic enzymes. This unfortunately means that valid comparisons can be made only between data obtained under strictly identical conditions, preferably at the same laboratory.

Since naturally occurring lipids are by definition insoluble in water, "interfacial activation" can be said in the light of the above overview to be little more than an artefact which has stimulated the imaginations of many biochemists, but which has not turned out to be of any very great physiological significance. Lipases might therefore be quite pragmatically re-defined as carboxyl-esterases which catalyse the hydrolysis of long-chain acylglycerols.

1.2. Monolayer techniques for studying lipase kinetics.

1.2.1. Why use lipid monolayers as lipase substrates ?

There are at least five major reasons for using lipid monolayers as substrates for lipolytic enzymes: the reader is referred to previous reviews for details [21-25]. (i) The monolayer technique is highly sensitive and very little lipid is needed to obtain kinetic measurements. This advantage can often be decisive in the case of synthetic or rare lipids. Moreover, a new phospholipase A_2 has been discovered using the monolayer technique as an analytical tool [26]. (ii) During the course of the reaction, it is possible to monitor one of several physicochemical parameters characteristic of the monolayer film: surface pressure, potential, radioactivity, etc. These variables often give unique information. (iii) With this technique, the lipid packing of a monomolecular film of substrate is maintained constant during the course of hydrolysis, and it is therefore possible to obtain accurate presteady state kinetic measurements with minimal perturbation caused by increasing amounts of reaction products. (iv) Probably the most important basic reason is that it is possible with lipid monolayers to vary and modulate the "interfacial quality", which depends on the nature of the lipids forming the monolayer, the orientation and conformation of the molecules, the molecular and charge densities, the water structure, the viscosity, etc. One further advantage of the monolayer technique as compared to bulk methods is that with the former, it is possible to transfer the film from one aqueous subphase to another. (v) Inhibition of lipase activity by water-insoluble substrate analogues can be precisely estimated using a "zero order" trough and mixed monomolecular films in the absence of any synthetic, non-physiological detergent. The monolayer technique is therefore suitable for modelling *in vivo* situations.

1.2.2. Zero-order trough

Several types of troughs have been used to study enzyme kinetics. The simplest of these is made of Teflon™, rectangular-shaped (Figure 4a) but gives nonlinear kinetics [27]. To obtain rate constants, a semilogarithmic transformation of the data is required. This drawback was overcome by a new trough design ("zero order" trough, Figure 4b), consisting of a substrate reservoir and a reaction compartment containing the enzyme solution [28]. The two compartments are connected to each other by a narrow surface canal. The kinetic recordings obtained with this trough are linear, unlike the nonlinear plots obtained with the usual one-compartment trough. The surface pressure can be maintained constant automatically by the surface barostat method described elsewhere [28]. Fully automatized monolayer systems of this kind are now commercially available (KSV - Helsinki - Finland).

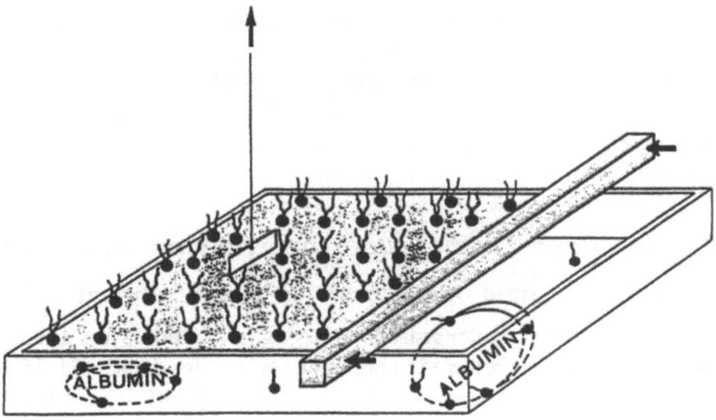

Figure 4A: Method for studying the hydrolysis of long-chain lipid monolayers with controlled surface density. A large excess of serum albumin has to be present in the aqueous subphase in order to solubilize the lipolytic products.

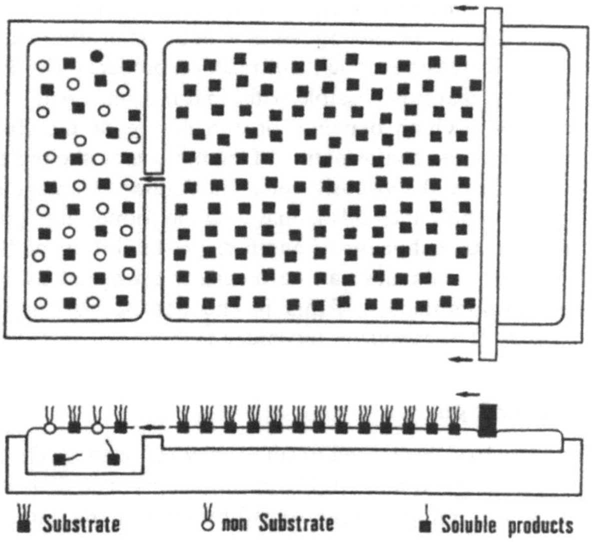

Figure 4B: Principle of the method for studying enzymatic lipolysis of mixed monomolecular films (from Piéroni *el al.* [29]).

1.2.3. Interfacial binding and film recovery

Using the monomolecular film technique, several investigators have reported that an optimum occurs in the velocity-surface pressure profile. The exact value of the optimum varies considerably with the particular enzyme/substrate combination used. Qualitative interpretations have been given to explain this phenomenon. The first hypothesis, proposed by Hughes [30] and supported by later workers [31, 32] was that a packing dependent orientation of the substrate may be one of the factors on which the regulation of lipolysis depends. Of special interest is the recent approach by Peters *et al* [33] on the structure and dynamics of 1,2-sn-dipalmitoylglycerol monolayers undergoing two phase transitions at 38.3 and 39.8Å^2 per molecule. The first transition is unique for diglyceride molecules and is driven by a reorganization of the headgroups causing a change in the hydrophobicity of the oil-water interface. X-ray diffraction studies of different mesophases shows that in the two highest pressure phases, the alkyl chains pack in an hexagonal structure relaxing to a distorted-hexagonal lattice in the lowest pressure phase with the alkyl chains titled by ~14° in a direction close to a nearest neighbour direction.

Another interpretation was put forward by Esposito *et al.* [34], who explained the surface pressure optimum in terms of changes in the lipase conformation upon adsorption at the interface, resulting in an optimal conformation at intermediate values of the interfacial free energy (film pressure). It was suggest that lower and higher values of surface pressure would lead to inactive forms either because of denaturation or because the conformational changes in the enzyme structure are not sufficiently marked. This view was challenged by Verger *et al.* [21] and Pattus *et al.* [35]. Using radiolabelled enzymes, these authors showed that the observed maxima in the velocity-surface pressure profile disappear when they are correlated with the interfacial excess of enzyme. Indeed, the main difference between the monolayer and the bulk system lies in the interfacial area to volume ratios which differ from each other by several orders of magnitude. In the monolayer system, this ratio is usually about 1 cm^{-1}, depending upon the depth of the trough, whereas in the bulk system it can be as high as 10^5 cm^{-1}, depending upon the amount of lipid used and the state of lipid dispersion. Consequently, under bulk conditions, the adsorption of nearly all the enzyme occurs at the interface, whereas with a monolayer only one enzyme molecule out of hundred may be at the interface [27]. Owing to this situation, a small but unknown amount of enzyme, responsible for the observed hydrolysis rate, is adsorbed on the monolayer. In order to circumvent this limitation, different methods were proposed for recovering and measuring the quantity of enzymes adsorbed at the interface [36-39].

After performing velocity measurements, Momsen and Brockman [38] transferred the monolayer to a piece of hydrophobic paper and the adsorbed enzyme was then assayed titrimetrically. After correcting for blank rate and subphase carry-over, the moles of adsorbed enzyme were calculated from the net velocity and the specific enzyme activity.

In assays performed with radioactive enzymes [36], the film was aspirated by inserting the end of a bent glass capillary into the liquid meniscus emerging above the ridge of the Teflon™ compartment walls. The other end of the same capillary was dipped into a 5-ml counting vial connected to a vacuum pump. As radioactive molecules dissolved in the subphase were unavoidably aspirated with the film constituents, the results had to be corrected by counting the radioactivity in the same volume of aspirated subphase. The difference between the two values, which actually expressed an excess radioactivity existing at the surface, was attributed to the enzyme molecules bound to the film.

The surface bound enzyme includes not only those enzyme molecules directly involved in the catalysis but also an unknown amount of protein present close to the monolayer. These enzyme molecules were not necessarily involved in the enzymatic hydrolysis of the film. Since it is possible with the monolayer technique to measure the enzyme velocity expressed in $\mu mol/cm^2/min$ and the interfacial excess of enzyme in mg/cm^2, it is easy to obtain an enzymatic specific activity value, which can be expressed as usual in µmol/min/mg.

Using radiolabelled 5,5'-dithiobis(2-nitrobenzoic acid) (DTNB), Gargouri *et al.* [40] investigated the interactions between covalently labelled [^{14}C]TNB-human gastric lipase ([^{14}C]TNB-HGL) and monomolecular lipid films. It is worth noting that [^{14}C]TNB-HGL is an inactive enzyme, and moreover, to facilitate detection, that the [^{14}C]TNB-HGL concentrations used by Gargouri *et al.* [40] to study its binding to monomolecular films were about 40 times higher than the usual catalytic concentrations of HGL. Under these conditions, the existence of a correlation between the increase in surface pressure and the amount of protein bound to the interface was the only possible conclusion which could be drawn by the authors. They showed that in the presence of egg PC films, the total amounts of surface bound [^{14}C]TNB-HGL decreased linearly as the initial surface pressure increased; whereas, using lipase substrates such as dicaprin films, the amounts of surface bound inactive [^{14}C]TNB-HGL remained constant at variable surface pressures [40].

The ELISA/biotin-streptavidin system has been found to be as sensitive as the use of radiolabelled proteins [41]. In addition, biotinylation preserves the biological activities of many proteins. Using this labelling procedure, Aoubala *et al.* [42] recently developed a specific double sandwich ELISA to measure the human gastric lipase (HGL) levels of duodenal contents. An other application of this method was to develop a sensitive sandwich ELISA, using the biotin-streptavidin system, and to measure the amount of surface-bound HGL and anti-HGL monoclonal antibodies (mAbs) adsorbed to monomolecular lipid films. HGL and two anti-HGL mAbs (denoted mAbs 4-3 and 218-13) were biotinylated without any significant loss of their biological activities occurring. They were further detected by ELISA using either anti-HGL or anti-mouse IgG polyclonal antibodies as specific captors before being revealed using a streptavidin-peroxidase conjugate as tracer. The detection limit was 25 pg and 85 pg in the case of HGL and mAb, respectively.

By combining the above sandwich ELISA technique with the monomolecular film technique, it was possible to measure the enzymatic activity of HGL on 1,2 didecanoyl-*sn*-glycerol monolayers as well as to determine the corresponding interfacial excess of the enzyme [39]. The HGL turnover number increased steadily with the lipid packing. The specific activities determined on dicaprin films spread at 35 mN/m were found to be in the range of the values measured under optimal bulk assay conditions, using tributyrin emulsion as substrate (*i.e.*, 1000 µmoles per min per mg of enzyme). At a given lipase concentration in the water subphase, the interfacial binding of HGL to the non hydrolyzable egg yolk phosphatidylcholine monolayers was found to be ten times lower than in the case of dicaprin monolayers [39].

1.2.4. Pure lipid monolayers as lipase substrates

A new field of investigation was opened in 1935 when Hughes [30] used the monolayer technique for the first time to study enzymatic reactions. He observed that the rate of the phospholipase A-catalyzed hydrolysis of a lecithin film, measured in terms of the decrease in surface potential, decreased considerably when the number of lecithin molecules per square centimeter increased. Since this early study, several laboratories have used the monolayer technique to monitor lipolytic activities, mainly with glycerides and phospholipids as substrates. These studies can be tentatively divided into four groups. Either long-chain lipids were used and their surface density (number of molecules/cm2) was not controlled, or short-chain lipids were applied, again without controlling the surface density, or short-chain lipids were used at constant surface density. As shown in Figure 4A, we have developed a fourth method at our laboratory for studying the hydrolysis of long chain phospholipid monolayers by various phospholipases A_2 involving controlled surface density [23, 24].A large excess of serum albumin has to be present in the aqueous subphase in order to solubilize the lipolytic products. This step was found not to be rate limiting. The linear kinetics obtained by these authors with natural long chain phospholipids were quite similar to those previously described in the case of short chain phospholipids using a "zero order" trough with the barostat technique [28]. Another way of using long chain lipid monolayers is based on ß-cyclodextrin. A study on the desorption rate of insoluble monomolecular films of oleic acid, monoolein, 1,2-diolein, 1,3-diolein and triolein at the argon/water interface by water-soluble ß-cyclodextrin has recently been published [43]. The desorption of the water insoluble reaction products (oleic acid and monoolein) probably involves the complexation of the single acyl chain into the ß-cyclodextrin cavity and sequestration of the soluble oleic acid/ß-cyclodextrin and monoolein/ß-cyclodextrin complexes formed from the argon/water interface [43]. In the case of monolayers of multiple acyl chain molecules such as diolein and triolein, no detectable change in the surface pressure occurred after ß-cyclodextrin injection. The surface rheological dilatational properties of the monolayers of diolein and triolein in the presence of ß-cyclodextrin in the subphase were studied. The elasticity of the diolein monolayer remained unchanged, whereas the decrease in the surface elasticity of the triolein film was attributed to the formation of a water-insoluble triolein/ß-cyclodextrin complex. With the 'tuning fork' model, one acyl chain of triolein can be included in the ß-cyclodextrin cavity. The triolein/ß-cyclodextrin complex formed at the interface retarded the propagation of the dilatational deformation along the plane of the monolayer. Schematic models have been proposed to attempt to explain the complexation of these lipids by ß-cyclodextrin at the argon/water interface (see Figure 5). In addition to the above results, the presence of ß-cyclodextrin in the water subphase makes it possible for the first time to perform kinetic measurements of the lipase hydrolysis rates of long-chain glycerides forming monomolecular films.

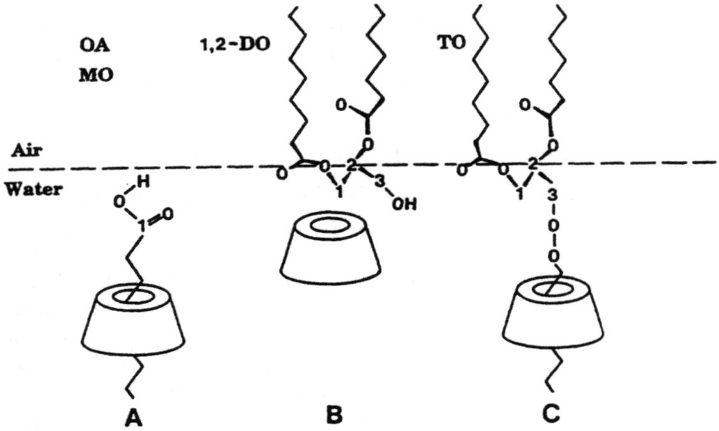

Figure 5: A: Diagram of the water-soluble oleic acid/ß-cyclodextrin inclusion complex. B: No complex formation between 1,2-diolein and ß-cyclodextrin. C: Inclusion of a single acyl chain having the tuning fork type conformation of triolein into the ß-cyclodextrin cavity. (From Laurent *et al.* [43]).

1.2.5. Stereoselectivity of lipases is controlled by surface pressure.

Lipases, which are lipolytic enzymes acting at the lipid/water interface, display stereoselectivity towards glycerides and other esters [44-46]. Biological lipids, which self-organize and orientate at interfaces, are chiral molecules, and their chirality is expected to play an important role in the molecular interactions between proteins and biomembranes. The most unusual aspect of glyceride hydrolysis catalyzed by pure lipases is its particular stereochemistry [47-55]. Under physiological conditions, many other hydrolases, such as phospholipases, proteases, glycosidases and nucleases, encounter only one optical antipode of their substrates. Lipases, on the contrary, can encounter both chiral forms of their substrates as well as molecules which are prochiral. This unique situation calls for some fundamental clarification, since the physiological consequences of lipases' stereopreferences are not known. Membrane-like lipid structures, such as monolayers, provide attractive model systems for investigating to what extent lipolytic activities depend upon the chirality and other physicochemical characteristics of the lipid/water interface.

The mechanism whereby an enzyme differentiates between two antipodes of a chiral substrate may be influenced by physicochemical factors such as the temperature [56, 57], the solvent hydrophobicity [58-61] or the hydrostatic pressure [62], which can affect the stereoselectivity of the reaction. To achieve a measurable impact of hydrostatic pressure on a protein in a bulk solution, however, high pressures of the order of 3 kbar need to be used and monitoring the enzyme activity under these conditions is difficult. On the other hand, the surface pressure is easy to manipulate and its effects on the enzyme activity can be readily controlled. Rogalska *et al.* [51] investigated the assumption that the stereoselectivity, which is one of the basic factors involved in enzymic catalysis, may be pressure-dependent. When working with bulk solutions, the external pressure is not a practical variable because liquids are highly incompressible, whereas the monolayer surface pressure is easy to manipulate. To establish the effects of the surface pressure on the stereochemical course of the enzyme action, during which optical activity is generated in a racemic substrate insoluble in water, the authors developed a method with which the enantiomeric excess of the residual substrate can be measured in monomolecular films. With all four lipases tested (*Mucor miehei* lipase, lipoprotein lipase, *Candida antarctica* B lipase and human gastric lipase), low surface pressures enhanced the stereoselectivity while decreasing the catalytic activity. This finding, which to our knowledge is unprecedented, should help to elucidate the mode of action of water-soluble enzymes on water-insoluble substrates.

In another study, Rogalska *et al.* [63] presented the results of an extensive comparative study on the stereoselectivity of 23 lipases of animal and microbial origin. Contrary to previous studies on lipase-glyceride chiral recognition where racemic or prochiral substrates were used, the substrates chosen here were three optically pure dicaprin isomers. The monomolecular film technique was chosen as a particularly appropriate method for use in chiral recognition studies [64-68] To establish the effects of the surface pressure on the stereochemical course of the enzyme action, Rogalska *et al.* [63] used as lipase substrates optically pure dicaprin enantiomers 1,2-*sn*-dicaprin, 2,3-*sn*-dicaprin and 1,3-*sn*-dicaprin, spread as monomolecular films at the air/water interface. The two former isomers are optically active antipodes (enantiomers), forming stable films up to 40 mN/m, while the latter one is a prochiral compound, which collapses at a surface pressure of 32 mN/m. To our knowledge, this is the first report on the use of three diglyceride isomers as lipase substrates under identical, controlled physico-chemical conditions. In this study, the authors showed that the regioselectivity, as well as the stereoselectivity, which are main factors involved in lipolytic catalysis of glycerides, are surface-pressure-dependent. The lipases tested displayed highly typical behaviour, characteristic of each enzyme, which allowed them to classify the lipases into groups on the basis of i). enzyme velocity profiles as a function of the surface pressure, ii). their preferences for a given diglyceride isomer, quantified using new parameters coined as: stereoselectivity index, vicinity index and surface pressure threshold. The general finding which was true of all the enzymes tested, was that the three substrates are clearly differentiated, and the differentiation is more pronounced at high interfacial energy (low surface pressure). This finding supports the hypothesis that lipase conformational changes resulting from the enzyme-surface interaction affect the enzymes' specificities. Generally speaking, the stereopreference for either position *sn*-1 or *sn*-3 on glycerides is maintained in the case of both di- and triglycerides.We link this effect to the assumed lipase conformational changes involving the active site, occuring upon the enzyme-interface interaction.

1.2.6. Mixed monolayers as lipase substrates

Most studies on lipolytic enzyme kinetics have been carried out *in vitro* with pure lipids as substrates. Virtually all biological interfaces are composed of complex mixtures of lipids and proteins. The monolayer technique is ideally suited for studying the mode of action of lipolytic enzymes at interfaces using controlled mixtures of lipids. There exist two methods of forming mixed lipid monolayers at the air-water interface: either by spreading a mixture of water-insoluble lipids from a volatile organic solvent, or by injecting a micellar detergent solution into the aqueous subphase covered with preformed insoluble lipid monolayers.

A new application of the "zero-order" trough was proposed by Piéroni and Verger [29] for studying the hydrolysis of mixed monomolecular films at constant surface density and constant lipid composition (Figure 4b). A Teflon™ barrier was placed transversely over the small channel of the "zero-order" trough in order to block surface communications between the reservoir and the reaction compartment. The surface pressure was first determined by placing the platinum plate in the reaction compartment, where the mixed film was spread at the required pressure. Surface pressure was then measured after switching the platinum plate to the reservoir compartment, where the pure substrate film was subsequently spread. The surface pressure of the reservoir was equalized to that of the reaction compartment by moving the mobile barrier. The barrier between the two compartments was then removed in order to allow the surfaces to communicate. The enzyme was then injected into the reaction compartment and the kinetics was recorded as described [28]. The main purpose of this study was to describe the influence of lecithin upon lipolysis of mixed monomolecular films of trioctanoylglycerol/didodecanoyl phosphatidylcholine by pancreatic lipase in order to mimick some physiological situations. The authors used a radiolabeled pancreatic lipase (5-thio-2-nitro [^{14}C] benzoyl lipase) to determine the influence of the film composition on the enzyme penetration (adsorption) and/or turnover at the interface. Of special interest is the concept of Scow *et al.* [69, 70], according to which triglyceride hydrolysis can be taken to be the first step in lateral flow lipid transport in cell membranes. Lipolytic activity was enhanced 3- to 4-fold in the presence of colipase, an effect which was attributed to the increase in the enzyme turnover number. When a pure triglyceride film was progressively diluted with lecithin, the minimum specific activity of lipase exhibited a bell-shaped curve : a mixed film containing only 20 % trioctanoylglycerol was hydrolyzed at the same rate as a monolayer of pure triglyceride. The main conclusion drawn from this study was that considerable activation or inhibition may result as a function of the lipid composition, lipid packing, and surface defects in mixed films.

1.2.7. Inhibition of lipases acting on mixed substrate/inhibitor monomolecular films

It is now becoming clear from the abundant lipolytic enzyme literature that any meaningful interpretation of inhibition data has to take into account the kinetics of enzyme action at the lipid-water interface. As shown in figure 6, Ransac *et al.* [71] have devised a kinetic model, which is applicable to water insoluble competitive inhibitors, in order to quantitatively compare the results obtained at several laboratories. Furthermore, with the kinetic procedure developed, it was possible to make quantitative comparisons with the same inhibitor placed under various physico-chemical situations, *i.e.*, micellar or monolayer states. Adding a potential inhibitor to the reaction medium can lead to paradoxical results. Usually, variable amounts of inhibitor are added, at a constant volumic concentration of substrate. The specific area and the interfacial concentration of substrate are continuously modified accordingly. In order to minimize modifications of this kind, it was proposed to maintain constant the sum of inhibitor and substrate (I+S) when varying the inhibitory molar fraction ($\alpha = I/(I+S)$), *i.e.*, to progressively substitute a molecule of inhibitor for each molecule of substrate [72]. This is the minimal change that can be made at increasing inhibitor concentrations. Of course, with this method, the classical kinetic procedure based upon the Michaelis-Menten model is not valid because both inhibitor and substrate concentrations vary simultaneously and inversely. However, by measuring the inhibitory power (Z), as described by Ransac *et al.* [71] or Xi(50) as used by Jain *et al.* [73], it is possible to obtain a normalized estimation of the relative efficiency of various potential inhibitors.

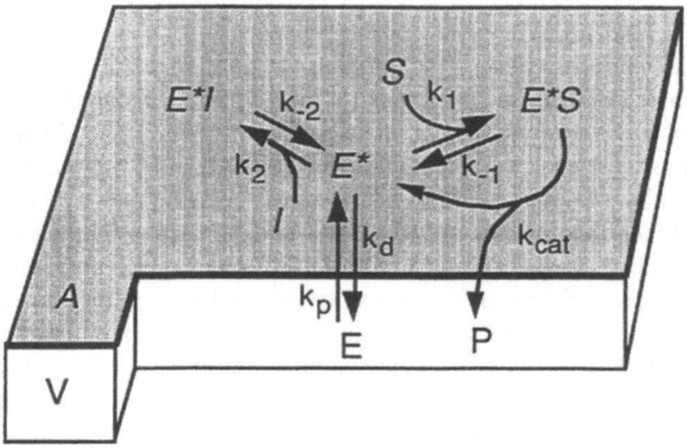

Figure 6: Proposed model for competitive inhibition at interfaces (from Ransac *et al.* [71]).

Human gastric lipase (HGL) and rabbit gastric lipase (RGL) are enzymes which have been found to be inactivated on the one hand by classical sulfhydryl reagents (5,5'-dithiobis(2-nitrobenzoic acid) (DTNB); 4,4'-dipyridyl disulfide) and by a new hydrophobic disulfide compounds: dodecyl dithio-5-(2-nitrobenzoic acid) (12:0-TNB) [74, 75]. On the other hand Hadváry *et al.* [76] and Borgström [77] have established that tetrahydrolipstatin, an hydrophobic lactone derived from lipstatin which is produced by

Streptomyces toxytricini, acts *in vitro* as a potent inhibitor of pancreatic and gastric lipases as well as cholesterol ester hydrolase. These authors suggested that a stoichiometric enzyme-inactivator complex of an acyl-enzyme type was formed and was slowly hydrolyzed with water as the final acceptor, leaving an intact enzyme and the inactive form of tetrahydrolipstatin (open lactone ring) [76, 77]. Achieving specific and covalent inactivation of lipolytic enzyme is a difficult task because of non-mutually exclusive processes such as interfacial denaturation, changes in "interfacial quality" and surface dilution phenomena [21, 78]. Furthermore, the interfacial enzyme binding and/or the catalytic turnover can be diversely affected by the presence of potential amphipatic inactivators [71, 79]. With emulsified systems, it is not possible to control the "interfacial quality", and to easily assess the distribution of soluble *versus* adsorbed amphiphilic molecules. Ransac *et al.* [80] studied the covalent inhibition of lipases by the monolayer technique. The authors described the inactivation of porcine pancreatic and human and rabbit gastric lipases, acting on mixed monomolecular films of dicaprin containing tetrahydrolipstatin or hydrophobic disulfide compounds such as 12:0-TNB,, 16:0-TNB and 18:1-TNB. The stoichiometry of the interfacial situation can be described as follows: one lipase molecule embedded among 105 substrate molecules will be inactivated to half its initial velocity by the presence of 10 tetrahydrolipstatin molecules. This inactivation was independent of the surface pressure. When tested in the form of mixed films, all the disulfide compounds investigated specifically reduced the hydrolysis of dicaprin films by gastric lipase, but did not affect hydrolysis by pancreatic lipase. With this "poisoned interface" system, tetrahydrolipstatin was found to be the most potent inactivator, whereas disulfide compounds showed a higher degree of selectivity than tetrahydrolipstatin.

Marguet *et al.* [81] recently studied the inhibition of human gastric and pancreatic lipases by the monomolecular film technique using mixed films of chiral organophosphorus compounds which are triacylglycerol analogs (see Figure 7) and dicaprin. Interfacial lipase binding has been evaluated by means of ELISA tests with biotinyled lipases, with which it was possible to measure the surface density of enzymes in the ng range. With both enzymes, kinetic experiments were performed at various molar ratios of dicaprin premixed with each of the four chiral inhibitors. All the four stereosiomers investigated reduced the hydrolysis of dicaprin by human gastric and pancreatic lipases. With human pancreatic lipase, the four stereoisomers exhibited a rather weak inhibition capacity, and no significant differences were observed among them. With each inhibitor tested, interfacial binding experiments using ELISA tests showed no significant difference in the surface density of human pancreatic lipase, which confirmed the low stereoselectivity of pancreatic lipases using either triglycerides or triglycerides analogs. With respect to gastric lipase however, the enzyme adsorbed less on each stereoisomeric inhibitor that on the dicaprin substrate. Furthermore, the various organophosphorus enantiomers displayed differential inhibitory effects. The inhibition was much dependent upon the chirality on the *sn*-2 carbon of the glycerol backbone, while the chirality on the phosphorus atom had no influence. The R_CS_P and R_CR_P, which both contain the phosphorus moiety at the *sn*-3 position, were found to be the best inhibitors. This latter finding correlates well with the *sn*-3 preference, during the hydrolysis of triglycerides catalyzed by gastric lipases. Moreover, the levels of surface density of gastric lipase differed significantly with each enantiomeric inhibitor used. A clear correlation was observed between the molar ratio (α50) of inhibitor leading to half inhibition and the surface concentration of gastric lipase: the highest enzymatic inhibition was observed with films containing the enantiomeric inhibitor to which the human gastric lipase was best adsorbed.

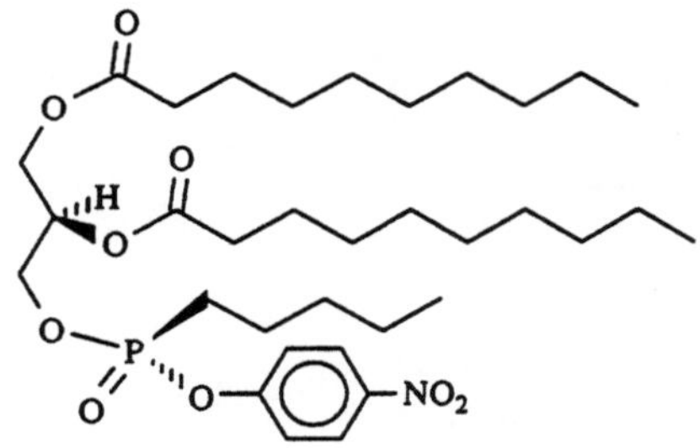

A (R_c, R_p)

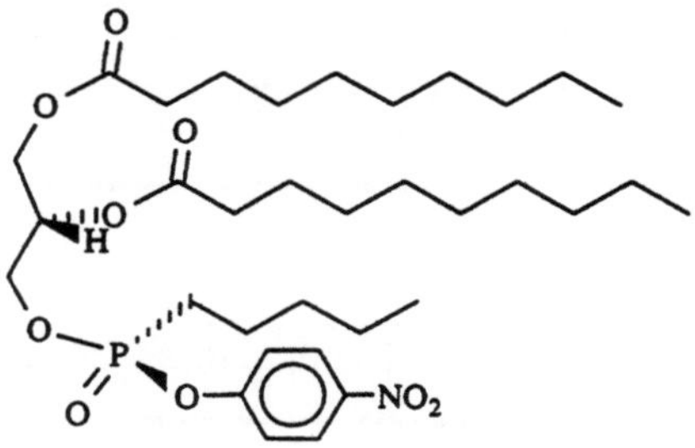

A' (S_c, S_p)

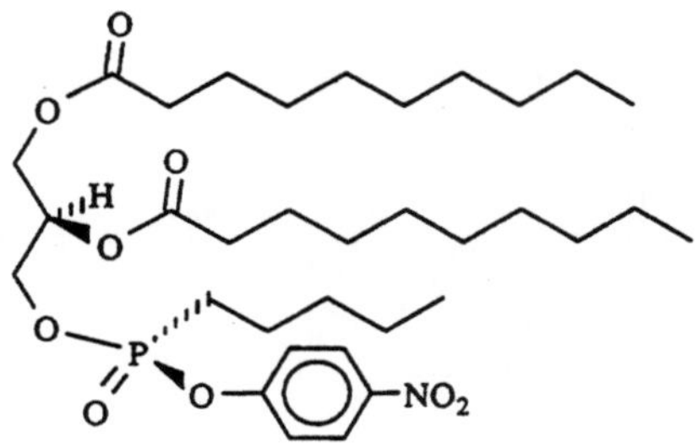

B (R_c, S_p)

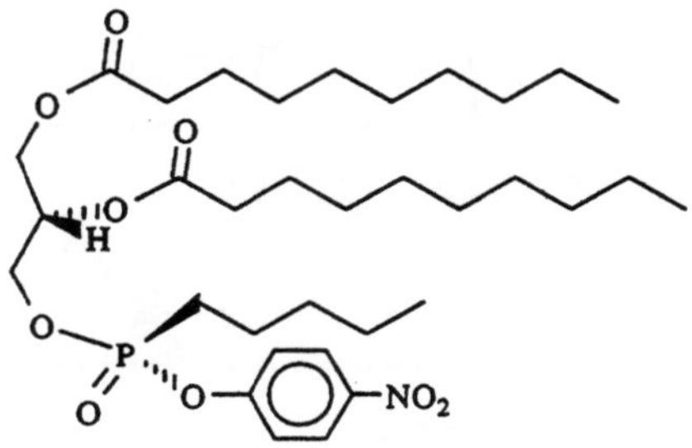

B' (S_c, R_p)

Figure 7: Organophosphorus inhibitors which are triacylglycerol analogs.

1.2.8. Glyceride synthesis catalysed by cutinase using the monomolecular film technique

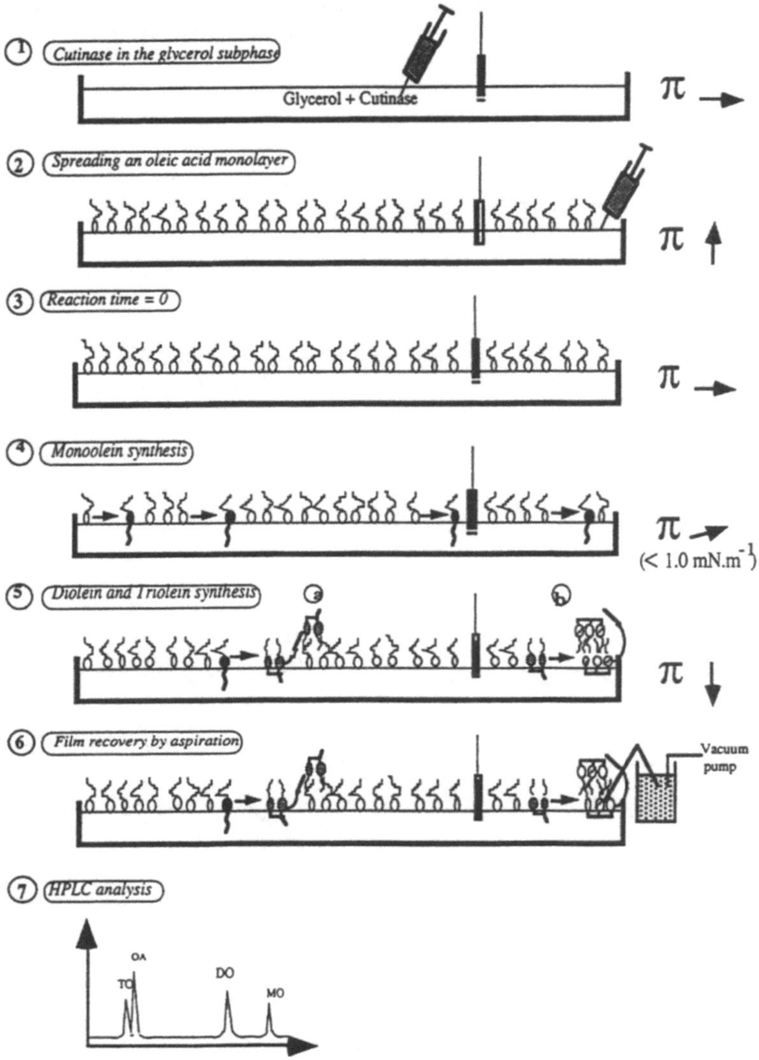

Figure 8: Steps occurring in a first-order trough after the spreading of an oleic acid monolayer (2) over a glycerol subphase containing cutinase (1). Monoolein synthesis (4).Diolein and triolein synthesis (5). Film recovery (6) and HPLC analysis (7). Open ovals stand for oleic acid; solid ovals, monoolein; stippled ovals, diolein; diagonally striped ovals, triolein. Surface pressures are indicated by the slopes of the arrows on the right-hand side. The platinum plate is depicted as a thick vertical bar. Acylglycerol molecules are pointed by an arrow. (From Melo *et al.* [82]).

When enzymatic catalysis takes place in systems with a low water content, the resulting thermodynamic equilibrium favours synthesis over hydrolysis. Hydrolytic enzymes can therefore be used to catalyze the formation of ester bonds in these systems. Among the hydrolytic enzymes, lipases have been widely used to perform esterification or transesterification reactions. Lipid monolayers have been used recently as lipase substrates with the monomolecular film technique. Melo *et al.* [82] adapted the monomolecular film technique for use in the synthesis of oleoyl glycerides (monoolein, diolein and triolein) as schematically shown in figure 8. The water subphase was replaced by glycerol and a stable film of oleic acid was initially spread on its surface. This method makes it possible to study and perform ester synthesis in a new and original system involving a specific array of self organized lipid substrate molecules. The authors continuously recorded the surface pressure and furthermore estimated the glyceride synthesis after film recollection and HPLC analysis, at the end of the experiment. A purified recombinant cutinase from *Fusarium solani* [83] was used as a biocatalyst in this study. More than 50% of the oleic acid film was acylated after seven minutes of reaction. The surface pressure applied to the monomolecular film acts as a physical selectivity factor, since glyceride synthesis can be steered so as to produce either diolein or triolein.

1.3. The oil-drop tensiometer for studying lipase action

As can be seen from the literature [44, 84-86], numerous techniques are available for measuring lipase activity. These can be classified into three groups on the basis of either the substrate consumption, the product formation, or the changes with time of one physical property, such as the conductivity, turbidity or interfacial tension. Among the interfacial methods, monomolecular film technology at the air-water interface has been extensively developed and used. With this technique, it is possible to measure and control some important interfacial parameters. One prerequisite of the monolayer technique however is that the insoluble monomolecular film of substrate should generate water-soluble products during the reaction process. This is why synthetic medium-acyl chain lipids are mainly used as substrates for lipolytic enzymes [27, 28, 87]. Nevertheless, the question remains as to whether the behaviour of the lipid film at the air-water interface is actually representative of what is occurring either at an oil-water interface or in a complex biomembrane. In 1987, Nury *et al.* [88] established that one can gain unique information by measuring the variations in the oil-water interfacial tension ($\gamma_{o/w}$) as a function of time during the lipase hydrolysis of natural long chain triacylglycerols. These authors adapted the so-called "oil-drop method" for use in studying the rate of lipase hydrolysis of natural long chain triacylglycerols. The accumulation of insoluble hydrolysis products at the surface of the drop is responsible for the $\gamma_{o/w}$ decrease, which in turn is correlated with changes with time in the oil drop profile. The theoretical basis of the calculation of $\gamma_{o/w}$ from a hanging drop profile, using the Laplace equation, has been extensively described in the physics literature since 1911 [89-99]. Nury *et al.* [88] demonstrated that this method constitutes a reliable, sensitive and convenient means of investigating lipase kinetics by taking oil drop pictures in order to determine the interfacial tension from the accurately measured diameters of the drops. The main drawbacks of this technique are however the lengthy film processing and profile analysis and the fact that it does not yield real-time measurements. On the basis of this initial study, Nury *et al.* [98], Grimaldi *et al.* [99] and Cagna *et al.* [100] developed a new set-up whereby the oil drop profile was automatically digitised and analysed by image processing; the interfacial tension was calculated in real time using the Laplace equation (see Figure 9).Fully automatized Oil Drop Tensiometer (ODT) systems of this kind are now commercially available (IT Concept-AXONE 69930 St Clément les Places-France)

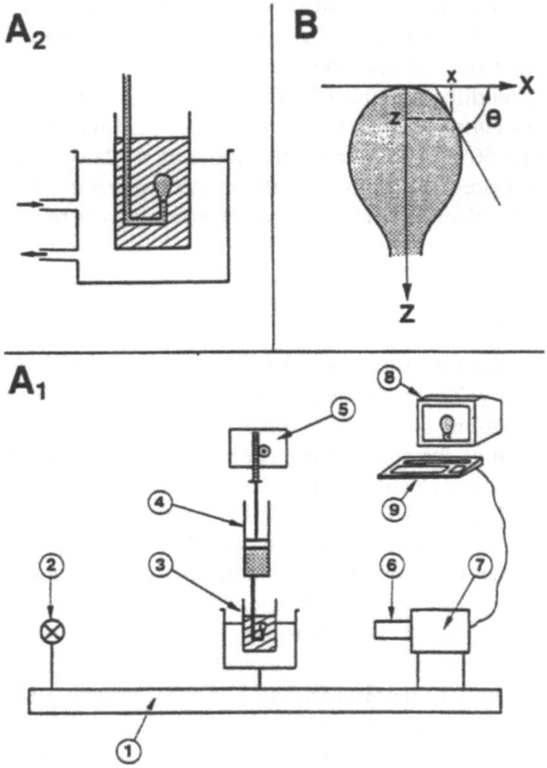

Figure 9: A1: Diagram of the experimental set-up. (1) Optical bench. (2) Integrated sphere light source. (3) Drop formation device with thermostated cuvette. (4) Syringe. (5) DC motor driving the piston of the syringe. (6) Telecentric gaging lens. (7) CCD camera. (9) Personal computer. A2: Blow-up of the drop formation device. B: Principle of the analysis of the profile of a mounting drop using the Laplace equation. (From Grimaldi *et al.* [99] and Labourdenne *et al.* [101]).

Furthermore, by keeping the oil-water interfacial tension at a fixed end point value, the enzyme kinetics can be monitored using the change with time in the area of the oil drop. Recently, Labourdenne *et al.* [101] described some potential applications of the oil drop tensiometer to lipolytic enzyme kinetics. The following three main aspects were discussed: (i) By increasing the drop volume in order to maintain $\gamma_{O/W}$ constant, it is possible to accurately monitor the lipase kinetics. (ii) A new method is proposed for phospholipase A_2 assay. (iii) The oil drop method can also be used to investigate lipase kinetics under high hydrostatic pressure in order to modulate catalysis by inducing possible enzyme conformational changes.

1.4. Kinetic models.

One of the main assumptions implicitely underlying the Michaelis-Menten model is the fact that the enzymatic reaction must take place in an isotropic medium (*i.e.* both the enzyme and the substrate must be in the same phase). This model therefore cannot be used as it stands to study lipolytic enzymes acting mainly at the interface between a water phase and an insoluble lipid phase. In 1973 [102], a kinetic model was developed, based on the idea that an enzyme/substrate complex might be formed at the interface, which would therefore have a surface density as one of its dimensions (Figure 6). The enzymatic reaction can be subdivided into two elementary catalytic stages. First the enzyme is fixed at the interface by an unknown molecular mechanism (adsorption, penetration, etc.). One formal consequence of this stage is the dimensional change in the enzyme concentration. This stage sometimes, but not always, involves enzyme activation (*via* the opening of the amphiphilic lid covering the active site, for example, see § III). The second stage is similar to a Michaelis-Menten pseudo-reaction occurring at a surface, and not within a volume. All the concentrations used for this purpose (those of the enzyme, the substrate, the product, and the enzyme/substrate complex) are therefore expressed as surface densities.

In mathematical terms, this model can be accounted for by the following expression, which gives the various products released in the course of time :

$$P = k_{cat} \cdot E_0 \frac{S}{K_M^*} \frac{t + \dfrac{\tau_1^2}{\tau_1 - \tau_2}\left(e^{-t/\tau_1} - 1\right) - \dfrac{\tau_2^2}{\tau_1 - \tau_2}\left(e^{-t/\tau_2} - 1\right)}{1 + \dfrac{k_d}{k_p \, A/V} + \dfrac{S}{K_M^*}}$$

The rate of production of these same products in the stationary state was written as follows :

$$v = \frac{dP}{dt} = k_{cat} \cdot E_0 \frac{S}{K_M^*\left(1 + \dfrac{k_d}{k_p \, A/V}\right) + S}$$

For definitions of the parameters in the above equations, see references [71, 103]. τ_1 and τ_2 are the characterictic times required for the adsorption at the interface and the catalysis in the interfacial plane (*i.e.* stage 1 and stage 2) respectively, to be completed.

Kinetic models accounting for the competitive inhibition in the presence and absence of detergent [104] and for irreversible inactivation [80] of the process have also been developed.

1.5. The mechanisms underlying lipase action.

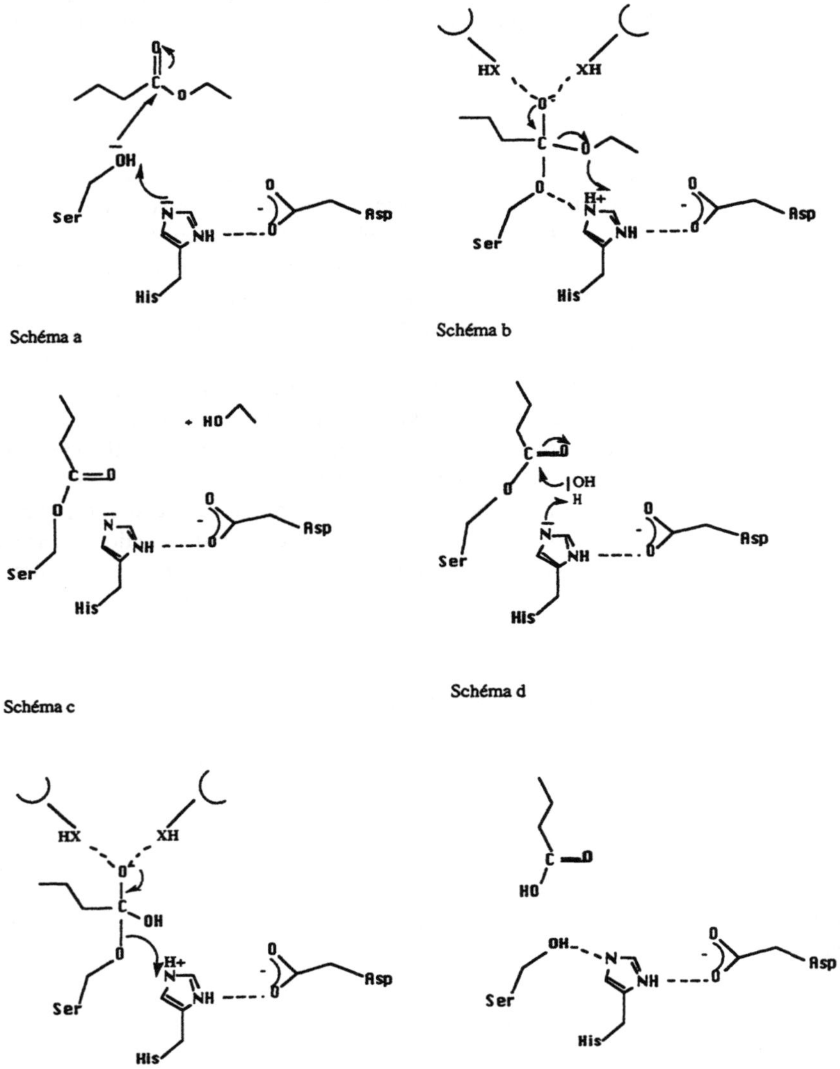

Figure 10: Hydrolysis mechanism of a carboxylic ester by the catalytic triad of a serine enzyme. a) Native conformation of the enzyme and substrate binding. b) Formation of the first tetrahedral intermediate and stabilisation by the oxyanion hole. c) Alcohol release and formation of the acyl-enzyme. d) Nucleophilic attack by a water molecule. e) Formation of the second tetrahedral intermediate and stabilisation by the oxyanion hole. f) Carboxylic acid release and regeneration of the enzyme in the native form.

Lipolytic enzymes, lipases and phospholipases A_2 catalyse the hydrolysis of lipidic substrates (triacylglycerols and phospholipids) and possess an active site consisting of a triad of amino-acids. These catalytic residues are Asp-His-Asp in the case of phospholipases A_2, and Ser-His-Asp/Glu in that of lipases. Contrary to what happens in the case of phospholipases A_2, the mechanism involved in the action of lipases was extrapolated from that previously proposed in the case of chymotrypsin (Figure 10). The hydroxyl group of serine is activated via the charge relay network and attacks the carbon atom in the substrate's carboxylic group (a), thus forming the first tetrahedral intermediate (b). Hydrogen bonds between the negatively charged oxygen (oxyanion) of the tetrahedral intermediate and N-H peptidic groups stabilize the negative charge on the oxyanion. This atomic and electronic environment has been called the oxyanionic hole. It stabilizes the tetrahedral structure and thus lowers the energy required to activate the reaction. During the overall acido-basic catalysis, the carboxylic ester bond is broken and the alcohol group is released, taking with it one proton from the imidazolium ion of the histidine residue. The acyl chain of the ester bond remains covalently bonded to the enzyme in the form of an acyl-enzyme intermediate (c). This intermediate is in turn attacked by a water molecule (d), resulting in a second tetrahedral intermediate (e). The latter disintegrates, releasing the removed acyl group, and thus the lipase is regenerated back into its initial state (f).

II - Specificities.

The specificity of lipases' action on triacylglycerols can be defined at the following three levels :

- in terms of their typoselectivity towards a given fatty acid;
- in terms of their regioselectivity, *i.e.*, the ability of a lipase to preferentially hydrolyze carboxylic ester bonds located in external positions *sn*-1 and *sn*-3 (primary esters) as opposed to those in the internal position *sn*-2 (secondary ester);
- and in terms of their stereoselectivity, *i.e.*, the ability to discriminate between two enantiomers in the case of a racemic substrate, and between two stereoheterotopic but homomorphic (enantiotopic) groups in the case of prochiral acylglycerols (position *sn*-1 *versus* *sn*-3).

2.1. Stereoselectivity.

Chirality is a general attribute of objects and figures, and by definition does not depend in any way on the material world, although is has important links with it. In 1904, Lord Kelvin said "I call any geometrical figure, or group of points, *chiral*, and say it has *chirality* if its image in a plane mirror, ideally realized, cannot be brought to coincide with itself" [105]. Chirality therefore means the absence of miror symmetry.

2.1.1. Enantiomeric differentiation (enzymatic resolution).

Several examples have been described which illustrate enzymes' ability to recognize enantiomers [106-108]. This phenomenon can be explained by looking at the various diastereoisomeric interactions which occur between enantiomers and the active (chiral) site of an enzyme.

Although many enzymes have a high degree of enantioselectivity, this is not always the case [107]. As far as lipases are concerned, the two enantiomers present in a racemic mixture usually behave as suitable substrates, and the enantioselectivity will depend on their chemical properties and on the origin of the lipases and the experimental conditions used [51, 63, 109, 110].

2.1.2. Enantiotopic differentiation.

The most spectacular aspect of enzymes' stereospecificity is the fact that some enzymes are able to differentiate between the two enantiotopic groups or enantiomeric faces present in a prochiral molecule. All the lipases tested so far on prochiral triacylglycerols have been found to have this discriminatory ability, apart from porcine pancreatic lipase, which did not distinguish between the prochiral carboxylic ester groups of trioctanoin and triolein [50, 54].

2.2. The mechanisms underlying lipases' stereoselectivity.

Some of the stereochemical aspects of the lipasic hydrolysis of glycerides are quite unusual. Under normal physiological conditons, many other hydrolases, such as phospholipases, proteases, glycosidases and nucleases, metabolize only one of the substrate's antipodes, whereas lipases are often capable of recognizing enantiomers as well as enantiotopic groups of prochiral molecules. The crystallographic structures which have been recently determined so far [6, 7, 9, 111-128] do not help to explain the stereochemical data [54]. Although all these lipases have very different stereoselectivities, their basic architecture is quite similar : the residues constituting the catalytic triad and the structural motif β–εSer-α characteristic of the nucleophilic centre generally match [116-119]. Cygler *et al.* recently described the mechanism possibly explaining how the Candida rugosa lipase is able to distinguish between the two enantiomers of a menthol phosphonic ester [126]. This was the first time that a structure of a lipase crystallized with two chiral inhibitors had been resolved, although the *Candida rugosa* lipase was previously determined without any inhibitors by the same group [122, 128]. As regards the stereochemistry of lipases, Cygler *et al.* reached the conclusion that the enantiopreference shown by Candida rugosa lipase among the secondary alcohol esters is determined not by an alcohol binding site other than the catalytic site, but by conformational changes in the catalytic triad. The inhibitors used in the study in question were structurally very differenct from the natural substrates of lipases, however. Egloff *et al.* [127] recently used a racemic mixture of methyl *n*-undecyl-phosphonate and *p*-nitrophenyl, as well as each of the two enantiomers in the medium prepared for co-crystallizing the human pancreatic lipase-colipase complex. The two enantiomers were differently oriented in the cleft of the active site (see §III, 2). The results of these two above studies unfortunately cannot be extrapolated to all lipases and to all experimental conditions.

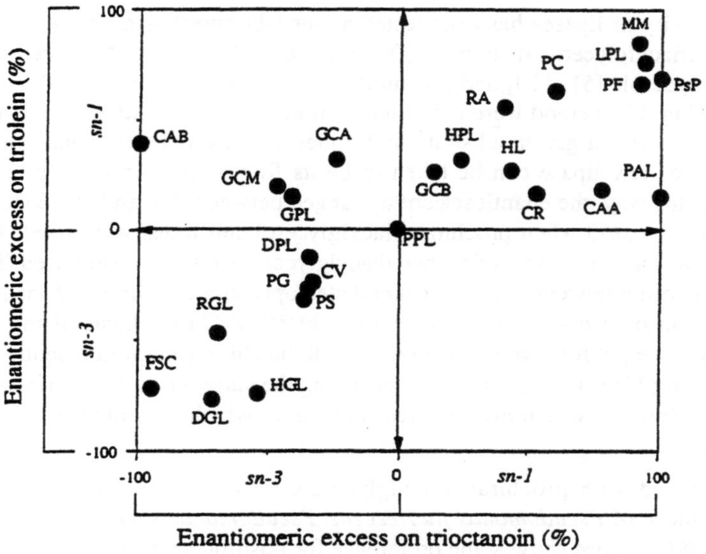

Figure 11a: Lipase stereoselectivity on trioctanoin and triolein. Correlation between the stereoselectivities on the two substrates (from Rogalska *et al.* [54]).

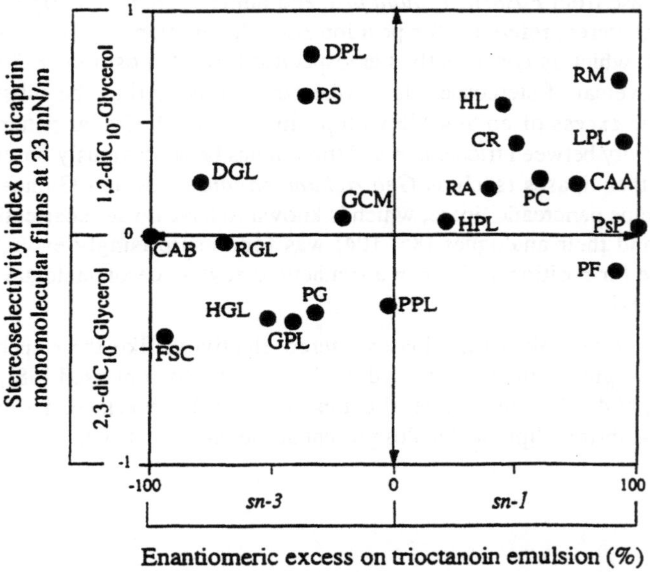

Figure 11b: Lipase stereoselectivity on trioctanoin and 1,2(2,3)-dicaprin enantiomers. Correlation between the stereoselectivities on the tri and di-glycerides (from Rogalska *et al.* [63]).

The stereoselectivity of lipases has been tested at our laboratory using monomolecular films of di- and triacylglycerol analogues [104], with optically pure diacylglycerols [63], racemic diacylglycerols [51, 110] and prochiral triacylglycerol emulsions [50, 54]. The data obtained using this method were independant of the typoselectivity of the lipases. It was observed that with a given substrate and under specific reaction conditions, the stereoselectivity of any lipase can be taken to be its fingerprint. Among the 25 pure lipases tested in this way, the enantioselectivity ranged between 0 % and 100 % towards positions *sn*-1 versus *sn*-3 when prochiral triacylglycerol substrates were used (Figure 11a), and that it was also very variable when diacylglycerol substrates were used (Figure 11b). It is worth noting however that the general stereopreference of most of the lipases tested is similar on both tri- and diacylglycerol substrates. On the basis of the above experimental data, we put forward the hypothesis that the chiral recognition centre might undergo enantiomorphic organization processes during the stage when the interfacial E*S complex is being formed, which may be precisely that at which the induced fit process takes place [54].

The results obtained with prochiral triacylglycerols [54] show that although some lipases, such as those of *Pseudomonas fluorescens*, *Pseudomonas species*, *Rhizomucor miehei* and *Candida rugosa* show some preference for position *sn*-1 of triacylglycerols, considerable differences exist from one lipase to another when triolein is used as the substrate: an enantiomeric excess of 70 % was recorded in *Pseudomonas species* lipase, as compared with 16 % in the case fo *Candida rugosa* lipase. On the other hand, many enzymes, such as cutinase from *Fusarium solani pisi*, and human, rabbit and dog gastric lipases, show a distinct stereopreference for position *sn*-3. In addition, the lipase from *Candida antarctica* B, which is consistently stereospecific towards position *sn*-3 with trioctanoin, shows a reversal of stereospecificity when triolein is used as the substrate (giving an enantiomeric excess of up to 40 % with position *sn*-1). A similar pattern of reversal of stereoselectivity between trioctanoin and triolein has been previously observed in the case of some other lipases (such as *Geotrichum candidum* A and Guinea pig pancreatic lipase). Porcine pancreatic lipase, which is known to have no stereoselectivity with triacylglycerols and their analogues [85, 104] was found surprisingly enough to have a stereopreference for position sn-3 when a synthetic diacyl glycerol analogue was used as the substrate [104].

The fact should not be overlooked that lipases' stereoselectivtiy, like their specific activity, can depend to a great extent on the hydrophicity of the solvent used [58], the interfacial tension [51, 104, 110], the chemical composition of the interface, and even the chirality of the non-substrate lipid molecules present at the interface [110].

III. Structural aspects : Relationships between structures and kinetic properties of pancreatic lipases

Pancreatic lipase is the major lipolytic enzyme involved in the digestion of dietary triglycerides [129], and it amounts to around 3 % of the total proteins secreted by the exocrine pancreas [130]. Pancreatic lipase hydrolyzes primary ester bonds of tri- and diglycerides thus generates 2-monoglycerides and fatty acids, which are absorbed through the intestinal barrier. In contrast to most of the other pancreatic enzymes, which are secreted as proenzymes and further activated by proteolytic cleavage in the small intestine, pancreatic lipase is directly secreted as a 50 kDa active enzyme consisting of 449 amino acid residues [131, 132]. The structural and catalytic properties of classical human pancreatic lipase (HPL) will be described below, based on X-ray structural data.

In addition to "interfacial activation", pancreatic lipases are also characterized by their behavior in the presence of bile salts [133]. They are inactive on an emulsified triglyceride substrate in the presence of micellar concentrations of bile salts, such as those found in the small intestine during digestion. Bile salts are amphiphilic molecules which are mainly to be found absorbed to the lipid/water interface or dispersed as micelles in solution. It was proposed that the bile salt coating of triglyceride globules has a negatively charged surface that inhibits pancreatic lipase adsorption and activation. To counteract this effect, a specific lipase-anchoring protein, colipase, is present in the exocrine pancreatic juice. It forms a 1:1 complex with the lipase that facilitates its adsorption to bile salt-covered lipid-water interfaces [115, 134, 135]. The inhibition by bile salts and the reactivation by colipase of HPL have been clearly observed using tributyrin as substrate and various concentrations of sodium taurodeoxycholate. When increasing the bile salt concentration, HPL activity towards tributyrin is lost above the critical micellar concentration (1-2 mM) in the absence of colipase. HPL is desorbed from the lipid/water interface and no significant activity towards tributyrin monomers present in solution is observed, probably because the enzyme is in an inactive conformation in solution. Consequently, HPL can be reactivated by adding colipase to the assay medium. Alternatively, HPL alone is irreversibly denaturated at the high energy tributyrin/water interface and further addition of colipase does not restore the activity. When colipase was added prior to HPL, lipolytic activity was observed but it remained unstable with time. It is only by adding bile salts and increasing their concentration that the HPL-colipase complex can be made to reach a high and stable specific activity. This behavior can be explained by the lowering of tributyrin/water interfacial tension induced by bile salts.[136]

All the previous properties clearly show how HPL adapts to the physiological conditions present in the small intestine, where the pancreatic and biliary secretions are mixed with dietary lipids. Through the concerted action of colipase and "interfacial activation", HPL displays a high specificity towards insoluble triglycerides.

Pancreatic lipolytic enzymes in the guinea pig have previously been found to differ significantly from those present in other mammals. After a partial purification that revealed a high phospholipase activity still associated with lipase activity, two cationic lipases with high phospholipase A_1 activity from guinea pig pancreas were purified by Durand *et al.* [137] and Fauvel *et al.* [138]. These authors observed an unusually high phospholipase/lipase activity ratio of 1, which was higher by 3-5 orders of magnitude to than with lipases from other sources. Furthermore, no evidence suggesting the existence of a classical secretory phospholipase A_2 in guinea pig pancreas was found [139]. The high phospholipase A_1 activity of guinea pig pancreatic (phospho)lipase might thus be of physiological significance for the degradation of dietary phospholipids and a detailed study of this enzyme seemed to be worth undertaking.

3.1. Functional organization of pancreatic lipase in several structural domains

The resolution of the 3D structure of HPL by Winkler *et al.* [7] confirmed the existence of two distinct domains in pancreatic lipase: a larger N-terminal domain comprising residues 1-336, and a smaller C-terminal domain made up of residues 337-449. This scheme had previously been suggested by Bousset-Risso *et al.* [140]. The high degree of amino acid sequence homology observed within the lipase gene family supports the view that this particular architecture is also common to lipoprotein lipase and to hepatic lipase [141-145].

The large N-terminal domain is a typical α/β structure dominated by a central parallel β-sheet [118]. It contains the active site with a catalytic triad formed by Ser 152, Asp 176, His 263, all of which are conserved in lipoprotein lipase and hepatic lipase. This catalytic triad is chemically analogous to that originally described in serine proteases such as chymotrypsin [146], but is structurally distinct. The β-strand/εSer/α-helix structural motif including the Gly-X-Ser-X-Gly consensus sequence has been detected only in lipases and esterases so far [142, 147]. The structure of HPL clearly demonstrated that Ser 152 is the nucleophilic residue essential for catalysis, in agreement with the chemical modification of Ser 152 in porcine pancreatic lipase [148], and in contradiction with results suggesting a function of Ser 152 in interfacial recognition [149].

In the structure resolved by Winkler *et al.* [7], the active site is covered by a surface loop between the disulfide-bridge Cys 237 and 261. This surface loop includes a short one-turn α-helix with a tryptophan residue (Trp 252) completely buried and sitting directly on top of the active site Ser 152. Under this closed conformation, the "lid" prevents the substrate from having access to the active site. Spectroscopic studies of tryptophan fluorescence have shown that large spectral changes are induced by acylation of pancreatic lipase with the inhibitor tetrahydrolipstatin in the presence of bile salt micelles [150]. By crystallizing the pancreatic lipase-procolipase complex in the presence of mixed lipid micelles, it was shown that the "lid" was shifted to one side, exposing both the active site and a larger hydrophobic surface [120]. This motion is induced when the binding to the lipid occurs and is probably the structural basis for "interfacial activation" of pancreatic lipase, as illustrated in figure 12.

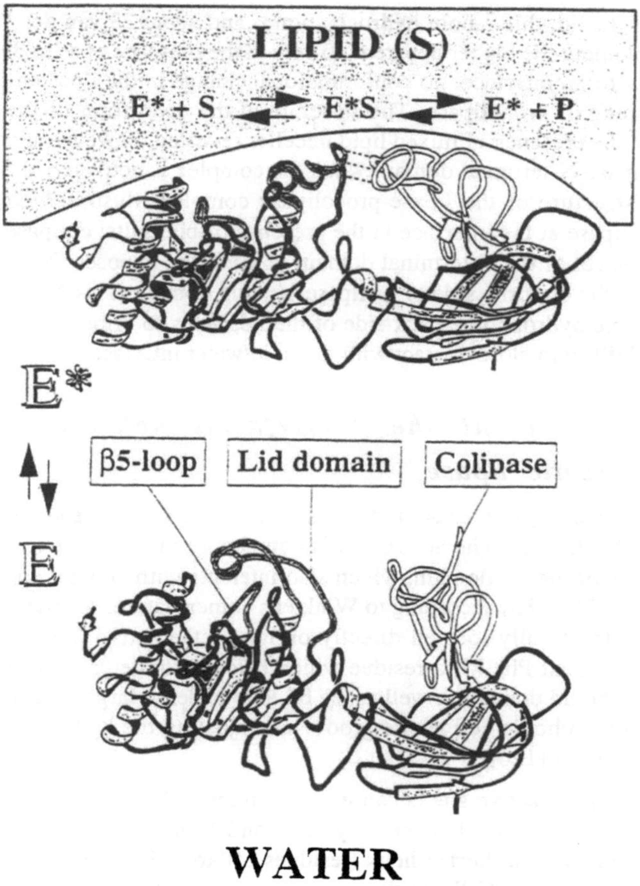

Figure 12: Structure of the HPL-procolipase complex in the closed conformation (E), and structure of the HPL-procolipase complex in the open conformatioin (E*S). These two figures show the conformational changes in the lid, the β5-loop, and the colipase during "interfacial activation".

The ß-sandwich C-terminal domain of pancreatic lipase is necessary for colipase binding to occur, as shown in the 3D structure of the HPL-porcine procolipase complex [115]. Procolipase is a "three finger" protein which is topologically comparable to snake toxins, even though these proteins do not share any sequence homology. Colipase lacks any well-defined secondary structure elements. This small protein seems to be stabilized mainly by an extended network of five disulfide bridges that runs throughout the flatly shaped molecule, reticulating its four finger-like loops. The colipase surface can be divided into a rather hydrophilic part, interacting with lipase, and a more hydrophobic part, formed by the tips of the fingers which are very mobile and constitute the lipid interaction surface. The interaction between colipase and the C-terminal domain of lipase is stabilized by eight hydrogen bonds and about 80 van der Waals contacts. Upon the opening of the lid, three more hydrogen bonds and about 28 van der Waals contacts are added, explaining the higher apparent affinity in the presence of a lipid/water interface. In the absence of an interface, no conformational change in the lipase molecule is induced by the binding of procolipase. The structure of the open form of the HPL-procolipase complex in the presence of mixed lipid micelles revealed, however, that the "lid" binds to the procolipase N-terminal domain when the complex is activated at an interface [120]. The open structure of the lipase-procolipase complex illustrates how colipase might anchor the lipase at the interface in the presence of bile salts: colipase binds to the non-catalytic ß-sheet of the C-terminal domain of HPL and exposes the hydrophobic tips of its fingers at the opposite side of its lipase-binding domain. This hydrophobic surface, in addition to the hydrophobic back side of the lid, helps to bring the catalytic N-terminal domain of HPL into close contact with the lipid/water interface.

3.2. Details about the "interfacial activation" of human pancreatic lipase

In solution (absence of interface), the classical HPL is in a closed conformation (E form in Figure 12) [7, 115]. The active site is completely inaccessible to solvent due to the conformation of the lid domain, which also interacts with another surface loop, the ß5-loop (residues 76 to 85) according to Winkler's nomenclature. A residue in the lid region, Trp 252, is structurally located directly on top of the active Ser 152, with the indole ring packed against Phe 77, a residue which belongs to the ß5-loop. In the presence of an interface, the lid domain as well as the ß5-loop undergo large conformational changes, creating a hydrophobic active-site groove and ajusting the hydrolytic machinery (E*S form in Figure 12) [120].

The opening of the active site is caused by a structurally complicated reorganization of the complete lid domain between Cys 237 and Cys 261. As mentioned above, this surface loop contains a short α-helix (residues 248 to 255) covering the active site in the closed conformation of HPL. In the presence of lipids, this helix partially unwinds and two new helices are formed (residues 241 to 246, and residues 251 to 259). The conformational change in the lid results in a maximal main-chain movement of 29 Å for Ile 248. The substrate now has access to the active site. The Trp 252 which filled the active-site pocket in the closed conformation is now conveyed to the surface of the molecule and is involved in a new interaction with the core of the protein.

Another consequence of the lid reorganization is the conformational change of the ß5-loop. In the closed conformation of the HPL-procolipase complex, this loop makes van der Waals contact exclusively with the lid. In the open conformation, these interactions are lost and the ß5-loop is lifted back onto the core of the protein. This movement creates an electrophilic region close to the active site Ser 152. This is probably the so called oxyanion hole which stabilizes the transition-state intermediate formed during catalysis. The main-chain nitrogen of Phe 77 from the ß5-loop moves to an ideal position to stabilize the negative charge developed during ester hydrolysis. The main-chain nitrogen of Leu 153 is also involved in the formation of the oxyanion hole.

In the open conformation of the HPL-procolipase complex, procolipase also binds to the lid domain. This interaction is mediated by three direct hydrogen bonds between residues of the lid (Val 246, Ser 243, Asn 24) and residues of the procolipase N-terminal domain (Arg 38, Leu 16, Glu 15), respectively. The latter domain undergoes a considerable conformational change from the closed to the open conformation of the complex.

As the result of all these conformational changes, the open lid and the extremities of the procolipase fingers form an impressive continuous hydrophobic plateau, extending over more than 50 Å. This surface might be able to interact strongly with a lipid/water interface, and this might explain the effects of colipase in the presence of bile salts.

A racemic mixture of the C11 alkyl phosphonate inhibitor was recently used in the cristallization of the pancreatic lipase-colipase complex, and there is evidence that both enantiomers are bound at the active site [127]. The C11 alkyl chain of the first enantiomer fits into a hydrophobic groove and this is thought to mimic the interaction of the leaving fatty acid of a triglyceride substrate. The alkyl chain of the second enantiomer also has an elongated conformation and interacts with hydrophobic patches on the surface of the open amphipathic lid. This binding mode may be indicative of an interaction with a second alkyl chain of a triglyceride substrate [127]. The most challenging problem results, however, from the fact that lipases differ dramatically in their specific activities and stereopreferences, and generalizing the above results obtained with just one or two enzymes is simply not possible.

3.3. A subfamily of pancreatic lipases with new kinetic properties

Until quite recently, pancreatic lipase was thought to be well characterized as far as its structural and kinetic properties were concerned, and to be a separate member of the lipase gene superfamily. Several new members of the pancreatic lipase family have now been cloned, sequenced and partly characterized [13-15, 151, 152]. Primary structure analysis of these new lipases revealed that the pancreatic lipase family can be divided into three subgroups: (i) classical pancreatic lipases, (ii) pancreatic lipase-related proteins 1 (PLRP1) and (iii) pancreatic lipase-related proteins 2 (PLRP2) as shown in figure 13. Among the RP1 subfamily, only HPL-RP1 has been expressed *in vitro* but it was not found to display any enzymatic activity under the assay conditions used [13]. In the case of the RP2 subfamily however, the kinetic properties of two lipases detected in the guinea pig (GPL-RP2 [14]) and in the coypu (CoPL-RP2 [15]) have been studied. Both enzymes have an atypical enzymatic behaviour and challenge the classical distinction between lipases, esterases and phospholipases.

In contrast to the classical HPL, CoPL-RP2 and GPL-RP2 display no "interfacial activation" *i.e.*, both enzymes are already fully active on monomers of partly soluble triglycerides and their active site is probably accessible to a water-soluble substrate. In GPL-RP2, the lid domain is shortened to a "mini-lid" and the active centre is thought to possibly be freely accessible, as suggested by the model based on HPL 3D structure [14].

In CoPL-RP2, the activity in solution is probably due to a permanently open conformation of a normal sized lid domain. The lid domain sequence was found to be poorly conserved, however, upon comparing CoPL-RP2 and HPL. In particular, replacing Trp 252 by a Leu in CoPL-RP2 may weaken the interactions of the lid with the ß5-loop, thus facilitating the spontaneous opening of the active site.

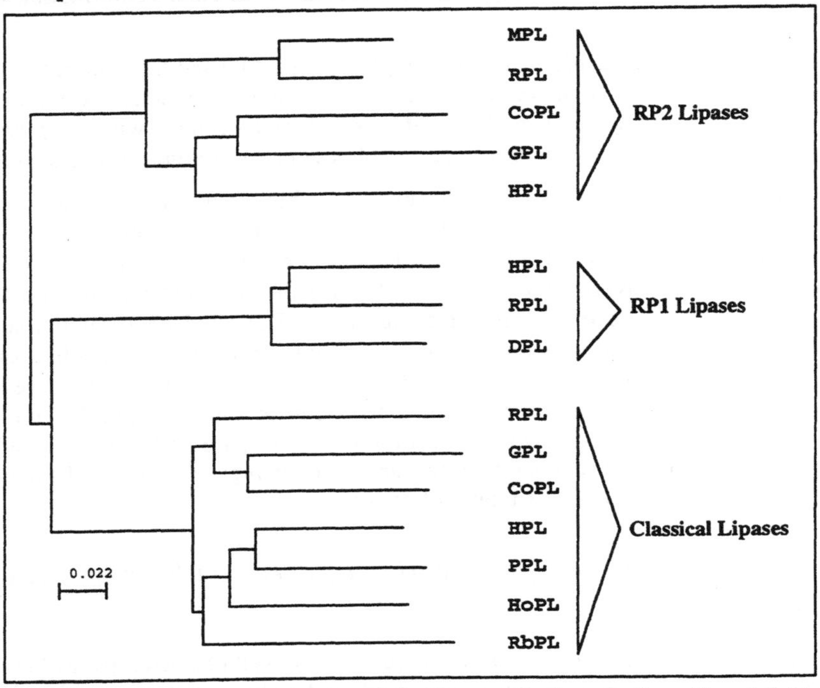

Figure 13: Dendogram of sequence alignment within the pancreatic lipase family and its subdivision onto three subfamilies. Lipases are those from rat (RPL), guinea-pig (GPL), coypu (CoPL), human (HPL), pig (PPL), horse (HoPL), rabbit (RbPL), and mouse (MPL)

It was also demonstrated that colipase does not reactivate GPL-RP2 and CoPL-RP2 at high bile salt concentrations [15]. Although colipase has absolutely no effect on GPL-RP2 activity whatever the bile salt concentration, the colipase binding to CoPL-RP2 is not completely abolished, since a clearly detectable effect of colipase was observed below the bile salt critical micellar concentrations, at 1 mM sodium taurodeoxycholate. However, the affinity of colipase for CoPLRP2 is probably very weak. The behaviour of GPL-RP2 can be easily explained: among the 12 residues of the HPL C-terminal domain involved in colipase binding [115], 9 are different in GPL-RP2, including the important Lys 399. Moreover, colipase also interacts with the open lid of HPL [120]. These interactions cannot exist in a lipase with a "mini-lid" such as GPL-RP2. In CoPL-RP2, the absence of colipase effect at bile salt concentrations above 2 mM is not easily interpretable. Almost all the residues involved in the interaction with colipase are conserved (both in the C-terminal domain and the lid). One striking exception however is residue 403, which is conserved as a tyrosine within all the classical lipases but is different in all RP2 lipases. Tyr 403 interacts strongly with colipase through several van der Waals contacts in the HPL-colipase complex [115]. The stacking of Tyr 403 of lipase and Arg 65 of colipase mainly confers the apolar component of the binding energy. The point mutation of residue 403 in CoPL-RP2 may induce a low affinity for colipase. Also the lid/colipase interactions might be suppressed in CoPL-RP2 due to an open conformation of the lid-domain different from the one observed in HPL. The open conformation of HPL lid is stabilized by interactions with the core of the protein, the colipase and the ß5-loop [120]. The lid domain interacts with the core of the protein through a salt bridge (Asp 257...Lys 268) and a hydrogen bond (Arg 256...Tyr 267). In the RP2 subfamily, these residues are almost all different, whereas they are completely conserved in the other two pancreatic lipase subfamilies [15, 153]. The lid domain also interacts with the ß5-loop through a salt bridge (Arg 256...Asp 79), and a hydrogen bond (Trp 252...Glu 83). Here again, most of these residues are different in the RP2 subfamily. In CoPL-RP2, all the stabilizing interactions observed in HPL between the lid domain, the protein core and the ß5-loop, are missing. One can assume that the lid domain in CoPL-RP2 possesses a higher degree of freedom and that the interactions with colipase are weakened if not completely absent.

Finally, both GPL-RP2 and CoPL-RP2 display high phospholipase activities of 500 U/mg and 180 U/mg, respectively (1 unit = 1 µmole of fatty acid released per minute), using egg yolk lecithin as substrate. For comparison, the classical porcine pancreatic phospholipase A_2 displays a specific activity of 700 U/mg under similar conditions. In contrast, HPL showed no measurable activity on this substrate. From the structural point of view, the reason for the high phospholipase activity of the RP2 lipases remains unclear. These RP2 enzymes display a high phospholipase activity towards lecithins but also towards other phospholipid classes [14, 15], whereas classical pancreatic lipases can hydrolyze negatively charged phospholipids only at very low specific activity levels [154]. From the GPL-RP2 modelling based on HPL 3D structure, the core of the N-terminal domain appears to be conserved overall, with the exception of the lid-domain. Furthermore, within a 10 Å radius sphere of the active site serine, there are no insertions/deletions and only four minor residue changes pointing away from the active site. Thus, it was speculated previously that the absence of the lid domain in GPL-RP2 could be possibly responsible for its activity towards phospholipids [14]. Conversely, the lack of high phospholipase activity in classical pancreatic lipases might be viewed as a depressed action on phospholipids due to the presence of the lid-domain [14]. Now that CoPL-RP2 has been characterized, it is clear there is no direct relationship between the absence of the lid domain and phospholipase activity within the RP2 subfamily. One can further speculate that the low affinity of classical pancreatic lipases for phospholipids is due to the peculiar open conformation of the lid domain in HPL and the large hydrophobic area exposed around the active site entrance. This hydrophobic environment, stabilized in the presence of a lipid/water interface, might induce a repulsion towards phospholipid polar heads. Thus pancreatic lipases of the RP2 type can therefore be said to be "natural mutants" challenging the classical distinctions between lipases, esterases and phospholipases.

References

1. Small D.M. (1968) A classification of biological lipids based upon their interaction in aqueous systems. *J. Amer. Oil Chemists Soc.* **45**, 108-119.
2. Sarda L. and Desnuelle P. (1958) Action de la lipase pancréatique sur les esters en émulsion. *Biochim. Biophys. Acta* **30**, 513-521.
3. Holwerda K., Verkade P.E. and de Willigen A.H.A. (1936) Vergleichende Untersuchungen über die Verseifungsgeschwindigkeit einiger einsäuriger Triglyceride unter Einfluss von Pankreasextrakt. *Rec. Trav. Chim. Pays-Bas* **55**, 43-57.
4. Schønheyder F. and Volqvartz K. (1945) On the affinity of pig pancreas lipase for tricaproin in heterogenous solution. *Acta Physiol. Scand.* **9**, 57-67.
5. Desnuelle P., Sarda L. and Ailhaud G. (1960) Inhibition de la lipase pancréatique par le diéthyl-*p*-nitrophényl phosphate en émulsion. *Biochim. Biophys. Acta* **37**, 570-571.
6. Brady L., Brzozowski A.M., Derewenda Z.S., Dodson E., Dodson G., Tolley S., Turkenburg J.P., Christiansen L., Huge-Jensen B., Norskov L., Thim L. and Menge U. (1990) A serine protease triad forms the catalytic centre of a triacylglycerol lipase. *Nature* **343**, 767-770.
7. Winkler F.K., d'Arcy A. and Hunziker W. (1990) Structure of human pancreatic lipase. *Nature* **343**, 771-774.
8. Deveer A. (1992) Mechanism of activation of lipolytic enzymes. Thesis, University of Utrecht, Netherlands.
9. Noble M.E.M., Cleasby A., Johnson L.N., Egmond M.R. and Frenken L.G.J. (1993) The crystal structure of triacylglycerol lipase from *Pseudomonas glumae* reveals a partially redundant catalytic aspartate. *FEBS Lett.* **331**, 123-128.
10. Jaeger K.-E., Ransac S., Koch H.B., Ferrato F. and Dijkstra B.W. (1993) Topological characterization and modeling of the 3D structure of lipase from *Pseudomonas aeruginosa*. *FEBS Lett.* **332**, 143-149.
11. Jaeger K.-E., Ransac S., Dijkstra B.W., Colson C., Vanheuvel M. and Misset O. (1994) Bacterial lipases. *FEMS Microbiol. Rev.* **15**, 29-63.
12. Uppenberg J., Hansen M.T., Patkar S. and Jones T.A. (1994) Sequence, crystal structure determination and refinement of two crystal forms of lipase B from *Candida antarctica*. *Structure* **2**, 293-308.
13. Giller T., Buchwald P., Blum-Kaelin D. and Hunziker W. (1992) Two novel human pancreatic lipase related proteins, hPLRP1 and hPLRP2. Differences in colipase dependence and in lipase activity. *J. Biol. Chem.* **267**, 16509-16516.
14. Hjorth A., Carrière F., Cudrey C., Wöldike H., Boel E., Lawson D.M., Ferrato F., Cambillau C., Dodson G.G., Thim L. and Verger R. (1993) A structural domain (the lid) found in pancreatic lipases is absent in the guinea pig (phospho)lipase. *Biochemistry* **32**, 4702-4707.
15. Thirstrup K., Verger R. and Carrière F. (1994) Evidence for a pancreatic lipase subfamily with new kinetic properties. *Biochemistry* **33**, 2748-2756.
16. van Oort M.G., Deveer A.M.T.J., Dijkman R., Leuveling Tjeenk M., Verheij H.M., de Haas G.H., Wenzig E. and Götz F. (1989) Purification and substrate specificity of *Staphylococcus hyicus* lipase. *Biochemistry* **28**, 9278-9285.
17. Pieterson W.A., Vidal J.C., Volwerk J.J. and de Haas G.H. (1974) Zymogen-catalysed hydrolysis of monomeric substrates and the presence of a recognition site for lipid-water interfaces in phospholipase A_2. *Biochemistry* **13**, 1455-1460.
18. van Eijk J.H., Verheij H.M., Dijkman R. and de Haas G.H. (1983) Interaction of phospholipase A_2 from *Naja melanoleuca* snake venom with monomeric substrate analogs. Activation of the enzyme by protein-protein or lipid-protein interactions? *Eur. J. Biochem.* **132**, 183-188.

19. van den Berg B., Tessari M., Boelens R., Dijkman R., Kaptein R., de Haas G.H. and Verheij H.M. (1995) Solution structure of porcine pancreatic phospholipase A_2 complexed with micelles and a competitive inhibitor. *Journal of Biomolecular NMR* **5**, 110-121.

20. Soares de Araujo P., Rosseneu M.Y., Kremer J.M.H., van Zoelen E.J.J. and de Haas G.H. (1979) Structure and thermodynamic properties of the complexes between phospholipase A_2 and lipid micelles. *Biochemistry* **18**, 580-586.

21. Verger R. and de Haas G.H. (1976) Interfacial enzyme kinetics of lipolysis. *Annual Review Biophys. Bioeng.* **5**, 77-117.

22. Verger R. (1980) Enzyme kinetics of lipolysis. *Methods Enzymol.* **64**, 340-392.

23. Verger R. and Pattus F. (1982) Lipid-protein interactions in monolayers. *Chem. Phys. Lipids* **30**, 189-227.

24. Ransac S., Moreau H., Rivière C. and Verger R. (1991) Monolayer techniques for studying phospholipase kinetics. *Methods Enzymol.* **197**, 49-65.

25. Piéroni G., Gargouri Y., Sarda L. and Verger R. (1990) Interactions of lipases with lipid monolayers. Facts and questions. *Adv. Colloid Interface Sci.* **32**, 341-378.

26. Verger R., Ferrato F., Mansbach C.M. and Piéroni G. (1982) Novel intestinal phospholipase A_2: Purification and some molecular characteristics. *Biochemistry* **21**, 6883-6889.

27. Zografi G., Verger R. and de Haas G.H. (1971) Kinetic analysis of the hydrolysis of lecithin monolayers by phospholipase A. *Chem. Phys. Lipids* **7**, 185-206.

28. Verger R. and de Haas G.H. (1973) Enzyme reactions in a membrane model. 1: A new technique to study enzyme reactions in monolayers. *Chem. Phys. Lipids* **10**, 127-136.

29. Piéroni G. and Verger R. (1979) Hydrolysis of mixed monomolecular films of triglyceride/lecithin by pancreatic lipase. *J. Biol. Chem.* **254**, 10090-10094.

30. Hughes A. (1935) The action of snake venoms on surface films. *Biochem. J.* **29**, 437-444.

31. Smaby J.M., Muderhwa J.M. and Brockman H.L. (1994) Is lateral phase separation required for fatty acid to stimulate lipases in a phosphatidylcholine interface? *Biochemistry* **33**, 1915-1922.

32. Muderhwa J.M. and Brockman H.L. (1992) Lateral lipid distribution is a major regulator of lipase activity. Implications for lipid-mediated signal transduction. *J. Biol. Chem.* **267**, 24184-24192.

33. Peters G.H., Toxvaerd S., Larsen N.B., Bjørnholm T., Schaumburg K. and Kjaer K. (1995) Structure and dynamics of lipid monolayers: implications for enzyme catalysed lipolysis. *Structural Biology* **2**, 395-401.

34. Esposito S., Sémériva M. and Desnuelle P. (1973) Effect of surface pressure on the hydrolysis of ester monolayers by pancreatic lipase. *Biochim. Biophys. Acta* **302**, 293-304.

35. Pattus F., Slotboom A.J. and de Haas G.H. (1979) Regulation of Phospholipase A_2 Activity by the Lipid -Water Interface: a Monolayer Approach. *Biochemistry* **13**, 2691-2697.

36. Rietsch J., Pattus F., Desnuelle P. and Verger R. (1977) Further studies of mode of action of lipolytic enzymes. *J. Biol. Chem.* **252**, 4313-4318.

37. Bhat S.G. and Brockman H.L. (1981) Enzymatic Synthesis/Hydrolysis of Cholesteryl Oleate in Surface Films. *J. Biol. Chem.* **256**, 3017-3023.

38. Momsen W.E. and Brockman H.L. (1981) The adsorption to and hydrolysis of 1,3-didecanoyl glycerol monolayers by pancreatic lipase. Effect of substrate packing density. *J. Biol. Chem.* **256**, 6913-6916.

39. Aoubala M., Ivanova M., Douchet I., de Caro A. and Verger R. (1995) Interfacial binding of Human Gastric Lipase to lipid monolayers, measured with an ELISA. *Biochemistry* **accepted**,

40. Gargouri Y., Moreau H., Piéroni G. and Verger R. (1989) Role of a sulfhydryl group in gastric lipases. A binding study using the monomolecular-film technique. *Eur. J. Biochem.* **180**, 367-371.

41. Guesdon J.L., Térnynck T. and Avrameas S. (1979) *J. Histoch. Cytochem.* **8**, 1131-1139.

42. Aoubala M., Douchet I., Laugier R., Hirn M., Verger R. and de Caro A. (1993) Purification of human gastric lipase by immunoaffinity and quantification of this enzyme in the duodenal contents using a new ELISA procedure. *Biochim. Biophys. Acta* **1169**, 183-188.

43. Laurent S., Ivanova M.G., Pioch D., Graille J. and Verger R. (1994) Interactions between β-Cyclodextrin and insoluble glyceride monomolecular films at the argon/water interface: application to lipase kinetics. *Chem. Phys. Lipids* **70**, 35-42.

44. Brockerhoff H. and Jensen R.G. (1974) *Lipolytic Enzymes* . (Academic Press, New York).

45. Borgström B. and Brockman H.L. (1984) *Lipases* . (Elsevier, Amsterdam).

46. Chen C.-S. and Sih C.J. (1989) *Angew. Chem. Int. Engl.* **28**, 695-707.

47. Morley N., Kuksis A. and Buchnea D. (1974) Hydrolysis of synthetic triacylglycerols by pancreatic and lipoprotein lipase. *Lipids* **9**, 481-488.

48. Akesson B., Gronowitz S., Herslöf B., Michelsen P. and Olivecrona T. (1983) Stereospecificity of different lipases. *Lipids* **18**, 313-318.

49. Jensen R.G., Galluzzo D.R. and Bush V.J. (1990) Selectivity is an important characteristic of lipase (acylglycerol hydrolases). *Biocatalysis* **3**, 307-316.

50. Rogalska E., Ransac S. and Verger R. (1990) Stereoselectivity of lipases. II. Stereoselective hydrolysis of triglycerides by gastric and pancreatic lipases. *J. Biol. Chem.* **265**, 20271-20276.

51. Rogalska E., Ransac S. and Verger R. (1993) Controlling lipase stereoselectivity *via* the surface pressure. *J. Biol. Chem.* **268**, 792-794.

52. Hult K. and Norin T. (1992) Enantioselectivity of some lipases : control and prediction. *Pure and Appl. Chem.* **64**, 1129-1134.

53. Holmquist M., Martinelle M., Berglund P., Clausen I.G., Patkar S., Svendsen A. and Hult K. (1993) Lipases from *Rhizomucor miehei* and *Humicola lanuginosa*. Modification of the lid covering the active site alters enantioselectivity. *J. Protein Chem.* **12**, 749-757.

54. Rogalska E., Cudrey C., Ferrato F. and Verger R. (1993) Stereoselective hydrolysis of triglycerides by animal and microbial lipases. *Chirality* **5**, 24-30.

55. Cernia E., Delfini M., Magri A.D. and Palocci C. (1994) Enzymatic catalysis by lipase from *Candida cylindracea*, enantiomeric activity evaluation by H^1 and C^{13} NMR. *Cell. Mol. Biol.* **40**, 193-199.

56. Holmberg E. and Hult K. (1991) Temperature as an enantioselective parameter in enzymic resolutions of racemic mixtures. *Biotechnol. Lett.* **13**, 323-326.

57. Lam L.K., Hui R.A.H.F. and Jones J.B. (1986) Enzymes in Organic Synthesis. 35. Stereoselective Pig Liver Esterase Catalysed Hydrolyses of 3-Substituted Glutarate Diesters. Optimization of Enantiomeric Excess Via Reaction Conditions Control. *J. Org. Chem.* **51**,

58. Parida S. and Dordick J.S. (1991) Substrate Structure and Solvent Hydrophobicity Control Lipase Catalysis and Enantioselectivity in Organic Media. *J. Am. Chem. Soc.* **113**, 2253-2259.

59. Wu S.H., Guo Z.W. and Sih C.J. (1990) Enhancing the enantioselectivity of Candida lipase catalyzed ester hydrolysis via noncovalent enzyme modification. *J. Am. Chem. Soc.* **112**, 1990-1995.

60. Matori M., Asahara T. and Ota Y. (1991) Reaction conditions influencing positional specificity index (PSI) of microbial lipases. *J. Ferment. Bioeng.* **72**, 413-415.

61. Makamura K., Takebe Y., Kitayama T. and Ohno A. (1991) Effect of solvent structure on enantioselectivity of lipase catalyzed transesterification. *Tetrahedron Lett.* **32**, 4941-4944.

62. Kamat S.V., Beckman E.J. and Russell A.J. (1993) Control of enzyme enantioselectivity with pressure changes in supercritical fluoroform. *J Am Chem Soc* **115**, 8845-8846.

63. Rogalska E., Nury S., Douchet I. and Verger R. (1995) Lipase stereo- and regioselectivity towards three isomers of dicaprin, a kinetic study by the monomolecular film technique. *Chirality* **accepted**,

64. Böhm C., Möhwald M., Leiserowitz L., Als-Nielsen J. and Kjaer K. (1993) Influence of chirality on the structure of phospholipid monolayers. *Biophys. J.* **64**, 553-559.

65. Andelman D. and Orland H. (1993) Chiral discrimination in solutions and in langmuir monolayers. *J. Am. Chem. Soc.* **115**, 12322-12329.

66. Harvey N.G., Mirajovsky D., Rose P.L., Verbiar R. and Arnett E.M. (1989) Molecular recognition in chiral monolayers of stearoylserine methyl ester. *J. Am. Chem. Soc.* **111**, 1115-1122.

67. Dvolaitzky M. and Guedeau-Boudeville M.-A. (1989) Chiral discrimination in the monolayer packing of hexadecylthiophospho-2-phenylglycinol with two chiral centers in the polar head group. *Langmuir* **5**, 1200-1205.

68. Landau E.M., Levanon L., Leiserowitz L., Lahav M. and Sagiv J. (1985) Transfer of structural information from Langmuir monolayers of three dimensional growing crystals. *Nature* **318**, 353-356.

69. Scow R.O., Blanchette-Mackie E.J. and Smith L.C. (1976) Role of capilary endothelium in the clearance of chylomicrons. A model for lipid transport from blood by lateral diffusion in cell membranes. *Circ. Res.* **39**, 149-163.

70. Scow R.O., Desnuelle P. and Verger R. (1979) Lipolysis and lipid movement in a membrane model. Action of lipoprotein lipase. *J. Biol. Chem.* **254**, 6456-6463.

71. Ransac S., Rivière C., Gancet C., Verger R. and de Haas G.H. (1990) Competitive inhibition of lipolytic enzymes. I. A kinetic model applicable to water-insoluble competitive inhibitors. *Biochim. Biophys. Acta* **1043**, 57-66.

72. de Haas G., van Oort M., Dijkman R. and Verger R. (1989) Phospholipase A_2 inhibitors: Monoacyl, monoacylamino-glycero-phosphocholines. *Biochemical Society Transactions* **625**, 274-276.

73. Jain M.K. and Berg O.G. (1989) The kinetics of interfacial catalysis by phospholipase A2 and regulation of interfacial activation: hopping versus scooting. *Biochim. Biophys. Acta* **1002**, 127-156.

74. Gargouri Y., Moreau H., Piéroni G. and Verger R. (1988) Human gastric lipase: A sulfhydryl enzyme. *J. Biol. Chem.* **263**, 2159-2162.

75. Moreau H., Gargouri Y., Piéroni G. and Verger R. (1988) Importance of sulfhydryl group for rabbit gastric lipase activity. *FEBS Lett.* **236**, 383-387.

76. Hadvàry P., Lengsfeld H. and Wolfer H. (1988) Inhibition of pancreatic lipase *in vitro* by the covalent inhibitor tetrahydrolipstatin. *Biochem. J.* **256**, 357-361.

77. Borgström B. (1988) Mode of action of tetrahydrolipstatin: A derivative of the naturally occuring lipase inhibitor lipstatin. *Biochim. Biophys. Acta* **962**, 308-316.

78. Deems R.A., Eaton B.R. and Dennis E.A. (1975) Kinetic analysis of phospholipase A2 activity toward mixed micelles and its implications for the study of lipolytic enzymes. *J. Biol. Chem* **250**, 9013-9020.

79. Kurganov B.I., Tsetlin L.G., Malakhova E.A., Chebotareva N.A., Lankin V.Z., Glebova G.D., Berezovsky V.M., Levashov A.V. and Martinek K. (1985) A novel approach to study of action of water-insoluble inhibitors of enzyme reactions. *J. Biochem. Biophys. Methods* **11**, 117-184.

80. Ransac S., Gargouri Y., Moreau H. and Verger R. (1991) Inactivation of pancreatic and gastric lipases by tetrahydrolipstatin and alkyl-dithio-5-(2-nitrobenzoic acid). A kinetic study with 1,2-didecanoyl-sn-glycerol monolayers. *Eur. J. Biochem.* **202**, 395-400.

81. Marguet F., Douchet I., Verger R. and Buono G. (1995) Molecular or interfacial chiral recognition by digestive lipases of chiral organophosphorus triglycerides analogs? *J. Am. Chem. Soc.* **Submited,**

82. Melo E.P., Ivanova M.G., Aires-Barros M.R., Cabral J.M.S. and Verger R. (1995) Glyceride synthesis catalyzed by cutinase using the monomolecular film technique. *Biochemistry* **34**, 1615-1621.

83. Lauwereys M., de Geus P., de Meutter J., Stanssens P. and Matthyssens G. (1991) Cloning, expression and characterization of cutinase, a fungal lipolytic enzyme. *GBF monogr.* **16**, 243-251.

84. Brockman H.L. (1981) Triglyceride lipase from porcine pancreas. *Methods Enzymol.* **71**, 619-627.

85. Jensen R.G., Dejong F.A. and Clark R.M. (1983) Determination of lipase specificity. *Lipids* **18**, 239-252.

86. Verger R. (1984) Pancreatic lipases *In Lipases* (eds. Borgström B. and Brockman H.L.) 83-149 (Elsevier, Amsterdam).

87. Dervichian D.G. (1971) Méthode d'étude des réactions enzymatiques sur une interface. *Biochimie* **53**, 25-34.

88. Nury S., Piéroni G., Rivière C., Gargouri Y., Bois A. and Verger R. (1987) Lipase kinetics at the triacylglycerol-water interface using surface tension measurements. *Chem. Phys. Lipids* **45**, 27-37.

89. Ferguson A. (1911) Photographic measurements of pendent drops. *Phil. Mag. S.* **23**, 417-431.

90. Andreas J.M., Hauser E.A. and Tucker W.B. (1938) Boundary tension by pendant drop. *J. Phys. Chem.* **42**, 1001-1019.

91. Lin M. (1981) Transition de phase d'alcools aliphatiques à l'interface liquide/liquide sous différentes pressions hydrostatiques. Thèse de Doctorat d'Etat, University de Provence.

92. Girault H.H.J., Schiffrin D.J. and Smith B.D.V. (1982) Drop Image Processing for Surface and Interfacial Tension Measurments. *J. Electroanal. Chem.* **137**, 207-217.

93. Girault H.H.J., Schiffrin D.J. and Smith B.D.V. (1984) The Measurment of Interfacial Tension of Pendant Drop Using a Video Image Profile Digitizer. *J. Coll. Interface Sci.* **101**, 257-266.

94. Anastasiadis S.H., Chen J.K., Koberstein J.T., Siegel A.F., Sohn J.E. and Emerson J.A. (1987) Determination of Interfacial Tension by Video Image Processing of Pendant Fluid Drop. *J. Coll. Interface. Sci.* **119**, 55-66.

95. Cheng P., Li D., Boruvka L., Rotenberg Y. and Neumann A.W. (1990) Automation of Axis symmetric drop shape analysis for measurments of interfacial tensions and contact angles. *Colloids and Surfaces* **43**, 151-167.

96. Pallas N.R. and Harisson Y. (1990) An automated drop shape apparatus and the surface tension of pure water. *Colloids and Surfaces* **43**, 169-194.

97. Satherley J., Girault H.H.J. and Schiffrin D.J. (1989) The Measurment of Ultralow Interfacial Tension by Video Image Digital Techniques. *J. Coll. Interface Sci.* **136**, 574-580.

98. Nury S., Gaudry-Rolland N., Rivière C., Gargouri Y., Bois A., Lin M., Grimaldi M., Richou J. and Verger R. (1991) Lipase kinetics at the triacylglycerol-water interface. *GBF Monogr.* **16**, 123-127.

99. Grimaldi M., Bois A., Nury S., Rivière C., Verger R. and Richou J. (1991) Analyse de la forme du profil d'une goutte pendante par traitement d'images numériques. (Mesure en temps réel de la tension interfaciale). *Opto.* **91**, 104-110.

100. Cagna A., Esposito G., Rivière C., Housset S. and Verger R. (1992) 33rd International Conference on the Biochemistry of Lipids, Lyon, France.

101. Labourdenne S., Gaudry-Rolland N., Letellier S., Lin M., Cagna A., Esposito G., Verger R. and Rivière C. (1994) The oil-drop tensiometer: potential applications for studying the kinetics of (phospho)lipase action. *Chem. Phys. Lipids.* **71**, 163-173.

102. Verger R., Mieras M.C.E. and de Haas G.H. (1973) Action of phospholipase A at interfaces. *J. Biol. Chem.* **248**, 4023-4034.

103. Ransac S. (1991) Modulation des activités (phospho)lipasiques. Mise en œuvre de la technique des films monomoléculaire pour l'étude d'inhibiteur spécifiques and la détermination de la stéréo-sélectivité des enzymes lipolytiques. Thesis, University of Aix-Marseille II.

104. Ransac S., Rogalska E., Gargouri Y., Deveer A.M.T.J., Paltauf F., de Haas G.H. and Verger R. (1990) Stereoselectivity of lipases. I. Hydrolysis of enantiomeric glyceride analogues by gastric and pancreatic lipases. A kinetic study using the monomolecular film technique. *J. Biol. Chem.* **265**, 20263-20270.

105. Kelvin W.T. (1904) Baltimore lectures on molecular dynamics and the wave theory of light *In* (eds. Clay C.J.) 618-619 London).

106. Alworth W.A. (1972) *Stereochemistry and its application in biochemistry* . (Wiley-Interscience, New York).

107. Dixon M. and Webb E.C. (1964) *Enzymes* . (Longmans, London).

108. Fersht A. (1985) *Enzyme structure and mechanism* . (Freeman W.H. and Company, New York).

109. Ransac S., Rogalska E., Gargouri Y., Deveer A.M.T.J., Paltauf F., Gancet C., Dijkman R., De Haas G.H. and Verger R. (1991) Stereoselectivity of lipases. Hydrolysis of enantiomeric glyceride analogs by gastric and pancreatic lipases, a kinetic study using the monomolecular film technique. *GBF Monogr.* **16**, 117-122.

110. Rogalska E., Ransac S., Douchet I. and Verger R. (1994) Lipase stereoselectivity depends on the "interfacial quality". Closing meeting of the BRIDGE lipase T-project, International Workshop - Bendor Island, Bandol, France.

111. Brzozowski A.M., Derewenda U., Derewenda Z.S., Dodson G.G., Lawson D.M., Turkenburg J.P., Bjorkling F., Huge-Jensen B., Patkar S.A. and Thim L. (1991) A model for interfacial activation in lipases from the structure of a fungal lipase-inhibitor complex. *Nature* **351**, 491-494.

112. Schrag J.D., Li Y., Wu S. and Cygler M. (1991) Ser-His-Glu triad forms the catalytic site of the lipase from *Geotrichum candidum. Nature* **351**, 761-764.

113. Derewenda Z.S., Derewenda U. and Dodson G.G. (1992) The crystal and molecular structure of the Rhizomucor miehei triacylglyceride lipase at 1.9 Å resolution. *J. Mol. Biol.* **227**, 818-839.

114. Martinez C., de Geus P., Lauwereys M., Matthyssens G. and Cambillau C. (1992) *Fusarium solani* cutinase is a lipolytic enzyme with a catalytic serine accessible to solvent. *Nature* **356**, 615-618.

115. van Tilbeurgh H., Sarda L., Verger R. and Cambillau C. (1992) Structure of the pancreatic lipase-procolipase complex. *Nature* **359**, 159-162.

116. Cygler M., Schrag J.D. and Ergan F. (1992) Advances in structural understanding of lipases. *Biotechnol. Genet. Eng. Rev.* **10**, 143-184.

117. Dodson G.G., Lawson D.M. and Winkler F.K. (1992) Structural and evolutionary relationships in lipase mechanism and activation. *Faraday Discuss.* **93**, 95-105.

118. Ollis D.L., Cheah E., Cygler M., Dijkstra B., Frolow F., Franken S.M., Harel M., Remington S.J., Silman I., Schrag J., Sussman J.L., Verschueren K.H.G. and Goldman A. (1992) The α/β hydrolase fold. *Protein Eng.* **5**, 197-211.

119. Cygler M., Schrag J.D., Sussman J.L., Harel M., Silman I., Gentry M.K. and Doctor B.P. (1993) Relationship between sequence conservation and three-dimensional structure in a large family of esterases, lipases, and related proteins. *Protein Sci.* **2**, 366-382.

120. van Tilbeurgh H., Egloff M.-P., Martinez C., Rugani N., Verger R. and Cambillau C. (1993) Interfacial activation of the lipase-procolipase complex by mixed micelles revealed by X-Ray crystallography. *Nature* **362**, 814-820.

121. Grochulski P., Bouthillier F., Kazlauskas R.J., Serreqi A.N., Schrag J.D., Ziomek E. and Cygler M. (1994) Analogs of reaction intermediates identify a unique substrate binding site in *Candida rugosa* lipase. *Biochemistry* **33**, 3494-3500.

122. Grochulski P., Li Y., Schrag J.D. and Cygler M. (1994) Two conformational states of *Candida rugosa* lipase. *Protein Sci.* **3**, 82-91.

123. Uppenberg J., Hansen M.T., Patkarr S. and Jones T.A. (1994) The sequence, crystal structure determination and refinement of two crystal forms of Lipase-B from *Candida antarctica* (Vol 2, pg 293, 1994). *Structure* **2**, 453-454.

124. Derewenda U., Swenson L., Wei Y.Y., Green R., Kobos P.M., Joerger R., Haas M.J. and Derewenda Z.S. (1994) Conformational lability of lipases observed in the absence of an Oil-Water interface. Crystallographic studies of enzymes from the fungi *Humicola lanuginosa* and *Rhizopus delemar. J. Lipid Res.* **35**, 524-534.

125. Rubin B. (1994) Grease pit chemistry exposed. *Nature Struct. Biology* **1**, 568-572.

126. Cygler M., Grochulski P., Kazlauskas R.J., Schrag J.D., Bouthillier F., Rubin B., Serreqi A.N. and Gupta A.K. (1994) A structural basis for the chiral preferences of lipases. *J. Am. Chem. Soc.* **116**, 3180-3186.

127. Egloff M.-P., Marguet F., Buono G., Verger R., Cambillau C. and van Tilbeurgh H. (1995) The 2.46 Å resolution structure of the pancreatic lipase-colipase complex inhibited by a C11 alkyl phosphonate. *Biochemistry* **34**, 2751-2762.

128. Grochulski P., Li Y., Schrag J.D., Bouthillier F., Smith P., Harrison D., Rubin B. and Cygler M. (1993) Insights into interfacial activation from an open structure of *Candida rugosa* lipase. *J. Biol. Chem.* **268**, 12843-12847.

129. Carrière F., Barrowman J.A., Verger R. and Laugier R. (1993) Secretion and contribution to lipolysis of gastric and pancreatic lipases during a test meal in humans. *Gastroenterology* **105**, 876-888.

130. Scheele G. and Kern H. (1986) The exocrine pancreas. *Molecular and cellular basis of digestion.* (eds. Desnuelle P.) 173-192 (Elsevier, Amsterdam).

131. de Caro J., Boudouard M., Bonicel J., Guidoni A., Desnuelle P. and Rovery M. (1981) Porcine pancreatic lipase. Completion of the primary structure. *Biochim. Biophys. Acta* **671**, 129-138.

132. Lowe M.E., Rosemblum J.L. and Strauss A.W. (1989) Cloning and characterization of human pancreatic lipase cDNA. *J. Biol. Chem.* **264**, 20042-20048.

133. Borgström B. and Erlanson-Albertsson C. (1973) Pancreatic lipase and colipase. Interactions and effects of bile salts and other detergents. *Eur. J. Biochem.* **37**, 60-68.

134. Maylié M.F., Charles M., Gache C. and Desnuelle P. (1971) Isolation and partial identification of a pancreatic colipase. *Biochim. Biophys. Acta* **229**, 286-289.

135. Borgström B. and Erlanson-Albertson C. (1971) Pancreatic juice colipase: Physiological importance. *Biochim. Biophys. Acta* **242**, 509-513.

136. Gargouri Y., Bensalah A., Douchet I. and Verger R. (1995) Kinetic behaviour of five pancreatic lipases using emulsion and monomolecular films of synthetic glycerides. *Biochim. Biophys. Acta* **accepted**,

137. Durand S., Clemente F., Thouvenot J.P., Fauvel-Marmouyet J. and Douste-Blazy L. (1978) A lipase with high phospholipase activity in guinea pig pancreatic juice. *Biochimie* **60**, 1215-1217.

138. Fauvel J., Bonnefis M.J., Sarda L., Chap H., Thouvenot J.P. and Douste-Blazy L. (1981) Purification of two lipases with high phospholipase A1 activity from guinea-pig pancreas. *Biochim. Biophys. Acta* **663**, 446-456.

139. Fauvel J., Bonnefis M.J., Chap H., Thouvenot J.P. and Douste-Blazy L. (1981) Evidence for the lack of classical secretory phospholipase A2 in guinea-pig pancreas. *Biochim. Biophys. Acta* **666**, 72-79.

140. Bousset-Risso M., Bonicel J. and Rovery M. (1985) Limited proteolysis of porcine pancreatic lipase. *FEBS Lett.* **182**, 323-326.

141. Kirchgessner T.G., Chuat J.C., Heinzmann C., Etienne J., Guilhot S., Svenson K., Ameis D., Pilon C., d'Auriol L., Andalibi A., Schotz M.C., Galibert F. and Lusis A.J. (1989) Organization of the human lipoprotein lipase gene and evolution of the lipase gene family. *Proc. Natl. Acad. Sci. USA* **86**, 9647-9651.

142. Derewenda Z.S. and Cambillau C. (1991) Effects of gene mutations in lipoprotein and hepatic lipases as interpreted by a molecular model of the pancreatic triglyceride lipase. *J. Biol. Chem.* **266**, 23112-23119.

143. van Tilbeurgh H., Roussel A., Lalouel J.M. and Cambillau C. (1994) Lipoprotein lipase. Molecular model based on the pancreatic lipase X-Ray structure: consequences for heparin binding and catalysis. *J. Biol. Chem.* **269**, 4626-4633.

144. Hide W.A., Chan L. and Li W.H. (1992) Structure and evolution of the lipase superfamily. *J. Lipid Res.* **33**, 167-178.

145. Persson B., Bentsson-Olivecrona G., Enerbäck S., Olivecrona T. and Jörnvall H. (1989) Srtuctural features of lipoprotein lipase. Lipase family relationships, binding interactions, non-equivalence of lipase cofactors, vitellogenin similarities and functional subdivision of lipoprotein lipase. *Eur. J. Biochem.* **179**, 39-45.

146. Blow D.M. (1971) The structure of chymotrypsin *The enzymes* (eds. Boyer P.D.) 185-212

147. Derewenda Z.S. and Sharp A.M. (1993) News from the interface: the molecular structures of triacylglyceride lipases. *Trends Biochem. Sci.* **18**, 20-25.

148. Guidoni A., Benkouka F., de Caro J. and Rovery M. (1981) Characterization of the serine reacting with diethyl *p*-nitrophenylphosphate in porcine pancreatic lipase. *Biochim. Biophys. Acta* **660**, 148-150.

149. Chapus C. and Sérémiva M. (1976) Mechanism of pancreatic lipase action. 2. Catalytic properties of modified lipases. *Biochemistry* **15**, 4988-4991.

150. Lüthi-Peng Q. and Winkler F.K. (1992) Large spectral changes accompany the conformational transition of human pancreatic lipase induced by acylation with the inhibitor tetrahydrolipstatin. *Eur. J. Biochem.* **205**, 383-390.

151. Grusby M.J., Nabavi N., Wong H., Dick R.F., Bluestone J.A., Schotz M.C. and Glimcher L.H. (1990) Cloning of an interleukin-4 inducible gene from cytotoxic T lymphocytes and its identification as a lipase. *Cell* **60**, 451-459.

152. Wishart M.J., Andrews P.C., Nichols R., Blevins G.T. Jr., Logsdon C.D. and Williams J.A. (1993) Identification and cloning of GP-3 from rat pancreatic acinar zymogen granules as a glycosylated membrane-associated lipase. *J. Biol. Chem.* **268**, 10303-10311.

153. Carrière F., Thirstrup K., Boel E., Verger R. and Thim L. (1994) Structure-function relationships in naturally occurring mutants of pancreatic lipase. *Protein Eng.* **7**, 563-569.

154. Verger R., Rietsch J. and Desnuelle P. (1977) Effects of colipase on hydrolysis of monomolecular films by lipase. *J. Biol. Chem.* **252**, 4319-4325.

Phospholipases A_2 and the Production of Bioactive Lipids

Henk van den Bosch, Casper Schalkwijk, Margriet J.B.M. Vervoordeldonk, Arie J. Verkleij*
and Johannes Boonstra*
Centre for Biomembranes and Lipid Enzymology
Institute for Biomembranes
Utrecht University
Padualaan 8
3584 CH Utrecht
The Netherlands

Phospholipases A_2 and their regulation have received much attention during the last decades because they have been implicated in the release of precursors for bioactive lipid production from structural membrane phospholipids. As such they are considered to be the rate limiting enzymes in the pathways leading to the formation of various eicosanoids and platelet-activating factor from released arachidonate and lyso-platelet-activating factor, respectively. These bioactive lipids are usually secreted from cells and activate specific receptors on neighbouring cells (Negishi *et al.*, 1993). As such they serve in many physiological processes and after excessive formation also in pathophysiological processes such as inflammation, fever, pain and allergic conditions. Hence, not only lipid mediators but also phospholipase A_2 has been implicated in inflammation (Pruzanski and Vadas, 1991). Arguments in favour of this view are several fold. In the early stages of inflammatory conditions activated monocytes/macrophages secrete pro-inflammatory cytokines such as interleukin-1β (IL-1β) and tumor necrosis factor (TNF). These in turn activate phospholipases A_2 in many target cells and frequently lead to secretion of phospholipases A_2. In line with this notion, highly elevated levels of circulating phospholipase A_2 have been found under inflammatory conditions such as rheumatoid arthritis, peritonitis and septic shock (Pruzanski and Vadas, 1991). Recently, these observations have been extended to typhoid infections (Keuter *et al.*, 1995), meningitis and acute phases of sickle cell anaemia (unpublished observations).

In this paper we summarize our studies on the long-term regulation of phospholipases A_2 by IL-1β and TNF in rat mesangial cells. This involves the identification of the secreted enzyme

---

* Department of Molecular Cell Biology, Institute for Biomembranes, Utrecht University, Padualaan 8, 3584 CH Utrecht

NATO ASI Series, Vol. H 96
Molecular Dynamics of Biomembranes
Edited by Jos A. F. Op den Kamp

in response to these cytokines, that induce *de novo* synthesis of mainly a 14 kDa phospholipase A_2. We then describe the suppressing effects of anti-inflammatory agents such as corticosteroids and transforming growth factor-β2 (TGF-β2) on this induced phospholipase A_2 synthesis in relation to their effects on prostaglandin synthesis. These results have been published and have recently been reviewed (van den Bosch *et al.*, 1994, Vervoordeldonk *et al.*, 1995). Therefore, they will be only briefly described here without showing the data on which the statements are based. The reader is referred to these previous reviews for this. Thereafter, new data concerning the short-term regulation of an 85 kDa high molecular weight or cytosolic phospholipase A_2 ($cPLA_2$) in fibroblasts will be presented.

Cytokine induction of 14 kDa group II phospholipase A_2 ($sPLA_2$)

Treatment of rat glomerular mesangial cells with IL-1β leads, after a lag period of 6–8 hours, to secretion of a PLA_2 activity into the culture medium in parallel to enhanced prostaglandin E_2 (PGE_2) production. Gel filtration indicated the PLA_2 to be of the 14 kDa class and no $cPLA_2$ activity appeared to be secreted. Using monoclonal antibodies that we had previously prepared with specificity for 14 kDa group II PLA_2, the secreted enzyme was identified to belong to this class (Schalkwijk *et al.*, 1991a). Immunoaffinity chromatography to purify the enzyme and sequence determination confirmed this identification (Schalkwijk *et al.*, 1992a). In rat platelets this enzyme was shown to be stored in α-granules by immunogold electronmicroscopy (Aarsman *et al.*, 1989) from where it could be released upon thrombin activation. This led us to investigate whether also in mesangial cells the enzyme was present in a pre-existing cellular pool from which it would be released upon IL-1β stimulation. This did not appear to be the case. The IL-1β induced secretion followed *de novo* synthesis of the enzyme, which hardly could be detected in non-stimulated cells (Schalkwijk *et al.*, 1991a). Double-label immunofluorescence experiments indicated that the enzyme that was not yet secreted was mainly present in the Golgi area and in punctate fluorescent structures throughout the cytoplasm, presumably representing secretory granules (Vervoordeldonk *et al.*, 1994). Using pulse-chase techniques the half-life of the enzyme in mesangial cells was found to be approximately 2 hours.

Suppression of $sPLA_2$ induction.

It has long been known that glucocorticosteroids inhibit prostaglandin formation (Flower, 1988) and they do so also in mesangial cells (Schalkwijk *et al.*, 1991b). Such inhibitions have previously been ascribed to corticosteroid-induced synthesis of PLA_2 inhibitory proteins called lipocortins or annexins (Flower, 1988). An alternative explanation was revealed when we found that the synthetic corticosteroid analog dexamethasone not only inhibited the IL-1β induced $sPLA_2$ activity but also the induced synthesis of $sPLA_2$ protein (Schalkwijk *et al*, 1991b). Subsequent experiments indicated that dexamethasone failed to affect annexin I levels in mesangial cells under the conditions where it attenuated induced $sPLA_2$ activity (Vervoordeldonk *et al.*, 1994). We concluded from these findings that the dexamethasone-induced inhibition of $sPLA_2$ activity was not mediated by annexin and was solely due to dexamethasone-induced suppression of $sPLA_2$ synthesis as induced by the pro-inflammatory cytokine IL-1β, thus providing a new mechanism for the molecular action of dexamethasone as an anti-inflammatory agent.

The availability of a cDNA for $sPLA_2$ (Van Schaik *et al.*, 1993) then allowed us to determine the site of action of dexamethasone by following $sPLA_2$ mRNA and protein levels. It could be shown that the IL-1β-induced $sPLA_2$ protein synthesis followed a many-fold increase in mRNA level. This mRNA level was not affected by pretreatment with dexamethasone, while $sPLA_2$ protein levels were strongly reduced to near control levels. This indicated that dexamethasone exerted its inhibition in this case at the post-transcriptional level. By contrast, when $sPLA_2$ activity was induced by the cAMP elevating agent forskolin, dexamethasone prevented accumulation of mRNA, indicative of transcriptional regulation (Vervoordeldonk *et al.*, 1995).

Similar effects as described in the previous paragraph were also obtained when the cells were pre-treated with TGF-β2, except that the inhibition of PGE_2 formation was not complete (Schalkwijk *et al.*, 1992, 1992b) and that both the IL-1β and the forskolin-induced $sPLA_2$ expression were attenuated at the transcriptional level. The observation that PGE_2 production was only partially inhibited could be explained by the finding that TGF-β2 increased the activity of $cPLA_2$ as was also found for IL-1β (Schalkwijk *et al.*, 1992b). The latter increase was also caused by *de novo* synthesis of $cPLA_2$ and this was also prevented by dexamethasone (Schalkwijk *et al.*, 1993).

The above results are schematically summarized in figure 1. TGF-β2 overrules the induction of $sPLA_2$ by IL-1β and completely suppresses $sPLA_2$ synthesis and secretion, thereby blocking the propagation of inflammatory reactions and the destructive action of this $sPLA_2$. The formation of PGE_2 is inhibited but not completely blocked, as in the case of dexamethasone, because under conditions where TGF-β2 completely prevents induced $sPLA_2$ synthesis it still causes enhanced PGE_2 formation when compared to control cells because TGF-β2 leads to increased $cPLA_2$ activity to the same extent as caused by IL-1β. The fact that IL-1β activates $cPLA_2$ to a comparable level as observed for TGF-β2, yet causes more PGE_2 formation, suggests that this is due to the fact that IL-1β additionally induces $sPLA_2$

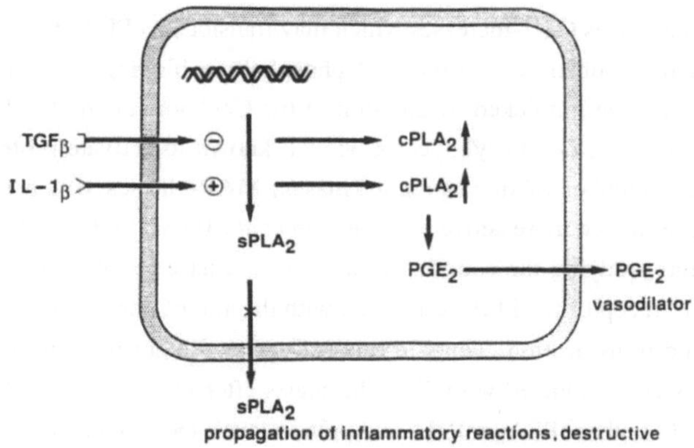

Figure 1. Schematic representation of cytokine effects on PLA_2 activities and PGE_2 formation in rat mesangial cells (Reproduced with permission from van den Bosch *et al.*, 1994.

which may contribute to arachidonate release for PGE_2 formation in an extracellular loop when this enzyme becomes activated by the mM concentration of extracellular Ca^{2+}. Additional evidence for the contribution of such an extracellular loop was obtained in experiments in which the cells were stimulated with IL-1β in the presence of either neutralizing anti-$sPLA_2$ antibodies or a selective inhibitor (Pfeilschifter *et al.*, 1993). This resulted in 50 to 70% decrease in PGE_2 formation. It is likely, however, that such contributions of $sPLA_2$ to arachidonate release for PGE_2 formation in the case of enzyme circulating in plasma do only take place with cells that in the course of inflammation become primed for an enhanced susceptibility to the action of this enzyme. It remains a major goal for future research to identify such priming mechanisms (Kudo *et al.*, 1993).

Short-term regulation of $cPLA_2$ activity

The regulatory aspects that were discussed above all constitute long-term effects involving induction of enzyme synthesis. There are a number of other, short-term, regulatory mechanisms as well participating in the complex and multifold regulation of PLA_2 activities as well. The latter involve mainly $cPLA_2$, which is regulated via receptor-mediated mechanisms. Activation of phospholipase C (PLC), either PLC-β via G-protein coupled or PLC-γ1 via tyrosine kinase, receptor activations causes increased formation of intracellular second messengers inositoltrisphosphate and diacylglycerol (Liscovitch, 1992). In turn, inositoltrisphosphate causes Ca^{2+}-increases which may translocate $cPLA_2$ from the cytosol to membranes to reach its substrates via its Ca^{2+}-phospholipid binding domain (Clark *et al.*, 1991). Such events can be mimicked by addition of the Ca^{2+}-ionophore A23187. The other product of PLC activation, *i.e.* diacylglycerol, is well-known to activate protein kinases C, and indirectly via a number of downstream kinases, MAP-kinases which phosphorylate $cPLA_2$ to give a several-fold more active enzyme (Lin *et al.*, 1993). In the next paragraphs we describe experiments studying the coupling of a tyrosine kinase receptor, *i.e.* the epidermal growth factor (EGF) receptor to $cPLA_2$ activation with the aim of unravelling the sequence of phosphorylation and translocation events in this activation. Rather than in mesangial cells, these experiments were conducted with 3T3 fibroblasts after we established that these cells express about 30-fold higher $cPLA_2$ mRNA and protein levels (Schalkwijk *et al.*, 1993).

EGF-induced activation of $cPLA_2$ in fibroblasts

Initial experiments showed that an NIH3T3 mouse fibroblast cell line, devoid of endogenous EGF-receptor and transfected with the cDNA for the human EGF-receptor, responded to EGF-treatment with an approximately 2.5-fold increase in $cPLA_2$ activity measured in cytosolic fractions prepared from these cells (figure 2) (Spaargaren *et al.*, 1992). Partially purified enzyme preparations obtained from EGF-treated cells retained the higher activities indicating a stable modification. However, treatment of such fractions with alkaline

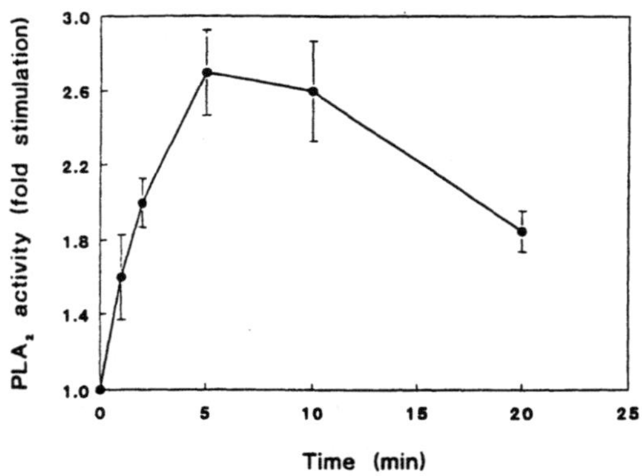

Figure 2. Time dependency of EGF-induced $cPLA_2$ activation.

phosphatase reduced the activity to control levels, indicating activation by phosphorylation. In view of the intrinsic tyrosine kinase activity of the EGF-receptor, we investigated the nature of this phosphorylation in more detail. In co-immunoprecipitation experiments we showed that $cPLA_2$ did not co-precipitate with the EGF-receptor, whereas such an association between the EGF-receptor and PLC-γ1, which becomes activated by tyrosine phosphorylation, could be demonstrated. This suggested that $cPLA_2$ did not become phosphorylated on tyrosine residues and this was confirmed by phosphoamino acid analysis of enzyme isolated from ^{32}P-labeled cells. Whereas increased $cPLA_2$ phosphorylation by EGF treatment was shown, this appeared to be solely on serine residues (Schalkwijk *et al.*, 1995). This indicates an indirect phosphorylation mechanism, presumable via EGF-receptor mediated activation of MAP-kinase. This is in line with the presently only identified activating phosphorylation of $cPLA_2$ via MAP-kinase on serine 505 of the enzyme (Lin *et al.*, 1993).

Induced phosphorylations represent one part of the EGF-signalling pathway, the other part being caused by the increases in intracellular Ca^{2+} concentrations. Increased Ca^{2+} concentrations following EGF treatment in the fibroblasts employed were confirmed using fluorescence indicators. To more directly determine the source of the Ca^{2+} required to activate $cPLA_2$ in the cells, the latter were prelabeled with ^3H-arachidonate and stimulated with EGF in media with and without extracellular Ca^{2+}. EGF induced only a substantial increase in ^3H-arachidonate release through $cPLA_2$ activation in media containing Ca^{2+}. This indicated that Ca^{2+} release from intracellular stores only marginally affected $cPLA_2$ activity and that the Ca^{2+} required for $cPLA_2$ activation had to come mainly from outside via entry through Ca^{2+} channels. In separate experiments it was shown that Ca^{2+} influx in itself,

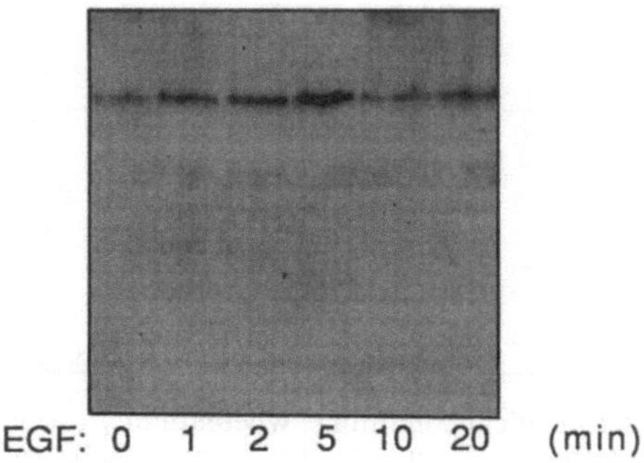

Figure 3. EGF-induced translocation of $cPLA_2$ to membranes.

realized by incubating the cells without EGF in the presence of Ca^{2+} ionophore and various buffered Ca^{2+} concentrations, was sufficient to induce ^{3}H-arachidonate release (Schalkwijk *et al.*, 1995).

It was then investigated whether the increase in intracellular Ca^{2+} concentrations following EGF exposure was indeed sufficient to affect $cPLA_2$ activation through translocation. To answer this question, cells were stimulated with EGF for various times. After washing, the cells were disrupted, membrane fractions were prepared and separated in SDS-PAGE whereafter blots were analyzed for $cPLA_2$ content by immunodetection. The data showed (figure 3) that the Ca^{2+} mobilization induced by EGF is sufficient to cause $cPLA_2$ translocation, which is optimal after about 5 min and thereafter declines. This is the same time scale as observed for $cPLA_2$ activation through phosphorylation (figure 2) as measured in the presence of a fixed added amount of Ca^{2+} in *in vitro* assays with cytosolic fractions.

Thus EGF leads to both phosphorylation and translocation of $cPLA_2$ with both phenomena being optimal after about 5 min and thus allowing no conclusions about their exact sequence at this stage.

Sequence of phosphorylation and translocation in $cPLA_2$ activation.

The question about the sequence of events leading to full $cPLA_2$ activation could be addressed by using fibroblasts expressing different numbers of EGF receptors. Table 1 shows the various cells used in these experiments. The cells contained identical amounts of $cPLA_2$, both as determined by activity measurements and Western blotting. As expected, this activity

Table 1: EGF-induced activation of $cPLA_2$ in fibroblast cell lines expressing different numbers of EGF-receptors*

Cell line	Receptor number	$cPLA_2$ activity	
3T3(0)	0	100 ± 12	(n=6)
HERC13	70000	196 ± 10	(n=5)
NEF	300000	202 ± 13	(n=5)
HER14	300000	201 ± 15	(n=6)

* The cells were treated with or without EGF under identical conditions, whereafter cytosolic fractions were prepared for $cPLA_2$ activity measurements. Results for EGF-treated cells are expressed as percentage (± S.D. for the indicated number of determinations) of the corresponding non-treated cells.

did not increase upon EGF treatment in cells devoid of EGF receptors. All other cell lines gave an about 2-fold increase in $cPLA_2$ activity, irrespective of whether they expressed 70000

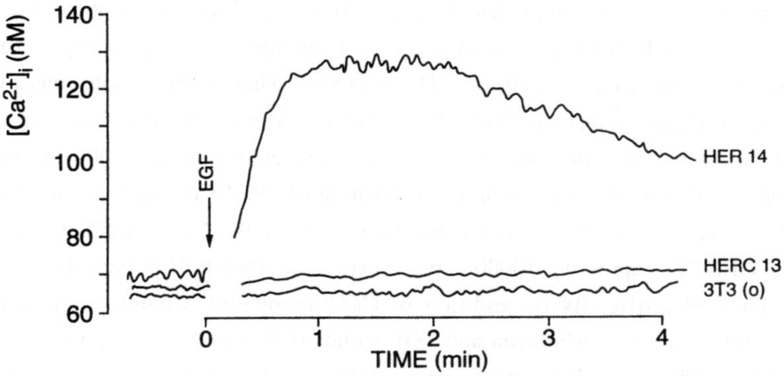

Figure 4. Changes in calcium levels induced by EGF in various fibroblasts. Similar data as indicated for HER14 cells were obtained for NEF cells (compare table 1 for cells).

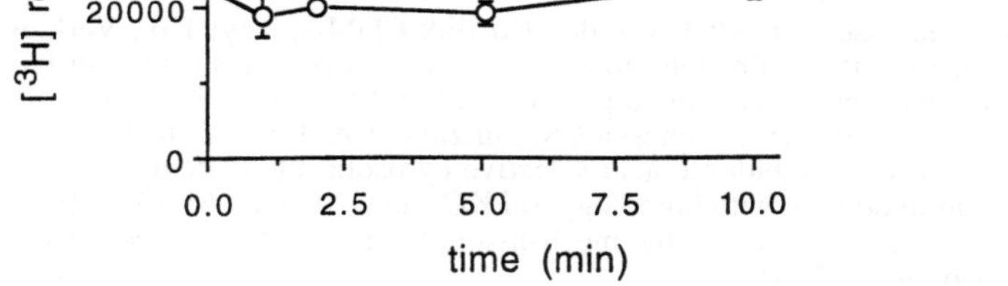

Figure 5. Arachidonate release from fibroblasts. Cells were stimulated with ionophore for 5 minutes followed by EGF for the indicated periods or with EGF followed by ionophore as shown.

or 300000 receptors. This indicates that the EGF-induced phosphorylation of $cPLA_2$ proceeded normally in all cell lines (Table 1). By contrast, EGF-induced increases in Ca^{2+}-levels (figure 4) were only observed in cells expressing 300000 copies of the receptor and were not seen in 3T3(0) cells expressing zero or HERC13 cells expressing 70000 copies. It follows from these observations that the cells expressing 70000 copies of the EGF receptor responded to EGF treatment with only half of the EGF signalling pathway, *i.e.* phosphorylation but no Ca^{2+} transients. This appeared to have considerable consequences for arachidonate release in response to EGF treatment. This was stimulated in cells expressing 300000 copies but remained unaffected in cells expressing either 70000 or zero copies. Obviously, both phosphorylation and translocation of $cPLA_2$ is required for EGF-induced arachidonate release and if there is no increase in intracellular Ca^{2+} for $cPLA_2$ translocation, as in the HERC13 expressing 70000 copies, there is no release. However, this translocation can be induced artificially by addition of Ca^{2+} ionophore. Combination of ionophore stimulation to induce translocation and EGF stimulation to induce phosphorylation in cells expressing 70000 copies then gave a clear insight into the sequence of events. As shown in figure 5, the result depended on the sequence of the additions. When the cells were first treated with ionophore, $cPLA_2$ becomes translocated to membranes and a subsequent time course of EGF addition showed no further increase in arachidonate release. By contrast, when the cells were pre-treated with EGF to induce $cPLA_2$ phosphorylation, a time-dependent increase after ionophore addition to translocate the enzyme was observed (figure 5). Thus, for an effective EGF response it appears that the phosphorylating activation of $cPLA_2$ has to proceed first in the cytosol whereafter the phosphorylated enzyme has to translocate to the membrane. Apparently, when the enzyme is first translocated it is no longer available for EGF-induced phosphorylation.

References

Aarsman AJ, Leunissen-Bijvelt J, van den Koedijk CDMA, Neys FW, Verkley AJ and van den Bosch H (1989) Phospholipase A_2 activity in platelets. Immuno-purification and localization of the enzyme in rat platelets. J. Lipid Mediators 1:49-61

Clark JD, Lin L-L, Kriz RW, Ramesha CS, Sultzman LA, Lin AY, Milona N and Knopf JL (1991) A novel arachidonic acid-selective cytosolic PLA_2 contains a Ca^{2+}-dependent translocation domain with homology to PKC and GAP. Cell 65:1043–1051

Flower RJ (1988) Lipocortin and the mechanism of action of the glucocorticosteroids. Br. J. Pharmacol. 94:987–1015

Keuter M, Dharmana D, Kullberg BJ, Schalkwijk C, Gasem MH, Seuren L, Djokomoeljanto R, Dolmans WMV, van den Bosch H and van der Meer JWM (1995) Phospholipase A_2 is a circulating mediator in Typhoid fever. J. Infect. Dis., in press

Kudo I, Murakami M, Hara S and Inoue K (1993) Mammalian non-pancreatic phospholipase A_2. Biochim. Biophys. Acta 1170:217–231

Lin L-L, Wartmann M, Lin AY, Knopf JL, Seth A and Davis RJ (1993) $cPLA_2$ is phosphorylated and activated by MAP kinase. Cell 72:269–278

Liscovitch M (1992) Crosstalk among multiple signal-activated phospholipases. Trends Biochem. Sci. 17:393–399

Negishi M, Sugimoto Y and Ichikawa A (1993) Prostanoid receptors and their biological actions. Prog. Lipid Res. 32:417–432

Pfeilschifter J, Schalkwijk C, Briner VA and van den Bosch H (1993) Cytokine-stimulated secretion of group II phospholipase A_2 by rat mesangial cells: Its contribution to arachidonic acid release and prostaglandin synthesis by cultured rat glomerular cells. J. Clin Invest. 92:2516-2523

Pruzanski W and Vadas P (1991) Phospholipase A_2 — a mediator between proximal and distal effectors of inflammation. Immunol. Today 12:143–146

Schalkwijk C, Pfeilschifter J, Märki F and van den Bosch H (1991a) Interleukin 1-β, tumor necrosis factor and forskolin stimulate the synthesis and secretion of group II phospholipase A_2 in rat mesangial cells. Biochem. Biophys. Res. Commun. 174:268-275

Schalkwijk CG, Vervoordeldonk M, Pfeilschifter J, Märki F and van den Bosch (1991b) Cytokine- and forskolin-induced synthesis of group II phospholipase A_2 and prostaglandin E_2 in rat mesangial cells is prevented by dexamethasone. Biochem. Biophys. Res. Commun. 180:46-52

Schalkwijk C, Pfeilschifter J, Märki F and van den Bosch H (1992a) Interleukin-1β- and forskolin-induced synthesis and secretion of group II phospholipase A_2 and prostaglandin E_2 in rat mesangial cells is prevented by transforming growth factor-β2. J. Biol. Chem. 267:8846-8851

Schalkwijk CG, de Vet E, Pfeilschifter J and van den Bosch H (1992b) Interleukin-1β and transforming growth factor β2 enhance cytosolic high molecular weight PLA_2 activity and induce prostaglandin E_2 formation in rat mesangial cells. Eur. J. Biochem. 210:169-176

Schalkwijk CG, Vervoordeldonk M, Pfeilschifter J and van den Bosch (1993) Interleukin-1β-induced cytosolic phospholipase A_2 activity and protein synthesis is blocked by dexamethasone in rat mesangial cells. FEBS Lett. 333:339-343

Schalkwijk CG, Spaargaren M, Defize LHK, Verkleij AJ, van den Bosch H and Boonstra J (1995) EGF induces serine phosphorylation, phosphorylation-dependent activation and calcium-dependent translocation of the cytosolic phospholipase A_2. Eur. J. Biochem., in press

Spaargaren M, Wissink S, Defize LHK, de Laat SW and Boonstra J (1992) Characterization and identification of an epidermal growth factor-activated phospholipase A_2. Biochem. J. 287:37–43

Van den Bosch H, Vervoordeldonk MJBM , Sanchéz RM, Pfeilschifter J and Schalkwijk CG (1994) Phospholipases A_2 and prostaglandin formation in rat glomerular mesangial cells. In: Mackness MI and Clerc M (eds.) Esterases, Lipases and Phospholipases. From Structure to Clinical Significance. NATO ASI Series Life Sciences 266:193–202, Plenum Press, New York

Van Schaik RHN, Verhoeven NM, Neys FW, Aarsman AJ and van den Bosch H (1993) Cloning of the cDNA coding for 14 kDa group II phospholipase A_2 from rat liver. Biochim. Biophys. Acta 1169:1-11

Vervoordeldonk MJBM, Schalkwijk CG, Vishwanath BS, Aarsman AJ and van den Bosch H (1994) Levels and localization of group II phospholipase A_2 and annexin I in interleukin- and dexamethasone-treated rat mesangial cells: evidence against annexin mediation of the dexamethasone-induced inhibition of group II phospholipase A_2. Biochim. Biophys. Acta 1224:541-550

Vervoordeldonk MJBM, Schalkwijk CG, Sanchéz RM, Pfeilschifter J and van den Bosch H (1995) Regulation of 14 kDa group II PLA_2 in rat mesangial cells. In: Packer L and Wirtz KWA (eds.) Signalling mechanisms from transcription factors to oxidative stress. NATO ASI Series Cell Biology 92:383-393, Springer Verlag, Heidelberg

PHOSPHOLIPASES IN THE YEAST SACCHAROMYCES CEREVISIAE

M. Fido, S. Wagner, H. Mayr, S.D. Kohlwein and F. Paltauf
Institut für Biochemie und Lebensmittelchemie
Technische Universität Graz
Petersgasse 12/II
8010 Graz
Austria

Phospholipases are ubiquitous enzymes capable of hydrolyzing acylester or phosphodiester bonds of glycerophospholipids. The physiological functions of phospholipases are diverse. Degradation of extracellular phospholipids by digestive phospholipases allows organisms to utilize phospholipid components, such as fatty acids, choline or inositol. Other phospholipases take part in intracellular signal transmission systems by catalyzing the controlled release of second messengers or their precursors. Phospholipases A allow the formation of the typical phospholipid molecular species pattern of membranes by a deacylation-reacylation cycle. Intracellular phospholipases participate in the disintegration of effete organelle membranes. It can be assumed that most of these functions of phospholipases must be fulfilled in *Saccharomyces cerevisiae*. All the more surprising is the paucity of information available on yeast phospholipases. On the other hand, studies on yeast phospholipases including the exploitation of the ease of genetic manipulation of this simple eukaryotic organism are expected to shed more light on the biological relevance of the different phospholipases.

Phospholipases A_1 act on the primary acylester group of diacylglycerophospholipids. They occur in a variety of tissues and subcellular structures (Waite, 1991). The occurrence of phospholipase A_1 activity in yeast has been postulated (Yost *et al.*, 1991), but the enzyme has not yet been characterized, nor has its physiological function been described. Results obtained from studies on the incorporation of ^{18}O into acyl ester groups of yeast glycerophospholipids are compatible with the participation of phospholipase A_1 in the generation of phospholipid molecular species (see below).

NATO ASI Series, Vol. H 96
Molecular Dynamics of Biomembranes
Edited by Jos A. F. Op den Kamp

In vertebrates phospholipases A_2 have been found in almost any tissue (Waite, 1991). They are most intensively studied because of their role in the release of arachidonic acid in the initial step of agonist-stimulated eicosanoid biosynthesis. This reaction, however, cannot be of significance in yeast which does not contain polyunsaturated fatty acids. On the other hand, participation of phospholipase A_2 in the generation of phospholipid molecular species as described for mammalian cells may well be of significance in yeast. Results from our own work indicate that such a reaction is an active process in yeast as described in detail below, together with the characterization of a mitochondria-associated phospholipase A_2.

True phospholipases B which remove both acyl chains from glycerophospholipids have only scarcely been identified in animal cells. Examples described so far include phospholipase B from guinea pig (Gassama-Diagne *et al.*, 1992) or rat and rabbit (Pind *et al.*, 1987; Pind *et al.*, 1989) small intestine. Phospholipases B have also been found in microorganisms, such as *Penicillium notatum* (Saito, 1991) or the yeasts, *S. cerevisiae* (Witt *et al.*, 1984) and *Torulaspora delbrueckii* (Watanabe *et al.*, 1994). Results obtained on phospholipase B from *S. cerevisiae* are described in detail below.

Phosphatidylinositolbisphosphate- specific phospholipase C plays a key role in cellular signalling. The structural gene encoding an enzyme with high homology to mammalian phospholipases C has been cloned from yeast (Yoko-o *et al.*, 1993; Payne *et al.*, 1993; Flick *et al.*, 1993). The gene product was found to be essential for cellular growth under certain conditions, such as elevated temperature, high osmolarity and certain nutrients.

Increasing attention is being paid to the role of phospholipases D in cellular signalling sequences. The existence of phospholipase D activity in yeast has been reported and a possible function has been ascribed to this enzyme in the disintegration of mitochondrial membranes during glucose repression (Grossman *et al.*, 1973) and in the onset of sporulation (Ella *et al.*, 1995). A more detailed characterization of yeast phospholipase D, including subcellular localization, substrate specificity and activation, is given below.

Phospholipase A_2

Previous work suggested phospholipase A_2 and A_1 activities in yeast mitochondria (Yost *et al.*, 1991). We found a membrane bound phospholipase A_2 with maximum enzymatic acitivity in the inner membrane and contact sites of mitochondria. No phospholipase A_2 activity could be detected in the cytosol or the mitochondrial matrix. The enzyme was partially purified from crude mitochondria by solubilization with detergent and a combination of chromatographic steps. The size of the enzyme was estimated by ultrafiltration to be around 30 kDa. During purification a 30 kDa protein (SDS-PAGE) was enriched. The enzyme showed, in accordance to many described phospholipases A_2 highest activity at pH 8.0 (Fig. 1) and was dependent on Ca^{2+} ions (Fig. 2). The partially purified enzyme showed similar specificity for phosphatidylcholine (PtdCho) and phosphatidylethanolamine (PtdEtn) while phosphatidylserine (PtdSer) was accepted less well.

In the yeast system phospholipase A_2 catalyzed arachidonate release cannot play a role because yeast lacks polyunsaturated fatty acids. However, yeast phospholipids undergo a rapid and selective acyl turnover, pointing to an extensive deacylation-reacylation process (Wagner *et al.*, 1994). Striking differences in the fatty acid distribution of phospholipids that are metabolically closely related (e.g. phosphatidylserine and phosphatidylethanolamine, phosphatidylethanolamine and phosphatidylcholine, and phosphatidylinositol and phosphatidylserine) suggest that pathways must exist for the generation of distinct phospholipid molecular species within the different phospholipid classes. The highly selective incorporation of exogenously supplied fatty acids (e.g. 90% for palmitic acid, 99% for oleic acid) into the *sn*-2 position of phosphatidylcholine, and the preferential incorporation of these fatty acids into the *sn*-2 position of phosphatidyl-ethanolamine (70% and 90%, respectively, for palmitic acid and oleic acid) are compatible with the postulate that phospholipase A_2-mediated deacylation followed by reacylation of the lysophospholipids is involved in the generation of phospholipid molecular species in yeast (Wagner *et al.*, 1994).

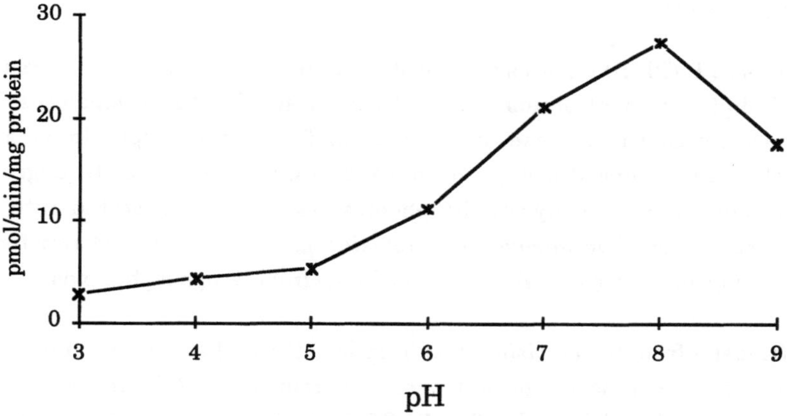

Fig. 1: pH-profile of yeast mitochondrial phospholipase A_2 activity

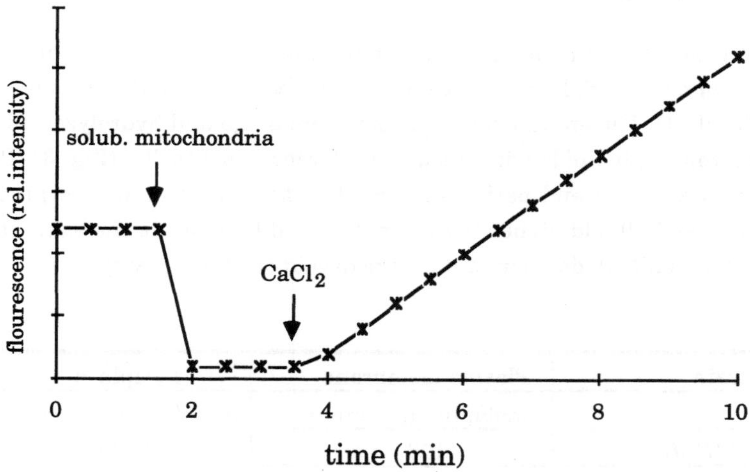

Fig. 2: Effect of Ca^{2+} on phospholipase A_2-catalyzed hydrolysis of 1,2-dipyrene-decanoyl-*sn*-glycero-3-phosphocholine

Phospholipase B

Phospholipase B (PL B) releases fatty acids from both the *sn-1* and *sn-2* position of phospholipids without accumulation of lysophospholipid intermediates. The enzyme catalyzes the release of fatty acids from lysophospholipids more effectively than from diacylglycerophospholipids. Thus PL B degrades phospholipids producing glycerophosphodiesters that are secreted into the medium. PL B from *Saccharomyces cerevisiae* also possesses acyltransferase activity, catalyzing the synthesis of phospholipids from lysophospholipids.

The *PLB1*-gene from *S. cerevisiae* has been isolated and sequenced (Lee *et al.*, 1994). The gene mapps to the left arm of chromosome VIII and encodes a polypeptide of 664 amino acids. The *PLB1*-gene shows high homology to the *Torulaspora delbrückii* (Watanabe *et al.*, 1994) and *Penicillium notatum* (Masuda *et al.*, 1983) *PLB*-genes. The encoded proteins are similar in size (*T.delbrückii*: 628 aa and *P.notatum* : 603 aa) and share 72% and 45% sequence identity with the *S. cerevisiae* gene.

Phospholipase B occurs at two different sites in yeast cells, the plasma membrane and the periplasmic space. We examined PL B activity from both sites *in vitro* (Table 1). The enzyme has a pH-optimum of 4,5 and hydrolyzes all major yeast glycerophospholipids with a slight preference for PtdCho (Fig. 3). Plasma membrane associated and periplasmic PL B activities from an overproducing strain are 4- and 10-fold higher than from the wild type, while the same cellular fractions from a *plb1Δ* deletion mutant are devoid of PL B-activity.

Strain	Plasma membrane	Periplasm
	nmol/min/mg protein	nmol/min/g wet cells
YEp352 (*PLB1*)	196	7144
Wild type	62	742
plb1Δ::URA3	0	0

Table 1: Phospholipase B activity in plasma membrane and periplasmic space

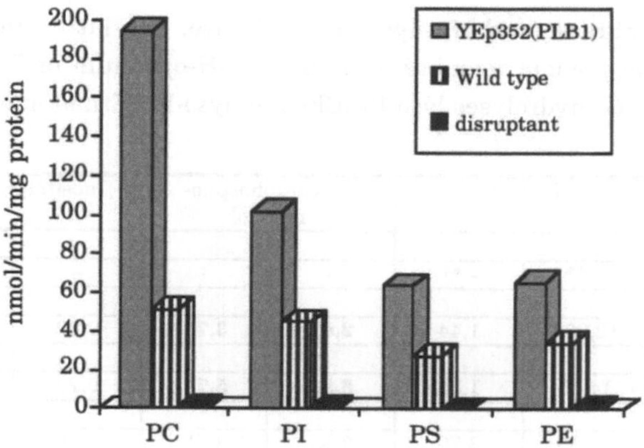

Fig. 3: Substrate-specifity of PL B, as determined using mixed vesicles

Highest enzyme specific activities in both fractions were found in the exponential phase of growth on glucose. Many enzymes of the phospholipid biosynthetic pathway show reduced activities in cells grown on media supplemented with inositol and choline. Wild type cells grown in the presence of 75 µM inositol show a reduction in phospholipase B specific activity to about 25% of control cells grown without supplementation (data not shown). These data together with mRNA analyses of *PLB1*-expression suggest that PL B is regulated at the transcriptional level by the soluble lipid precursor inositol, but not by choline. The 5' region of the *PLB1*-gene contains several copies of an UAS_{INO} element that might be responsible for the regulation of expression in response to supplementation with lipid precursors.

Despite a lack in phospholipase B activity, *plb1Δ* mutants are devoid of any growth phenotype, and still release substantial amounts of glycerophosphodiesters into the medium. It is rather unlikely that intracellularly produced glycerophosphodiesters could traverse the plasma membrane. Therefore we assume that other phospholipases are responsible for the formation of glycerophosphodiesters. While Plb1p contributes to the production of secreted glycerophosphodiesters, as can be seen from the significant increase of these products upon overproduction of the enzyme, an alternative glycerophosphodiester producing enzyme activity observed with *plb1Δ* mutant cells is almost as effective (Tables 2 and 3). It consists most likely of a combination of phospholipase A_2, which we observed in mitochondria (see preceding paragraph) and plasma membrane, and of a lysophospholipase. The latter enzyme activity was found in the plasma membrane. It has a pH-optimum of 7.5, it does not depend on Ca^{2+} and hydrolyses lyso-PtdCho and lyso-PtdEtn at similar rates.

	Inositol		Glycerophospho-inositol		Inositolphosphate	
	+ **	- **	+	-	+	-
wild type	12,33*	1,44	**2,85**	**3,73**	0,32	0,23
YEp352[*PLB1*]	14,35	1,55	**6,96**	**5,75**	0,28	0,23
plb1Δ::URA3	30,66	3,76	**0,83**	**1,32**	0,28	1,32

Table 2: Release of radioactive products into the medium by cells labeled with ^{3}H-inositol. Numbers indicate percentage of total cellular radioactivity released from cells within 360 minutes.

** + media supplemented with 75 μM inositol
- without supplementation

	Choline	Glycerophospho-choline	Cholinephosphate
wild type	1,24	**1,08**	0,69
YEp352[*PLB1*]	1,41	**5,56**	1,40
plb1Δ::URA3	1,35	**2,70**	0,91

Table 3: Release of radioactive products into the medium by cells labeled with ^{3}H-choline. Numbers indicate percentage of total cellular radioactivity released from cells within 360 minutes. Media were supplemented with 1 mM choline.

Phospholipase B from yeast is highly glycosylated (molecular weight between 145 and 200 kDa; Witt *et al.*, 1984); the protein sequence obtained from the cloned gene contains numerous potential glycosylation sites (Fig. 4). Another interesting feature of the *PLB1* primary translation product is a sequence at the C-terminus which is highly homologous to a sequence identified in glycosylphosphatidylinositol (GPI)- anchored yeast proteins (Fig. 5). It can be speculated that the plasma membrane associated form of PL B is tethered to the membrane by a GPI- anchor, and that this anchor can be cleaved leading to the soluble, periplasmic form of PL B. The mechanism and regulation of the release of the Plb1p is currently under investigation.

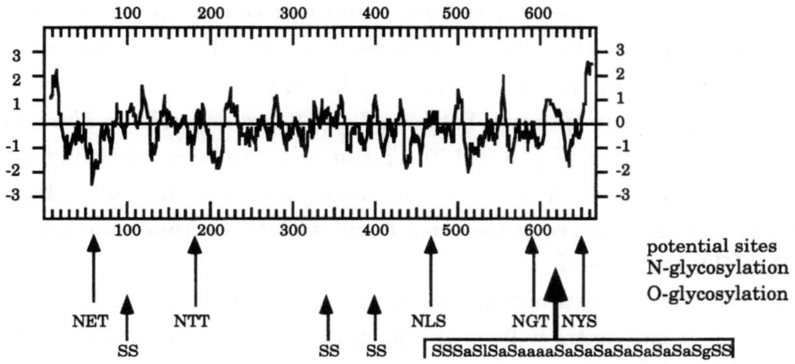

Fig. 4: Kyte & Doolittle plot and putative glycosylation sites of PL B from *S. cerevisiae*

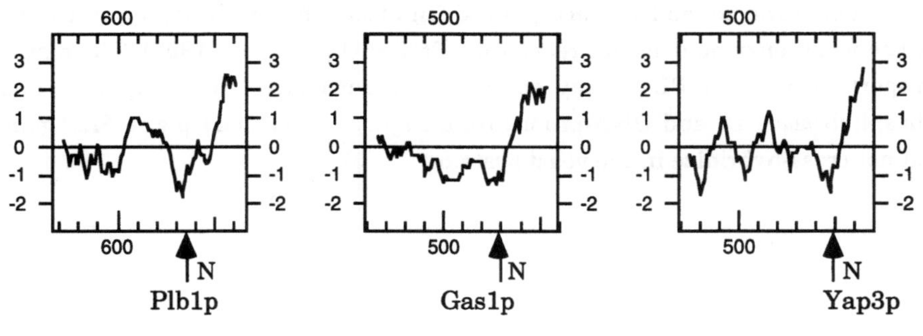

"Consensus": S x S x S x x x A x S x S x S x S x S A S x S S S S K K N

Fig. 5: Consensus sequence for glycolipid anchor attachment and homologies with other glycolipid anchored proteins in the yeast *S. cerevisiae* (N is the anchor attachment site).

PLB1 (PHOSPHOLIPASE B)
GAS1 (GLYCOLIPID ANCHORED SURFACE PROTEIN)
YAP3 (YEAST ASPARTYL PROTEASE)

Phospholipase D

In raw extracts from yeast cells cultivated on glucose we found PL D activity associated with the membrane fraction but no evidence for a soluble form of this enzyme. Further characterization showed that maximal enzyme activities were obtained from cells grown in non-fermentable lactate-medium. Analyzing the subcellular distribution of PL D the highest specific activities were measured in the inner mitochondrial membrane, somewhat less in contact sites between mitochondrial membranes, even less in microsomes, very little in the outer mitochondrial membrane, and nothing in the mitochondrial matrix.
Mitochondrial PL D has a pH-optimum between 7 and 8 and depends on divalent cations, preferentially calcium (Fig. 6 and 7). The enzyme can be solubilized from mitochondrial membranes with a variety of detergents but also with high concentrations of salt. Similar properties were found with PL D from mammalian membranes of a HL60 cell line (Brown *et al.*, 1993).
The size of the enzyme was estimated to be around 75 kDa using gel filtration chromatography of a salt extract from mitochondrial membranes. Substrate specificity determined with partially purified enzyme shows the following preference: PtdSer ≥ PtdEtn > PtdCho >> PtdIns (phospholipids were purified from yeast and supplied in form of detergent mixed micelles). This preference seems to be a remarkable feature of yeast PL D. The high activity towards PtdSer has not been reported for phospholipases D from other eukaryotic sources. PL D from the plant *Ricinus communis*, which occurs in soluble as well as in membrane-bound form, accepts PtdEtn rather than PtdCho, but in contrast to the yeast enzyme does not react with PtdSer (Dyer *et al.*, 1994). The distinct properties of yeast PL D may explain that antibodies raised against plant phospholipases D, and DNA-probes from highly homologous plant *PLD*-genes, did not cross-hybridize in the yeast system.

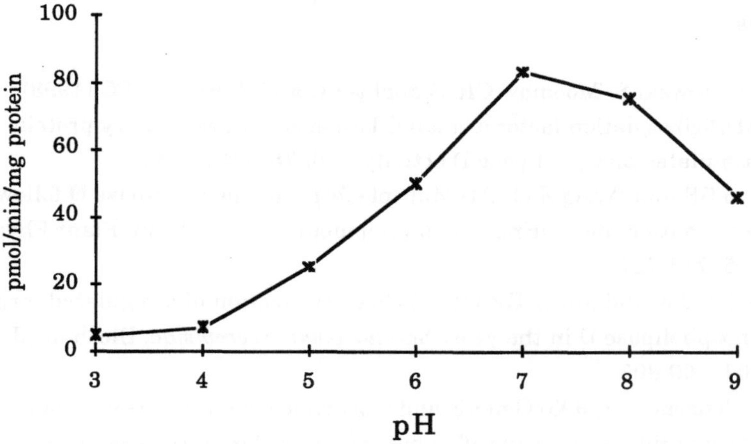

Fig. 6: pH-profile of yeast mitochondrial phospholipase D activity

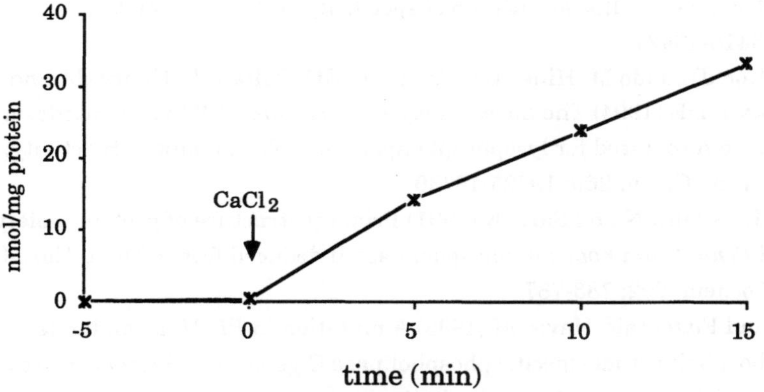

Fig. 7: Ca^{2+}-dependence of partially purified phospholipase D from yeast

References

Brown HA, Gutowski S, Moomaw CR, Slaughter C and Sternweis PC (1993) ADP-ribosylation factor a small GTP-dependent regulatory protein stimulates phospholipase D activity. Cell 75: 1137-1144

Dyer JH, Ryu SB and Wang X (1994) Multiple forms of phospholipase D following germination and during leaf development of castor bean. Plant Physiol. 105: 715-724

Ella KM, Dolan JW and Meier KE (1995) Characterization of a regulated form of phospholipase D in the yeast *Saccharomyces cerevisiae*. Biochem. J. 307: 799-805

Flick JS and Thorner J (1993) Genetic and biochemical characterization of a phosphatidylinositol-specific phospholipase C in *Saccharomyces cerevisiae*. Mol. Cell. Biol. 13: 5861-5876

Gassama-Diagne A, Rogalle P, Fauvel J, Willson M, Klaébé and Chap H (1992) Substrate specificity of phospholipase B from guinea pig intestine. A glycerol ester lipase with broad specificity. J. Biol. Chem. 267: 13418-13424

Lee KS, Patton JL, Fido M, Hines LK, Kohlwein SD, Paltauf F, Henry SA and Levin DE (1994) The *Saccharomyces cerevisiae PLB1* gene encodes a protein required for lysophospholipase and phospholipase B activity. J. Biol. Chem. 269: 19725-19730

Masuda N, Kitamura N and Saito K (1991) Primary structure of protein moiety of *Penicillium notatum* phospholipase B deduced from c-DNA. Eur. J. Biochem. 202: 783-787

Payne WE and Fitzgerald-Hayes M (1993) A mutation in *PLC1*, a candidate phosphoinositide-specific phospholipase C gene from *Saccharomyces cerevisiae*, causes aberrant mitotic chromosome segregation. Mol. Cell. Biol. 13: 4351-4364

Pind S and Kuksis A (1989) Association of the intestinal brush-border membrane phospholipase A_2 and lysophospholipase activities (phospholipase B) with a stalked membrane protein. Lipids 24: 357-362

Pind S and Kuksis A (1987) Isolation of purified brush-border membranes from rat jejunum containing Ca^{2+}-independent phospholipase A_2 activity. Biochim. Biophys. Acta 901: 78-87

Saito K, Sugatani J and Okumura T (1991) Phospholipase B from *Penicillium notatum*. Methods Enzymol. 197: 446-456

Wagner S and Paltauf F (1994) Generation of glycerophospholipid molecular species in the yeast *Saccharomyces cerevisiae*. Fatty acid pattern of phospholipid classes and selective acyl turnover at *sn*-1 and *sn*-2 positions. Yeast 10: 1429-1437

Waite M (1991) Phospholipases. In: Biochemistry of Lipids, Lipoproteins and membranes, Vance DE and Vance J, eds., 269-295

Watanabe Y, Yashiki Y, Sultana GN-N, Maruyama M, Kangawa K and Tamai Y (1994) Cloning and sequencing of phospholipase B gene from the yeast *Torulaspora delbrueckii*. FEMS Microbiol. Letters 124: 29-34

Witt W, Schweingruber ME and Mertsching A (1984) Phospholipase B from the plasma membrane of *Saccharomyces cerevisiae*. Separation of two forms with different carbohydrate content. Biochim. Biophys. Acta 795: 108-116

Yoko-o T, Matsui Y, Yagisawa H, Nojima H, Uno I and Toh-e A (1993) The putative phosphoinositide-specific phospholipase C gene, *PLC1*, of the yeast *Saccharomyces cerevisiae* is important for cell growth. Proc. Natl. Acad. Sci. USA 90: 1804-1808

Yost RW, Grauvickel SJ, Cantwell R, Bomalaski JS and Hudson AP (1991) Yeast mitochondria of *Saccharomyces cerevisiae* contain Ca^{2+}-independent phospholipase A_1 and A_2 activities: effect of respiratory state. Biochem. Int. 24: 199-208

Acknowledgement

This work is supported by grants S-5813 (to F. P.) and S-5812 (to S. D. K.) of the Fonds zur Förderung der wissenschaftlichen Forschung in Österreich.
We would like to thank David E. Levin for providing yeast strains and plasmids and Claudia Kalaus for communicating information prior to publication.

FUNCTIONAL ANALYSIS OF PHOSPHATIDYLINOSITOL TRANSFER PROTEINS.

Brian G. Kearns, James G. Alb, Jr., Robert T. Cartee, and Vytas A. Bankaitis
Department of Cell Biology
University of Alabama at Birmingham
Birmingham, Alabama, 35294-0005
U.S.A.

All eukaryotic cells contain a battery of cytosolic proteins that catalyze the energy-independent transfer of PLs, as monomers, between membrane bilayers *in vitro* (Rueckert and Schmidt, 1990; Wirtz, 1991). These proteins were initially detected in assays designed to identify polypeptides that could be involved in intracellular lipid sorting/ trafficking and, based solely upon this operational definition, such proteins are referred to as phospholipid transfer proteins (PL-TPs). These PL-TPs are, in turn, categorized into three general classes on the basis of their unique catalytic activities which reflect the differing PL headgroup specificities of these proteins. For example, the monospecific PL-TPs, exemplified by the phosphatidylcholine transfer protein, exhibit an absolute specificity for one PL species. On the other hand, the nonspecific transfer proteins are able to mobilize most PL species, glycolipids, and sterols in the *in vitro* transfer reaction. Finally, the oligospecific transfer proteins, exemplified by the phosphatidylinositol (PI)/phosphatidylcholine (PC) transfer proteins (PI-TPs) are able to transfer only a few PLs. In the case of PI-TPs these substrates are PI and PC.

With few exceptions, PL-TPs execute efficient exchange reactions but are at best very inefficient catalysts of PL-transfer via the net transfer mode. For this reason, PL-TPs are sometimes referred to as phospholipid exchange proteins. Moreover, PL-TPs can utilize virtually any natural or synthetic membrane as PL-donor, or PL-acceptor, in the *in vitro* transfer reaction (Rothman, 1990; Wirtz, 1991). This promiscuity, when coupled with the energy-independence of the transfer reaction and the clear propensity of PL-TPs to catalyze PL-exchange reactions over net transfer reactions, poses a

NATO ASI Series, Vol. H 96
Molecular Dynamics of Biomembranes
Edited by Jos A. F. Op den Kamp

challenge to the validity of the general assumption that PL-TPs function to catalyze PL-transfer *in vivo.*

I. PHOSPHATIDYLINOSITOL/ PHOSPHATIDYLCHOLINE TRANSFER PROTEINS.

PI-TPs have the capability to transfer either PI or PC between membrane bilayers *in vitro.* Competition experiments have demonstrated that PI is clearly the preferred ligand in the transfer reaction. The rate of PI-transfer is some 16- to 20-fold greater for PI than it is for PC (Van Paridon *et al.,* 1987). On the basis of such a substrate preference, these PL-TPs are generally referred to as PI-TPs rather than PI/PC-TPs which is technically the more precise term. PI-TPs exhibit two outstanding features that identify them as a particularly interesting class of PL-TP. First, PI-TPs are unique in that, while these polypeptides contain one PL-binding site per protein monomer, this PL-binding site exhibits a dual headgroup specificity. Thus, PI-TPs have the ability to accomodate the binding of two dissimilar PLs (i.e. PI and PC) in a mutually exclusive binding reaction. This unique feature suggests that binding of ligand might potentially be utilized as a molecular switch by which some other activity of a PI-TP can be regulated. Second, PI-TPs represent a class of proteins that exhibits a high level of primary sequence conservation. It is noteworthy that no other class of PL-TP, not even the monospecific PL-TPs, exhibits significant primary sequence conservation -- even when analagous PL-TPs from related organisms are compared (Rueckert and Schmidt, 1990; Wirtz, 1991). PI-TPs of mammalian, amphibian, and insect origin are all approximately 35 kD in molecular weight, are recognized by polyclonal antisera raised against rat PI-TP, and mammalian PI-TPs share very high primary sequence identity (Dickeson *et al.,* 1989; unpublished data). This conservation of epitopes suggests substantial primary sequence similarity of PI-TPs from humans to *Drosophila.* We have found that fungal PI-TPs are also approximately 35kD in molecular weight and are highly similar to each other at the primary sequence level (Bankaitis *et al.,* 1989; Bankaitis *et al.,* 1990; Salama *et al.,* 1990; Carmen-Lopez *et al.,* 1994). However, these proteins share no primary sequence similarity at all with mammalian

PI-TPs (Bankaitis *et al.*, 1989; Dickeson *et al.*, 1989). Instead, fungal PI-TPs share significant primary sequence homology with the mammalian retinaldehyde binding protein, a polypeptide that plays a crucial role in mammalian vision (Salama *et al.*, 1990). The lack of primary sequence similarity between mammalian and fungal PI-TPs notwithstanding, these proteins exhibit very similar *in vitro* transfer properties. One distinction between these PI-TPs is that mammalian PI-TP activity is sensitive to N-ethylmaleimide (NEM) challenge, whereas fungal PI-TP activity is not (Helmkamp, 1985; Daum and Paltauf, 1984). As it is widely assumed that the PI- and PC-binding/transfer activities of rPI-TP are somehow relevant to the *in vivo* function of these proteins, it becomes necessary to consider how the cell would impose a suitable regulation on the phospholipid transfer/binding activities of PI-TP, so as to effectively harness the *in vitro* PL-transfer activity to *in vivo* function. An illustration of such regulation in the context of vectorial PI-transfer is given in Figure 1.

Currently, there exists no structural understanding of how PI-TPs execute their PI- and PC-binding/transfer activities. Thus, the necessity for a detailed functional analysis of the PL-transfer activities of rPI-TP is emphasized. In particular, the definition of what rPI-TP domains are specifically dedicated to effecting the transfer of PI and PC, respectively, defines one of the outstanding issues in the field that require resolution. Such knowledge would be of great use as it would identify potential sites through which each PL-transfer activity might be specifically regulated; thereby providing the ultimate means for individually testing the role of each PL-transfer activity in PI-TP function *in vivo*.

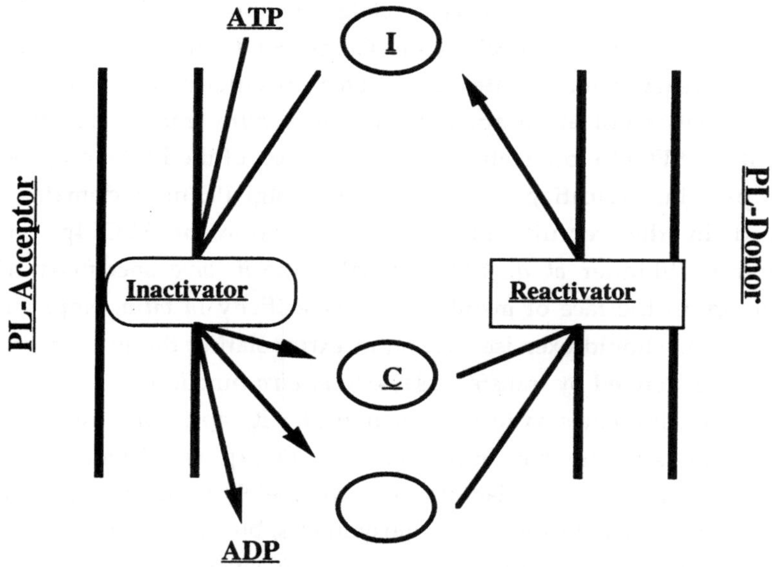

Figure 1. A speculative energy utilizing cycle for vectorial transfer of PI from a donor to an acceptor membrane is depicted. Upon discharge of PI (I) into the acceptor membrane a resident regulator inactivates PI-TP in an ATP-dependent manner so that PI-TP is incapable of reloading with PI. PI-TP subsequently disengages from the acceptor membrane either as a PL-free protein, or as a PC-bound species. Upon re-engagement of PI-TP with the PL-donating membrane, a resident regulator reactivates PI-TP and permits reloading with PI for the next round of transfer.

II. PI-TP FUNCTION IN BUDDING YEAST.

The first opportunity for a detailed *in vivo* analysis of PI-TP function, and indeed the function of any PL-TP, was identified by the finding that the *Saccharomyces cerevisiae SEC14* gene product (SEC14p), whose function is essential for protein transport from the yeast Golgi complex (Bankaitis *et al.*, 1989; Novick *et al.*, 1980), is the yeast PI-TP (Bankaitis *et al.*, 1990). Subsequent work has also established that SEC14p exhibits a specific association with yeast Golgi membranes *in vivo*, and that this association is stable as evidenced by copurification of SEC14p with yeast Golgi membranes (Cleves *et al.*, 1991b). The specific localization of SEC14p to Golgi membranes, a minor membrane system in yeast, demonstrates an *in vivo* specificity of PI-TP membrane targeting that is not at all apparent in the *in vitro* transfer reaction. This specificity of SEC14p localization is not a function of the PL-binding capability of this protein, indicating that specialized Golgi PL microdomains are not involved in the recruitment and/or retention of SEC14p on Golgi membranes (Skinner *et al.*, 1993). Finally, the *in vivo* specificity of Golgi localization, in the face of membrane nonspecificity *in vitro*, emphasizes the point that one should exercise caution in extrapolating the *in vitro* properties of PL-TPs, as defined by transfer assays, to *in vivo* function.

Penetrating clues as to the function of SEC14p *in vivo* were obtained from genetic studies that indicated the SEC14p requirement for Golgi secretory function and cell viability to be completely bypassed by inactivation of the CDP-choline pathway for PC biosynthesis, but not by inactivation of the PE-methylation pathway for PC biosynthesis (Cleves *et al.*, 1991). The CDP-choline pathway is comprised of three reactions (catalyzed by CKIase, CCTase, and CPTase, respectively) that result in the incorporation of free choline into PC. A biochemical rationale for this specific genetic relationship between SEC14p function and the CDP-choline pathway was forthcoming from the demonstration that the primary *in vivo* consequence of SEC14p dysfunction is a specific CDP-choline pathway-driven increase in Golgi PC content (McGee *et al.*, 1994). Our demonstration that SEC14p functions to maintained reduced Golgi membrane PC levels can be reconciled by one of two mechanisms. First, a PL-transfer mechanism could operate where SEC14p mobilizes PC from the Golgi. The difficulties with this idea are two-fold. It fails to account for why the methylation pathway is irrelevant to SEC14p function, and it posits a net transfer mode for SEC14p -- a transfer reaction SEC14p cannot efficiently perform *in vitro*. An alternative mechanism is that SEC14p is a PL-sensor that functions to repress the CDP-choline pathway. The weakness of this model is that it also fails to explain why the CDP-choline pathway is relevant while the methylation pathway is not. Nevertheless, both models are consistent with all of the genetic data and most of the biochemical data. Two distinguishing predictions of the sensor model are that: **(i)** SEC14p mediates a specific regulation directed at the activity of at least one of the three structural enzymes of the CDP-choline pathway, and **(ii)** the inhibitory activity of SEC14p is a function of the particular PL (PI or PC) bound to SEC14p. Both *in vivo* and *in vitro* data have been obtained that are consistent with these predictions.

[^{14}C]-Choline and [^{32}P]-orthophosphate pulse-radiolabeling experiments independently revealed that overproduction of SEC14p leads to a significant depression in bulk cellular CDP-choline pathway activity, but not in methylation pathway activity (McGee *et al.*, 1994; Skinner *et al.*, 1994). This *in vivo* effect on the CDP-choline pathway exhibits the signature of cholinephosphate cytidylyltransferase (CCTase) deficiency, and in vitro experiments have indicated that SEC14p effects a potent and specific inhibition of CCTase (Skinner *et al.*, 1994). The finding that CCTase is inhibited by SEC14p supports the first prediction of the sensor model, and is reasonable since CCTase is the rate-determining enzyme of the CDP-choline pathway. Finally, the potency of SEC14p-mediated inhibition of CCTase appears to be a function of membrane PC content. This suggests that SEC14p~PC is the CCTase inhibitor (Skinner *et al.*, 1994). The indication that the regulatory function of SEC14p is controlled by the ligand-bound state of SEC14p provides additional support for the sensor model.

The data of McGee *et al.* (1994) and Skinner *et al.* (1994) have led us to entertain the following refinement of the sensor model, and a working model is illustrated in Figure 2 below. The data are consistent with a mechanism where SEC14p functions to downregulate CDP-choline pathway activity in yeast Golgi membranes *in vivo* by inhibiting CCTase, and that PC-bound form of SEC14p (SEC14p~PC) is responsible for inhibition of CCTase. Thus, a sensitive feedback loop is created that couples the action of CCTase to Golgi PC levels. We consider the CCTase-directed activity of SEC14p to itself be regulated through a constitutive PL-exchange reaction executed by SEC14p on Golgi membranes; an activity that manifests itself as a PL-exchange activity *in vitro*. Finally, the PI-binding activity of SEC14p is suggested to constitute a competitive PL-binding reaction that constantly regulates the amount of SEC14~PC present in response to the PI/PC ratio present in Golgi membranes.

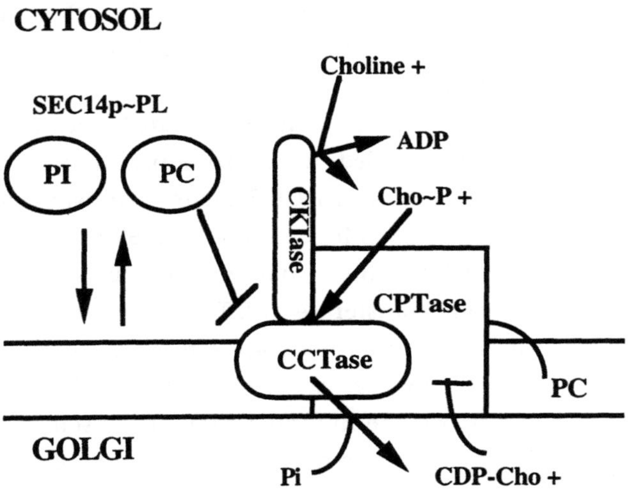

Figure 2. Sensor model for SEC14p function. SEC14p is localized to the cytosolic surface of yeast Golgi membranes where it is proposed to execute a constitutive PL-exchange reaction that generates a pool of PL-bound SEC14p comprised of both SEC14p~PC and SEC14p~PI, as indicated. SEC14p~PC serves as a negative effector of CCTase activity in yeast Golgi membranes. Diacylglycerol is abbreviated to DAG.

This proposed mechanism of SEC14p function is analagous to that established for *ras*-like GTP-binding proteins where ligand occupancy regulates signal transduction from a sensor component to a downstream effector. For SEC14p, the inherent PL-exchange activity associated with this protein provides a mechanism for controlling its CCTase-directed regulatory activity. These data emphasize the potential for an indirect relevance of the *in vitro* PL-transfer activity of a protein to *in vivo* function, and indicate that the mechanism of function determined for one PL-TP cannot be confidently extrapolated to that of another: even to one that exhibits similar PL-substrate specificities *in vitro*.

Another interesting possibility raised by the sensor model is that, as in the case of *ras*-like GTP binding proteins, there may exist multiple members of a SEC14p family in cells; each dedicated to regulating a specific cellular process. Thus, one could conceivably encounter a situation where characterization of the *in vivo* function of two related PI-TPs might reveal diverse *in vivo* functions for each. Studies of the *in vivo* function of the

SEC14p from the dimorphic yeast *Yarrowia lipolytica* (SEC14p^{YL}) suggest this is likely the case. *Y. lipolytica* is widely diverged from both *S. cerevisiae* and *S. pombe* and is typified by two distinct developmental forms, the yeast and the mycelial forms, whose predominance can be controlled at the level of the growth medium. The data indicate a considerable level of both primary sequence homology and functional homology between SEC14p and SEC14p^{YL}, as evidenced by the 65% identity shared between these proteins and ability of the latter to efficiently substitute for the essential function of the former in *S. cerevisiae* (Carmen-Lopez *et al.*, 1994) Also, in a manner entirely analagous to the SEC14p paradigm, SEC14p^{YL} is a PI-TP that localizes to what are likely to be *Y. lipolytica* Golgi bodies. However, in contrast to the case of SEC14p in *S. cerevisiae*, SEC14p^{YL} is neither required for the cellular viability of *Y. lipolytica* nor is it required for efficient secretory pathway function. SEC14p^{YL} is required for *Y. lipolytica* to undergo the yeast-mycelial transition that is typical of this species, however.

The collective data demonstrate that SEC14p and SEC14p^{YL} are involved in controlling distinct physiological processes in their respective host organisms. Thus, the elements of conservation between these PI-TPs notwithstanding, SEC14p^{YL} and SEC14p either: (i) play mechanistically divergent roles in their respective organisms, or (ii) play mechanistically similar roles in their respective host organisms but the coupling of their function to downstream regulatory circuits has diverged. Indeed, both functional complementation studies, data base searches and genome sequencing efforts have revealed SEC14p homologs in *Saccharomyces cerevisiae* and in several plant species. Thus, we believe it likely that the SEC14p/PI-TP protein families will ultimately represent new classes of intracellular sensor/regulatory proteins with a diversity and ubiquity approaching that of the small GTP-binding proteins. It promises to be an exciting future for the study of this class of proteins.

III. MOLECULAR ANALYSES OF PI-TP FUNCTION IN A YEAST SURROGATE SYSTEM.

Recent studies have implicated mammalian PI-TP (mPI-TP) as a cofactor in various membrane signalling events; e.g. the priming of the regulated fusion of dense core granules and the stimulation of PI-specific phospholipase C activity (Hay and Martin, 1993; Thomas *et al.*, 1993). These findings have largely been interpreted in the context of mPI-TP effecting vectorial transfer of PI to signalling membranes. As described above, vectorial PI-transfer in cells requires the superimposition of regulation on mPI-TP function (Figure 1). Recent data obtained by Alb *et al.* (1995) now suggest a means by which such a regulation could be imposed, and what the identity of the components of such a candidate machinery may be. Alb *et al.* have demonstrated that the S25F, T59I, P78L, and E248K mutations in the alpha-isoform of rat PI-TP are distinguished by their *in vitro* manifestation of specific PI-transfer defects, in the face of largely unadulterated PC-transfer capabilities. These findings provide the first demonstration that the PI- and PC-transfer activities can be uncoupled in the context of the PI-TP itself. The specificity of the PI-transfer defects associated with these mutations suggests that S25, P78, T59, and E248 either cooperate to form a PI headgroup recognition/binding site in rPI-TP, or that these residues neighbor such a site. Thus, these data also provide the first structural clues as to the potential identity of rPI-TP residues dedicated to headgroup binding/recognition.

The finding that two of these affected residues (i.e. T59 and E248) lie within consensus PKC phosphorylation motifs suggests that at least one consensus protein kinase C (PKC) phosphorylation site could potentially serve as a regulatory site on this particular mPI-TP. The sensitivity of PI-transfer to the state of T59 predicts that phosphorylation of T59 will manifestly downregulate the PI-binding/transfer activity of PI-TP (alpha-isoform) in mammalian cells, and it suggests a regulatory circuit by which mPI-TP could be regulated so as to effect vectorial PI-transfer to a signalling membrane. A speculative model for how a PKC/protein phosphatase cycle might operate in the context of vectorial PI-transfer by rPI-TP *in vivo* is depicted in Figure 3.

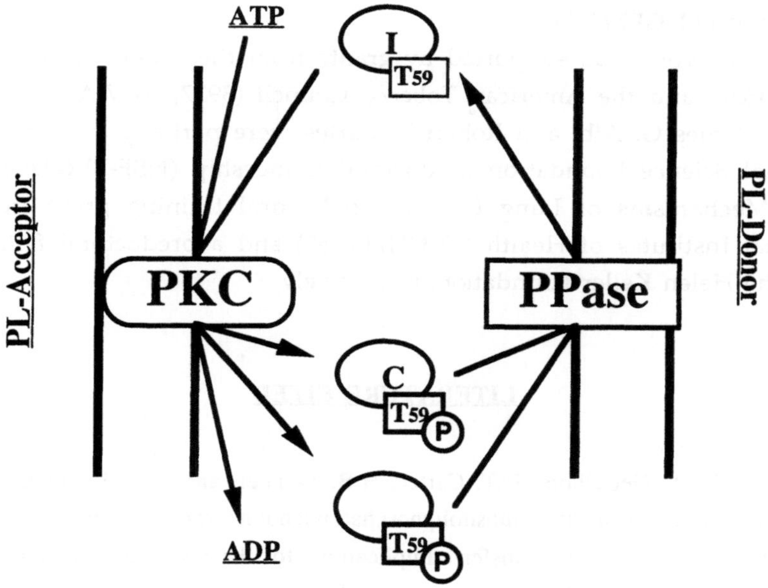

Figure 3. A speculative PKC/PPase cycle for vectorial transfer of PI from a donor to an acceptor membrane in mammals is depicted. Upon discharge of PI (I) into the acceptor membrane, a resident PKC phosphorylates mPI-TP on T59; thereby rendering mPI-TP incapable of reloading with PI. mPI-TP subsequently disengages from the acceptor membrane either as a PL-free protein, or as a PC-bound species. Re-engagement of mPI-TP with the PL-donor membrane introduces mPI-TP to a resident protein phosphatase (PPase) that dephosphorylates PI-TP and allows it to reload with PI for the next round of transfer.

Supporting evidence for the cycle depicted in Figure 3 comes from the demonstration that: (i) phosphorylation of rodent PI-TP is stimulated by PKC agonists *in vivo* (Snoek *et al.*, 1992), (ii) that rat PI-TP (alpha-isoform) is a substrate for PKC *in vitro* (Snoek *et al.*, 1993), and (iii) that the T59D and T59E mutants are specifically defective in PI-transfer (Alb *et al.*, 1995). This model is also appealing because it predicts a physiologically reasonable directionality to the vectorial PI-transfer reaction. Since activated PKC resides on membranes engaged in phosphoinositide driven signalling events, the active kinase is properly situated to effect an inactivation of mPI-TP at membranes serving as efficient PI-acceptors in cellular signalling reactions.

ACKNOWLEDGEMENTS

This work was supported by grants from the Public Health Service (GM-44530) and the American Tobacco Council (3937) to V.A.B. Brian G. Kearns, James G. Alb, and Robert T. Cartee were partially supported by a National Science Foundation predoctoral traineeship (NSF-BIR-9256853), a Basic Mechanisms of Lung Disease predoctoral training grant from the National Institutes of Health (5T32HL07553) and a predoctoral fellowship from the Helen Keller Foundation, respectively.

LITERATURE CITED

Alb, Jr., J.G., A. Gedvilaite, R.T. Cartee, H.B. Skinner, and V.A. Bankaitis. 1995. Mutant rat phosphatidylinositol/phosphatidylcholine transfer proteins defective in phosphatidylinositol transfer: Implications for the regulation of phospholipid transfer activity. *Proc. Natl. Acad. Sci. U.S.A.* (In Press).

Bankaitis, V.A., J.R. Aitken, A.E. Cleves, and W. Dowhan. 1990. An essential role for a phospholipid transfer protein in yeast Golgi function. *Nature* **347**::561-562.

Bankaitis, V.A., D.E. Malehorn, S.D. Emr, and R. Greene. 1989. The *Saccharomyces cerevisiae SEC14* gene encodes a cytosolic factor that is required for transport of secretory proteins from the yeast Golgi complex. *J. Cell Biol.* **108**: 1271-1281.

Cleves, A. E., T. P. McGee, E. A. Whitters, K. M. Champion, J. R. Aitken, W. Dowhan, M. Goebl, and V. A. Bankaitis. 1991. Mutations in the CDP-choline pathway for phospholipid biosynthesis bypass the requirement for an essential phospholipid transfer protein. *Cell* **64**: 789-800.

Daum, G. and Paltauf, F. (1984). Isolation and partial characterization of a phospholipid transfer protein from yeast cytosol. *Biochim. Biophys. Acta* **794**: 385-391.

Dickeson, S.K., Lim, C.N., Schuyler, G.T., Dalton, T.P., Helmkamp, G.M., Jr. and Yarbrough, L.R. (1989). Isolation and sequence of cDNA clones encoding rat phosphatidylinositol transfer protein. *J. Biol. Chem.* **264**: 16557-16564.

Hay, J.C. and Martin, T.F.J. (1993). Phosphatidylinositol transfer protein required for ATP-dependent priming of Ca^{2+}-activated secretion. *Nature* **366**: 572-575.

Helmkamp, G.M., Jr. (1985). Phospholipid transfer proteins: Mechanisms of action. *Chem. Phys. Lipids* **38**: 3-16.

McGee, T. P., H. B. Skinner, E. A. Whitters, S. A. Henry, and V. A. Bankaitis. 1994. A phosphatidylinositol transfer protein controls the phosphatidylcholine content of yeast Golgi membranes. *J. Cell Biol.* **124**: 273-287.

Novick, P., Field, C. and Schekman, R. (1980). Identification of 23 complementation groups required for post-translational events in the yeast secretory pathway. *Cell* **21**: 205-215.

Rothman, J.E. (1990). Phospholipid transfer market. *Nature* **347**: 519-520.

Rueckert, D.G. and Schmidt, K. (1990). Lipid transfer proteins. *Chemistry and Physics of Lipids* **56**: 1-20.

Salama, S.R., A.E. Cleves, D.E. Malehorn, E.A. Whitters, and V.A. Bankaitis. 1990. Cloning and characterization of *Kluyveromyces lactis SEC14*, a gene whose product stimulates Golgi secretory function in *Saccharomyces cerevisiae*. *J. Bacteriol.* **172**: 4510-4521.

Skinner, H.B., T.P. McGee, C.R. McMaster, M.R. Fry, R.M. Bell, and V.A. Bankaitis. 1995. The Saccharomyces cerevisiae phosphatidylinositol-transfer protein effects a ligand-dependent inhibition of choline-phsphate cytidylyltransferase activity. *Proc. Nat. Acad. Sci. U.S.A.* **92**: 112-116.

Snoek, G.T., de Wit, I.S.C., van Mourik, J.H.G. and Wirtz, K.W.A. (1992). The phosphatidylinositol transfer protein in 3T3 mouse fibroblast cells is associated with the Golgi system. *J. Cellular Biochem.* **49**: 339-348.

Snoek, G.T., Westerman, J., Wouters, F.S., and Wirtz, K.W.A. (1993). Phosphorylation and redistribution of the phosphatidylinositol-transfer protein in phorbol 12-myristate 13 acetate- and bombesin-stimulated Swiss mouse 3T3 fibroblasts. *Biochem. J.* **291**: 649-656.

Thomas, G.M.H., E. Cunningham, A. Fensome, A. Ball, N.F. Totty, O. Truong, J. Hsuan, and S. Cockroft. 1993. An essential role for phosphatidylinositol transfer protein in phopsholipase C-mediated inositol lipid signaling. *Cell* **74**: 919-928.

Wirtz, K.W.A. (1991). Phospholipid transfer proteins. *Annu. Rev. Biochem.* **60**: 73-99.

Phosphatidylcholine Biosynthesis in *Saccharomyces cerevisiae*: Effects on Regulation of Phospholipid Synthesis and Respiratory Competence

Peter Griac[1] and Susan A. Henry
Department of Biological Sciences
Carnegie Mellon University
4400 Fifth Avenue
Pittsburgh, PA 15213
USA

Phosphatidylcholine (PC) can be synthesized via two distinct pathways. One pathway involves three sequential methylations of phosphatidylethanolamine (PE) (Bremer et al., 1960). Alternatively, PC can be synthesized from free choline via the CDP-choline pathway (Kennedy and Weiss, 1956). These two pathways are found in all eukaryotes so far investigated, including mammals (Bjornstad and Bremer, 1966) and yeast (Steiner and Lester, 1972; Waechter and Lester, 1973). Phospholipid biosynthesis is highly regulated in yeast. Much of the regulation occurs at the level of gene transcription in response to the soluble precursors of phospholipid biosynthesis, inositol and choline (for review see Carman and Henry, 1989; Paltauf et al., 1992). In the presence of inositol, transcription of the co-regulated biosynthetic genes (*INO1*-Inositol-1-phosphate synthase, *CHO1*-Phosphatidylserine synthase, *CHO2/PEM1* - Phosphatidylethanolamine methyltransferase and *OPI3/PEM2* - Phospholipid methyltransferase) (Figure 1) is repressed. In the absence of inositol, transcription of these genes is derepressed. If choline is added to medium in which inositol is already present, these genes are further repressed. However, if choline is present in the growth medium by itself, it has little or no effect on transcription of the co-regulated genes.

Interestingly, with the exception of the *CHO1* gene (Atkinson et al. 1980; Kovac et al. 1980) each of the structural genes leading to synthesis of phosphatidylcholine via the methylation pathway has a "backup mechanism" enabling the yeast cells to survive, when a particular gene is deleted, even in the absence of choline (or ethanolamine). Thus, there are two genes (*PSD1* and *PSD2*) (Trotter et al., 1993; Clancey et al., 1993, Trotter and Voelker, 1995) the products of which are able to catalyze the decarboxylation reaction of phosphatidylserine to phosphatidylethanolamine. Therefore, attempts to isolate yeast strains defective in phosphatidylserine decarboxylase (PSD) as ethanolamine auxotrophs failed and

[1]Present address:
Institute of Animal Biochemistry and Genetics
Slovak Academy of Sciences
900 28 Ivanka pri Dunaji
Slovakia

NATO ASI Series, Vol. H 96
Molecular Dynamics of Biomembranes
Edited by Jos A. F. Op den Kamp

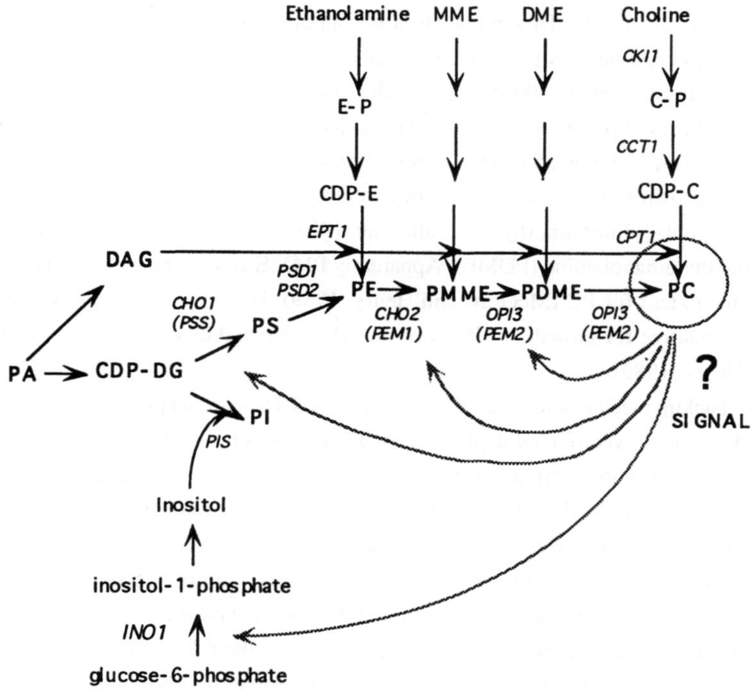

FIGURE 1. Phospholipid metabolic pathways in yeast. Abbreviations: PA - phosphatidic acid, DAG - *sn*-1,2 diacylglycerol, CDP-DG - cytidinediphosphate diacylglycerol, PI - phosphatidylinositol, PS - phosphatidylserine, PE - phosphatidylethanolamine, PMME - phosphatidylmonomethylethanolamine, PDME - phosphatidyldimethylethanoamine, PC - phosphatidylcholine, C-P - choline phosphate, CDP-C - cytidinediphosphate choline, E-P - ethanolamine phosphate, CDP-E - cytidinediphosphate ethanolamine. Structural genes: *INO1* - inositol-1-phosphate synthase, *PIS* - PI synthase, *CHO1* - PS synthase, *PSD1, PSD2* - PS decarboxylases, *CHO2/PEM1* - PE *N*-methyl-transferase, *OPI3/PEM2* - phospholipid *N*-methyltransferase, *CKI1* - choline kinase, *CCT1* - choline phosphate:CTP cytidylyltransferase, *CPT1* - *sn*-1,2-diacylglycerol choline phosphotransferase, *EPT1* - *sn*-1,2-diacylglycerol ethanolamine phosphotransferase.

alternative methods (Trotter et al., 1993; Clancey et al., 1993) had to be used to isolate strains defective in PSD activity and to clone the genes. Similarly, *cho2* disruption mutants are not choline auxotrophs (Summers et al., 1988). They have grossly altered membrane phospholipid composition and grow considerably slower than wild type strains in the absence of choline or its methylated precursors. The ability of the *cho2* strains to grow in the absence of choline is probably due to the presence of a second phospholipid methyltransferase, product of the *OPI3* gene, that to some extent can substitute for the metabolic defect in *cho2* mutants and allows a small amount of residual PC biosynthesis in the absence of exogenous choline (Summers et al., 1988). This explanation is supported by the observation that the *OPI3* gene on a high copy number plasmid can suppress the phenotypes associated with the *cho2* mutation (Preitschopf et al., 1993). Yeast strains carrying *opi3* mutations, defective in the two terminal phospholipid *N*-methyltransferase activities, are also choline prototrophs (Greenberg et al., 1983; McGraw and Henry, 1989). The tightest *opi3* mutants, such as gene disruptants, when grown without choline do not contain any detectable amount of PC but accumulate phosphatidylmonomethylethanolamine (PMME) and a small amount of phosphatidyldimethylethanolamine (PDME). Apparently PMME and the residual PDME can substitute to some extent for PC (McGraw and Henry, 1989). However, tight *opi3* mutants are temperature sensitive for growth at 37°C and quickly loose viability at stationary phase (McGraw and Henry, 1989).

Similar "backup mechanisms" are also present in the CDP-choline pathway. Thus, the function of *sn*-1,2-diacylglycerol cholinephosphotransferase (product of the *CPT1* gene) can be substituted by ethanolaminephosphotransferase (product of the *EPT1* gene) (Hjelmstad and Bell, 1991; McMaster and Bell, 1994). Analyzing a mutant defective in choline kinase (product of the *CKI1* gene) Hosaka et al. (1989) concluded that the *CKI1* gene encodes the only choline kinase in yeast and that the ethanolamine kinase activity is a second activity of choline kinase in yeast. We assessed incorporation of ^{14}C labeled choline into PC in the strain with disrupted *CKI1* gene and compared it to a strain in which the *CKI1* and *CHO2* genes were both disrupted. We found a large increase of incorporation of the label in the *cho2 cki1* double mutant as compared to the *cki1* single mutant (Figure 2).

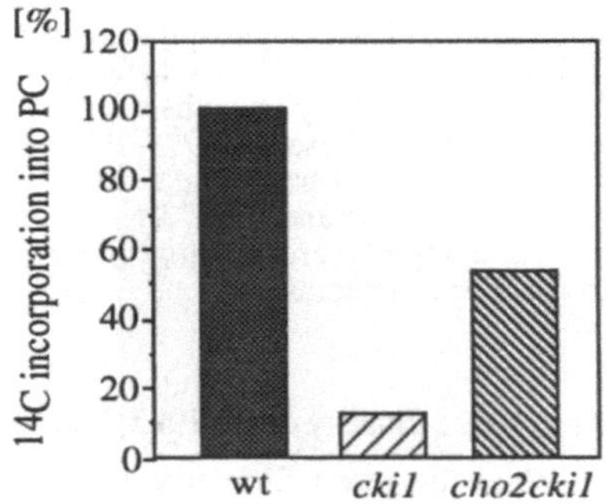

FIGURE 2. Incorporation of labeled choline into PC. Yeast strains (*wt*; *cki1*; *cho2 cki1*) were grown in the presence of [methyl-^{14}C] choline for at least 5 generations in vitamin-defined medium with 1 mM choline and 75 µM inositol. Lipids were extracted and resolved as described in Steiner and Lester (1972). Spots corresponding to PC were quantified by liquid scintillation. All values were standardized to the final turbidity of cultures and are an average from two independent experiments.

We interpret this increase as a regulatory response of yeast cells containing a block in methylation pathway that is caused by disruption of the *CHO2* gene. This result also indicates that the product of a gene other than the *CKI1* gene is able, under special circumstances, to phosphorylate exogenously supplied choline.

Overall, these studies suggest that yeast are very flexible in satisfying their phospholipid requirements and that the cell has evolved multiple mechanisms for insuring the synthesis of vital lipids.

Phenotypes of phospholipid deficient mutants. Despite the fact that most disruptions of structural genes in the PC biosynthetic pathway do not lead to auxotrophic requirements for ethanolamine or choline, the mutant cells are not wild type in all their properties. Two major general abnormalities in these mutants are defects in transcriptional regulation of coordinately regulated phospholipid genes and a tendency to produce respiratory deficient cells in excessive amounts.

Respiratory deficiency. The *cho1* mutants, for example, produce high proportions of respiratory deficient cells when grown on complete media (Atkinson et al., 1980). Yeast cells defective in phosphatidylserine decarboxylase (*PSD1*) (Trotter et al., 1993; Trotter and Voelker, 1995) also have a pronounced tendency to form petite cells. Similarly, a triple mutant strain, blocked simultaneously in the CDP-choline pathway by a combination of *cpt1* and *ept1* mutations and in the methylation pathway by the *cho2* mutation, displayed high tendency to form petites (Figure 3).

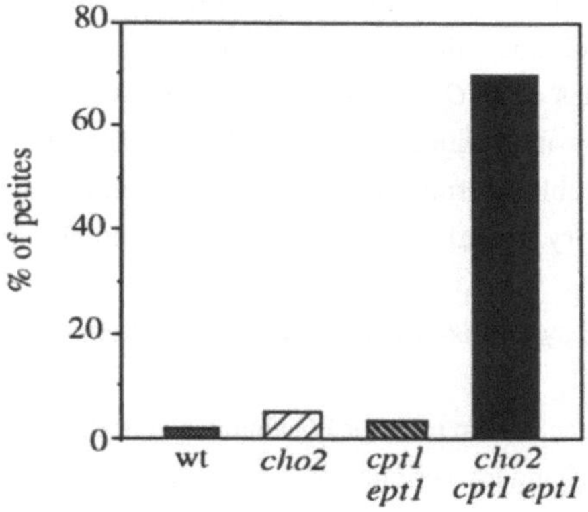

FIGURE 3. The accumulation of respiratory deficient cells during exponential phase of growth. Cultures (*wt*; *cho2*; *cpt1 ept1*; *cho2 cpt1 ept1*) were grown from a respiratory competent colony in YEPD liquid medium . Cultures were successively diluted with fresh YEPD to maintain log-phase of growth. After two days samples were taken, diluted and spread on YEP medium containing 0.1% glucose and 3% glycerol, in order to distinguish between respiratory competent and deficient colonies. After 7 days at 30°C petites were scored as percentage of total colonies.

Our results indicate that, at least in the case of the *cho2*, *cptl*, *eptl* triple mutant strain, these non-respiring cells are mitochondrial *rho-* petites that are not able to properly maintain their mitochondrial genome. Most likely, the generation of petites under these conditions is due to the markedly changed membrane phospholipid composition. All three of the above mentioned yeast strains which have a tendency to accumulate petits: *chol*; *psdl* as well as the *cho2*, *cptl*, *eptl* triple mutant, show abnormalities in the phospholipid composition of their membranes. However, due to the different nature of the mutations, the phospholipid profiles of these mutants differ dramatically: For example, *chol* cells lack any detectable PS (Atkinson et al. 1980), while *psdl* cells have decreased levels of PE (Trotter et al., 1993) and *cho2*, *cptl*, *eptl* cells display significantly decreased levels of PC (6-10% of total phospholipids). Apparently, mitochondrial biogenesis and/or maintenance is more sensitive to compositional changes in the membranes than other cellular functions. The mechanisms leading from compositional changes in mitochondrial membranes to mitochondrial abnormalities are under investigation.

Defects in transcriptional regulation. Yeast strains carrying mutations in the *CHO1*, *CHO2* and *OPI3* structural genes (Figure 1) are not able to repress properly the transcription of the coordinately regulated phospholipid genes in response to inositol. Normal regulation is restored if a metabolite, which enters the PC biosynthetic pathway downstream of the genetic block, is supplied exogenously. Thus, *INO1* regulation is restored in *cho1* mutants if ethanolamine, monomethylethanolamine (MME), dimethylethanolamine (DME) or choline is supplied. *cho2* mutants respond to each of the three methylated species, and *opi3* mutants to DME or choline (Letts and Henry, 1985; McGraw and Henry, 1989, Summers et al. 1988). These studies suggest that PC biosynthesis is somehow required for correct regulation of phospholipid biosynthesis. However, the exact nature of this signal remains elusive. Our recent studies of *INO1* expression in mutants blocked in the CDP-choline pathway, alone or in combination with the *cho2* mutation indicate that:

1. None of the soluble intermediates in the CDP-choline pathway is responsible for generation of a regulatory signal necessary for proper transcription of phospholipid biosynthetic genes.

2. The route of PC synthesis (methylation or CDP-choline pathway) is not important in the regulation.

3. Yeast cells may be able to monitor PC abundance or the actual formation of PC and respond accordingly, modifying transcription of coordinately regulated genes.

Search for regulatory mutants. Recently, we have begun a search for mutants that are able to regulate correctly the transcription of the *INO1* gene even when a defect in PC biosynthesis, caused by a *cho2* mutation, is present in the strain. A fully regulated gene fusion containing the *INO1* promoter and a portion of the coding sequence fused in frame to the *lacZ* gene, has been integrated in single copy at the *URA3* locus of a wild type strain (Lopes and Henry 1991). Subsequently, the *CHO2* gene was disrupted in this strain. The colonies of the parental *cho2* strain containing the fusion construct are blue when grown in the absence of inositol and choline on medium containing the chromogenic substrate, X-gal. While wild-type colonies are white in the presence of inositol alone, the colonies of a *cho2* strain are blue under the same conditions as a result of a misregulation of the *INO1* gene. This misregulation is corrected by addition of choline in the medium resulting in a restoration of PC biosynthesis via

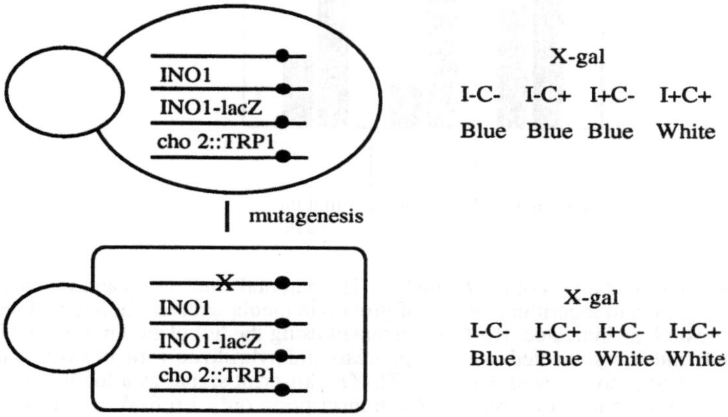

FIGURE 4. Mutagenic strategy to isolate mutants that are able to regulate correctly the transcription of phospholipid biosynthetic genes in a *cho2* genetic background. A strain harboring an *INO1-lacZ* fusion as well as a disrupted *CHO2* gene was mutagenized with EMS. While the parental *cho2* strain is not able to repress transcription of the *INO1* gene in the presence of inositol alone (I+C-) resulting in a blue color of colonies under these conditions, potential mutants are able to do so, displaying a white phenotype on the medium with inositol and without choline (I+C-).

the CDP-choline pathway. Using the blue-white phenotype, we screened for mutants able to repress the fusion construct in the presence of inositol and absence of choline in a *cho2* background. This repression would be indicated by a white phenotype on plates containing inositol alone (Figure 4). So far we have identified 2 mutants in this screen that are dark blue on medium without inositol and choline and only light blue on medium with inositol and without choline, indicating at least partial ability to repress the *INO1* gene in the presence of inositol alone. This ability to repress the *INO1* gene in a *cho2* background, in response to inositol alone, was confirmed by Northern blot analysis of *INO1* mRNA (Figure 5). The two mutants have the same phospholipid composition as a parental *cho2* strain, indicating that the new mutations do not bypass or suppress the *cho2* block but rather somehow modify the requirement for ongoing PC biosynthesis for the correct regulation of *INO1*.

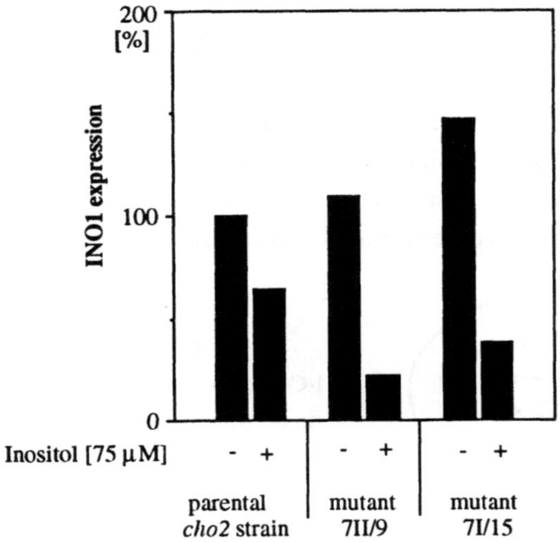

FIGURE 5. Northern blot analysis of *INO1* mRNA. The parental *cho2* strain and two mutants (7II/9; 7I/15) were grown to logarithmic phase of growth in media without choline and with or without 75 μM inositol, as indicated. RNA was isolated using the hot phenol method of Elion and Warner (1984) and was subjected to electrophoresis and hybridized with an *INO1* specific probe. *INO1* mRNA was normalized using the *TCM1* transcript signal as a loading control. The parental *cho2* strain is able to repress *INO1* transcription only 1.6 fold as a response to the presence of inositol in medium, while both mutants respond by repressing *INO1* transcription approximately 5 fold.

Further analysis of *INO1* regulation in these new mutants in combination with existing mutants defective in other aspects of signal transduction, should enable us to gain new insights into the complex mechanisms controlling phospholipid biosynthesis in yeast.

ACKNOWLEDGEMENTS

We wish to thank Dr. V.Bankaitis (University of Alabama at Birmingham) for providing the *cki1* mutant and Dr. R. Bell (Duke University Medical Center) for providing the *cpt1, ept1* mutant. This work was supported by grant GM 19629 from the National Institutes of Health to SAH.

REFERENCES

Atkinson K, Fogel S and Henry SA (1980) Yeast mutant defective in phosphatidylserine synthase. J. Biol. Chem. 255: 6653-6661.

Bjornstad P and Bremer J (1966) *In vivo* studies on pathways for the biosynthesis of lecithin in the rat. J. of Lipid Res. 7: 38-45.

Bremer J, Figard P and Greenberg D (1960) The biosynthesis of choline and its relation to phospholipid metabolism. Biochim. Biophys. Acta 43: 477-488.

Carman GM and Henry SA (1989) Phospholipid biosynthesis in yeast. Ann. Rev. Biochem. 58: 635-669.

Clancey CJ, Chang S and Dowhan W (1993) Cloning of a gene (*PSD1*) encoding phosphatidylserine decarboxylase from *Saccharomyces cerevisiae* by complementation of an *Escherichia coli* mutant. J. Biol. Chem. 268: 24580-24590.

Elion EA and Warner JR (1984) The major promoter element of rRNA transcription in yeast lies 2kb upstream. Cell 39: 663-673.

Greenberg ML, Klig LS, Letts VA, Loewy BS and Henry SA (1983) Yeast mutant defective in phosphatidylcholine synthesis. J. Bacteriol. 153: 791-799.

Hjelmstad RH and Bell RM (1991) *sn*-1,2-diacylglycerol choline- and ethanolaminephosphotransferases in *Saccharomyces cerevisiae*. Nucleotide sequence of the *EPT1* gene and comparison of the *CPT1* and *EPT1* gene products. J. Biol. Chem. 266: 5094-5103.

Hosaka K, Tsumomu K and Yamashita S (1989) Cloning and characterization of the yeast *CKI* gene encoding choline kinase and its expression of *Escherichia coli*. J. Biol. Chem. 264: 2053-2059.

Kennedy EP and Weiss SB (1956) The function of cytidine coenzymes in the biosynthesis of phospholipids. J. Biol. Chem. 222: 193-214.

Kovac L, Gbelska I, Poliachova V, Subik J and Kovacova V (1980) Membrane mutants: A yeast mutant with a lesion in phosphatidylserine biosynthesis. Eur. J. Biochem. 111: 491-501.

Letts VA and Henry SA (1985) Regulation of phospholipid synthesis in phosphatidylserine synthase-deficient (*cho1*) mutants of *Saccharomyces cerevisiae*. J. Bacteriol. 163: 560-567.

McGraw P and Henry SA (1989) Mutations in the *Saccharomyces cerevisiae opi3* gene: Effects on phospholipid methylation, growth and cross-pathway regulation of inositol synthesis. Genetics 122: 317-330.

McMaster CR and Bell RM (1994) Phosphatidylcholine biosynthesis in *Saccharomyces cerevisiae*: Regulatory insights from studies employing null and chimeric *sn*-1,2-diacylglycerol choline- and ethanolamine- phosphotransferases. J. Biol. Chem. 269: 28010-28016.

Paltauf F, Kohlwein SD and Henry SA (1992) Regulation and compartmentalization of lipid synthesis in yeast. Molecular Biology of the Yeast *Saccharomyces cerevisiae* Eds. J. Broach, E. Jones and J. Pringle. Cold Spring Harbor Press. 415-500.

Preitschopf W, Luckl H, Summers E, Henry SA, Paltauf F and Kohlwein SD (1993) Molecular cloning of the yeast *OPI3* gene as a high copy number suppressor of the *cho2* mutant. Curr. Genet. 23: 95-101.

Steiner MR and Lester RL (1972) *in vitro* studies of phospholipid biosynthesis in *Saccharomyces cerevisiae*. Biochim. Biophys. Acta 260: 222-243.

Summers EF, Letts VA, McGraw P and Henry SA (1988) *Saccharomyces cerevisiae cho2* mutants are deficient in phospholipid methylation and cross-pathway regulation of inositol synthesis. Genetics 120: 909-922.

Trotter PJ, Pedretti J and Voelker DR (1993) Phosphatidylserine decarboxylase from *Saccharomyces cerevisiae*. J. Biol. Chem. 268: 21416-21424.

Trotter PJ and Voelker DR (1995) Identification of a non-mitochondrial phosphatidylserine decarboxylase activity (PSD2) in the yeast *Saccharomyces cerevisiae*. J. Biol. Chem. 270: 6062-6070.

Waechter C and Lester R (1973) Differential regulation of the *N*-methyltransferases responsible for phosphatidylcholine synthesis in *Saccharomyces cerevisiae*. Arch. Biochem. Biophys. 158: 401-410.

RESYNTHESIS OF THE CELL SURFACE POOL OF PHOSPHATIDYLINOSITOL

Daniel J. Sillence and Martin G. Low
Department of Physiology,
University College London,
LONDON, WC1E 6JJ,
UK
ucgbdaj@ucl.ac.uk

Relatively little information is available about the transbilayer distribution of phosphatidylinositol (PtdIns) in plasma membranes. In order to address this question, *B. thuringiensis* PtdIns phospholipase C was used as a probe for cell surface PtdIns in a variety of mammalian cell lines. Addition of PtdIns phospholipase C to [^{3}H] inositol labelled bovine aortic endothelial cells hydrolysed ~9% of the total PtdIns levels after 5 minutes (similar results were obtained for J774 macrophages, NIH3T3 fibroblasts, NRK fibroblasts and rat cortical astrocytes).

Although inositol phospholipids are distributed widely throughtout the cell, only inositol phospholipids on the inner leaflet of the plasma membrane serve as substrates for an agonist activated phospholipase C. Such action generates the intracellular second messengers, diacylglycerol and Ins(1,4,5)P_3 (Berridge and Irvine 1989) The replenishment of agonist sensitive phosphoinositides during continued stimulation requires the recycling of the inositol moiety of Ins(1,4,5)P_3 (Berridge and others 1989). In most cell types inner leaflet diacylglycerol is also efficiently recycled to reform PtdIns (Kirk and others 1981). However the mechanisms which regulate PtdIns resynthesis are not well understood. The first step in these recycling reactions involves the phosphorylation of diacylglycerol which yields phosphatidic acid (Allan and others 1978). The recycling reactions continue with the activation of phosphatidic acid which is metabolised to CDP-Diacylglycerol. The recycling reactions culminate in the combination of CDP-Diacylglycerol and inositol to yield PtdIns (Sillence and Downes 1993).

Previously reported work has indicated that the rapid resynthesis that occurs following agonist stimulation is not directly dependent on the presence of agonist (Monaco and Adelson 1991). Therefore it seems probable that the changes in the level of the substrates or products of PtdIns-phospholipase C activity regulate PtdIns resynthesis rather than the production of second

NATO ASI Series, Vol. H 96
Molecular Dynamics of Biomembranes
Edited by Jos A. F. Op den Kamp

messengers. If this is the case then artificial degradation of cell surface PtdIns by *B.thuringiensis* PtdIns-phospholipase C provides an independent method for studying the control of PtdIns resynthesis. Cleavage of cell surface PtdIns leads to the generation of cell surface diacylglycerol, since diacylglycerol has been reported to undergo rapid transbilayer movement (Ganong and Bell 1984; Allan and others, 1978) it was expected that this would constitute an immediate signal for PtdIns synthesis. In order to measure PtdIns synthesis we used cells acutely labelled with [^{3}H]inositol. Under these conditions the labelling of inositol lipids and phosphates can reflect changes in both mass and specific radioactivity. Changes in specific activity can be useful to measure the flux of the inositol moiety through the signalling pathway.

B.thuringiensis PtdIns phospholipase C hydrolyses cell surface PtdIns in intact cells

Addition of *B. thuringiensis* phospholipase C to equilibrium labelled bovine aortic endothelial cells led to the clear accumulation of cyclic inositol monophosphate in the medium accounting for 9±1 % of the total [^{3}H]PtdIns after 5 minutes. Trypan blue exclusion indicated >99% cell viability suggesting that adherent dead cells are not a significant source of [^{3}H]PtdIns. Even long term exposure to the phospholipases used in this study did not increase the permeability of the cells to trypan blue or cause the release of the cytosolic marker lactate dehydrogenase.

The action of PtdIns phospholipase C on intact cells was characterised by a rapid rise in the level of medium cyclic inositol monophosphate which was accompanied by a corresponding decrease in cellular PtdIns. After the initial rapid rise further small but significant increases in medium inositol monophosphate could be detected. The further increases in medium cyclic inositol monophosphate were dependent on the slow transport of intracellular PtdIns to the cell surface via the secretory patway since brefeldin A inhibited these increases without any detectable effect of the total PtdIns levels (Figure 1; Miller and others 1992).
Next, it was of interest to examine he effects of PtdIns phospholipase C on PtdIns synthesis. It was thought that if PtdIns derived diacylglycerol was a key trigger for PtdIns resynthesis then production of cell surface diacylglycerol should lead to a rapid increase in intracellular PtdIns synthesis (Allan and others 1978; Ganong and Bell 1984) .

Figure 1 Medium inositol monophosphate accumulation in Bovine Aortic Endothelial cells

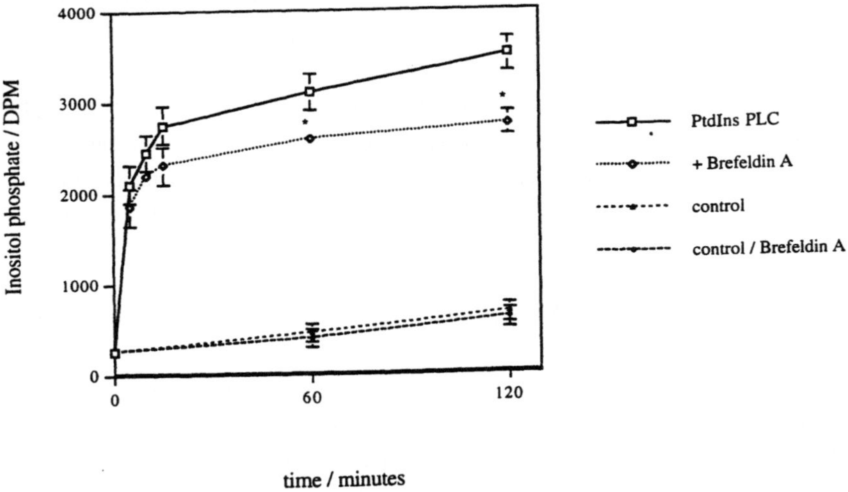

BAE cells were grown to confluence in MEM containing 10% FBS and 2μCi [3H]inositol for 60 hours. Monolayers were incubated in the presence of 0.5U/ml PtdIns-PLC and 10μg/ml brefeldin A as indicated. All incubations in this experiment were in the presence of 100μg/ml BSA. Total [3H]PtdIns levels were 38 852±2539 d.p.m./well. *Significantly different from PtdIns-PLC $P<0.05$.

Addition of phospholipase C to intact cells does not lead to rapid PtdIns resynthesis.

Bovine aortic endothelial cells express a purinergic receptor of the P_2-type which is linked to an intracellular phospholipase C (Purkiss and others 1992). Activation of this receptor by the addition of the agonist ATPgS resulted in a rapid increase in the labelling of inositol phospholipids in acutely labelled cells. This effect is due to the diacylglycerol generated from receptor activation being rapidly phosphorylated and resynthesised into PtdIns (Sillence and Downes 1992). In contrast, the generation of diacylglycerol from the cell surface PtdIns by *B.thuringiensis* phospholipase C did not lead to a rapid increase in PtdIns labelling (Table 1). This result is consistent with the hypothesis that the rapid resynthesis of PtdIns that occurs during agonist stimulation may be dependent on the activation of diacylglycerol kinase. However, addition of PtdIns-PLC and ATPgS together did not lead to more PtdIns labelling when compared to the addition of these agents alone (results not shown). The lack of phosphorylation of outer leaflet diacylglycerol could also be due to the cell surface PtdIns being of a different fatty acid composition to the intracellular PtdIns. Indeed, a diacylglycerol kinase has been characterised in swiss 3T3 cells which preferentially phosphorylated arachidonic acid rich diacylglycerol. This diacylglycerol kinase was also shown to be important in PDGF stimulated PtdIns turnover (MacDonald and others 1988a; MacDonald and others 1988b). It is of interest to note however, that *B. thuringiensis* PtdIns-PLC, when added to intact Jurkat T-cells stimulates the production of arachidonic acid rich diacylglycerol which is not phosphorylated by the intracellular kinase (van der Bend and others 1994). Consistent with the above, the metabolic fate of exogenously added 1-steoryl,2-arachidonyl diacylglycerol was predominantly PtdCho and triacylglycerol in intact NIH 3T3 fibroblasts (Florin Christensen and others 1992). It is possible that rather than being phosphorylated these diacylglycerols are substrates for the reverse reaction of a cell surface or endosomal sphingomyelin synthase (van Helvoort and others 1994).

PtdIns labelled in response to the addition of propranolol is available to ATPgS stimulated phospholipase C

Propranolol is an inhibitor of phosphatidate phosphohydrolase and leads to the accumulation of phosphatidate which is then channelled to PtdIns synthesis (Eichberg and others 1979). Propranolol stimulated the labelling of PtdIns in acutely labelled bovine aortic endothelial cells. Propranolol-induced increases in labelling of intracellular inositol phosphate were observed in the presence and absence of the purinergic agonist ATPgS (Table 2). The above result suggests that the PtdIns labelled in response to propranolol was available to the intracellular agonist activated phospholipase C. In 1321 N1 cells addition of propranolol lead to the accumulation of CMP-phosphatidate in the endoplasmic reticulum (Sillence and Downes 1993). Hence, it could be argued that the above result suggests that endomplasmic recticular PtdIns can take place in the PtdIns cycle. However, phosphatidate phosphohydrolase has also been localised to the plasma membrane in some cells (Day and Yeaman 1992). It is therefore possible that propranolol induces the synthesis of PtdIns at locations other than the endoplasmic reticulum.

Table 1

Stimulation of PtdIns resynthesis in BAE cells

Incubation	Time/(min)	Radioactivity (d.p.m/well)
Control	(1.5)	2307± 353
PtdIns-PLC	(1.5)	1998± 268
ATPgS	(5)	27 140± 3742
PtdIns-PLC	(5)	2818± 139
ATPgS	(10)	46 022± 2520
PtdIns-PLC	(10)	3705± 136
Control	(10)	3893± 323

Monolayers were labelled for 1 hour in the presence of 5μCi [^{3}H]inositol. ATPgS (75 μM), PtdIns-PLC (0.36 U/ml) were added at time 0. All incubations contained 5mM LiCl. Results are expressed ±S.E.

Table 2

ATPgS and propranolol-stimulated PtdIns and intracellular inositol phosphate labelling

Incubation	Radioactivity (d.p.m./well) in intracellular InsP	Radioactivity (d.p.m./well) in cellular PtdIns
Control	5109± 214	7290± 479
ATPgS	10 484± 616	54 438± 4454
Propranolol	10 521± 860	136 937±10236
Propranolol/ATPgS	21 777± 1320	160 110±11259

Monolayers were labelled for 1 hour in the presence of 5μCi [^{3}H]inositol. ATPgS (75 μM) was added for 5 min. Preincubations with propranolol (250 μM) were for the second 30 minutes of the incubation period. All incubations contained LiCl (5 mM). The results are expressed ±S.E.

PtdIns labelled in response to the addition of propranolol or ATPgS slowly becomes available to extracellular phospholipase C

In the light of the results in figure 1 we would predict that newly synthesised PtdIns would be transported to the cell surface at a similar rate to the second phase of PtdIns phospholipase C induced PtdIns hydrolysis. Table 3 shows that the PtdIns pools labelled in response to the addition of propranolol or ATPgS were inaccessible to *B. thuringiensis* phospholipase C and was transported to the cell surface at a similar rate to vesicular transport (1-2%/hour; Table 3).

Table 3

Sensitivity of PtdIns synthesised in response to propranolol and ATPgS to extracellular PtdIns-PLC

Incubation	Medium InsP (d.p.m./well)	[^{3}H]PtdIns (d.p.m./well)
Control	190± 23	9997± 275
ATPgS	252± 12	115 956± 6641
Propranolol	241± 9	68 075± 6215
PtdIns-PLC	286± 40	11 594± 1776
ATPgS/PtdIns-PLC	657± 51*	93 933± 4691
Propranolol/PtdIns-PLC	1047± 83*	81 930± 3424

Cells were labelled in the presence of 5μCi of [^{3}H]inositol for 1 hour. ATPgS (75 μM) or Propranolol (250 μM) were then added as indicated. After 20 min PtdIns-PLC (0.36 U/ml) and then incubated for another 20 min. Results are expressed ±S.E. * Significantly different from relevant control P<0.05.

Chronic treatment with B. thuringiensis PtdIns phospholipase C does lead to enhanced PtdIns synthesis

Table 4 compares the effects of *B. cereus* phosphatidylcholine phospholipase C and *B. thuringiensis* PtdIns phospholipase C on PtdIns labelling. It was found that extended incubations with *B. thuringiensis* PtdIns phospholipase C but not *B. cereus* PtdCho phospholipase C led to the increased labelling of PtdIns. This suggests that PtdIns synthesis could only be observed during prolonged treatment under conditions were significant uptake of the phospholipase may have occurred. It is possible that the activation of PtdIns synthesis correlates with the slow depletion of intracellular PtdIns levels that may take place during prolonged incubations with PtdIns phospholipase C.

Table 4

Effects of chronic treatment with PtdIns-PLC and PtdCho-PLC on the labelling of PtdIns.

Incubation	$[^3H]$PtdIns	Medium $[^3H]$InsP	Intracellular $[^3H]$inositol
Control	13 677± 931	1319± 92	17 839±1135
PtdIns-PLC	22 694± 1148*	4699± 117	19 667±1338
PtdCho-PLC	12 791± 1903	1780± 305	28 956±1332
PC+PI-PLC	23 911± 1788*	5490± 286	37 371± 385

Monolayers were incubated in the presence of 2% FBS and 1µCi of $[^3H]$inositol and 1U/ml *B.cereus* PtdCho-PLC or 1U/ml PtdIns-PLC for 16 hours. After this time 300µl of medium was taken for the analysis of inositol phosphates. Results are expressed d.p.m./well±S.E. * Significantly different from control $P<0.05$.

Conclusions

1) A cell surface pool of PtdIns was found in a variety of mammalian cell types. Such a cell surface pool of PtdIns could be a substrate for a cell surface PtdIns-PLC which has been found in several mouse fibroblast cell lines and in rat kidney fibroblasts (Ting and Pagano 1990; Ting and Pagano 1991).

2) Newly synthesised PtdIns is resistant to the action of the phospholipase and appears on the surface of the cell at a similar rate to bulk flow in the secretory pathway.

3) The metabolic fate of the most of the cell surface diacylglycerol produced by the hydrolysis of cell surface PtdIns is different from diacylglycerol formed from the agonist activated phospholipase C.

4) Prolonged exposure to PtdIns-PLC but not PtdCho-PLC does lead to enhanced PtdIns synthesis as judged by an increase in the labelling of PtdIns under non-equilibrium conditions. It is suggested that the stimulation of PtdIns synthesis may be more closely linked to intracellular decreases in PtdIns rather than to the generation of PtdIns-derived diacylglycerol.

References.

Allan D, Thomas P, and Michell RH (1978) Rapid transbilayer diffusion of 1,2-diacylglycerol and its relevance to control of membrane curvature. Nature 276 5685: 289-90.

Berridge MJ, Downes CP and Hanley MR (1989) Neural and developmental actions of lithium: a unifying hypothesis. Cell 59: 411-9.

Berridge MJ and Irvine RF (1989) Inositol phosphates and cell signalling. Nature 341: 197-205.

Day CP and Yeaman SJ (1992) Identification of a plasma membrane phosphatidate phosphohydrolase. Biochim. Biophys. Acta. 1127 87-94.

Eichberg J, Gates J and Hauser G (1979) The mechanism of modification by propranolol of the metabolism of phosphatidyl-CMP (CDP-diacylglycerol) and other lipids in the rat pineal gland. Biochim. Biophys. Acta. 573: 90-107.

Florin Christensen J, Florin Christensen M, Delfino JM, Stegmann T and Rasmussen H (1992) Metabolic fate of plasma membrane diacylglycerols in NIH 3T3 fibroblasts. J. Biol. Chem. 267: 14783-9.

Ganong BR and Bell RM (1984) Transmembrane movement of phosphatidylglycerol and diacylglycerol sulfhydryl analogues. Biochemistry 23: 4977-83.

Kirk CJ, Michell RH and Hems DA (1981) Phosphatidylinositol metabolism in rat hepatocytes stimulated by vasopressin. Biochem. J. 194: 155-65.

MacDonald ML, Mack KF, Richardson CN, and Glomset JA (1988a) Regulation of diacylglycerol kinase reaction in Swiss 3T3 cells. Increased phosphorylation of endogenous diacylglycerol and decreased phosphorylation of didecanoylglycerol in response to platelet-derived growth factor. J. Biol. Chem. 263: 1575-83.

MacDonald ML, Mack KF, Williams BW, King WC and Glomset JA (1988b) A membrane-bound diacylglycerol kinase that selectively phosphorylates arachidonoyl-diacylglycerol. Distinction from cytosolic diacylglycerol kinase and comparison with the membrane-bound enzyme from Escherichia coli. J. Biol. Chem. 263: 1584-92.

Miller SG, Carnell L and Moore HH (1992) Post-Golgi membrane traffic: brefeldin A inhibits export from distal Golgi compartments to the cell surface but not recycling. J. Cell Biol. 118: 267-83.

Monaco ME and Adelson JR (1991) Evidence for coupling of resynthesis to hydrolysis in the phosphoinositide cycle. Biochem J. 279: 337-341

Purkiss J, Owen PJ, Jones JA and Boarder MR (1992) Stimulation of phosphatidic acid synthesis in bovine aortic endothelial cells in response to activation of P2-purinergic receptors. Biochem. Pharmacol. 43: 1235-42.

Sillence DJ and Downes CP (1992) Lithium treatment of affective disorders: effects of lithium on the inositol phospholipid and cyclic AMP signalling pathways. Biochim. Biophys. Acta 1138: 46-52.

Sillence DJ and Downes CP (1993) Subcellular distribution of agonist-stimulated phosphatidylinositol synthesis in 1321 N1 astrocytoma cells. Biochem. J. 290: 381-387

Ting AE and Pagano RE (1990) Detection of a phosphatidylinositol-specific phospholipase C at the surface of Swiss 3T3 cells and its potential role in the regulation of cell growth. J. Biol. Chem. 265: 5337-40.

Ting AE and Pagano RE (1991) Density-dependent inhibition of cell growth is correlated with the activity of a cell surface phosphatidylinositol-specific phospholipase C. Eur. J. Cell Biol. 56: 401-6.

van der Bend R, de Widt J, Hilkmann H and van Blitterswijk W (1994) Diacylglycerol kinase in receptor-stimulated cells converts its substrate in a topologically restricted manner. J. Biol. Chem. 269: 4098-102.

van Helvoort A, van't Hof W, Ritsema T, Sandra A and van Meer G (1994) Conversion of diacylglycerol to phosphatidylcholine on the basolateral surface of epithelial (Madin-Darby canine kidney) cells. Evidence for the reverse action of a sphingomyelin synthase. J Biol Chem 269: 1763-9.

THE GlcNAc-PI DE-N-ACETYLASE OF GLYCOSYLPHOSPHATIDYLINOSITOL (GPI) BIOSYNTHESIS IN *TRYPANOSOMA BRUCEI*

Deepak K. Sharma and Michael A.J. Ferguson

Department of Biochemistry

University of Dundee

Dundee DD1 4HN

Introduction

Glycosyl-phosphatidylinositol (GPI) membrane anchors are present in organisms at most stages of eukaryotic evolution, including protozoa, yeast, slime moulds, invertebrates and vertebrates, and are found on a diverse range of proteins. They are primarily responsible for the anchoring of cell-surface proteins in the plasma membrane (or in some cases to the topologically equivalent lumenal surface of secretory vesicles) and may be considered as an alternative to the hydrophobic transmembrane polypeptide anchor of type-1 integral membrane proteins (Fig. 1). Many other functions have been proposed (though some remain controversial) for GPI anchors, including roles in intracellular sorting, transmembrane signalling, and novel endocytic processes. Most of these proposed functions are dependent on the hypothesis that a GPI anchor might allow proteins to associate in specialised membrane microdomains. These putative functions, as well as the structure, biosynthesis and distribution of GPI anchors, have been extensively reviewed (Englund, 1993; McConville and Ferguson, 1993; Brown, 1993; Ferguson, 1994).

The mRNAs of GPI-anchored proteins encode an N-terminal signal sequence, to direct the protein to the endoplasmic reticulum (ER), and a C-terminal GPI-attachment signal sequence. This sequence is cleaved with the concomitant addition of a preassembled GPI precursor. The anchor attachment site (ω) may be one of six amino acid residues, all of which have small side-chains (Ala, Asn, Asp, Cys, Gly or Ser) (Moran et al., 1991; Gerber et al., 1992). In addition, the residue at the $\omega + 2$ position is restricted to Ala, Gly or Ser (Gerber et al., 1992). The transfer of GPI precursor to protein occurs in the ER and GPI-anchored proteins are subsequently transported through the Golgi stacks to (generally) the plasma membrane.

NATO ASI Series, Vol. H 96
Molecular Dynamics of Biomembranes
Edited by Jos A. F. Op den Kamp

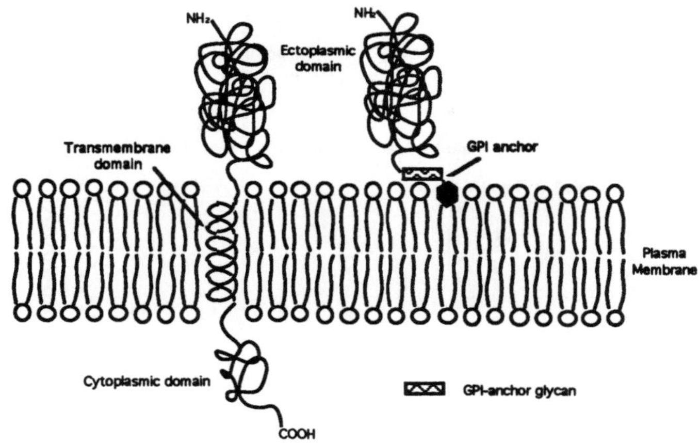

Fig. 1 Comparison of a type-1 transmembrane protein (left) with a GPI-anchored protein (right)

Many of the details of the biosynthesis of GPI precursors have come from studying African trypanosomes. The tsetse-fly transmitted African trypanosomes, which cause human sleeping sickness and a variety of livestock diseases, are able to survive in the mammalian bloodstream by virtue of their dense cell-surface coat. This coat consists of 10 million copies of a 55 kDa GPI anchored glycoprotein called the variant surface glycoprotein (VSG) (Cross, 1990). The relative abundance of the VSG protein in *Trypanosoma brucei* has made this organism extremely useful for the study of GPI anchor biosynthesis. The structure of the VSG GPI anchor is known (Ferguson et al., 1988), see Fig. 2, and the principal features of the GPI biosynthetic pathway in trypanosomes were elucidated using a cell-free system based on washed trypanosome membranes (Masterson et al., 1989, 1990; Menon et al., 1990b). The pathway is summarised in Fig. 3. The first step in the pathway involves the transfer of GlcNAc from UDP-GlcNAc to endogenous phosphatidylinositol (PI), via a sulphydryl-dependent GlcNAc-transferase (Milne et al., 1992), to form GlcNAc-PI which is rapidly de-N-acetylated (Doering et al., 1989; Milne et al., 1994) to form glucosaminyl-PI (GlcN-PI). Three α-mannose residues are sequentially transferred onto GlcN-PI from dolichol-phosphate-mannose (Dol-P-Man) (Menon et al., 1990a) to form the intermediate Manα1-2Manα1-6Manα1-4GlcN-PI. At least this much of the pathway is believed to occur on the cytoplasmic face of the endoplasmic reticulum (Vidugiriene and Menon, 1993; 1994). Ethanolamine phosphate (EtNP) is then transferred from phosphatidylethanolamine (Menon and Stevens, 1992; Menon et al., 1993) to the terminal mannose residue to form EtNP-6Manα1-2Manα1-6Manα1-4GlcN-PI (known as glycolipid A'). This species then undergoes a series of fatty acid remodelling

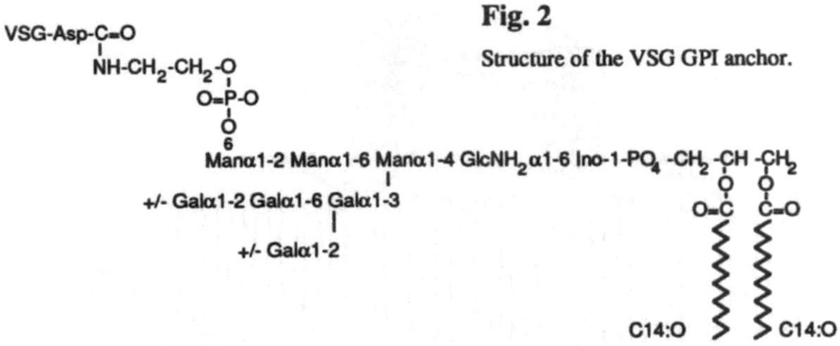

Fig. 2

Structure of the VSG GPI anchor.

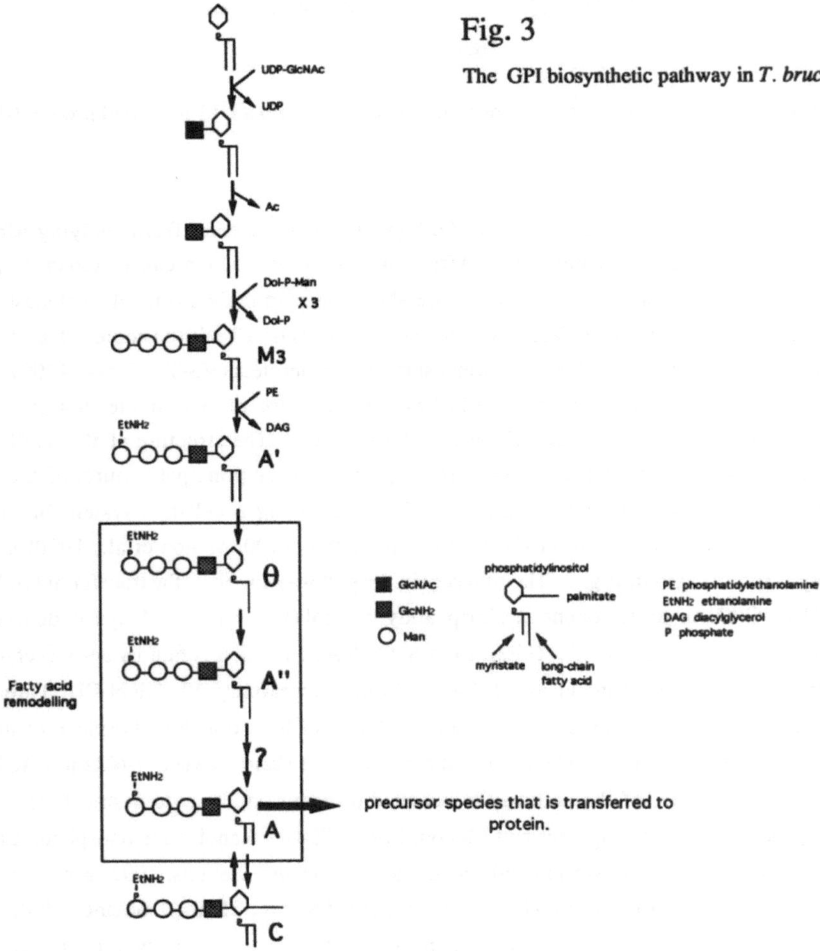

Fig. 3

The GPI biosynthetic pathway in *T. brucei*.

reactions (Masterson et al., 1990), whereby the fatty acids of the PI moiety are removed and replaced with myristate, to yield the mature GPI precursor glycolipid A. Concomitant with the formation of glycolipid A is the formation of glycolipid C (the inositol-acylated version of glycolipid A). Both glycolipid A and glycolipid C have been shown to be competent for transfer to VSG polypeptide when added exogenously to a trypanosome cell-free system (Mayor et al., 1991), although there is no evidence for the transfer of glycolipid C *in vivo*.

The GPI biosynthetic pathways in mammalian cells, reviewed in Englund (1993) and McConville and Ferguson (1993), and in yeast (Costello and Orlean, 1992; Sipos et al., 1994), as well as in other protozoan organisms such as *Toxoplasma* (Tomavo et al., 1992a,b) and *Plasmodium falciparum* (Gerold et al., 1994), appear to be broadly similar to that described above for the bloodstream form of *T.brucei*. Some notable differences in the mammalian GPI pathway include the almost quantitative inositol-acylation of all GPI intermediates from GlcN-PI onwards and the addition of extra ethanolamine phosphate groups. Similarly, the yeast GPI intermediates are also inositol-acylated from GlcN-PI onwards. The fatty acid remodelling reactions, as described above, appear to be unique to bloodstream form African trypanosomes. However, there is evidence for the exchange of diacyglycerol for alkylacylglycerol in mammalian cells (Singh et al., 1994) and for the exchange of diacyglycerol for ceramide in yeast (Conzelmann et al., 1992; Sipos et al., 1994).

The GlcNAc-PI de-N-acetylase

The second step of the GPI biosynthetic pathway involves the removal of the acetyl group from GlcNAc-PI to form $GlcNH_2$-PI (Doering et al., 1989), see Fig. 4. The enzyme resposible for this step, GlcNAc-PI de-N-acetylase, has been partially purified from *T.brucei* membranes and partially characterised with respect to substrate specificity (Milne et al., 1994). This study suggested that the enzyme has little or no specificity for the absolute configuration (D or L) of the *myo*-inositol ring and that the N-acetamido group is essential for substrate recognition. Unlike the mammalian enzyme (Stevens, 1993), the trypanosome de-N-acetylase showed no evidence of stimulation by GTP as a partially purified enzyme (Milne et al., 1994) or when *in situ* in trypanosome membranes (Sharma and Ferguson, unpublished data).

In order to improve the ananlysis of the trypanosome de-N-acetylase, a new assay was developed based on that described in (Milne et al., 1994). In the new assay synthetic GlcN-PI (Cottaz et al., 1993) is N-acetylated with $[^{14}C]$acetic anhydride to form GlcN$[^{14}C]$Ac-PI. The synthetic substrate was characterised by negative-ion electrospray mass spectrometry (Fig. 5)

Fig. 4

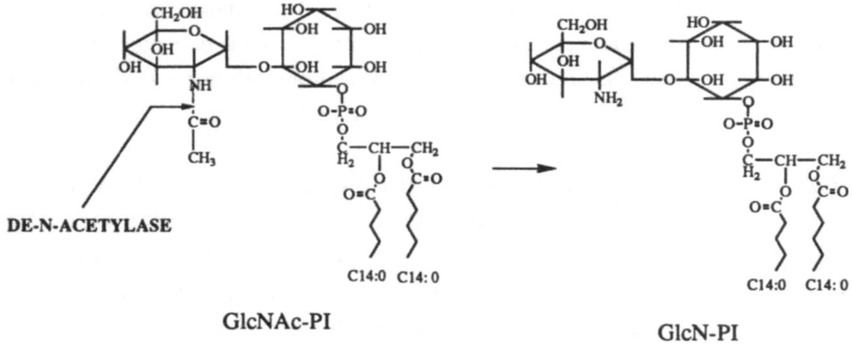

Fig. 5

ESMS scan of synthesisedGlcN[^{14}C]Ac-PI.
Major peek corresponds to synthesised substrate.

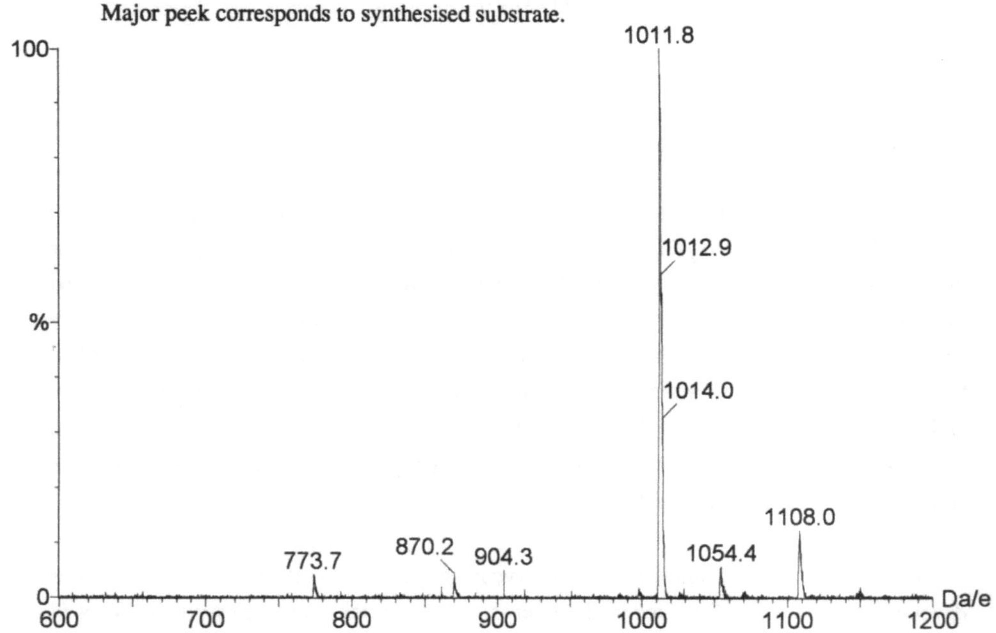

Fig. 6a

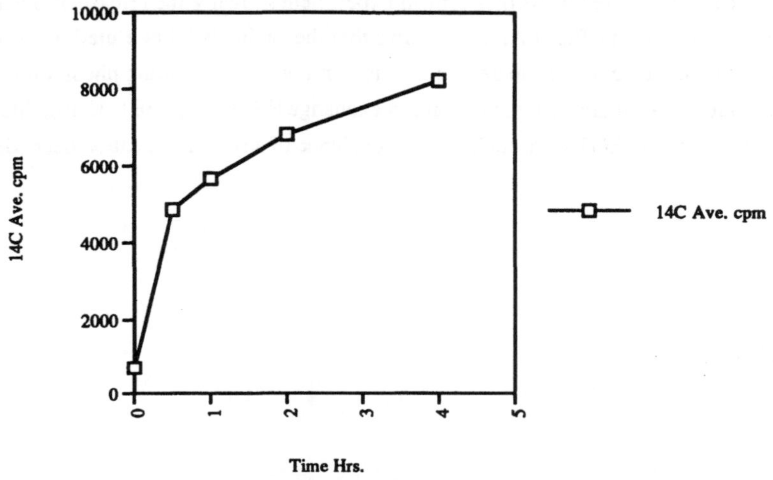
10000
8000
6000
4000
2000
0
14C Ave. cpm
0
1
2
3
4
5
Time Hrs.
14C Ave. cpm

Fig. 6b

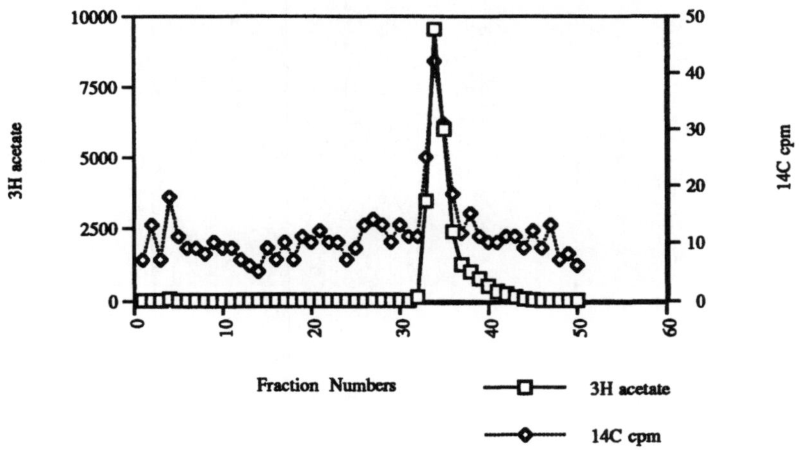
10000
7500
5000
2500
0
3H acetate
50
40
30
20
10
0
14C cpm
0
10
20
30
40
50
60
Fraction Numbers
3H acetate
14C cpm

and the specific activity was calculated following accurate measurement of the inositol content by GC-MS (Ferguson 1993). The release of [^{14}C]acetic acid from the substrate by the action of the de-N-acetylase was detected by measuring the radiolabel that no longer bound to a C8 reverse-phase cartridge. The results of a typical experiment showing the release of acetic acid against time can be seen in Fig. 6A. To confirm that the radiolabel that eluted from the C8 cartridge was due to the action of the de-N-acetylase, and not contaminating phospholipase(s), the labelled product was analysed by strong anion exchange HPLC (Fig. 6B). Using this assay in presence of EDTA and EGTA we could find no evidence of any divalent cation dependency.

Fig. 7

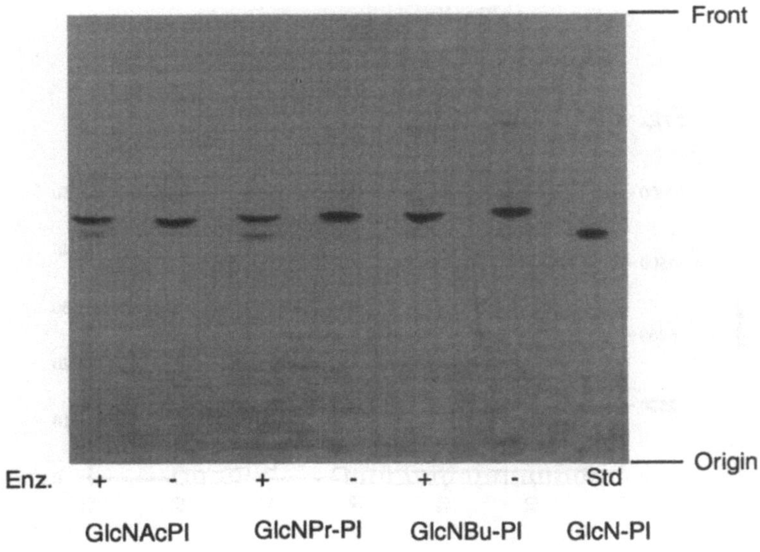

In order to probe the substrate specificity of the de-N-acetylase with respect to the nature of the N-acyl group, three GlcNR-PI substrates (radiolabelled in the fatty acid residues of the PI moiety) were prepared. To make these substrates, living trypanosomes were labelled with [^{3}H]myristic acid in the presence of PMSF, as described in (Milne et al., 1992). The presence of PMSF inhibits the addition of ethanolamine phosphate to the Man_3GlcN-PI intermediate (Masterson and Ferguson, 1991) and therefore label accumulates in Man_3GlcN-PI. This compound was converted to GlcN-PI by exhaustive digestion with jack bean α–mannosidase. Finally, the GlcN-PI was converted to GlcNAc-PI, GlcNPr-PI and GlcNBu-PI (where Ac, Pr and Bu indicate N-acetyl, N-propyl and N-butyl groups, respectively) by acylation with the appropriate anhydide (acetic, propionic and butyric anhydride).The formation of the correct products was confirmed by HPTLC. These three substrates were incubated with the crude de-N-acetylase preparation and the products were analysed by HPTLC and fluorography (Fig. 7). The results indicate that the enzyme can remove both N-acetyl and N-propyl groups but not N-butyl groups. These results suggest that the enzyme active site may be constrained with respect to the size of the acyl group that can be accommodated. Future studies will concentrtate on further defining the substrate specificity of the enzyme and on exploiting this information towards inhibitor design.

Acknowledgements

This work is supported by the Wellcome Trust. DKS thanks the MRC for a PhD studentship. MAJF is a Howard Hughes International Research Scholar. We thank Terry Smith and Lucia Güther for helpful comments and advice.

References

Brown, D. (1993) The tyrosine kinase connection - how GPI-proteins activate T-cells. Current Opinion in Immunol. **5**, 349-354.

Conzelman, A., Puoti, A., Lester, R. L. and Desponds, C. (1992) Two different types of lipid moieties are present in glycophosphoinositol-anchored membrane proteins of Saccharomyces cerevisiae. EMBO. **11**, 457-466.

Costello, L. C. a. O., P. (1992) Inositol acylation of a potential glycosyl phosphatidylinositol anchor precursor from yeast requires acyl coenzyme A. J. Biol. Chem. 8599-8603.

Cottaz, S., Brimacombe, J.S. and Ferguson, M.A.J. (1993) Parasite glycoconjugates. Part 1. The synthesis of some early and related intermediates in the biosynthetic pathway of glycosyl-phosphatidylinositol membrane anchors. J. Chem . Soc. Perkin Trans. **1** , 2945-2951

Cross, G. A. M. (1990a) Cellular and genetic aspects of antigenic variation in Trypanosomes. Annu. Rev. Cell Biol. **8**, 83-100.

Doering, T. L., Masterson, W. J., Englund, P. T. and Hart, G. W. (1989) Biosynthesis of the glycosyl-phosphatidylinositol membrane anchor of the trypanosome variant surface glycoprotein: origin of the non-acetylated glucosamine. J. Biol. Chem. **264**, 11168-11173.

Englund, P. T. (1993) The structure and biosynthesis of glycosyl phosphatidylinositol protein anchors. Annu. Rev. Biochem. **62**, 121-138.

Ferguson, M. A. J. (1993) Lipid modifications of proteins: A practical approach. (Hooper, N.M. and Turner, A.J., eds.) pp191-230, IRL Press, Oxford.

Ferguson, M. A. J. and Homans, S. W. (1988) Parasite glycoconjugates: towards the exploitation of their structure. Parasite Immunology. **10**, 465-479.

Ferguson, M.A.J. (1994) What can GPI do for you? Parasitology Today. **10**, 48-52.

Gerber, L. D., Kodukula, K. and Udenfriend, S. (1992) Phosphatidylinositol Glycan (PI-G) anchored membrane proteins: Amino acid requirements adjacent to the site of cleavage and PI-G attachment in the COOH-terminal signal peptide. J. Biol. Chem. **267**, 12168-12173.

Gerold, P., Dieckmannschuppert, A., Schwarz, R.T. (1994) Glycosyl-phosphatidylinositols synthesized by asexual erythrocytic stages of the malarial parasite, plasmodium-falciparum-candidates for plasmodial glycosylphosphatidylinositol membrane anchor precursors and pathogenicity factors. J. Biol. Chem. **269**, 2597-2606.

Masterson, W. J., Doering, T. L., Hart, G. W. and Englund, P. T. (1989) A novel pathway for glycan assembly: biosynthesis of the glycosyl-phosphatidylinositol anchor of the trypanosome variant surface glycoprotein. Cell. **56**, 793-800.

Masterson, W. J., Raper, J., Doering, T. L., Hart, G. W. and Englund, P. T. (1990) Fatty acid remodeling: A novel reaction sequence in the biosynthesis of trypanosome glycosyl Phosphatidylinositol membrane anchors. Cell. **62**, 73-80.

Mayor, S., Menon, A. K. and Cross, G. A. M. (1991) Transfer of glycosyl-phosphatidylinositol membrane anchors to polypeptide acceptors in cell-free system. J. Cell Biol. **114**, 61-71.

McConville, M. J. and Ferguson, M. A. J. (1993) The structure, biosynthesis and function of glycosylated-phosphatidylinositols in the parasitic protozoa and higher eukaryotes. Biochem J. **294**, 305-324.

Menon, A. K., Mayor, S. and Schwarz, R. T. (1990a) Biosynthesis of glycosyl-phosphatidylinositol lipids in *Trypanosoma* brucei: involvement of mannosyl-phosphoryldolichol as the mannose donor. EMBO J. **9**, 4249-4258.

Menon, A., Schwarz, R., Mayor, S. and Cross, G. A. M. (1990b) Cell-free synthesis of glycosyl-phosphatidylinositol precursors for the glycolipid membrane anchor of *Trypanosoma brucei* variant surface glycoproteins. J. Biol. Chem. **265**, 9033-9042.

Menon, A.K., Eppinger, M., Mayor, S., Schwarz, R.T. (1993) Phosphatidylethanolamine is the donor of the terminal phosphoethanolamine group in trypanosome glycosylphosphatidylinositols. EMBO J. **12**, 1907-1914.

Menon, A.K., Stevens, V.L. (1992) Phosphatidylethanolamine is the donor of the ethanolamine residue linking a glycosylphosphatidylinositol anchor to protein. J. Biol. Chem. **267**, 15277-15280.

Singh, N., Zoeller, R.A., Tykocinski, M.L., Lazarow, R.B., Tartakoff, A.M. (1994) Addition of lipid substitutents of mammalian protein glycosylphsophoinositol anchors. **14**, 21-31.

Sipos, G., Puoti, A., Conzelmann, A. (1994) Glycosylphosphatidylinositol membrane anchors in saccharomyces-cerevisiae - absence of ceramides from complete precursor glycolipids. EMBO J. **13**, 2789-2796.

Milne, K. G., Ferguson, M. A. J. and Masterson, W. J. (1992) Inhibition of the GlcNAc transferase of glycosylphosphatidylinositol anchor biosynthesis in African Trypanosomes. Eur. J. Biochem. **208**, 309-314.

Milne, K.G., Field, R.A., Masterson, W.J., Cottaz, S., Brimacombe, J.S. and Ferguson, M.A.J. (1994) Partial purification and characterization of the N-acetylglucosaminyl-phosphatidylinositol De-N-acetylase of glycosylphosphatidylinositol anchor biosynthesis in African Trypanosomes. J. Biol. Chem. **269**, 16403-16408.

Moran, P., Raab, H., Kohr, W.J., Caras, I.W. (1991) Glycophospholipid membrane anchor attachment - Molecular analysis of the cleavage attachment site. J. Biol. Chem. **266**, 1250-1257.

Tomavo, S., Dubrametz, J.-F. and Schwarz, R.T. (1992) Biosynthesis of glycolipid precursors for glycosylphosphatidylinositol membrane anchors in a *Toxoplasma gondii* cell-free system. J. Biol. Chem. **267**, 21446-21458.

Tomavo, S., Dubremetz, J.-F. and Schwarz, R. T. (1992) A family of glycolipids from *Toxoplasma gondii*. Identification of candidate glycolipid precursor(s) for *Toxoplasma gondii* glycosylphosphatidylinositol membrane anchors. J. Biol. Chem. **267**, 11721-11728.

Stevens, V.L. (1993) Regulation of glycosylphosphatidylinositol biosynthesis by GTP. J. Biol. Chem. **268**, 9718-9724.

Vidugiriene, J. and Menon, A.K. (1993) Early lipid intermediates in glycosyl-phosphatidylinositol anchor assembly are synthesized in the ER and located in the cytoplasmic leaflet of the ER membrane bilayer. J. Cell Biology **121**, 987-996.

Vidugiriene, J. and Menon, A.K. (1994) The GPI anchor of cell-surface proteins is synthesized on the cytoplasmic face of the endoplasmic reticulum. J. Cell Biol. **127**, 333-341.

PHOSPHOLIPID FLIPPASES: NEITHER EXCLUSIVELY, NOR ONLY INVOLVED IN MAINTAINING MEMBRANE PHOSPHOLIPID ASYMMETRY

Ben Roelofsen, Esther Middelkoop, Willem P. Vermeulen, Alexander J. Smith* and J.A.F. Op den Kamp
Centre for Biomembranes and Lipid Enzymology
Institute of Biomembranes
Utrecht University
P.O. Box 80.054
3508 TB Utrecht, The Netherlands

Shortly after its discovery in the early seventies, the phenomenon of an asymmetric distribution of phospholipids in the red cell and other plasma membranes raised questions regarding its generation and maintenance.
Although transbilayer phospholipid asymmetry, exhibiting the dominating presence of the choline-containing sphingomyelin (SM) and phosphatidylcholine (PC) in the outer and most of the phosphatidylethanolamine (PE) and all of the phosphatidylserine (PS) in the inner leaflet, is widely accepted as a general characteristic of plasma membranes (Op den Kamp, 1979), most of the pioneering work has been – and still is – done on the membrane of the (human) red blood cell. This typically applies to attempts to address the above two questions.
Enhanced accessibility of purified phospholipases A_2 towards the aminophospholipids in intact human erythrocytes, observed after mild oxidative cross-linking of their membrane skeletal proteins, led Haest and colleagues (Haest *et al.*, 1978) to propose an important role of this protein network – and more specifically its spectrin – in maintaining phospholipid asymmetry in the red cell membrane. Six years thereafter, a novel and intriguing mechanism for the maintenance of phospholipid asymmetry was discovered by Seigneuret and Devaux (1984), as they had observed an ATP-dependent rapid inward translocation of spin-labelled analogues of both aminophospholipids that had been previously inserted into the outer membrane leaflet of intact human erythrocytes. Subsequent studies by the groups of Devaux and others provided substantial evidence for the existence of this system in a number of different (plasma) membranes, unravelled quite a number of its characteristics (see Schroit

* Division of Molecular Biology, The Netherlands Cancer Institute, Plesmanlaan 121, 1066 CX Amsterdam, The Netherlands

NATO ASI Series, Vol. H 96
Molecular Dynamics of Biomembranes
Edited by Jos A. F. Op den Kamp

and Zwaal, 1991; Devaux, 1992; Roelofsen and Op den Kamp, 1994, for some recent reviews) and demonstrated its (almost) absolute specificity for diacyl-PE and -PS as suitable candidates for translocation in the human erythrocyte membrane (Morrot *et al.*, 1989).

No doubt, this so-called "flippase" plays a most prominent role in maintaining phospholipid asymmetry in the red cell membrane. However, and in contrast to what has been propagated by others (Devaux, 1992; Pradhan *et al.*, 1991; Gudi *et al.*, 1990), we feel confident that there is also a major contribution of the membrane skeleton in this phenomenon, as becomes evident from a variety of studies on model systems, native and modified normal, as well as pathologic, erythrocytes. Another point of controversial opinions concerns the possible occurrence of a loss of phospholipid asymmetry, either artificially induced in normal erythrocytes, or *in situ* existing in this cell type under pathologic conditions. Careful analysis of the conditions under which phospholipid asymmetry in the red cell membrane is either maintained or lost in such cases, provides strong support for the view that indeed the flippase and the membrane skeleton are both involved in maintaining proper transbilayer phospholipid asymmetry in this membrane. This aspect will provide a 'fil rouge' for this chapter.

The last part of this contribution will be devoted to briefly discuss some recent observations which indicate the existence of a phospholipid flippase that has a function different from that of maintaining membrane phospholipid asymmetry.

Red cell membrane phospholipid asymmetry: generation and maintenance

The transbilayer distribution of the four major phospholipid classes in the normal human erythrocyte has been established in the early seventies by using highly purified phospholipases (Zwaal *et al.*, 1975), and has been confirmed in a great many of studies that have been published ever since. The two choline containing phospholipids dominate the outer leaflet, where we find 76% of the PC and 82% of the SM. Both amino-phospholipids, on the other hand, are for the greater part (80% of the PE), or even exclusively (PS), located in the cytoplasmic half of the bilayer. More recent studies (Bütikofer *et al.*, 1990; Gascard *et al.*, 1991) showed that approximately 20% of each of the following phospholipids is present in the outer membrane leaflet of the human red cell membrane: phosphatidic acid (PA), phosphatidylinositol (PI), and phosphatidylinositol 4,5-bisphosphate (PIP_2). Interestingly, such a positive detection failed for phosphatidylinositol 4-monophosphate (PIP), which may imply its exclusive localisation in the cytoplasmic leaflet, a feature that PIP would share with PS.

The presence, in the erythrocyte membrane, of a system that mediates an active transport of both aminophospholipids from the outer towards the inner membrane leaflet, has been well documented in a variety of studies using different exogenous probes, viz. spin-labelled

(Seigneuret and Devaux, 1984) or fluorescently labelled (Connor and Schroit, 1989) phospholipid analogues, short chain diacyl glycerophospholipids (Daleke and Huestis, 1989), as well as radiolabelled long chain diacyl glycerophospholipids and SM (Tilley *et al.*, 1986). Although the exact nature of this flippase could not yet be established – the 120 kDa Mg^{2+}-ATPase (Morrot *et al.*, 1990) or a 32 kDa Rh antigen like polypeptide (Connor and Schroit, 1989), eventually in a concerted action with an endofacial protein (Connor and Schroit, 1990) – quite a number of its features have been characterised in great detail. The system exhibits a markedly high degree of specificity, not only regarding the chemical structure of the molecules it can accept for translocation (Fig. 1).

Inward translocation of both aminophospholipids was reported to be inhibited by extracellularly added vanadate (50 µM) (Seigneuret and Devaux, 1984). The intracellular presence of Mg^{2+} (2 mM) is most essential for flippase functioning, whereas cytosolic Ca^{2+} (1 µM) appeared to be strongly inhibitory to the translocation process. The extracellular presence of

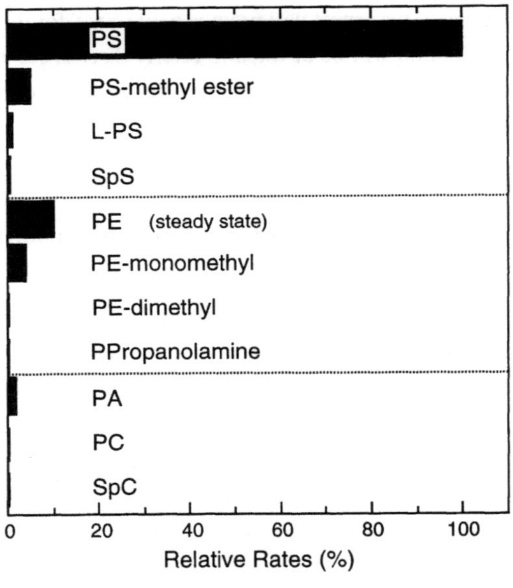

Fig. 1. Initial rates of flippase-mediated translocation of spin-labelled phospholipid analogues in human erythrocytes at 37°C. The data shown in this figure have been adopted from the work of Morrot, *et al.* (1989) after some modification. They have been corrected for non-flippase mediated translocations. Abbreviations: PS, PE, PC and PPropanolamine, phosphatidylserine, -ethanolamine, -choline and -propanolamine, resp.; L-PS, lyso-PS; PA, phosphatidic acid; SpC, sphingomyelin; SpS, sphingomyelin in which the choline moiety has been replaced by serine.
Note: As it is claimed that the spin-labelled PE is a genuine reporter of the fate of the endogenous PE, its translocation reflects the rate of the steady state translocation of PE from the outer monolayer pool, rather than its initial rate.

Ca^{2+}, as well as that of a series of other divalent cations, at concentrations of up to 1 mM has no effect (Bitbol *et al.*, 1987). Finally, the translocase has been shown to be inhibited by thiol-reagents (such as N-ethylmaleimide (NEM), pyridyldithioethylamide (PDA) and diamide (Connor and Schroit, 1989; Daleke and Huestis, 1989; Connor and Schroit, 1990)). It should be noted, however, that this information has been derived from studies involving short-chain diacyl-PE and -PS, or their fluorescently labelled analogues. Particularly as far as diamide is concerned, considerable inward translocation of radiolabelled long-chain diacyl-PE and -PS was still observed when inserted into the outer membrane leaflet of (fresh) human erythrocytes, previously treated with this reagent (Middelkoop *et al.*, 1989). This apparent discrepancy may raise some doubts as to the reliability of the unnatural probe molecules in their application as reporter molecules. It may be wondered whether they indeed will show a behaviour that is identical to that of their native counterparts under *all* circumstances.
More detailed information on the specific features of the aminophospholipid translocase can be found in a number of recent reviews (Devaux, 1992; Schroit and Zwaal, 1991; Roelofsen and Op den Kamp, 1994).

The involvement of the membrane skeleton, a two-dimensional protein network that underlies – and interacts with – the cytoplasmic side of the red cell membrane, in maintaining its phospholipid asymmetry, has been for the first time suggested by Haest and colleagues (Haest *et al.*, 1978). They observed a marked increase in the hydrolysis of both aminophospholipids when diamide (or tetrathionate) treated intact human erythrocytes were subsequently exposed to pure phospholipases A_2 under non-lytic conditions. Although it has been argued that this observation might reflect a transverse destabilisation of the lipid bilayer resulting in enhanced lipid translocation rates (Franck *et al.*, 1982), rather than an *in situ* change in phospholipid asymmetry (Franck *et al.*, 1986), it clearly indicates that a perturbation of the membrane skeleton, as caused by the diamide/tetrathionate-induced mild oxidative crosslinking of its components, strongly impairs its interaction with the inner monolayer phospholipids. Indeed, numerous studies, involving model systems such as lipid monolayers and liposomes, provided evidence for possible interactions of the aminophospholipids – most specifically PS – with components of the membrane skeleton, viz. spectrin (Sweet and Zull, 1970; Mombers *et al.*, 1980; Subbarao *et al.*, 1991) and protein 4.1 (Sato and Ohnishi, 1983; Rybicki *et al.*, 1988; Cohen *et al.*, 1988).

Studies on normal erythrocytes in which the membrane skeleton had been perturbed by diamide treatment provided strong indications for the involvement of the membrane skeleton in the maintenance of phospholipid asymmetry in the red cell membrane. Although mild oxidative crosslinking of membrane skeletal proteins by treatment of intact cells with either diamide or tetrathionate causes a destabilisation of the lipid bilayer, as reflected by a considerably enhanced flip-flop of PC molecules (Franck *et al.*, 1982), it appeared – and in contrast to earlier observations (Haest *et al.*, 1978) – not to give rise to an alteration in phospholipid asymmetry and an appearance of PS in the outer monolayer (Franck *et al.*, 1982; Middelkoop *et al.*, 1989). Particularly that latter fact, which is of considerable interest, was convincingly demonstrated by the negative response that diamide-treated fresh erythrocytes exhibited in the prothrombinase assay, as well as by the absence of any hydrolysis of PS when those cells were exposed to Pal-116-AMPA (Middelkoop *et al.*, 1989). Given the specificity and high sensitivity of both these techniques, it could be concluded that treatment of fresh intact human erythrocytes with those oxidative reagents does not give rise to an outward translocation of more than 1 mole % of the PS, if any. However, endogenous PS started to appear in the outer membrane leaflet after diamide-treated cells had been incubated for 4 hours at 37°C, a process that progressively increased during a further continuation of the incubation. Interestingly, the very first appearance of PS in the exofacial leaflet of these cells coincided markedly with a drop in cellular ATP down to some 10% of the original level. This agrees well with the immediate appearance of exofacial PS in diamide-treated cells that had been previously depleted of their ATP, a condition which – by itself – does not give rise to any appreciable change in the transbilayer distribution of PS (Middelkoop *et al.*, 1989). The above observations in fact suggest that in diamide-treated (fresh) erythrocytes, the aminophospholipid translocase is still in operation, a suggestion that appears to be in conflict with the reported inhibitory effect that this treatment supposedly exerts on the flippase (Daleke and Huestis, 1989; Connor and Schroit, 1990). Although this inhibition of an inward translocation of short-chain diacyl aminophospholipids or NBD-labelled analogues in diamide-treated red cells may tell us something about the suitability of the applied probe molecules (Connor and Schroit, 1990; Colleau *et al.*, 1991) under these conditions, rather than truthfully informing us about the condition of the flippase, the actual reason for this discrepancy is not immediately obvious. But whatever the reason, the application of radiolabelled long-chain diacyl glycerophospholipids as exogenous probe molecules, unambiguously demonstrated that the aminophospholipids – and in particular PS – are still translocated in favour of the inner membrane leaflet of fresh human erythrocytes previously exposed to diamide, although the efficiency of the process seems to be impaired when compared to that in control cells (Fig. 2). Of particular interest to note, however, is the observation that, after some 95% of the labelled PS had accumulated in the inner leaflet within the first 10 h of incubation, a fraction of it reappears in the outer leaflet during the subsequent 10 to 15 h. At the same time-point, the distribution of the labelled PE seems to have reached an equilibrium at which less than the normal 80% of this lipid is present in the inner leaflet. The reverse is observed for the labelled PC, since in the diamide-treated cell more than the normal 25% accumulates in the inner monolayer and does so at a higher flip rate (Fig. 2). It should be noted that, similarly as mentioned above with respect to the endogenous PS, the (re)appearance of part of the labelled PS in the outer leaflet coincides with an essentially complete deprivation of cellular ATP. Also, treatment of the cells with

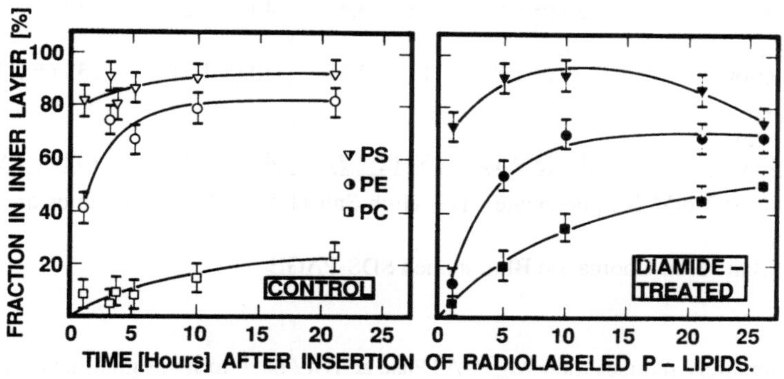

Fig. 2. Transbilayer migration of radiolabelled glycerophospholipids in control and diamide-treated human erythrocytes at 37°C. Trace amounts of radiolabelled phospholipids were inserted into the outer membrane leaflet of intact cells by using nonspecific lipid transfer protein from bovine liver. After a 30 min incubation, which started at zero time, cells were thoroughly washed and subsequently incubated at 37°C to enable a re-equilibration of the probe molecules. At the indicated time points, cells were treated with Pal-116-AMPA. The fractions of the radiolabelled phospholipids that could be degraded this way were assumed to be present in the exofacial membrane leaflet. The results shown are the means of 8 independent experiments; bars indicate corresponding SD values.

Pal-116-AMPA, 20 to 25 h after their exposure to diamide, caused a degradation of the endogenous phospholipids, under still non-lytic conditions, to the following extents: 10% of the PS, 30% of the PE and only 50% of the PC (Middelkoop *et al.*, 1989). Indeed, the resemblance between the ultimate distribution of the probe molecules and that of the endogenous glycerophospholipids in the membrane of the diamide-treated, ATP-depleted, erythrocyte is most striking.

Table I shows that removal of essential parts of the membrane skeleton by exposure of human red cell ghosts to either low (0.3 mM Na-phosphate) or high (1 M KCl) salt concentrations, renders these membranes a 30% enhanced capacity of hydrolysing Mg-ATP (Vermeulen *et al.*, 1995). This may reflect an increased functioning of the flippase, since the contribution of the membrane skeleton to retain the (greater part of) PS in the cytoplasmic leaflet has been lost. Indeed, reassociation of these membranes with a partially purified preparation of

Table I: Effect of Changes in the Amount of Membrane Skeletal Proteins on the Mg^{2+}-ATPase Activity in Human Erythrocyte Ghosts

		Residual Protein[a] (%)			**Mg^{2+}-ATPase**
Treatment	**(n)**	**Spectrin**	**Actin**	**4.1**	**Activity (%)**
None		100	100	100	100
LS-Extraction	(5)	30 (± 4.8)	33 (± 3.4)	45 (± 12.8)	130 (± 3.2)
• Subsequent HS-Extraction	(3)	9 (± 1.8)	14 (± 2.5)	40 (± 1.1)	133 (± 7.3)
• Subsequent Reassoc. with Conc. Extract	(4)	80 (± 9.1)	32 (± 3.2)	48 (± 6.1)	102 (± 1.9)

LS, Low Salt (0.3 mM Na-phosphate); HS, High Salt (1 M KCl); SEM values are given in parentheses.

[a]Relative to Band 3 on Coomassie Blue stained SDS-PAGE

spectrin, completely reduces their Mg^{2+}-ATPase activity back to its original level. Disturbed interactions between membrane bilayer and its skeleton, either or not accompanied by deficiencies in this protein network, which are found in erythrocytes from patients suffering of hereditary spherocytosis (HS), may similarly explain the considerably enhanced Mg^{2+}-ATPase activities which are invariably found in ghosts prepared from those abnormal cells (Vermeulen *et al.*, 1995). Furthermore, conversion of an appreciable fraction of the endogenous PS into PE, by treatment of ghosts with PS decarboxylase, results in a considerable enhancement of Mg^{2+}-ATPase activity (W.P. Vermeulen *et al.*, manuscript in preparation). In this case, the endogenous pools of PE have been enlarged, giving more work to the flippase to prevent an excessive accumulation of PE in the outer membrane leaflet. Conversely, labelling of 50-70% of the PE pool in the outer membrane leaflet by treatment of intact erythrocytes with TNBS, appears to cause a reduction of Mg^{2+}-ATPase activity in the subsequently prepared ghost membranes by not less than 60% (W.P. Vermeulen *et al.*, manuscript in preparation). In view of the extremely high specificity of the flippase (Fig. 1), TNP-PE will not act as a candidate to be translocated. Consequently, the pool of flippase-transportable (PE) molecules in the outer leaflet is considerably reduced. Since exposure of intact erythrocytes to TNBS has been shown not to affect the transport activity of the flippase (Connor and Schroit, 1989), the reduction in Mg^{2+}-ATPase activity may indeed be ascribed to a reduction of the pool of transportable molecules, *i.e.* unmodified PE. Also, an exclusive and complete hydrolysis of the PE in the outer membrane leaflet by treatment of intact erythrocytes with an appropriate phospholipase A_2 (plus sphingomyelinase C), results in an up to 80% inhibition of the Mg^{2+}-ATPase (W.P. Vermeulen *et al.*, manuscript in preparation). Hence, the above studies strongly suggest that the Mg^{2+}-ATPase activity as observed in red cell ghost membranes is – at least for the greater part – an expression of flippase functioning; the level of Mg^{2+}-ATPase activity appears to be closely related to the pool size of flippase-transportable molecules.

In summary, these studies clearly point to a function of the membrane skeleton in the stabilisation of phospholipid asymmetry, particularly as far as the cytofacial localisation of PS is concerned. Alternatively, an alteration ("loss") of phospholipid asymmetry in the erythrocyte membrane may occur only when *both* of the two following prerequisites are fulfilled: (i) a malfunctioning/inhibition of the aminophospholipid translocase, and (ii) a perturbation/destruction/removal of the membrane skeleton.

Changes in red cell phospholipid asymmetry

Alterations in red cell phospholipid asymmetry can be artificially induced *in vitro*, or may even occur *in situ* under certain pathologic conditions. Both classes of alterations have been intensively studied to gain a better insight into the conditions and mechanisms that govern the transbilayer distribution and dynamics of the red cell phospholipids.
Changes in asymmetry are usually referred to in terms such as 'a loss of asymmetry', or 'a scrambling of phospholipids'. Strictly spoken, this terminology implicitly refers to a complete randomising of all phospholipid classes over both leaflets, *i.e.* the creation of a phospholipid symmetric membrane. It may be strongly doubted, however, whether such a situation can ever be reached. Not only is there a highly static asymmetry in the transbilayer distribution of SM in the normal red cell membrane, a great many of studies have shown that this situation is fully maintained under a variety of conditions which give rise to accelerated flip-flop of the glycerophospholipids, or even to changes in their transbilayer distribution (see Roelofsen and Op den Kamp, 1994). This peculiar behaviour of SM is possibly best illustrated by the fact that SM asymmetry is completely maintained in vesicles that have been pinched off by heat treatment of erythrocytes, whilst asymmetry is lost for PE and PS (Dressler *et al.*, 1984). Hence, it can be concluded that a shift of aminophospholipids from inner to outer monolayer – as a consequence of whatever conditions – can be only compensated for by a shift of part of the PC into the opposite direction. In principle, there are two possibilities to consider for the ultimate (passive) redistribution of the three glycerophospholipids that may occur under conditions of a complete failure of the mechanisms that normally generate and maintain phospholipid asymmetry (Fig. 3). In both situations, I and II, the outward migration of aminophospholipids is compensated for by an inward migration of PC. Either PC re-equilibrates over both layers in a 1:1 ratio (Fig. 3, I), or all three glycerophospholipids mix up

and re-equilibrate on the basis of entire equality (Fig. 3, II). In the latter situation, one cannot expect to have more than 40% of each of the aminophospholipids in the outer leaflet, whereas in situation I this is even as little as 20% in the case of PS. Hence, and particularly when also considering SM, a complete loss of asymmetry, or scrambling of phospholipids, may not easily occur. The above also implies that reports which claim the appearance of more than 20% – not to mention over 40% – of the PS in the exofacial leaflet, under whatever conditions, should be considered with some reservation.

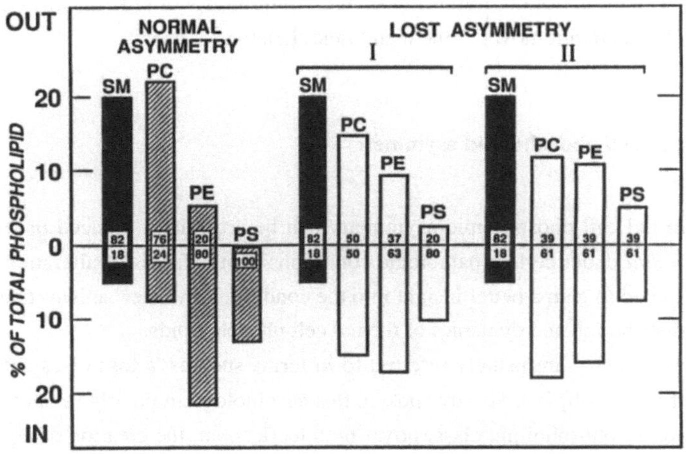

Fig. 3. Ultimate transbilayer redistributions of the three glycerophospholipids in the human erythrocyte membrane that one may expect to occur under complete failure of the mechanisms that normally maintain their asymmetric distribution. In either of the two possibilities shown, SM is not subject to transbilayer redistributions (see text). I, proportional outward migration of PE and PS, which is limited by a compensatory inward migration of PC to reach an equal distribution over both leaflets. II, identical transbilayer distributions of the three glycerophospholipids that will be established when the system does not discriminate among the different subclasses.

Activation of blood platelets triggers a number of rapidly proceeding processes, as for instance the release of Ca^{2+} from intracellular stores. At the same time, and within minutes, vast quantities of PS appear in the outer leaflet of the plasma membrane, the physiological significance of this event being obvious: the creation of a condition that is essential for the formation of the tenase and prothrombinase complexes (Schroit and Zwaal, 1991). It has been suggested that the sudden increase in cytosolic Ca^{2+} is closely related to the outward movement of PS in the plasma membrane and that in fact this process is triggered by Ca^{2+}. Indeed, elevation of the cytosolic Ca^{2+} concentration in erythrocytes by using the ionophore

A23187, similarly causes an alteration in phospholipid asymmetry, including the exofacial exposure of PS (Williamson *et al.*, 1992). The effects that enhanced intracellular Ca^{2+} concentrations may exert are manifold (see Roelofsen and Op den Kamp, 1994) and include activation of an endogenous phospholipase C to degrade polyphosphoinositides, thereby generating the bilayer-destabilising diacylglycerol; direct effects on membrane skeletal proteins and their interaction with the lipid bilayer; activation of transglutaminase to produce crosslinking of spectrin; activation of endogenous proteases that also can affect the membrane skeleton; activation of the Ca^{2+}-pump, leading to ATP depletion; inactivation of the flippase; and induction of non-bilayer structures. Hence, enhanced intracellular Ca^{2+} concentrations create ample conditions that inevitably will lead to considerably changes in phospholipid asymmetry, that is, a perturbation of membrane skeleton-bilayer interactions and an inhibition of the flippase.

Considerable changes in phospholipid asymmetry have been claimed to occur in a number of pathologic red cells. The most prominent examples are summarised in Table II which, for reasons of simplicity, only shows the changes in transbilayer localisation of the two aminophospholipids.

Schwartz and colleagues reported the accessibility of PS in the outer leaflet whereas, more recently, Kuypers and colleagues (Kuypers *et al.*, 1993) showed an absolutely normal phospholipid asymmetry in the cells of eight different HS patients, some of which containing as little as 34% of the normal spectrin content. Furthermore, these authors reported a normally

Table II: Reported Changes in Phospholipid Asymmetry in Abnormal Human Erythrocytes

	% Outside		
Erythrocyte	**PE**	**PS**	**Reference**
Normal	20	0	
Hereditary Spherocyte	26	16	Schwartz *et al.*, (1985)
Rh_{null}	40	0	Kuypers *et al.*, (1984)
Deoxygenated RSC	25	10	Lubin *et al.*, (1981)
Diabetes	26	15	Wali *et al.*, (1988)
Chronic Myeloid Leukaemia	26	32	Kumar *et al.*, (1987)
Malaria, *P. falciparum*	42	7	Schwartz,*et al.*, (1987)
(mature stages)	25	36	Joshi & Gupta, (1988)
	68	26	Maguire *et al.*, (1991)
	20	0	Moll *et al.*, (1990)

functioning flippase in those cells. Hence, the results obtained by Schwartz *et al.* might have been due to energy deprivation of the cells. The same could have been the reason for the considerably increased amount of PE that Kuypers *et al.* observed in Rh_{null} cells (Table II), at a time that we were not yet aware of the existence of the flippase and had not therefore taken the appropriate precautions to protect the cells against energy deprivation. Because of a (local) uncoupling of the lipid bilayer from the membrane skeleton in deoxygenated reversibly sicklable cells (RSCs), conditions are fulfilled again for the appearance of PS in the outer membrane layer under energy depriving conditions (Middelkoop *et al.*, 1988). The reasons for the exofacial exposure of PS that was claimed recently to occur in erythrocytes from diabetic patients (Table II), is not easily explained. It should be remembered, however, that *in vivo* the presence of only very little amounts of PS on the outer surface of an intact erythrocyte will be immediately recognised by the reticulo-endothelial system, causing the sequestration of the cell. This implies that the exposure of appreciable amounts of PS on the outer surface of the cells should have taken place during their handling after the blood had been drawn. This also applies, of course, to the results obtained with the red cells obtained from patients suffering from chronic myeloid leukaemia (CML), in which not less than one third of all PS had been claimed to be present in the outer leaflet (Table II). It should be added, however, that the standard techniques used in our laboratory completely failed to detect any abnormality in the phospholipid distribution in the erythrocytes of three different, clinically established, CML patients (M. van Linde-Sibenius Trip and B. Roelofsen, 1983, unpublished).

To conclude this brief account on abnormal phospholipid distributions in pathologic erythrocytes, it is worth paying special attention to the seemingly controversial situation as to such changes in malaria-infected red cells. Different groups have claimed rather dramatic changes to occur in the phospholipid distribution in the host cell membrane (Table II). These results, however, contrast markedly with those obtained in our laboratory, showing normal phospholipid asymmetry in the membrane of red cells that harbour mature stages of *P. falciparum* (Table II, Moll *et al.*, 1990). When critically comparing the experimental conditions that have been applied in the various studies, it appears that all those studies in which a change in phospholipid asymmetry is observed, exhibit one common feature: the absence of glucose in the media used during experimental handling of the cells. Withholding glucose from the cells may have a most dramatic effect, since the consumption of this fuel by the parasitised cells is known to increase up to two orders of magnitude. Therefore, starvation may easily cause a rapid and drastic drop in intracellular ATP and consequently an impaired functioning – or eventually a complete inhibition – of the aminophospholipid translocase. Since structural lesions are also known to occur in the parasitised erythrocyte, the two requirements for a change in phospholipid asymmetry are fulfilled again.

Maintenance of phospholipid asymmetry: concerted action of flippase and membrane skeleton

All of the above discussed studies point into one and the same direction and provide ample evidence for the view that the asymmetric distribution of the glycerophospholipids is maintained by the concerted action of the ATP-dependent aminophospholipid translocase and the membrane skeleton. The latter seems of doubtless importance for the maintenance of an exclusive localisation of PS in the cytoplasmic leaflet. On the other hand, there is a continuous – flippase mediated – pumping of PE molecules from outer towards inner membrane leaflet. Hence, the asymmetric distribution of PE may be considered to be the result of a steady-state dynamic process. Since the pronounced asymmetric distribution of SM appears to be a very static and permanent one, for which the mechanism/interactions are as yet unknown, PC has no other option than to passively occupy the space that is left open by SM and the aminophospholipids.

This concerted action of both these systems is certainly of physiological significance. First of all, it is quite clear that it is in the immediate interest of the cell's survival to prevent an appearance of PS on the outer surface of its membrane (Tanaka *et al.*, 1983; Schroit *et al.*, 1985; McEvoy *et al.*, 1986). Furthermore, it is conceivable that during the *in vivo* circulation of – even normal – erythrocytes, a (local) uncoupling of the membrane skeleton from the lipid bilayer may occur when the cells are squeezed through very narrow passages, *i.e.* the splenic endothelial slits. The flippase will then be in charge of an immediate back translocation of PS molecules which – consequent on the above uncoupling – may have flopped from inner to outer membrane leaflet. The flippase thus provides a repair mechanism under such conditions. The contribution of the membrane skeleton in retaining PS in the inner membrane leaflet, on the other hand, will save a lot of energy (ATP) that otherwise would be consumed by the flippase, if it were the only system responsible for the maintenance of phospholipid asymmetry in the red cell membrane.

A flippase with a task different from maintaining phospholipid asymmetry

The putative flippase concerned is a multi drug resistance (MDR) like P-glycoprotein (Pgp) that is abundantly present in the canalicular membrane of the liver. The Netherlands Cancer Institute has generated a strain of mice with a disruption in the *Mdr2* gene, and those mice are unable to secrete PC into their bile. This suggested that this Mdr2 Pgp and the human homomlogue MDR3 might be required for the transport of PC – the major phospholipid in bile – through the canalicular membrane of the hepatocyte. To get more information on the

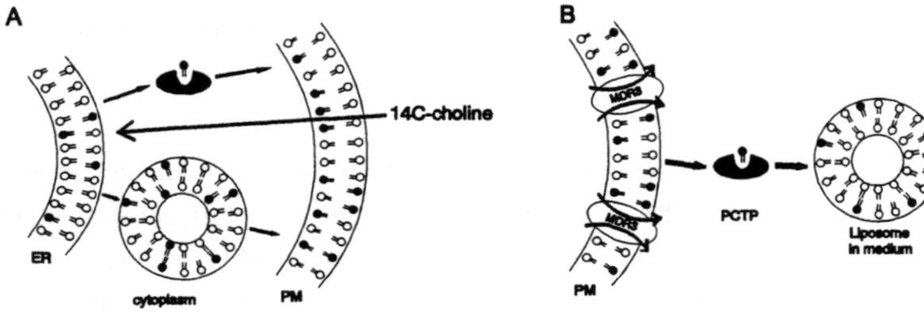

Fig. 4. Assay for MDR3-mediated translocation of PC through the plasma membrane of murine ear fibroblasts. A, incorporation of [^{14}C]choline labels PC (black symbols) in the endoplasmic reticulum (ER) and transport of labelled PC from the ER to the plasma membrane (PM), either by lipid transfer protein mediated monomeric diffusion (top) which inserts the labelled PC in the inner leaflet only, or by vesicle flow (bottom), delivering PC to both leaflets. B, flopping of labelled PC to the outer leaflet in the presence of MDR3 and exchange of the [^{14}C]PC by purified beef liver PC-specific transfer protein (PC-TP) from the outer PM leaflet to liposomes in the medium.

mechanism of this PC transport by this P-glycoprotein, we studied the transport of newly synthesised PC in mouse ear fibroblasts expressing the human *MDR3* gene at a high level. Normal mouse fibroblasts do not express the *Mdr2* gene and these were used as a control for PC transport in the absence of the P-glycoprotein. The principle of the assay is shown in Fig. 4. Control and MDR3 containing fibroblasts were incubated for 3 h in the presence of [^{14}C]choline to allow a radiolabelling of the PC that is synthesised in the ER. Thus labelled PC can be transferred to the plasma membrane by, either vesicular transport which introduces the label at both sides of the plasma membrane, by PC-TP mediated transport, introducing the [^{14}C]PC in the inner leaflet, or by both processes (Fig. 4A). Control and MDR3 containing fibroblasts were subsequently incubated together with a 15-fold excess of unlabelled acceptor PC/cholesterol liposomes in the presence of PC-TP, purified from beef liver (Fig. 4B). This transfer is known to mediate a 1:1 exchange of PC molecules between membranes. Consequently, [^{14}C]PC that appears in the outer leaflet of the fibroblasts' plasma membrane will be transported to the acceptor liposomes. Whenever MDR3 is indeed able to mediate the translocation of PC from inner to outer membrane leaflet, more of the radiolabelled PC will become available for translocation to the liposomes, compared to the controls in which such a protein is lacking. In the latter case, [^{14}C]PC can only reach the outer leaflet of the plasma membrane by intracellular vesicular transport (Fig. 4A) or by spontaneous flipping, a process known to proceed at a vary low rate. The presence of a translocator will result in a more rapid appearance of the [^{14}C]PC in the outer leaflet and, consequently, in a more rapid labelling of the acceptor liposomes in the presence of exogenous PC-TP. Indeed, this is exactly what was observed (Smith *et al.*, 1994): an appreciable transfer of [^{14}C]PC from the MDR3 containing fibroblasts which is markedly higher than in the case of control cells (Smith *et al.*, 1994). Also recently, Ruetz and Gros (1994) reported on the activity of the murine Mdr2 Pgp in yeast inside-out secretory vesicles which were found to accumulate a fluorescent PC analogue in an ATP-dependent manner. So it may be concluded that the results of both these independent studies are consistent with the hypothesis that the human MDR3 Pgp and its murine homologue Mdr2 Pgp are indeed able to mediate the transport of PC from the inner to the outer plasma membrane leaflet and, in this sense, should be considered as phospholipid flippases. Flippases that are specifically found in the canalicular membrane to facilitate the secretion of PC into the bile, rather than being involved in the maintenance of the asymmetric phospholipid distribution in this membrane.

References

Bitbol M, Fellmann P, Zachowski A, and Devaux PF (1987) Ion regulation of phosphatidylserine and phosphatidylethanolamine outside-inside translocation in human erythrocytes. Biochim. Biophys. Acta 904:268–282

Bütikofer P, Lin ZW, Chiu DT-Y, Lubin B and Kuypers FA (1990) Transbilayer distribution and mobility of phosphatidylinositol in human red blood cells. J. Biol. Chem. 265:16035–16038

Cohen AM, Liu SC, Lawler J, Derick L and Palek J (1988) Identification of the protein 4.1 binding site to phosphatidylserine vesicles. Biochemistry 27:614–619

Colleau M, Hervé P, Fellmann P and Devaux PF (1991) Transmembrane diffusion of fluorescent phospholipids in human erythrocytes. Chem. Phys. Lipids 57:29–37

Connor J and Schroit AJ (1989) Transbilayer movement of phosphatidylserine in erythrocytes. Evidence that the aminophospholipid transporter is a ubiquitous membrane protein. Biochemistry 28:9680–9685

Connor J and Schroit AJ (1990) Aminophospholipid translocation in erythrocytes: Evidence for the involvement of a specific transporter and an endofacial protein. Biochemistry 29:37–43

Daleke DL and Huestis W (1989) Erythrocyte morphology reflects the transbilayer distribution of incorporated phospholipids. J. Cell Biol. 108:1375–1385

Devaux PF (1992) Protein involvement in transmembrane lipid asymmetry. Annu. Rev. Biophys. Biomol. Struct. 21:417–439

Dressler V, Haest CWM, Plasa G, Deuticke B and Erusalimsky JD (1984) Stabilizing factors of phospholipid asymmetry in the erythrocyte membrane. Biochim. Biophys. Acta 775:189–196

Franck PFH, Roelofsen B and Op den Kamp JAF (1982) Complete exchange of phosphatidylcholine from intact erythrocytes after protein cross-linking. Biochim. Biophys. Acta 687:105–108

Franck PFH, Op den Kamp JAF, Roelofsen B and van Deenen LLM (1986) Does diamide treatment of intact human erythrocytes cause a loss of phospholipid asymmetry? Biochim. Biophys. Acta 857:127–130

Gascard P, Tran D, Sauvage M, Sulpice J-C, Fukami K, Takenawa T, Claret M and Giraud F (1991) Asymmetric distribution of phosphoinositides and phosphatidic acid in the human erythrocyte membrane. Biochim. Biophys. Acta 1069:27–36

Gudi SRP, Kumar A, Bhakuni V, Gokhale SM and Gupta CM (1990) Membrane skeleton-bilayer interaction is not the major determinant of membrane phospholipid asymmetry in human erythrocytes. Biochim. Biophys. Acta 1023:63–172

Haest CWM, Plasa G, Kamp D and Deuticke B (1978) Spectrin as a stabilizer of the phospholipid asymmetry in the human erythrocyte membrane. Biochim. Biophys. Acta 509:21–32

Joshi P and Gupta CM (1988) Abnormal membrane phospholipid organization in *Plasmodium falciparum*-infected human erythrocytes. Brit. J. Haematol. 68:255–259

Kumar A, Daniel S, Agarwal SS and Gupta CM (1987) Abnormal erythrocyte membrane phospholipid organisation in chronic myeloid leukemia. J. Biosci. 11:543–548

Kuypers FA, van Linde-Sibenius Trip M, Roelofsen B, Tanner MJA, Anstee DJ and Op den Kamp JAF (1984) Rh_{null} human erythrocytes have an abnormal membrane phospholipid organization. Biochem. J. 221, 931–934

Kuypers, FA, Lubin BH, Yee M, Agre P, Devaux PF and Geldwerth D (1993) The distribution of erythrocyte phospholipids in HS demonstrates a minimal role for erythrocyte spectrin on phospholipid diffusion and asymmetry. Blood 81:1051–1057

Lubin B, Chiu D, Bastacky J, Roelofsen B and van Deenen LLM (1981) Abnormalities in membrane phospholipid organization in sickled erythrocytes. J. Clin. Invest. 67:1643–1649

Maguire PA, Prudhome J and Sherman IW (1991) Alterations in erythrocyte membrane phospholipid organization due to the intracellular growth of the human malaria parasite *Plasmodium falciparum*. Parasitol. 102:179–186

McEvoy L, Williamson P and Schlegel RA (1986) Membrane phospholipid asymmetry as a determinant of erythrocyte recognition by macrophages. Proc. Natl. Acad. Sci. USA 83:3311–3315

Middelkoop E, Lubin BH, Bevers EM, Op den Kamp JAF, Comfurius P, Chiu DT-Y, Zwaal RFA, van Deenen LLM and Roelofsen B (1988) Studies on sickled erythrocytes provide evidence that the asymmetric distribution of phosphatidylserine in the red cell membrane is maintained by both ATP-dependent translocation and interaction with membrane skeletal proteins. Biochim. Biophys. Acta 937:281–288

Middelkoop E, van der Hoek EE, Bevers EM, Comfurius P, Slotboom AJ, Op den Kamp JAF, Lubin BH, Zwaal RFA and Roelofsen B (1989) Involvement of ATP-dependent phospholipid translocation in maintaining phospholipid asymmetry in diamide-treated human erythrocytes. Biochim. Biophys. Acta 981:151–160

Moll GN, Vial HJ, Bevers EM, Ancelin ML, Roelofsen B, Comfurius P, Slotboom AJ, Zwaal RFA, Op den Kamp JAF and van Deenen LLM (1990) Phospholipid asymmetry in the plasma membrane of malaria-infected erythrocytes. Biochem. Cell Biol. 68:579–585

Mombers C, de Gier J, Demel RA and van Deenen LLM (1980) Spectrin-phospholipid interaction. A monolayer study. Biochim. Biophys. Acta 603:52–62

Morrot G, Hervé P, Zachowski A, Fellmann P and Devaux PF (1989) Aminophospholipid translocase of human erythrocytes: phospholipid substrate specificity and effect of cholesterol. Biochemistry 28:3456–3462

Morrot G, Zachowski A and Devaux PF (1990) Partial purification and characterization of the human erythrocyte Mg^{2+}-ATPase. A candidate aminophospholipid translocase. FEBS Lett. 266:29–32

Op den Kamp JAF (1979) Lipid asymmetry in membranes. Annu. Rev. Biochem. 48, 47–71

Pradhan D, Williamson P and Schlegel RA (1991) Bilayer/cytoskeleton interactions in lipid-symmetric erythrocytes assessed by a photoactivatable phospholipid analogue. Biochemistry 30:7754–7758

Roelofsen B and Op den Kamp JAF (1994) Plasma membrane phospholipid asymmetry and its maintenance: The human erythrocyte as a model, Current Topics in Membranes 40:7–46

Rybicki AC, Heath R, Lubin B and Schwartz RS (1988) Human erythrocyte protein 4.1 is a phosphatidylserine binding protein. J. Clin. Invest. 81:255–260

Sato SB and Ohnishi S-I (1983) Interaction of a peripheral protein of the erythrocyte membrane, band 4.1, with phosphatidylserine-containing liposomes and erythrocyte inside-out vesicles. Eur. J. Biochem. 130:19–25

Schroit AJ, Madsen JW and Tanaka, Y (1985) *In vivo* recognition and clearance of red blood cells containing phosphatidylserine in their plasma membrane. J. Biol. Chem. 260:5131–5138

Schroit AJ and Zwaal RFA (1991) Transbilayer movements of phospholipids in red cell and platelet membranes. Biochim. Biophys. Acta 1071:313–329

Schwartz RS, Chiu DT-Y and Lubin B (1985) Plasma membrane phospholipid organization in human erythrocytes. Current Topics Hematol. 4:63–112

Schwartz RS, Olson JA, Raventos-Suarez C, Yee M, Heath RH, Lubin B and Nagel RL (1987) Altered plasma membrane phospholipid organization in *plasmodium falciparum*-infected human erythrocytes. Blood 69:401–407

Seigneuret M and Devaux PF (1984) ATP-dependent asymmetric distribution of spin-labeled phospholipids in the erythrocyte membrane: Relation to shape changes. Proc. Natl. Acad. Sci. USA 81:3751–3755

Smith AJ, Timmermans-Hereijgers JLPM, Roelofsen B, Wirtz KWA, van Blitterswijk WJ, Smit JJM, Schinkel AH and Borst P (1994) The human MDR3 P-glycoprotein promotes translocation of phosphatidylcholine through the plasma membrane of fibroblasts from transgenic mice. FEBS Lett. 354:263–266

Subbarao NK, MacDonald RJ, Takeshita K and MacDonald RC (1991) Characteristics of spectrin-induced leakage of extruded, phosphatidylserine vesicles. Biochim. Biophys. Acta 1063:147–154

Sweet C and Zull JE (1970) Interaction of the erythrocyte-membrane protein, spectrin, with model systems. Biochem. Biophys. Res. Commun. 41:135–141

Tanaka Y and Schroit AJ (1983) Insertion of fluorescent PS into the plasma membrane of red blood cells. Recognition by autologous macrophages. J. Biol. Chem. 258:11335–11343

Tilley L, Cribier S, Roelofsen B, Op den Kamp JAF and van Deenen LLM (1986) ATP-dependent translocation of aminophospholipids across the human erythrocyte membrane. FEBS Lett. 194:21–27

Vermeulen WP, Briedé JJ, Bunt G, Op den Kamp JAF, Kraaijenhagen RJ and Roelofsen B (1995) Enhanced Mg^{2+}-ATPase activity in ghosts from HS erythrocytes and in normal ghosts stripped of membrane skeletal proteins may reflect enhanced aminophospholipid translocase activity. Brit. J. Haematol. 90:56–64

Wali RK, Jaffe S, Kumar D and Kalra VY (1988) Alterations in organization of phospholipids in erythrocytes as factor in adherence to endothelial cells in diabetes mellitus. Diabetes 37:104–111

Williamson P, Kulick A, Zachowski A, Schlegel RA and Devaux PF (1992) Ca^{2+} induces transbilayer redistribution of all major phospholipids in human erythrocytes. Biochemistry 31:6355–6360

Zwaal RFA, Roelofsen B, Comfurius P and van Deenen LLM (1975) Organization of phospholipids in human red cell membranes as detected by the action of various purified phospholipases. Biochim. Biophys. Acta 406:83–96

A NEW EFFICIENT STRATEGY TO RECONSTITUTE MEMBRANE PROTEINS INTO LIPOSOMES : APPLICATION TO THE STUDY OF CA^{++}-ATPASES.

Jean-Louis Rigaud and Daniel Levy
Département de Biologie Cellulaire et Moléculaire
CEA Saclay
91191 Gif sur Yvette, Cedex .
France

Reconstitution of integral membrane proteins into unilamellar phospholipid vesicles has played and should stay a potentially powerful tool to analyse structural as well as functional areas of membrane protein research. Although it is difficult to list all informations published since the pioneering work of Racker and co-workers initiated 20 years ago, a number of reviews are available including specific classes of membrane proteins such as receptors, substrate carriers, energy-conserving enzymes involved in oxidative phosphorylation and ion motive ATPases (Racker, 1979; Eytan, 1982; Casey, 1984; Levitzki, 1985; Jain and Zakim, 1987; Villalobo, 1990; Cornelius, 1991; Rigaud *et al.,* 1995).

However despite the extensive use and diverse applications of proteoliposomes, it has to be stressed that the mechanism of their formation is in many ways surprisingly ill defined and that reconstitution has often seemed more like art than science. Nevertheless, in the last few years, important progress has been made in acquiring knowledge of the mechanisms of liposome formation as well as in understanding the physical behavior of detergent-lipid systems (Lichtenberg *et al.,*1983; Lasic, 1988, Silvius, 1992). This progress has made possible to build on a set of basic principles that limits the number of experimental variables and thus the empirical approach of reconstitution experiments. In this context, in order to analyse some aspects of energy-transducing membrane proteins through the use of well characterized proteoliposomes, we have developed in our laboratory a broad and systematic assessement of the reconstitution of different classes of membrane proteins including bacteriorhodopsin (Rigaud *et al.*, 1988), H^{+}-ATPase (Richard *et al.*, 1990) and Ca^{++}-ATPase (Levy *et al.*, 1992). To this end a new experimental strategy has been developed, providing

NATO ASI Series, Vol. H 96
Molecular Dynamics of Biomembranes
Edited by Jos A. F. Op den Kamp

more insight the mechanisms that trigger protein insertion into liposomes during the most employed technique, namely detergent-mediated reconstitutions. It is the purpose of this article to review our personal experience in the reconstitution field with the hope that it will promote the increase of integrated approach to this field. Besides the mechanisms of membrane protein reconstitution, the respective efficiencies and relevant advantages of the proteoliposomes reconstituted by this strategy will be considered and discussed in relation to some functional studies of Ca^{++} -ATPases in model systems.

I. MECHANISMS OF LIPID-PROTEIN ASSOCIATION DURING DETERGENT-MEDIATED RECONSTITUTIONS

The most successful and employed strategy to prepare proteoliposomes is that involving the use of detergents since most of membrane proteins are isolated and purified in the presence of detergents. In the standard procedure, membrane proteins are first cosolubilized with phospholipids in that appropriate detergent to form an isotropic solution of lipid-protein-detergent and lipid-detergent mixed micelles. Next the detergent concentration is lowered by dialysis, gel filtration, adsorption onto polystyrene beads or dilution, resulting in the progressive formation of bilayer vesicles with incorporated protein.

From the abundant litterature it appears that reconstitution from lipid-protein-detergent mixtures yield proteoliposomes of different sizes and compositions depending on the nature of the detergent, the particular procedure to remove it, the protein and lipid composition, the ionic conditions and also the precise conditions of initial solubilization. Therefore not surprisingly, each membrane protein responds differently to the various reconstitution procedures and the approach has been for a long time entirely empirical. This empirism is even more striking when one considers the ideal criteria imposed to fully optimize the potential use of reconstituted proteoliposomes in membrane protein research. Besides the need for conditions that preserve the integrity and the activity of the protein under study, the following important criteria should be fulfilled: the morphology and the size of the proteoliposomes, the homogeneity of their size and protein distribution, the number of proteins units incorporated, the final orientation of the incorporated protein and the permeability of the proteoliposomes.

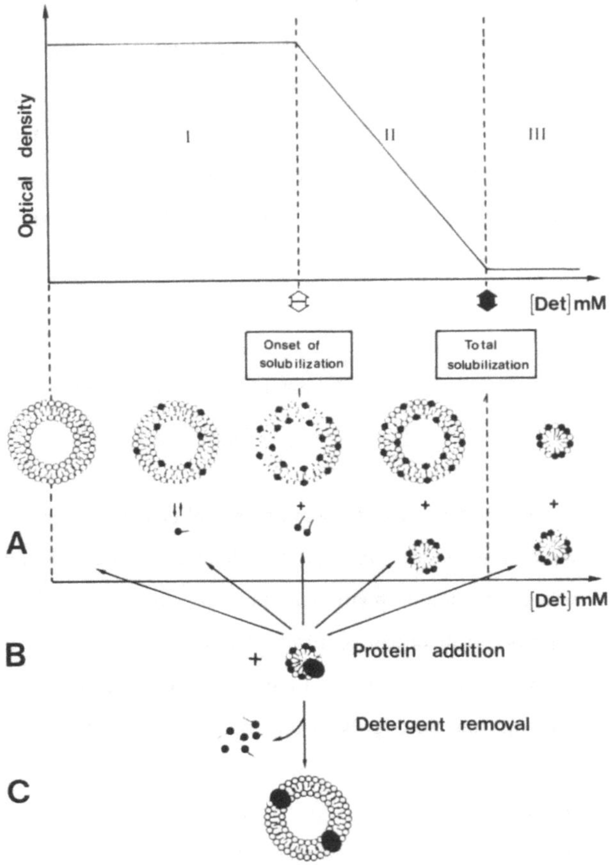

Fig. 1: Schematic representation of a new strategy for membrane protein reconstitution. The standard procedure for reconstituting membrane proteins was carried out in 3 different steps:

A - Stepwise solubilization of preformed liposomes: the solubilization process can be qualitatively analysed through turbidimetry; I, II, III correspond to the 3 stages of the solubilization process. White and black arrows correspond to the onset and total solubization respectively.

B - Protein addition at each step of the lamellar to micellar transition.

C - Detergent removal by hydrophobic adsorption onto polystyrene beads.

A/ A NEW RECONSTITUTION STRATEGY

To allow realistic experimental monitoring of the mechanisms by which proteins may associate to lipids during detergent-mediated reconstitutions, we developed a new strategy based on the idea that reassociation of lipids and proteins upon selective removal of detergent from a detergent-lipid-protein mixture is the mirror image of the solubilization process (Helenius and Simons, 1975; Levy *et al.*, 1990 a). Indeed addition of detergents to preformed liposomes destabilizes them, causing them to open and disintegrate into mixed micelles, following probably the same but reverse sequence of intermediate structures that occur upon detergent removal from micelles. Accordingly the standard procedure for studying the incorporation of membrane proteins was carried out in 3 different steps (Fig.1):

1) Stepwise solubilization of preformed liposomes: liposomes prepared by reverse phase evaporation (Rigaud et al, 1983) were resuspended at the desired concentration and treated with different amounts of detergent through all the range of lamellar to micellar transition. The process of liposome solubilization by different detergents has been the subject of considerable investigations. The methods used included turbidimetry, quasielastic light-scattering, centrifugation experiments, nuclear magnetic resonnance and electron microscopy (Silvius, 1992; Paternostre *et al.*, 1988; Levy *et al.*, 1990b). The results of these studies were related to a three-stage model describing the interactions of detergents with lipidic bilayers. For a given concentration of liposomes, three stages in the solubilization process are apparent depending on the nature and the concentration of the detergent (Fig. 1, upper part).

* In stage I, an increase in the total detergent concentration increases both the concentrations of monomeric detergent in the aqueous phase and the mole fraction of detergent in the bilayer. The incorporation of detergent into the bilayer phase can be described with a well defined partition coefficient. Stage I ends up when the bilayer becomes saturated with detergent.

* During stage II solubilization of detergent-saturated liposomes occurs with the appearance of a coexisting population of mixed micelles. The effective detergent to phospholipid ratios in the vesicles and in the mixed micelles are constant throughout most of stage II and correspond to the critical detergent to phospholipid ratios at which stage II begins (R_{sat}) and finishes (R_{sol}) respectively. During all the bilayer to micelle transition both of these amphiphilic structures coexist and only their relative proportion varies with increasing detergent concentrations.

* Finally in stage III, the phospholipids are completely solubilized into mixed micelles. As the total detergent concentration increases, the mole fraction of detergent in the micelles increases with concomitant decrease in the size of these micelles.

The results of systematic studies allowed to define quantitatively and accurately the process of solubilization by the equation:

$$(D_t) = D_w + R_{eff}(L)$$

where (L) and (D_t) are the total lipid and detergent concentrations, D_w the monomeric detergent concentration and R_{eff} the effective detergent to phospholipid ratios in mixed aggregates. R_{eff} will be referred as R_{sat} at the onset of solubilization and will correspond to the detergent to lipid ratio in detergent-saturated liposomes; R_{eff} will be referred to R_{sol} at total solubilization and will correspond to the detergent to phospholipid ratio in mixed micelles. According to the above equation, it may be possible to calculate R_{eff} from the dependence of the total detergent concentration at which phase transformation occurs over the lipid concentration. On a more technical stand point, one important general conclusion of our studies was that the 3 stage model could be easily visualized and quantitatively analysed through changes in turbidity (Paternostre *et al.*, 1988; Levy *et al.*, 1990 a; Rigaud *et al.*, 1995).

2) **Protein addition**: after incubation time sufficient for detergent equilibration, a solution of solubilized monomeric protein was added at each accurately adjusted step of the lamellar to micellar transition.

Such systematic studies were performed using essentially 3 prototypic energy transducing membrane proteins: bacteriorhodopsin (MW: 27 000; prototype of membrane proteins with 7 transmembrane α-helices; mainly hydrophobic; for a review see Oesterhelt *et al.*, 1992), Ca^{++}-ATPase (MW: 110 000; prototyppe of P-Type ATPases with 10 transmembrane α-helices and a large hydrophilic part; for a review see Bigelow and Inesi, 1992), H^{+}-ATPase from chloroplasts (MW: 550 000; prototype of F0F1 ATPases composed of a F0 hydrophobic part and a very large F1 hydrophilic part; both F0 and F1 are composed of many sub-units; for a review see Graber, 1990).

3) **Detergent removal**: this was generally performed by successive additions of SM_2 Biobeads. The efficient removal of detergent from the reconstituted proteoliposomes is an absolute necessity: residual detergent may either inhibit enzyme activity and/or increase drastically the passive permeability of the liposomes. The method we used to remove

detergent was based on the procedure originally described by Holloway (1973) namely detergent adsorption onto hydrophobic Bio-Beads SM_2. Systematic studies demonstrated the importance of the initial detergent concentration, the amount of beads, the nature and the commercial source of detergent, the temperature and the presence of phospholipids in determining the rates of detergent adsorption onto Bio-Beads. One of the main findings of our studies was that Bio-Beads allowed the almost complete removal of detergents (Triton X-100, $C_{12}E_8$, cholate and octyl glucoside, Levy *et al.*, 1990 a,b) whatever the initial experimental conditions.

B/ MECHANISMS OF PROTEIN INCORPORATION

After detergent removal the resulting proteoliposomes were analysed with respect to protein incorporation and orientation by freeze-fracture electron microscopy, sucrose density gradients and activity measurements. The results from our reconstitution studies allowed us to identify, depending upon the nature of the detergent, three mechanisms by which proteins can associate with phospholipid to give functional proteoliposomes (Fig. 2).

The results from the reconstitutions with sodium cholate demonstrated that proteoliposome formation only arose from detergent depletion of protein-lipid-detergent micelles. No protein incorporation into preformed liposomes, even destabilized by saturating levels of cholate could be detected. Proteoliposome formation was linearly related to the percentage of lipid solubilization in the starting lipid-detergent suspension. Maximal activities and homogeneous protein distribution were measured in samples reconstituted from isotropic micellar solutions.

In the case of Triton X-100-mediated reconstitutions, although no protein was found associated with phospholipids until the starting material contained mixed micelles, the efficiency of the reconstitution was not related to the amount of mixed micelles initially present in the incubation medium. Optimal reconstitutions were detected in samples reconstituted from Triton X-100-phospholipid-protein suspensions where about 60-70 % of the phospholipid was still present as Triton X-100 saturated liposomes. A time dependent incorporation of the proteins was observed suggesting a transfer of the protein molecules initially present in the micelles to detergent-saturated liposomes still present in the incubation

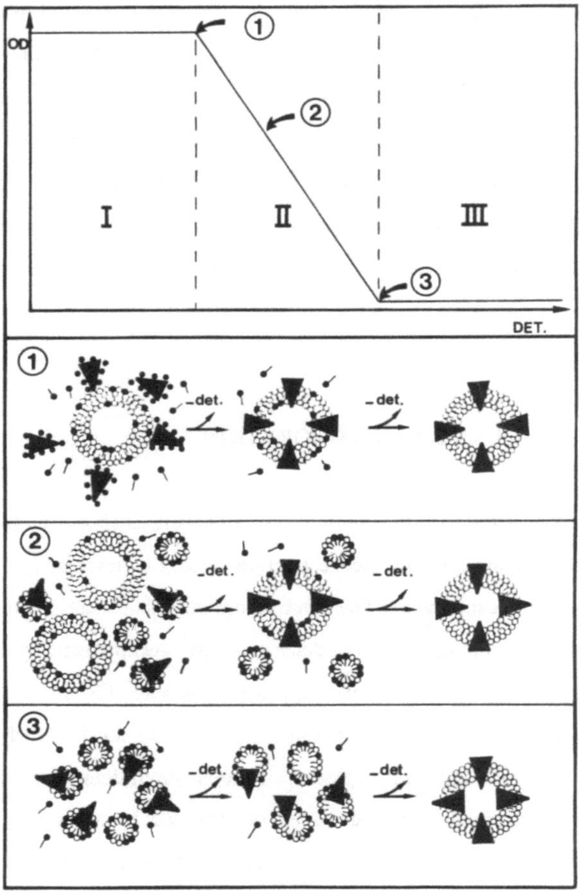

Fig. 2 : Schematic representation of the different mechanisms of protein-lipid association during step by step detergent-mediated reconstitutions. Proteoliposomes were reconstituted according to the general strategy described in Fig.1. The optical density of lipid-detergent-protein before detergent removal are schematically represented in the upper part of the figure. Number 1 refers to optimal octylglucoside- mediated reconstitutions (optimal proteoliposome formation arises from direct incorporation of the protein into preformed detergent saturated liposomes). Number 2 refers to optimal Triton X-100-mediated reconstitution (optimal proteoliposome formation arises from protein transfer from mixed micelles to detergent-saturated liposomes); Number 3 refers to optimal cholate-mediated reconstitution (optimal proteoliposome formation arises from detergent depletion of ternary mixed micelles).

medium. In the case of bacteriorhodopsin and/or H^+-ATPase, the transfer was total leading to the formation of an homogeneous proteoliposome preparation with a final lipid to protein ratio similar to the initial ratio.

The results from octylglucoside-mediated reconstitutions indicated that reconstitution was optimal when starting from a suspension at a detergent to phospholipid ratio around the onset of liposome solubilization. Thus proteoliposomes could be formed by direct incorporation of the protein into preformed liposomes provided they were destabilized by saturating levels of octylglucoside.

Noteworthy, these mechanisms of lipid-protein associations in the presence of different detergents were observed for bacteriorhodopsin, Ca^{++}-ATPases and H^+-ATPases, i.e for proteins with very different hydrophilic-hydrophobic balances. Thus the clear implication of our systematic studies is that the mode of optimal lipid-protein association relies more on the nature of the detergent than on the structure of the protein. Further extension and generalization of the proposed mechanisms is supported by many reports on other membrane proteins where detergent-mediated reconstitutions have been reported more efficient when starting from partially solubilized material in the presence of Triton X-100 or octylglucoside (Helenius and Simons, 1975; Curman *et al.*, 1980; Bullock and Cohen, 1986; Klein and Farenholz, 1994; Kramer and Heberger, 1986).

C/ EFFICIENCY OF THE NEW RECONSTITUTION PROCEDURE

Besides providing informations about the way by which proteins may associate to phospholipids during detergent-mediated reconstitutions, we believe that an important benefit of our study is the finding that the reconstitution method described is a method of choice for protein reconstitution, more suitable than the usual methods using detergents.

The advantage of the method described in this review, involving protein incubation in detergent-treated liposome suspensions at each step of the solubilization process, is to allow a "snapshot" on all situations that may occur in an usual detergent-mediated reconstitution. For comparison, the almost standard procedure for detergent-mediated reconstitution consists in cosolubilization of membrane proteins and lipids with detergent to form a suspension of mixed micelles, followed by detergent removal. The consequence is that, even if a protein can be inserted into a preformed liposome, this step can be missed during detergent removal from micellar solutions. This is clearly what is observed in octyglucoside and Triton X-100-mediated reconstitution where efficiencies of the proteoliposomes were found much higher at intermediate solubilizations than when starting from total solubilzation.

In this context, another important aspect of our study was that the mechanism by which proteins associated with phospholipid to give proteoliposomes was shown to critically affect the final orientation of the protein into the bilayer (Rigaud *et al.*, 1988; Levy *et al.*, 1992). Better asymmetric orientations and consequently higher activities were always observed for the samples reconstituted by incorporation of the protein into preformed liposomes. Optimal unidirectional orientations (85 % to 95 %) were obtained after direct incorporation of proteins into octylglucoside-saturated liposomes. Although less efficient, good asymmetric orientations (70 % to 80 %) were also observed in Triton X-100-mediated reconstitutions where the starting detergent-lipid mixtures contained a large amount of preformed detergent-doped liposomes. Noteworthy these results support the idea that the insertion of a protein into preformed liposomes leads to proteoliposomes with better asymmetric protein insertion than when proteoliposomes are formed by detergent removal from ternary phospholipid-detergent-protein micelles (Eytan, 1982; Jain and Zakim, 1987). A possible mechanism explaining unidirectional orientation of a membrane protein when it is incorporated into preformed liposomes is that protein inserts through the hydrophobic domain of the membrane always with its more hydrophobic moiety first. This seems obvious for asymmetric membrane proteins such as Ca^{++}-ATPase or H^{+}-ATPase. In the case of bacteriorhodopsin, its carboxylic tail is the most hydrophilic containing at least five COOH groups, while the NH_2-terminal region is more hydrophobic. Thus the latter will be first to insert the membrane leading to almost inside-out orientation of bacteriorhodopsin into the resulting proteoliposomes.

An additional advantage of this new reconstitution procedure relies on the batch procedure using SM_2 Biobeads as the detergent removing agent, because it provides an easy reproducible way of obtaining unilamellar, relatively large and impermeable vesicles. In particular ion passive permeability of proteoliposomes containing transport proteins such as bacteriorhodopsin, Ca^{++}-ATPase or H^{+}-ATPase is an important parameter in active transport processes since it partially determines the range of attainable $\Delta\mu H^{+}$ or $\Delta\mu Ca^{++}$ values as well as the rates of proton gradient-driven ADP phosphorylation by H^{+}-ATPases. Thus in all the reconstituted systems analysed in our laboratory, ion passive permeability was investigated in details. Proton and counterion fluxes generated by external acid pulses were monitored using the fluorescence of the pH-sensitive probe pyranine trapped inside reconstituted liposomes (Seigneuret *et al.,* 1985; Levy *et al.,* 1990 b, c and 1992). The most streaking features of our studies is that proteoliposomes obtained by the strategy described in this paper are relatively tight: impermeant to SO_4^{--}, PO_4^{--}, Ca^{++} and slighly permeant to H^+, K^+, Na^+. Such low permeabilities are clearly an improvement in the reconstitution procedures.

In conclusion, the method described in this paper which allows a rapid and easy determination of the experimental conditions for optimal detergent-mediated reconstitution has been shown useful for reconstitution of bacteriorhodopsin, Ca^{++}-ATPase or H^+-ATPase. Fairly high activities were observed for the 3 different proteins, thus proving the efficiency and general validity of the procedure in creating highly functional proteoliposomes. Furthermore we have been able to obtain succesful reconstitution of erythrocyte plasma membrane Ca^{++}-ATPase (Hao *et al.,* 1994) and of Photosystem I (manuscript in preparation). Finally we recently demonstrated that this strategy efficient for reconstitution of individual proteins was very successful for the co-reconstitution of two energy-coupled proteins: co-reconstitution of bacteriorhodopsin with F0F1 ATPases from different sources yields proteoliposomes able to sustain high ATP synthesis activities upon light activation of bacteriorhodopsin (Richard *et al.,* 1995). It is noteworthy that the activities of the reconstituted proteoliposomes are among the best reported up to now for each of these proteins and some of them relevant to physiological conditions.

II/ APPLICATION TO THE FUNCTIONAL STUDY OF CA^{++} -ATPASES

The membrane-bound Ca^{++}-ATPase of the sarcoplasmic reticulum (SR) isolated from skeletal muscle is among the most important and extensively studied of the transport enzymes. It catalyzes active Ca^{++} uptake into SR vesicles which is coupled to ATP hydrolysis. Mechanism, regulation and structure of the SR Ca^{++} pump have been and are still analysed in great details (for a review see Bigelow and Inesi, 1992).

An important question related to the mechanism of ATP-dependent Ca^{++} transport is whether a counterion is utilized by the Ca^{++}-ATPase in analogy to Na^+/K^+ or K^+/H^+ ATPases. Proton countertransport coupled to Ca^{++} uptake during the functioning of SR Ca^{++} pump has been suggested several times (Madeira, 1978; Chiesi and Inesi, 1980; Ueno and Sekine, 1981;

Yamaguchi and Kanazawa, 1985). Eventhough all these authors presented strong experimental support for their suggestions, conclusive evidence is hard to obtain and objections can be made. Indeed considerations has to be given to the fact that studies of Ca^{++} transport in their native membranes are hindered by difficulties related to their non ideal density of assembly within the membrane and to the presence of other ion channel proteins producing dispersion of electrolyte gradients (Haynes, 1982).

Therefore reconstitution of the Ca^{++} pump into impermeant proteoliposomes is a powerful tool for this kind of investigation. We endeavored to demonstrate Ca^{++}/H^{+} coutertransport by parallel measurements of H^{+} and Ca^{++} transport within well characterized proteoliposomes. Using our new reconstitution strategy, we obtained proteoliposomes which sustained very high rates of Ca^{++} transport between 1.0 and 2.0 μmol Ca^{++} per mg protein.min-1, reaching asymptotic levels of about 5-10 μmol Ca^{++} per mg protein which corresponds to a lumenal concentration of 20 mM Ca^{++} or higher (Levy *et al.*, 1992). In addition the proteins were found oriented almost unidirectionally and their density in the reconstituted membrane can be easily controled, thereby permitting ideal lumenal volume per ATPase molecule (i.e allowing prolonged time to reach the inhibitory level of lumenal Ca^{++} due to the high volume available per unit protein). Evenmore interesting for the purpose of the study was that the active enzymes were incorporated in the membranes of relatively impermeant proteoliposomes, whose observed permeabilities were in the range 5.10^{-5} cm.sec-1 (Levy *et al.*, 1990 c).

The first line of evidence for Ca^{++}/H^{+} countertransport during the operation of the Ca^{++}-ATPase came from Ca^{++} uptake measurements. In K_2SO_4 medium, a very low Ca^{++} uptake was obtained which was only slighly affected by the presence of valinomycin. On the contrary, Ca^{++} accumulation was increased 3-4-fold in the presence of the protonophore carbonyl-p-trifluoromethoxy phenylhydrazone, FCCP, indicating that a transmembrane pH gradient was built up during Ca^{++} uptake that inhibited the transport activity of the pump.

Additional evidence that the Ca^{++} pump operated as a Ca^{++}/H^{+} coutertransport was provided by measurements of ATP-dependent intraliposomal alkalinization using entrapped pyranine and accumulation of the weak acid acetate. A transmembrane pH gradient of abour 1 pH unit was generated whose kinetics parallel to those of the Ca^{++} uptake. Collapse of such internal alkalinization by FCCP but not valinomycin confirmed that it was due to outward H^{+} translocation through a Ca^{++}/H^{+} coutertransport.

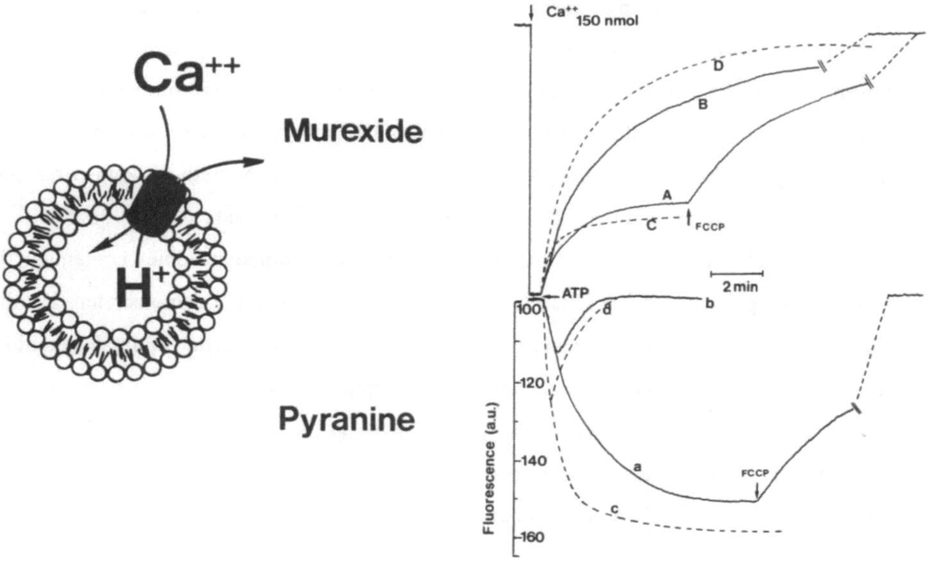

Fig. 3: Comparison of the time courses of Ca^{++} accumulation and internal alkalinization in Ca^{++}-ATPase proteoliposomes. Ca^{++} accumulation using murexide (upper traces) and changes in internal pH using pyranine (lower traces) were recorded on the same preparations in similar experimental conditions. Proteoliposomes at lipid:protein ratio of 40 w/w in the presence of FCCP (traces B, b) or in the absence of FCCP (traces A, a). Proteoliposomes at lipid protein ratio of 20 w/w in the presence of FCCP (traces D, d) or in the absence of FCCP (traces C, c).

The Ca^{++}/H^{+} stoechiometry was further determined by parallel recording on the same sample of Ca^{++} uptake and lumenal alkalinization (Fig.3). We found that the 2 time curves proceeded in parallel and with a soichiometric ratio very close to 1. Countertransport of 1 H^{+} per 1 Ca^{++} was corroborated by the net positive charge displacement observed during Ca^{++} transport. Using Oxonol VI absorption changes, a steady-state electrical potential was estimated to about 50 mV and was accounted for by the estimated charge transfer associated with a 1:1 Ca^{++}/H^{+} coutertransport (Yu *et al.*, 1993)

Studies of the effects of lumenal and medium pH variations on the Ca^{++}/H^{+} ratios (Yu *et al.*, 1994) have led to the suggestion that pK changes of acidic residues involved in Ca^{++} binding contributed to vectorial displacement of the bound Ca^{++}. These changes were produced by ATP through enzyme phosphorylation which induced cyclic inward displacement of Ca^{++} and outward displacement of H^{+}. The question of direct cation exchange was further discussed by these authors considering the structural and molecular biology information related to Ca^{++} binding and translocation. A diagram of a putative mechanism with a countertransport stoichiometry of two H^{+} and two Ca^{++} per ATPase cycle was proposed .

Finally it should be emphasized that similar studies using proteoliposomes containing the cadmodulin-dependent Ca^{++}-ATPase from the erythrocyte membrane demonstrated that this ATPase behaved as an electrogenic Ca^{++}/H^{+} exchanger and that under optimal conditions utilization of 1 mol of ATP was accompanied by uptake of one Ca^{++} by the vesicles and ejection of 1 H^{+} from the lumen of the vesicles (Hao *et al.*, 1994). Thus it is tempting to assume that H^{+} countertransport is a general feature of the Ca^{++} transporting ATPases which should be renamed Ca^{++}/H^{+} ATPases in analogy to Na^{+}/K^{+} and K^{+}/H^{+} ATPases.

CONCLUSION

Reconstitutions of membrane proteins into liposomes have played and should stay a potentially powerful tool that can be used to identify the mechanism of action of membrane proteins. As a prerequisite, a sound charaterization of the reconstitution is required to lead to optimal reconstituted proteoliposomes. As appears for this short review, it is obvious that an optimistic perspective can be foreseen provided the experimental analysis is systematically performed.

REFERENCES

Bigelow, D.J., and Inesi, G. (1992) Contributions of chemical derivatization and spectroscopic studies to the characterization of the Ca^{++} transport ATPase of sarcoplasmic reticulum, *Biochim. Biophys. Acta*, 1113, 323-338.

Bullock, J. O. and Cohen, F. S. (1986) Octylglucoside promotes incorporation of channels into neutral planar phospholipid bilayers. Studies with colicin Ia, *Biochim. Biophys. Acta*, 856, 101-108.

Casey, R. P. (1984) Membrane reconstitution of the energy-conserving enzymes of oxidative phosphorylation, *Biochim. Biophys. Acta*, 768, 319-347.

Cornelius, F (1991) Functional reconstitution of the sodium pump. Kinetics of exchange reactions performed by reconstituted Na / K-ATPase, *Biochim. Biophys. Acta*, 1071, 19-66.

Chiesi, M., and Inesi, G. (1980) Adenosine 5'-triphosphate dependent fluxes of manganese and hydrogen ions in sarcoplasmic reticulum vesicles. *Biochemistry*, 19, 2912-2918.

Curman, B., Klareskog, L., and Peterson, P.A. (1980). On the mode of incorporation of human transplantation antigens into lipid vesicles, *J. Biol. Chem.*, 255, 7820-7826.

Eytan, G. D.(1982) Use of liposomes for reconstitution of biological functions, *Biochim. Biophys. Acta*, 694, 185-202.

Gräber, P. (1990) Kinetics of proton transport coupled ATP synthesis in chloroplasts. *Bioelectrochemistry III*, Milazzo, G. and Blank, M., Plenum Press, New York, 277-310.

Hao, L., Rigaud, J.L., and Inesi, G. (1994) Ca^{++}/H^{+} coutertransport and electrogenicity in proteoliposomes containing erythrocyte plasma membrane Ca^{++}-ATPase and exogenous lipids. *J. Biol.Chem.* 269, 14268-14275

Haynes, D.H. (1982) Relationships between H^{-}, anion and monovalent cation movements and Ca^{++} transport in sarcoplasmic reticulum: further proof for a cation exchange mechanism for the Ca^{++}-ATPase pump. *Arch.Biochem.Biophys.*, 215, 444-461.

Helenius, A. and Simons, K.(1975) Solubilization of membranes by detergents, *Biochim. Biophys. Acta*, 415, 29-79.

Helenius, A., Sarvas, M. and Simons, K.(1981) Asymmetric and symmetric membrane reconstitution by detergent elimination, *Eur. J. Biochem.*, 116, 27-35.

Holloway, P.W. (1973) A simple procedure for removal of triton X-100 from protein samples, *Analytical Biochemistry*, 53, 304-307.

Jain, M. K. and Zakim, D.(1987) The spontaneous incorporation of proteins into preformed bilayers, *Biochim. Biophys. Acta*, 906, 33-68.

Klein, U., and Farenholz, F. (1994) Reconstitution of the myometrial oxytocin receptor into proteoliposomes. *Eur. J. Biochem.*, 220, 559-567.

Krämer, R., and Heberger, C. (1986) Functional reconstitution of carrier proteins by removal of detergent with hydrophobic ion exchange column. *Biochim. Biophys. Acta,* 863, 289-296.

Lasic, D. (1988) The mechanism of vesicle formation, *Biochem. J.*, 256, 1-11.

Levitzki, A.(1985) Reconstitution of membrane receptor systems, *Biochim. Biophys. Acta*, 822, 127-153.

Levy, D., Gulik, A., Seigneuret, M. and Rigaud, J.L (1990, a) Phospholipid vesicle solubilization and reconstitution by detergent. Symmetrical analysis of the two processes using octaethylene glycol mono-N-dodecyl ether ($C_{12}E_8$), *Biochemistry*, 29, 480-9488.

Levy, D., Bluzat, A., Seigneuret, M. and Rigaud, J.L. (1990, b) A systematic study of liposome and proteoliposome reconstitution involving Bio-Bead-mediated TX-100 removal, *Biochim. Biophys. Acta*, 1025, 179-190.

Levy, D., Seigneuret, M., Bluzat, A. and Rigaud, J.L. (1990, c) Evidence for proton coutertransport by the sarcoplasmic reticulum Ca^{++}-ATPase during calcium transport in reconstituted proteoliposomes with ionic permeability, *J. Biol. Chem.*, 265, 19524-19534.

Levy, D., Gulik, A., Bluzat, A. and Rigaud, J.L. (1992) Reconstitution of the sarcoplasmic reticulum Ca^{++}-ATPase: mechanisms of membrane protein insertion into liposomes during reconstitution procedures involving the use of detergents, *Biochim. Biophys. Acta*, 1107, 283-298.

Lichtenberg, D., Robson, R. J. and Dennis, E. A. (1983) Solubilization of phospholipids by detergents. Structural and kinetic aspects, *Biochim. Biophys. Acta*, 737, 285,

Madeira, V.M.C. (1978) Proton gradient formation during transport of Ca^{++} by sarcoplasmic reticulum. *Arch.Biochem.Biophys.*, 185, 316-325.

Oesterhelt, D., Tittor, J. and Bamberg, E. (1992) A unifying concept for ion translocation by retinal proteins, *Journal of Bioenergetics and Biomembranes*, 24, 181-191.

Paternostre, M. T., Roux, M. and Rigaud, J. L. (1988) Mechanisms of membrane protein insertion into liposomes during reconstitution procedures involving the use of detergents. 1. Solubilization of large unilamellar liposomes (prepared by reverse-phase evaporation) by triton X-100, octyl glucoside and sodium cholate, *Biochemistry*, 29, 2668-2677.

Racker, E.(1979) Reconstitution of membrane processes, *Methods in Enzymology*, 55, 699-711.

Richard, P., Rigaud, J.L. and Gräber, P.(1990) Reconstitution of CF0F1 into liposomes using a new reconstitution procedure, *Eur. J. Biochem*, 193, 921-925.

Richard, P., Pitard, B. and Rigaud, J.L. (1995) ATP synthesis by the F0F1 ATPase from the thermophilic Bacillus PS3 co-reconstituted with bacteriorhodopsin into liposomes. *J.Biol.Chem.*, in press

Rigaud, J. L. Bluzat, A. and Büschlen, S. (1983) Incorporation of bacteriorhodopsin into large unilamellar liposomes by reverse phase evaporation, *Biochem. Biophys. Res. Commun.*, 111, 373-382.

Rigaud, J. L., Paternostre, M. T. and Bluzat, A. (1988) Mechanisms of membrane protein insertion into liposomes during reconstitution procedures involving the use of detergents. 2. Incorporation of the light-driven proton pump bacteriorhodopsin, *Biochemistry*, 27, 2677-2688.

Rigaud, J.L, Pitard, B. and Levy, D. (1995) Reconstitution of membrane proteins into liposomes: application to energy transducing membrane proteins. *Biochim.Biophys. Acta,* reviews in Bioenergetic, in press.

Seigneuret, M. and Rigaud, J.L. (1985) Use of the fluorescent pH probe pyranine to detect heterogeneous directions of proton movement in bacteriorhodopsin reconstituted large liposomes, *FEBS*, 188, 101-106.

Seigneuret, M. and Rigaud, J.L. (1986, a) Analysis of passive and light-driven ion movements in large bacteriorhodospin liposomes reconstituted by reverse-phase evaporation. 1. Factors governing the passive proton permeability of the membrane, *Biochemistry*, 25, 6716-6723.

Seigneuret, M. and Rigaud, J.L. (1986, b) Analysis of passive and light-driven ion movements in large bacteriorhodopsin liposomes reconstituted by reverse-phase evaporation. 2. influence of passive permeability and back-pressure effects upon light-induced proton uptake, *Biochemistry*, 25, 6723-6729.

Seigneuret, M. and Rigaud, J.L. (1988) Partial separation of inwardly pumping and outwardly pumping bacteriorhodospin reconstituted liposomes by gel filtration, *FEBS*, 228, 79-84.

Silvius, J. R.(1992) Solubilization and functional reconstitution of biomembrane components, *Annu. Rev. Biophys. Biomol. Struct*, 21, 323-348.

Ueno, T. and Sekine, T. (1981) A role of H^+ flux in active Ca^{++} transport into SR vesicles. I. Effect of an artificially imposed H^+ gradient on Ca^{++} uptake. *J.Biochem.*, 89, 1239-1246.

Villalobo,A.(1990) Reconstitution of ion-motive transport ATPases in artificial lipids membranes. *Biochim. Biophys. Acta*, 1017, 1-48.

Yamaguchi, M. and Kanazawa, T. (1985) Coincidence of H^+ binding and Ca^{++} dissociation in the sarcoplasmic reticulum Ca^{++} ATPase during ATP hydrolysis. *J.Biol.Chem.*, 260, 4896-4900.

Yu, X., Hao, L. and Inesi, G. (1994) A pK change of acidic residues contributes to cation countertransport in the Ca-ATPase of sarcoplasmic reticulum. *J.Biol.Chem.*, 269, 16656-16661.

INTERACTION OF PULMONARY SURFACTANT-ASSOCIATED PROTEINS WITH PHOSPHOLIPID VESICLES

J. Pérez-Gil, A. Cruz, M.L.F. Ruano, E. Miguel, I. Plasencia, C. Casals

Dept. Bioquímica y Biología Molecular I, Fac. Biología
Universidad Complutense, 28040 Madrid, SPAIN

Introduction

The terminal airways and alveoli of lungs are stabilized and protected against collapse by the presence of a specialized material called pulmonary surfactant. In order to achieve its function, this system must be spread as a monomolecular layer at the air-water interface of the aqueous film covering the alveolar epithelium, substantially reducing its surface tension (Keough, 1992).

Surfactant is roughly composed of about 90% lipids and 10% proteins. The major lipid component is dipalmitoylphosphatidylcholine (DPPC). This phospholipid is mainly responsible for the tensoactive properties of pulmonary surfactant. Four proteins have been unequivocally associated to surfactant called SP-A, SP-B, SP-C and SP-D. SP-A and SP-D are hydrophilic, complex oligomeric macromolecules composed of monomers containing collagen and lectin domains. SP-B and SP-C are small peptides with an unusual hydrophobicity. All these proteins participate in the biophysical and physiological activities of surfactant. (Hawgood and Shiffer, 1991; Weaver and Whitsett, 1991).

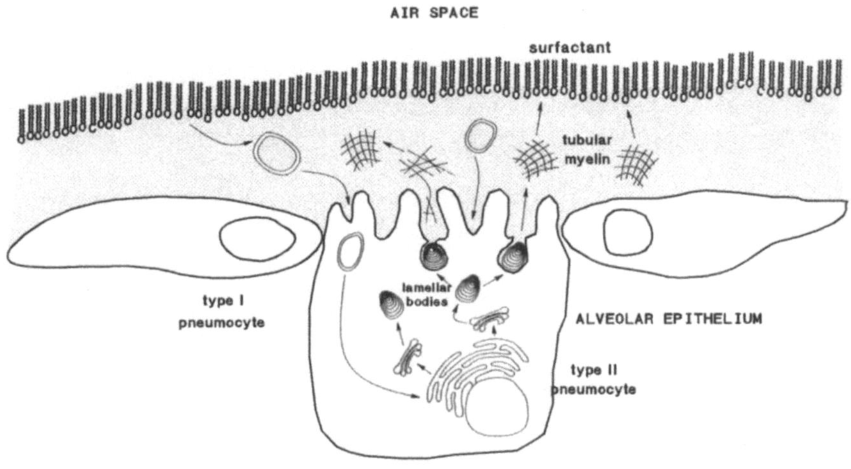

Figure 1. Pulmonary surfactant cycle in the alveolar spaces

This work has been supported by the grant PB92-0752 from Dirección General de Investigación Científica y Técnica (Spain)

NATO ASI Series, Vol. H 96
Molecular Dynamics of Biomembranes
Edited by Jos A. F. Op den Kamp

The lipid and protein components of pulmonary surfactant (Figure 1) are synthesized and assembled into lamellar bodies by alveolar type II cells. After secretion, the content of lamellar bodies expands in the hypophase and forms a special structure known as tubular myelin. This structure is composed of a network of regularly spaced crossing bilayers and is widely thought to be the immediate precursor of a monolayer at the alveolar air-liquid interface. During compression of alveoli, surfactant monolayer is somehow enriched in DPPC, the only component in surfactant able to produce the low surface tension values necessary to protect the alveoli against collapse (Goerke and Clements, 1986).

At least three of the surfactant-associated proteins (SP-A, SP-B and SP-C) potentiate the surface tension-reducing properties of the surfactant lipids (Hawgood and Shiffer, 1991; Weaver and Whitsett, 1991). One of the objectives of our research group is to characterize lipid-protein interactions between those surfactant proteins (SP-A, SP-B and SP-C) and the main surfactant phospholipids.

Interaction of SP-B with phospholipid vesicles

The extreme hydrophobicity of surfactant proteins SP-B and SP-C allows their purification from organic lipidic extract of surfactant according to the method described by Curstedt et al. (1987), optimized in our lab (Pérez-Gil et al., 1993).

SP-B is a small polypeptide of 79 residues with net positive charge and 7 cysteines forming three intrachain and one interchain disulfide bridges (Johansson et al., 1991). The protein is isolated as a homodimeric form. Predictive studies and characterization of the secondary structure have lead to propose that SP-B contains several amphipathical α-helical motifs which would respond for its affinity to bind to phospholipid bilayer and monolayers (Takahashi et al., 1990; Morrow et al., 1993). Studies on the effect of SP-B on the physical properties of DPPC or DPPG bilayers have concluded that SP-B probably interacts with the surface of the phospholipid bilayer, having little effects in deeper, hydrophobic, regions of the bilayer core (Morrow et al., 1993; Shiffer et al., 1993; Pérez-Gil et al., 1995).

SP-B molecule has a single tryptophan residue in all the species characterized so far. This residue can be used as intrinsic probe to monitor changes occurring in the protein molecule. We have characterized the tryptophan fluorescence spectrum of surfactant SP-B isolated from pig lungs and reconstituted in different environments (Pérez-Gil et al., 1993; Cruz et al., 1995). The interaction of SP-B with different phospholipid vesicles can be followed by observing the changes in the SP-B fluorescence emission spectrum (Figure 2). In this experiment, a fixed amount of SP-B dissolved in methanol was injected into aqueous solutions containing unilamellar vesicles of DPPC at different lipid concentrations. The fluorescence emission intensity of SP-B increased as a function of lipid concentration, reaching saturation at a phospholipid/protein weight ratio of 3:1 (Figure 2, insert). From these experiments, binding constants and lipid-protein stoichiometry can be determined according to the equation:

$$P + nL \underset{K_D}{\rightleftharpoons} P(L)_n$$

where n represents the number of phospholipid molecules saturating a single protein molecule and K_D

represents the apparent dissociation constant. A similar approach has been described for other proteins (Dufourc and Foucon, 1978; Batenburg et al., 1988; Gazit et al., 1994). The estimated K_D for interaction of SP-B with DPPC vesicles was of 5.5 ± 1.2 μM. The stoichiometry (n) was 24±3 mol of phospholipid/mol SP-B. These values are similar to those reported for other membrane-associated proteins such as melittin (Kuchinka and Seelig, 1989) or the myelin basic protein (Sankaram et al., 1989), which have similar membrane-interacting motifs.

Interaction of SP-B with phospholipid vesicles has been proposed to conduct to bilayer destabilization leading to vesicle aggregation and fusion (Williams et al., 1991; Poulain et al., 1992). These activities of SP-B have been related with the participation of SP-B in tubular myelin formation and bilayer-monolayer transitions. The bilayer perturbing properties of SP-B were analyzed by lipid mixing assays using resonance energy transfer between NBD-phosphatidylethanolamine (donor) and Rhodamine-phosphatidylethanolamine (acceptor) (Struck et al., 1981) (Figure 3).

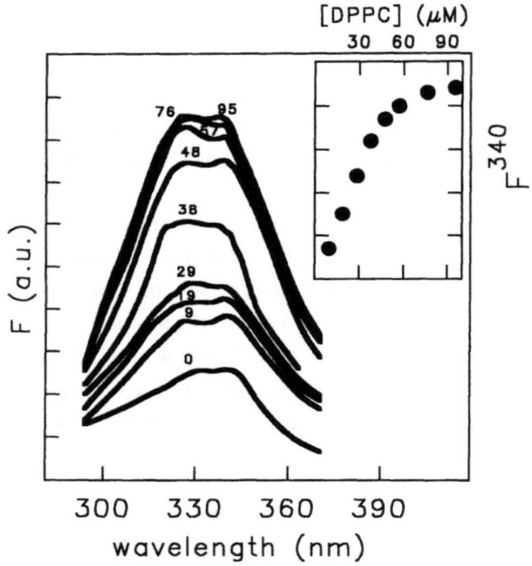

Figure 2. Fluorescence emission spectrum of porcine SP-B (14 μg/mL) in the absence and in the presence of the indicated concentrations of DPPC (μM). Insert: variation of SP-B fluorescence emission intensity at 340 nm versus DPPC concentration

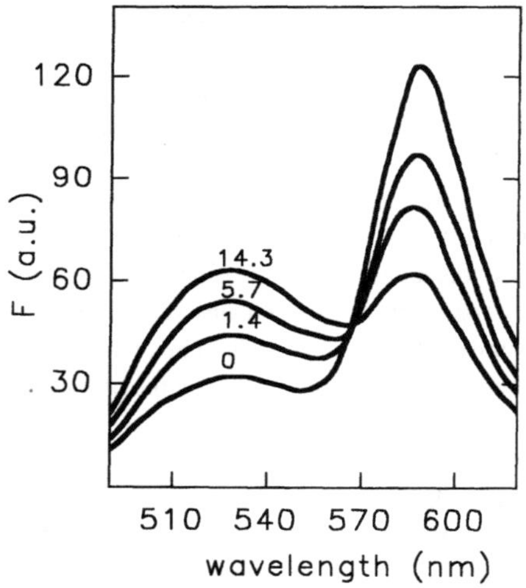

Figure 3. SP-B-induced lipid mixing assayed by resonance energy transfer between NBD-PE and Rh-PE. Figure presents the fluorescence emission spectrum of a mixture consisting of 50 μg/mL of DPPC/PG (7:3, w/w) vesicles and 4 μg/mL of DPPC/PG/NBD-PE/Rh-PE (7:3:0.1:0.1, w/w/w/w) vesicles, in the absence and in the presence of the indicated SP-B concentrations (μg/mL)

Increasing amounts of protein caused higher extent of lipid mixing between DPPC or DPPC/PG (7:3, w/w) vesicles, yielding maximal dilution of phospholipid probes in the bilayers. Further characterization of the process is necessary to unequivocally demonstrate the possible fusogenic activity of SP-B.

Reconstitution of SP-C in phospholipid vesicles

Surfactant SP-C is an small peptide of 35 amino acids with an unusual composition comprising about 60% of aliphatic, branched, hydrophobic residues that explains its extreme hydrophobicity (Johansson et al., 1988). The hydrophobicity of SP-C is further increased by the stoichiometric palmitoylation of the two cysteines in the sequence (Curstedt et al., 1990). The protein appears to form a transmembrane hydrophobic α-helix, which would be oriented perpendicular to the plane of phospholipid bilayer or monolayers (Vandenbussche et al., 1992). This has been confirmed by the determination of the three-dimensional conformation of SP-C in non-polar environments, by high-resolution NMR (Johansson et al., 1994).

Both SP-B and SP-C have an important role in promoting the adsorption of surfactant phospholipids from the aqueous hypophase to the air-liquid interface and in the formation of the tensoactive monolayer (Oosterlaken-Dijksterhuist et al., 1991; Pérez-Gil et al., 1992a). The biophysical properties of SP-C correlate with the effects induced by the protein on the physical behaviour of phospholipid monolayers (Pérez-Gil et al., 1992b) and bilayers (Pérez-Gil et al., 1995).

The primary sequence of SP-C does not comprise any tryptophan or tyrosine residue to be used as intrinsic probe to characterize lipid-protein interactions. We have used the approach of introducing an extrinsic fluorescent probe in the SP-C structure by selective chemical modification. This approach has been widely applied to characterize lipid-protein interactions with other proteins (i.e. Bardelle et al., 1993; Ferrandiz et al., 1994). We have optimized the procedure to selectively label the amino N-terminal group of SP-C with different probes, by using isothiocyanate-derived compounds and carrying out the labelling reaction at strictly controlled pH 7.8.

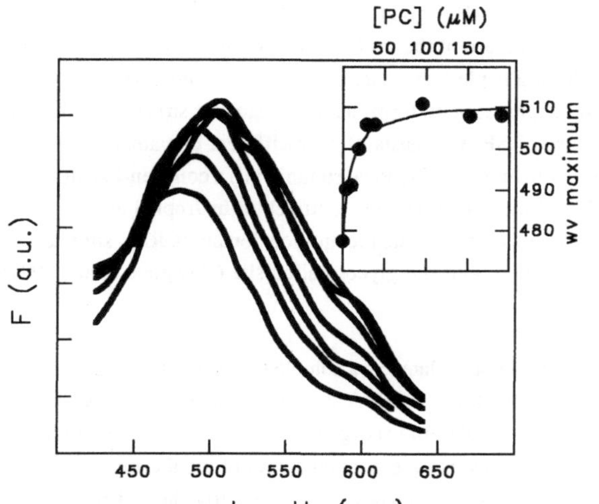

Figure 4. Fluorescence emission spectrum of N-t-dansyl-SP-C (14 μg/mL) in the absence and in the presence of different concentrations of egg-PC. Insert: shift of the wavelength of dansyl-SP-C fluorescence emission maximum versus the concentration of PC

Figure 4 shows the fluorescence emission spectrum of a N-terminal-dansyl labelled SP-C inserted in phosphatidylcholine bilayers. In this experiment, a fixed amount of dansyl-SP-C and different amounts of phosphatidylcholine were mixed in methanol and injected in a 50 mM Hepes buffer containing NaCl 150 mM (pH 7.0). The fluorescence emission spectrum of the dansyl group underwent some changes upon insertion of the protein in phospholipid vesicles. The maximum of the fluorescence emission spectrum of dansyl-SP-C shifted to larger wavelengths as lipid concentration increased. Also, the fluorescence emission intensity of the SP-C-attached dansyl increased upon insertion of the protein in lipid vesicles. However, this increase in fluorescence intensity was almost negligible compared to the increase observed in the fluorescence emission intensity of a free dansyl upon insertion in the same phospholipid vesicles (data not shown). This suggests that the probe attached to SP-C is exposed to solvent in both, lipid-free and lipid-inserted, dansyl-SP-C. This result is reasonable because the N-terminal end of SP-C is the polar, positively-charged, moiety of the protein. The shift in the position of the fluorescence emission maximum can be interpreted as a consequence of the change in the environment of the probe, from the aqueous bulk phase to the membrane surface, upon insertion of dansyl-SP-C into phospholipid bilayers. Similar effects have been described in other probes upon association with membrane surfaces (Wall et al., 1995). The described effects allow a quantitation of the interaction of dansyl-SP-C with phosphatidylcholine bilayers (Figure 4, insert). Saturation of dansyl-SP-C with phosphatidylcholine was estimated to occur at a protein:lipid molar ratio of about 1:15 which roughly agrees with the values determined by other techniques (Simatos et al., 1990; Pérez-Gil et al., 1995).

Interaction of SP-A with phospholipid vesicles

SP-A, the major surfactant-associated protein, is an oligomeric glycoprotein with a monomeric molecular weight of 36000 (Hawgood and Shiffer, 1991). Electron micrographs of human SP-A reveal the oligomeric structure of the protein, consisting of an octadecamer composed of six trimeric subunits. Trimers form a triple-helical stem stabilized by interchain disulfide bonds. Six trimers are assembled by lateral aggregation to form the octadecameric structure with a bouquet-like appearance. The primary structure of SP-A consists of four different domains: i) an amino-terminus containing a cysteine residue which dimerizes identical subunits; ii) a collagen-like domain consisting of Gly-X-Y repeats, iii) a neck region containing a short stretch of hydrophobic amino acids and an amphipathic helix, and iv) a C-terminal carbohydrate recognition domain (CRD), similar to several Ca^{2+}-dependent C-type lectins, containing a N-linked glycosylation site (Hawgood et al., 1991; Kuroki and Voelker, 1994).

Although SP-A can be isolated as a soluble protein, it is highly associated with phospholipids in the alveolar spaces. The ability of SP-A to bind and aggregate phospholipids is important for the intra alveolar surfactant phospholipid reorganization, i.e. the conversion of the multilamellar forms of the secreted lamellar bodies into the ordered arrays of tubular myelin which is the precursor of the DPPC monolayer at the air-water interface. It was recently showed that both the hydrophobic region of SP-A (neck domain) and likely the CRD domain are involved in the lipid binding properties of SP-A (Ogosawara et al., 1994; Kuroki et al., 1994).

We studied the effect of phospholipid on protein structure. The interaction of phospholipids with SP-A induced a conformational change in the protein molecule, detected by changes in intrinsic fluorescence and sensitivity to the digestion by trypsin (see Figures 5 and 6). The only two tryptophan residues of SP-A are located in the C-terminal 38 amino acids in all species studied so far. The observed increase in the intrinsic fluorescence of SP-A on interaction with DPPC vesicles (Figure 5) indicate a larger shielding of SP-A fluorophores from the polarizable groups of either the protein itself (disulfide, thiol or amine) or the solvent, which are responsible for tryptophan quenching. It is conceivable that this conformational change caused by phospholipid binding lead also to a reduced exposure to trypsin cleavage targets (Figure 6) located not only in the region in which phospholipids are bound but also in closed areas of the carbohydrate recognition domain (CRD).

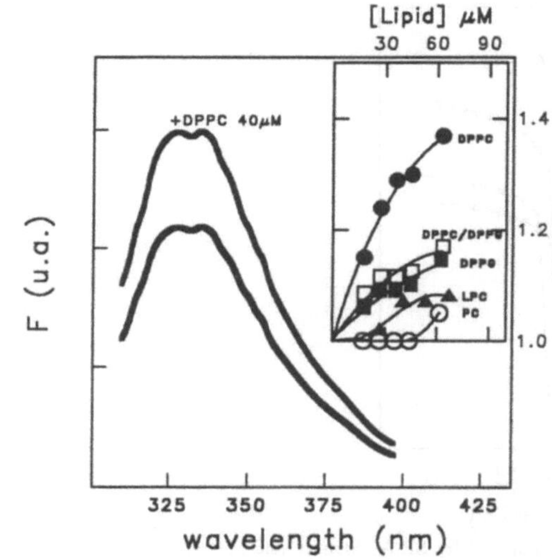

Figure 5. Fluorescence emission spectrum of porcine SP-A (10 µg/mL) in the absence and in the presence of DPPC vesicles (40 µg/mL). Insert: changes in fluorescence emission intensity of porcine SP-A on addition of phospholipid vesicles

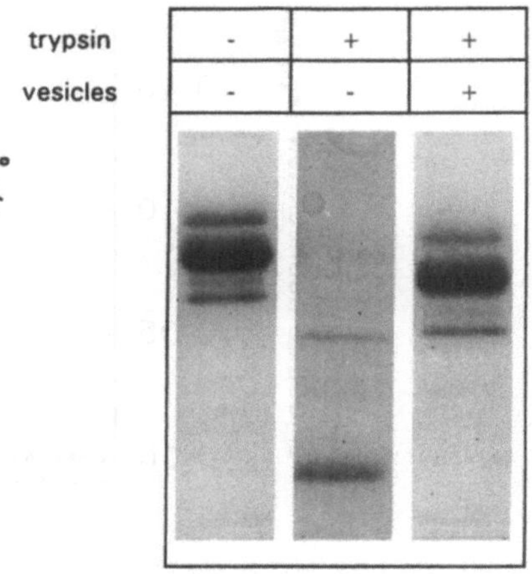

Figure 6. Susceptibility of canine SP-A to digestion by trypsin, in the absence and in the presence of DPPC vesicles. SP-A (160 µg/mL) was incubated 15 min with trypsin (20 µg/mL) in 5mM Tris/HCl buffer, pH 7.4, 150 mM NaCl, 1 µM Ca^{2+}, with or without DPPC vesicles (1.6 mg/mL)

On the other hand, we have determined that the interaction of SP-A with phospholipids was influenced by the nature of the polar head group of phospholipids as well as their physical state in aqueous systems (Casals et al., 1993). The interaction of SP-A with DPPC was much more pronounced than with other phospholipids such as DPPG (with the same acyl chains and Tm but different head group) or egg-PC (with the same polar group but different acyl chains and Tm) (Figure 1, insert). In addition we found that the interaction of SP-A with DPPC vesicles was stronger when vesicles were in the gel state than when they were in the liquid-crystalline state (Casals et al., 1993). SP-A seems to show preference for interacting with the specific head group/backbone conformation of highly ordered DPPC vesicles at 37°C. Wether SP-A can penetrate the bilayer or monolayer is currently unresolved. So far, there is little evidence for significant perturbation of bilayer structure when SP-A binds.

Interestingly, the interaction of negatively charged vesicles (DPPC/DPPG 7:3, w/w) with SP-A was abrogated at low ionic strength (Casals et al., 1993) as well as the aggregation of these vesicles induced by SP-A (Casals et al., 1994a). However, the interaction of neutral vesicles with SP-A and the subsequent aggregation was not dependent on the ionic strength. The lack of interaction of acidic vesicles with SP-A at low ionic strength is interpreted to arise from electrostatic repulsion between the negative charge of phospholipids and the negative surface charge on the protein (pI is 4.8-5.2).

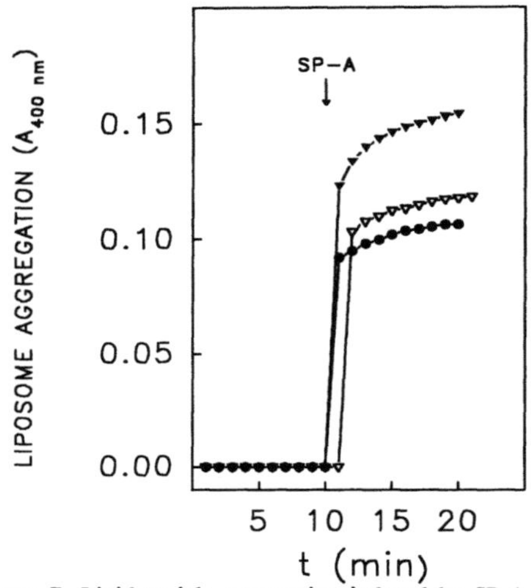

Figure 7. Lipid vesicle aggregation induced by SP-A. Sample and reference cuvettes were filled with either 100 μg/mL DPPC (●) or DPPC/SP-C (▾) or DPPC/SP-B (▿) (10 μg/mL SP-B or SP-C) in 50 mM Tris-HCl buffer, pH 7.4 containing 100 mM NaCl and 20 μM Ca^{2+}. After equilibration for 10 min at 37°C, SP-A was added to the sample cuvette (10 μg/mL, final concentration) and the change in optical density at 400 nm was monitorized

Little is known of the influence of hydrophobic surfactant proteins (SP-B or SP-C) on the biological activities of SP-A. We recently found that SP-C but not SP-B enhanced lipid aggregation activity of SP-A (Figure 7) (Casals et al., 1994b) at both neutral and acidic pH. On the other hand, Poulain et al. (1992) reported that SP-A increased the lipid mixing activity of SP-B. The study of possible protein-protein interactions between surfactant-associated proteins will be an important objective to understand the dynamics of that lipoprotein complex in the alveolar spaces.

Conclusions

Several methods are presented here to determine the interaction of isolated surfactant proteins, SP-A, SP-B and SP-C, with phospholipid vesicles, and the effects of those proteins on phospholipids.

The extensive characterization of lipid-protein interactions between those proteins and surfactant phospholipids will contribute to understand the role of surfactant proteins in the dynamics and function of the pulmonary surfactant system.

References

Bardelle, C., Furie, B., Furie, B.C. and Gilbert, G.E. (1993). J. Biol. Chem. 268, 8815-8824.
Batenburg, A.M., van Eschm J.H., Kruijff, B. (1988). Biochemistry 27, 2324-2331.
Casals, C., Herrera, L., Miguel, E., García-Barreno, P. and Municio, A.M. (1989). Biochim. Biophys. Acta 1003, 201-203.
Casals, C., Miguel, E. and Pérez-Gil, J. (1993). Biochem. J. 296, 585-593.
Casals, C., Miguel, E., Pérez-Gil, J. (1994a). Prog. Respir. Res. 27, 39-43.
Casals, C., Ruano, M.L.F., Miguel, E., Sanchez, P., Pérez-Gil, J. (1994b). Biochem. Soc. Trans 22, 370S.
Cruz, A., Casals, C. and Pérez-Gil, J. (1995). Biochim. Biophys. Acta 1255, 68-76.
Curstedt, T., Jornvall, H., Robertson, B., Bergman, T. and Berggren, P. (1987). Eur. J. Biochem. 168, 255-262.
Curstedt, T., Johansson, J., Persson, P., Eklund, A., Robertson, B., Lowenadler, B. and Jornvall, H. (1990). Pr. Natl. Acad. Sci. USA 87, 2985-2989.
Dufourc, J. and Faucon, J.-F. (1978). Biochemistry 17, 1170-1176.
Ferrandiz, C., Pérez-Payá, E., Braco, L. and Abad, C. (1994). Biochem. Biophys. Res. Comm. 203, 359-365.
Gazit, E., Lee, W.-J., Brey, P.T. and Shai, Y. (1994). Biochemistry 33, 10681-10692.
Goerke, J. and Clements, J.A. (1986). In *Alveolar Surface Tension and Lung Surfactant* (Macklend, P.T. and Mead, J., Eds.) pp 247-261- American Physiological Society, Washington D.C.
Haagsman, H.P., Hawgood, S., Sargeant, T., Buckley, D., White, R.T., Drickamer, K. and Benson, B.J. (1987). J. Biol. Chem. 262, 13877-13880.
Hawgood, S., Benson, B.J., Schilling, J., Damm, D., Clements, J.A. and White, R.T (1987). Pr. Natl. Acad. Sci. USA 84, 66-70.
Hawgood, S. and Shiffer, K. (1991). Annu. Rev. Physiol. 53, 375-394.
Johansson, J., Curstedt, T., Robertson, B. and Jornvall, H. (1988). Biochemistry 27, 3544-3547.
Johansson, J., Curstedt, T. and Jornvall, H. (1991). Biochemistry 30, 6917-6921.
Johansson, J., Szyperski, T., Curstedt, T. and Wuthrich, K. (1994). Biochemistry 33, 6015-6023.
Keough, K.M.W. (1992). In *Pulmonary Surfactant: from Molecular Biology to Clinical Practice* (Robertson, B., Van Golde, L.M.G. and Batenburg, J.J., Eds.) pp 109-164, Elsevier (Amsterdam).
Kuchinka, E. and Seelig, J. (1989). Biochemistry 28, 4216-4221.
Kuroki, Y. and Akino, T. (1991). J. Biol. Chem. 266, 3068-3073.
Kuroki, Y. and Voelker, D.R. (1994). J. Biol. Chem. 269, 25943-25946.
Kuroki, Y., McCormack, F.X., Ogosawara, Y., Mason, R.J., Voelker, D.R. (1994). J. Biol. Chem. 269, 29793-29800.
Morrow, M.R., Pérez-Gil, J., Simatos, G.A., Boland, C., Stewart, J., Absolom, D., Sarin, V. and Keough, K.M.W. (1993). Biochemistry 32, 4397-4402.
Ogosawara, Y., McCormack, F.X., Mason, R.J. and Voelker, D.R. (1994). J. Biol. Chem. 269, 29785-29792.
Pérez-Gil, J., Tucker, J., Simatos, G.A. and Keough, K.M.W. (1992a). Biochem. Cell Biol. 70, 332-338.
Pérez-Gil, J., Nag, K., Taneva, S. and Keough, K.M.W. (1992b). Biophys. J. 63, 197-204.
Pérez-Gil, J., Cruz, A. and Casals, C. (1993). Biochim. Biophys. Acta 1168, 261-270.
Pérez-Gil, J., Casals, C. and Marsh, D. (1995). Biochemistry 34, 3964-3971.
Poulain, F.R., Allen, L., Williams, M.C. , Hamilton, R.L. and Hawgood, S. (1992). Am. J. Physiol. 262, L730-L739.
Sankaram, M.B., Brophy, P.J. and Marsh, D. (1989). Biochemistry 28, 9699-9707.
Shiffer, K., Hawgood, S., Haagsman, H.P., Benson, B., Clements, J.A. and Goerke, J. (1993). Biochemistry 32, 590-597.
Simatos, G.A., Forward, K.B., Morrow, M.R. and Keough, K.M.W. (1990). Biochemistry 29, 5807-5814.
Struck, D.K., Hoekstra, D. and Pagano, R.E. (1981). Biochemistry 20, 4093-4099.
Suzuki, Y., Fujita, Y. and Kogishi, K. (1989). Am. Rev. Respir. Dis. 140, 75-81.
Takahashi, A., Waring, A.J., Amirkhanian, J., Fan, B. and Taeusch, H.W. (1990). Biochim. Biophys. Acta 1044, 43-49.
Vandenbussche, G., Clercx, A., Curstedt, T., Johansson, J., Jornvall, H. and Ruysschaert, J.M. (1992). Eur. J. Biochem. 203, 201-209.
Voss, T., Eistetter, H. and Schafer, K.P. (1988). J. Mol. Biol. 201, 219-227.
Wall, J., Golding, C.A., Van Veen, M. and O'Shea, P. (1995). Mol. Membr. Biol. 12, 183-192.
Weaver, T.E. and Whitsett, J.A. (1991). Biochem. J. 273, 249-264.
Williams, M.C., Hawgood, S. and Hamilton, R.L. (1991). Am. J. Respir. Cell Mol. Biol. 5, 41-50.
Wright, J.R. and Dobbs, L. (1991). Annu. Rev. Physiol. 53, 395-414.

Subject Index

NATO ASI Series H

Vol. 1: **Biology and Molecular Biology of Plant-Pathogen Interactions.** Edited by J.A. Bailey. 415 pages. 1986.

Vol. 2: **Glial-Neuronal Communication in Development and Regeneration.** Edited by H.H. Althaus and W. Seifert. 865 pages. 1987.

Vol. 3: **Nicotinic Acetylcholine Receptor: Structure and Function.** Edited by A. Maelicke. 489 pages. 1986.

Vol. 4: **Recognition in Microbe-Plant Symbiotic and Pathogenic Interactions.** Edited by B. Lugtenberg. 449 pages. 1986.

Vol. 5: **Mesenchymal-Epithelial Interactions in Neural Development.** Edited by J. R. Wolff, J. Sievers, and M. Berry. 428 pages. 1987.

Vol. 6: **Molecular Mechanisms of Desensitization to Signal Molecules.** Edited by T M. Konijn, P J. M. Van Haastert, H. Van der Starre, H. Van der Wel, and M.D. Houslay. 336 pages. 1987.

Vol. 7: **Gangliosides and Modulation of Neuronal Functions.** Edited by H. Rahmann. 647 pages. 1987.

Vol. 8: **Molecular and Cellular Aspects of Erythropoietin and Erythropoiesis.** Edited by I.N. Rich. 460 pages. 1987.

Vol. 9: **Modification of Cell to Cell Signals During Normal and Pathological Aging.** Edited by S. Govoni and F. Battaini. 297 pages. 1987.

Vol. 10: **Plant Hormone Receptors.** Edited by D. Klämbt. 319 pages. 1987.

Vol. 11: **Host-Parasite Cellular and Molecular Interactions in Protozoal Infections.** Edited by K.-P. Chang and D. Snary. 425 pages. 1987.

Vol. 12: **The Cell Surface in Signal Transduction.** Edited by E. Wagner, H. Greppin, and B. Millet. 243 pages. 1987.

Vol. 13: **Toxicology of Pesticides: Experimental, Clinical and Regulatory Perspectives.** Edited by L.G. Costa, C.L. Galli, and S.D. Murphy. 320 pages. 1987.

Vol. 14: **Genetics of Translation. New Approaches.** Edited by M.F. Tuite, M. Picard, and M. Bolotin-Fukuhara. 524 pages. 1988.

Vol. 15: **Photosensitisation. Molecular, Cellular and Medical Aspects.** Edited by G. Moreno, R. H. Pottier, and T. G. Truscott. 521 pages. 1988.

Vol. 16: **Membrane Biogenesis.** Edited by J.A.F Op den Kamp. 477 pages. 1988.

Vol. 17: **Cell to Cell Signals in Plant, Animal and Microbial Symbiosis.** Edited by S. Scannerini, D. Smith, P. Bonfante-Fasolo, and V. Gianinazzi-Pearson. 414 pages. 1988.

Vol. 18: **Plant Cell Biotechnology.** Edited by M.S.S. Pais, F. Mavituna, and J. M. Novais. 500 pages. 1988.

Vol. 19: **Modulation of Synaptic Transmission and Plasticity in Nervous Systems.** Edited by G. Hertting and H.-C. Spatz. 457 pages. 1988.

Vol. 20: **Amino Acid Availability and Brain Function in Health and Disease.** Edited by G. Huether. 487 pages. 1988.

NATO ASI Series H

Vol. 21: **Cellular and Molecular Basis of Synaptic Transmission.** Edited by H. Zimmermann. 547 pages. 1988.

Vol. 22: **Neural Development and Regeneration. Cellular and Molecular Aspects.** Edited by A. Gorio, J. R. Perez-Polo, J. de Vellis, and B. Haber. 711 pages. 1988.

Vol. 23: **The Semiotics of Cellular Communication in the Immune System.** Edited by E.E. Sercarz, F. Celada, N.A. Mitchison, and T. Tada. 326 pages. 1988.

Vol. 24: **Bacteria, Complement and the Phagocytic Cell.** Edited by F. C. Cabello und C. Pruzzo. 372 pages. 1988.

Vol. 25: **Nicotinic Acetylcholine Receptors in the Nervous System.** Edited by F. Clementi, C. Gotti, and E. Sher. 424 pages. 1988.

Vol. 26: **Cell to Cell Signals in Mammalian Development.** Edited by S.W. de Laat, J.G. Bluemink, and C.L. Mummery. 322 pages. 1989.

Vol. 27: **Phytotoxins and Plant Pathogenesis.** Edited by A. Graniti, R. D. Durbin, and A. Ballio. 508 pages. 1989.

Vol. 28: **Vascular Wilt Diseases of Plants. Basic Studies and Control.** Edited by E. C. Tjamos and C. H. Beckman. 590 pages. 1989.

Vol. 29: **Receptors, Membrane Transport and Signal Transduction.** Edited by A. E. Evangelopoulos, J. P. Changeux, L. Packer, T. G. Sotiroudis, and K.W.A. Wirtz. 387 pages. 1989.

Vol. 30: **Effects of Mineral Dusts on Cells.** Edited by B.T. Mossman and R.O. Begin. 470 pages. 1989.

Vol. 31: **Neurobiology of the Inner Retina.** Edited by R. Weiler and N.N. Osborne. 529 pages. 1989.

Vol. 32: **Molecular Biology of Neuroreceptors and Ion Channels.** Edited by A. Maelicke. 675 pages. 1989.

Vol. 33: **Regulatory Mechanisms of Neuron to Vessel Communication in Brain.** Edited by F. Battaini, S. Govoni, M.S. Magnoni, and M. Trabucchi. 416 pages. 1989.

Vol. 34: **Vectors asTools for the Study of Normal and Abnormal Growth and Differentiation.** Edited by H. Lother, R. Dernick, and W. Ostertag. 477 pages. 1989.

Vol. 35: **Cell Separation in Plants: Physiology, Biochemistry and Molecular Biology.** Edited by D. J. Osborne and M. B. Jackson. 449 pages. 1989.

Vol. 36: **Signal Molecules in Plants and Plant-Microbe Interactions.** Edited by B.J.J. Lugtenberg. 425 pages. 1989.

Vol. 37: **Tin-Based Antitumour Drugs.** Edited by M. Gielen. 226 pages. 1990.

Vol. 38: **The Molecular Biology of Autoimmune Disease.** Edited by A.G. Demaine, J-P. Banga, and A.M. McGregor. 404 pages. 1990.

NATO ASI Series H

Vol. 39: **Chemosensory Information Processing.**
Edited by D. Schild. 403 pages. 1990.

Vol. 40: **Dynamics and Biogenesis of Membranes.**
Edited by J. A. F. Op den Kamp. 367 pages. 1990.

Vol. 41: **Recognition and Response in Plant-Virus Interactions.**
Edited by R. S. S. Fraser. 467 pages. 1990.

Vol. 42: **Biomechanics of Active Movement and Deformation of Cells.**
Edited by N. Akkas. 524 pages. 1990.

Vol. 43: **Cellular and Molecular Biology of Myelination.**
Edited by G. Jeserich, H. H. Althaus, and T. V. Waehneldt. 565 pages. 1990.

Vol. 44: **Activation and Desensitization of Transducing Pathways.**
Edited by T. M. Konijn, M. D. Houslay, and P. J. M. Van Haastert. 336 pages. 1990.

Vol. 45: **Mechanism of Fertilization: Plants to Humans.**
Edited by B. Dale. 710 pages. 1990.

Vol .46: **Parallels in Cell to Cell Junctions in Plants and Animals.**
Edited by A. W Robards, W. J . Lucas, J . D. Pitts, H . J . Jongsma, and D. C. Spray. 296 pages. 1990.

Vol. 47: **Signal Perception and Transduction in Higher Plants.**
Edited by R. Ranjeva and A. M. Boudet. 357 pages. 1990.

Vol. 48: **Calcium Transport and Intracellular Calcium Homeostasis.**
Edited by D. Pansu and F. Bronner. 456 pages. 1990.

Vol. 49: **Post-Transcriptional Control of Gene Expression.**
Edited by J. E. G. McCarthy and M. F. Tuite. 671 pages. 1990.

Vol. 50: **Phytochrome Properties and Biological Action.**
Edited by B. Thomas and C. B. Johnson. 337 pages. 1991.

Vol. 51: **Cell to Cell Signals in Plants and Animals.**
Edited by V. Neuhoff and J. Friend. 404 pages. 1991.

Vol. 52: **Biological Signal Transduction.**
Edited by E. M . Ross and K . W. A. Wirtz. 560 pages. 1991.

Vol. 53: **Fungal Cell Wall and Immune Response.**
Edited by J. P. Latge and D. Boucias. 472 pages. 1991.

Vol. 54: **The Early Effects of Radiation on DNA.**
Edited by E. M. Fielden and P. O'Neill. 448 pages. 1991.

Vol. 55: **The Translational Apparatus of Photosynthetic Organelles.**
Edited by R. Mache, E. Stutz, and A. R. Subramanian. 260 pages. 1991.

Vol. 56: **Cellular Regulation by Protein Phosphorylation.**
Edited by L. M. G. Heilmeyer, Jr. 520 pages. 1991.

NATO ASI Series H

Vol. 57: **Molecular Techniques in Taxonomy.**
Edited by G. M. Hewitt, A. W. B. Johnston, and J. P. W. Young. 420 pages. 1991.

Vol. 58: **Neurocytochemical Methods.**
Edited by A. Calas and D. Eugene. 352 pages. 1991.

Vol. 59: **Molecular Evolution of the Major Histocompatibility Complex.**
Edited by J. Klein and D. Klein. 522 pages. 1991.

Vol. 60: **Intracellular Regulation of Ion Channels.**
Edited by M. Morad and Z. Agus. 261 pages. 1992.

Vol. 61: **Prader-Willi Syndrome and Other Chromosome 15q Deletion Disorders.**
Edited by S. B. Cassidy. 277 pages. 1992.

Vol. 62: **Endocytosis. From Cell Biology to Health, Disease and Therapie.**
Edited by P. J. Courtoy. 547 pages. 1992.

Vol. 63: **Dynamics of Membrane Assembly.**
Edited by J. A. F. Op den Kamp. 402 pages. 1992.

Vol. 64: **Mechanics of Swelling. From Clays to Living Cells and Tissues.**
Edited by T. K. Karalis. 802 pages. 1992.

Vol. 65: **Bacteriocins, Microcins and Lantibiotics.**
Edited by R. James, C. Lazdunski, and F. Pattus. 530 pages. 1992.

Vol. 66: **Theoretical and Experimental Insights into Immunology.**
Edited by A. S. Perelson and G. Weisbuch. 497 pages. 1992.

Vol. 67: **Flow Cytometry. New Developments.**
Edited by A. Jacquemin-Sablon. 1993.

Vol. 68: **Biomarkers. Research and Application in the Assessment of Environmental Health.** Edited by D. B. Peakall and L. R. Shugart. 138 pages. 1993.

Vol. 69: **Molecular Biology and its Application to Medical Mycology.**
Edited by B. Maresca, G. S. Kobayashi, and H. Yamaguchi. 271 pages. 1993.

Vol. 70: **Phospholipids and Signal Transmission.**
Edited by R. Massarelli, L. A. Horrocks, J. N. Kanfer, and K. Löffelholz. 448 pages. 1993.

Vol. 71: **Protein Synthesis and Targeting in Yeast.**
Edited by A. J. P. Brown, M. F. Tuite, and J. E. G. McCarthy. 425 pages. 1993.

Vol. 72: **Chromosome Segregation and Aneuploidy.**
Edited by B. K. Vig. 425 pages. 1993.

Vol. 73: **Human Apolipoprotein Mutants III. In Diagnosis and Treatment.**
Edited by C. R. Sirtori, G. Franceschini, B. H. Brewer Jr. 302 pages. 1993.

NATO ASI Series H

Vol. 74: **Molecular Mechanisms of Membrane Traffic.**
Edited by D. J. Morré, K. E. Howell, and J. J. M. Bergeron. 429 pages. 1993.

Vol. 75: **Cancer Therapy. Differentiation, Immunomodulation and Angiogenesis.**
Edited by N. D'Alessandro, E. Mihich, L. Rausa, H. Tapiero, and T. R.Tritton. 299 pages. 1993.

Vol. 76: **Tyrosine Phosphorylation/Dephosphorylation and Downstream Signalling.**
Edited by L. M. G. Heilmeyer Jr. 388 pages. 1993.

Vol. 77: **Ataxia-Telangiectasia.** Edited by R. A. Gatti, R. B. Painter. 306 pages. 1993.

Vol. 78: **Toxoplasmosis.** Edited by J. E. Smith. 272 pages. 1993.

Vol. 79: **Cellular Mechanisms of Sensory Processing. The Somatosensory System.**
Edited by L. Urban. 514 pages. 1994.

Vol. 80: **Autoimmunity: Experimental Aspects.**
Edited by M. Zouali. 318 pages. 1994.

Vol. 81: **Plant Molecular Biology. Molecular Genetic Analysis of Plant Development and Metabolism.**
Edited by G. Coruzzi, P. Puigdomènech. 579 pages. 1994.

Vol. 82: **Biological Membranes: Structure, Biogenesis and Dynamics.**
Edited by Jos A. F. Op den Kamp. 367 pages. 1994.

Vol. 83: **Molecular Biology of Mitochondrial Transport Systems.**
Edited by M. Forte, M. Colombini. 420 pages. 1994.

Vol. 84: **Biomechanics of Active Movement and Division of Cells.**
Edited by N. Akkaş. 587 pages. 1994.

Vol. 85: **Cellular and Molecular Effects of Mineral and Synthetic Dusts and Fibres.**
Edited by J. M. G. Davis, M.-C. Jaurand. 448 pages. 1994.

Vol. 86: **Biochemical and Cellular Mechanisms of Stress Tolerance in Plants.**
Edited by J. H. Cherry. 616 pages. 1994.

Vol. 87: **NMR of Biological Macromolecules.**
Edited by C. I. Stassinopoulou. 616 pages. 1994.

Vol. 88: **Liver Carcinogenesis. The Molecular Pathways.**
Edited by G. G. Skouteris. 502 pages. 1994.

Vol. 89: **Molecular and Cellular Mechanisms of H^+ Transport.**
Edited by B. H. Hirst. 504 pages. 1994.

Vol. 90: **Molecular Aspects of Oxidative Drug Metabolizing Enzymes: Their Significance in Environmental Toxicology, Chemical Carcinogenesis and Health.**
Edited by E. Arınç, J. B. Schenkman, E. Hodgson. 623 pages. 1994.

NATO ASI Series H

Vol. 91: **Trafficking of Intracellular Membranes: From Molecular Sorting to Membrane Fusion.** Edited by M. C. Pedroso de Lima, N. Düzgüneş, D. Hoekstra. 371 pages. 1995.

Vol. 92: **Signalling Mechanisms – from Transcription Factors to Oxidative Stress.** Edited by L. Packer, K. Wirtz. 466 pages. 1995.

Vol. 93: **Modulation of Cellular Responses in Toxicity.** Edited by C. L. Galli, A. M. Goldberg, M. Marinovich. 379 pages. 1995.

Vol. 94: **Gene Technology.** Edited by A. R. Zander, W. Ostertag, B. V. Afanasiev, F. Grosveld. 556 pages. 1996.

Vol. 95: **Flow and Image Cytometry.** Edited by A. Jacquemin-Sablon. 241 pages. 1996.

Vol. 96: **Molecular Dynamics of Biomembranes.** Edited by J. A. F. Op den Kamp. 424 pages. 1996.

Springer-Verlag and the Environment

We at Springer-Verlag firmly believe that an international science publisher has a special obligation to the environment, and our corporate policies consistently reflect this conviction.

We also expect our business partners – paper mills, printers, packaging manufacturers, etc. – to commit themselves to using environmentally friendly materials and production processes.

The paper in this book is made from low- or no-chlorine pulp and is acid free, in conformance with international standards for paper permanency.

MIX
Papier aus verantwortungsvollen Quellen
Paper from responsible sources
FSC® C105338

FSC
www.fsc.org

If you have any concerns about our products,
you can contact us on
ProductSafety@springernature.com

In case Publisher is established outside the EU,
the EU authorized representative is:
Springer Nature Customer Service Center GmbH
Europaplatz 3, 69115 Heidelberg, Germany

Printed by Libri Plureos GmbH
in Hamburg, Germany